Environmental Toxicology: Xenobiotics and Effects on Human Genetic Material

Kanu Priya · Shiv Kumar Giri
Editors

Environmental Toxicology: Xenobiotics and Effects on Human Genetic Material

Springer

Editors
Kanu Priya
Department of Life Science
Sharda University
Greater Noida, Uttar Pradesh, India

Shiv Kumar Giri
Department of Biotechnology
Maharaja Agrasen University
Barotiwala, Himachal Pradesh, India

ISBN 978-3-032-28618-5 ISBN 978-3-032-28619-2 (eBook)
https://doi.org/10.1007/978-3-032-28619-2

This Springer imprint is published by the registered company Springer Nature Switzerland AG
The registered company address is: Gewerbestrasse 11, 6330 Cham, Switzerland

This book is dedicated in the loving memory of my beloved father Late Sh. P.K. Gupta. Though you are no longer here to witness this milestone, your values, encouragement, and belief in the power of education have been the foundation of my journey.
Your love, and your belief in me is my greatest motivation.

Foreword

Environmental toxicology has emerged as one of the most critical interdisciplinary fields of contemporary science, as rapid industrialization, urbanization, and agricultural intensification have led to unprecedented human exposure to a wide range of xenobiotics. Among their most profound and long-lasting impacts are those exerted on human genetic material, leading to mutations, chromosomal aberrations, epigenetic alterations, and increased susceptibility to a range of diseases. Understanding their interaction with human genetic material is therefore not only of academic interest but also of immense relevance to public health, environmental safety, and sustainable development. In this context, the present book, *Environmental Toxicology: Xenobiotics and Effect on Human Genetic Material*, edited by Dr. Kanu Priya and Dr. Shiv Kumar Giri, assumes great importance and is highly relevant.

The editors have thoughtfully curated chapters that bridge classical concepts of toxicology with modern insights from molecular biology, genetics, and biotechnology. The book is well-organized, with each chapter complementing the others. Careful editorial planning is evident throughout the book.

This edited volume provides a comprehensive and multidisciplinary perspective on environmental toxicology, integrating concepts from toxicology, molecular biology, genetics, biotechnology, and environmental sciences. It elucidates the sources, classification, metabolism, and mechanisms of action of xenobiotics, with particular emphasis on genotoxicity, mutagenesis, DNA damage, epigenetic alterations, and their long-term implications for human health. Special emphasis on microplastics, nanoparticles, pesticides, heavy metals, industrial effluents, and emerging contaminants reflects the evolving nature of environmental risks in the modern world, underscoring the need for advanced research and informed risk assessment strategies. It further distinguishes itself by highlighting advanced experimental approaches and bioassays used to evaluate DNA damage, genetic polymorphisms, and life-cycle-based toxicity assessments.

This volume is particularly noteworthy as it outlines future research paths that emphasize interdisciplinary approaches, integration of genetics and environmental exposure data, development of predictive and alternative models, and stronger links between science, policy, and public health. Such perspectives are essential for advancing precision toxicology and for formulating evidence-based environmental regulations.

A notable strength of this book lies in its clear presentation of complex concepts, supported by relevant experimental approaches, biomarkers, and case studies. The integration of molecular techniques and toxicogenomic perspectives makes this work especially valuable for postgraduate students, research scholars, and early-career scientists.. The inclusion of topics such as occupational exposure, environmental risk assessment, and regulatory policies strengthens the practical relevance of this work academicians, environmental scientists, policymakers, and professionals involved in risk assessment, regulatory science, and environmental monitoring.

As a mentor, I take great pride in the scholarly maturity and scientific rigor demonstrated by my Ph.D. students, Dr. Kanu Priya and Dr. Shiv Kumar Giri, in editing this volume. Their dedication reflects not only their research training but also their commitment to addressing pressing environmental and genetic health challenges through education and research dissemination.

The editors, Dr. Kanu Priya and Dr. Shiv Kumar Giri, have demonstrated commendable academic vision and editorial rigor in bringing together contributions from experts working at the forefront of environmental toxicology and genetic research. Their efforts have resulted in a coherent and informative volume that bridges fundamental concepts with applied research, making it valuable to postgraduate students, doctoral researchers, academicians, and professionals engaged in environmental health and safety.

I am confident that this book will stimulate critical thinking, encourage interdisciplinary research, and contribute meaningfully to our understanding of how environmental xenobiotics influence human genetic integrity. This work will undoubtedly serve as an important reference source and stimulate further research into understanding and mitigating the genetic and health impacts of environmental xenobiotics.

I congratulate the editors and contributing authors for their scholarly contribution and wish this book a wide readership and impact in the scientific community.

Prof. Anita Yadav
Department of Biotechnology
Kurukshetra University
Kurukshetra, India

Preface

The rapid pace of industrialization, urban expansion, and agricultural intensification has profoundly altered the chemical landscape of the environment. Humans are increasingly exposed to a wide spectrum of foreign chemical substances—collectively known as xenobiotics—through air, water, food, and occupational settings. These substances, which include industrial chemicals, pesticides, pharmaceuticals, heavy metals, and persistent organic pollutants, pose significant risks to human health, particularly at the genetic and molecular levels.

Among the most concerning outcomes of environmental exposure is damage to human genetic material, manifesting as DNA mutations, chromosomal aberrations, epigenetic alterations, and disruptions in gene expression. Such genotoxic effects are closely linked to the development of cancer, reproductive disorders, congenital anomalies, and long-term heritable diseases.

This book, *Environmental Toxicology: Xenobiotics and Effects on Human Genetic Material*, is designed to provide a comprehensive and systematic overview of the sources, mechanisms, and consequences of xenobiotic-induced genetic toxicity. It explores the chemical nature and environmental distribution of xenobiotics, their absorption, metabolism, and bioaccumulation in the human body, and the molecular pathways through which they interact with DNA and cellular repair systems. Special emphasis is placed on mechanisms of genotoxicity, oxidative stress, mutagenesis, and epigenetic modulation.

The volume also highlights modern experimental and computational approaches used in environmental toxicology, including in vitro and in vivo genotoxicity assays, biomonitoring techniques, and emerging in silico tools for toxicity prediction and risk assessment. This book is intended for undergraduate and postgraduate students of toxicology, environmental sciences, biotechnology, pharmacy, and life sciences, as well as researchers, educators, and professionals engaged in environmental health and regulatory affairs.

It is our hope that this book will serve not only as an academic resource but also as a catalyst for further research and responsible decision-making in the protection of human genetic integrity in an increasingly chemical-driven world.

We extend our gratitude to the contributors who have shared their expertise, and to the readers whose curiosity and dedication continue to propel this field forward.

Greater Noida, India Kanu Priya
Barotiwala, India Shiv Kumar Giri

Acknowledgements

I extend my deepest gratitude to all those who contributed to the creation of this book, *Environmental Toxicology: Xenobiotics and Effects on Human Genetic Material*. Their support, guidance, and encouragement have been invaluable throughout this journey.

I am profoundly thankful to *Sharda University*, my mentors and academic colleagues, whose insight and expertise helped shape the scientific rigor and clarity of this work. Their constructive feedback and willingness to share knowledge strengthened every chapter of this book.

My sincere appreciation goes to the researchers, scholars, and authors whose foundational studies on environmental toxicology laid the groundwork for this text. Their dedication to advancing the understanding of xenobiotics and human genetics continues to inspire scientific exploration.

I am grateful to my scholars and technical staff who provided essential assistance during various phases of research and data collection. Their commitment to precision and excellence made many aspects of this work possible.

To my students and collaborators, thank you for your enthusiasm, curiosity, and contributions that enriched the discussions and broadened the perspectives presented in this book.

Finally, a heartfelt thank you to my husband, my family and friends for their unwavering support, patience, and encouragement. Their belief in me sustained my motivation during the most challenging moments.

Dr. Kanu Priya

Contents

About the Editors

Dr. Kanu Priya is working as an Associate Professor in the Department of Life Science in the School of Bioscience and Technology at Sharda University, Greater Noida, India. Her specialization is in animal biotechnology, nanobiotechnology, environmental biotechnology, and pharmacology and toxicology. She has been teaching graduate and postgraduate students of biotechnology with total experience of more than nine years. She had served as head of the department in her previous institution. She has taught topics related to animal biotechnology, microbiology, genetic engineering, and IPR. She has authored more than 35 scientific articles and six book chapters from international publishers. She has 14 patents published (four granted). She has received 40 lakhs project from Ministry of Ayush. She has also completed two small granted projects. She received a best women research award for excellence in research in 2025 by Springer Nature under "Her research, our future". She has presented her work on national and international platforms in conferences and workshops. She has served as an editorial board member and reviewer for several reputed international journals.

Dr. Shiv Kumar Giri is a renowned academic and researcher, currently serving as the Head of Department and Professor in the Department of Biotechnology at Maharaja Agrasen University, Baddi, Himachal Pradesh. With over 15 years of research expertise and a decade of teaching experience, Dr. Giri's areas of specialization include environmental toxicology, genomics, bioinformatics, and stem cell technology. Dr. Giri has successfully guided four Ph.D. scholars and completed a prestigious research project on stem cell funded by the DST-SERB, Government of India, in 2016, showcasing his ability to lead impactful research initiatives. He has also made significant contributions to scientific literature through numerous research papers, book chapters, and as an editor of three books within his field. Prior to his current role, Dr. Giri served as an Assistant Professor in the Department of Biotechnology at University College, Kurukshetra University. His career reflects a strong commitment to advancing biotechnology education and research innovation.

Abbreviations

8-OHdG	8-hydroxy-2′-deoxyguanosine
ABC	ATP-Binding Cassette
AChE	Acetylcholinesterase
AEP	Aggregate Exposure Pathway
AFFF	Aqueous Film-Forming Foam
AFM	Atomic Force Microscope
AGEs	Advanced Glycation End
AhR	Aryl Hydrocarbon Receptor
AKR	Aldo-Keto Reductase
AML	Acute Myeloid Leukemia
AMS	Accelerator Mass Spectrometry
AOP	Adverse Outcome Pathway
AP	Apurinic/Apyrimidinic
APIs	Active Pharmaceutical Ingredients
ARGs	Antibiotic Resistance Genes
ARNT	AhR Nuclear Translocator
AS3MT	Arsenite Methyltransferase
AST	Aspartate Transaminase
BCRP	Breast Cancer Resistance Protein
BER	Base Excision Repair
BFRs	Brominated Flame Retardants
BMF	Bio-magnification Index
BPA	Bisphenol A
BPDE	Benzo[a]pyrene diol epoxide
CAT	Catalase
CDC	Centers for Disease Control and Prevention
CNTs	Carbon Nanotubes
COPD	Chronic Obstructive Pulmonary Diseases
CRC	Colorectal Cancer
CTD	Comparative Toxicogenomic Database
CYP	Cytochrome P450

DDD	Dichlorodiphenyldichloroethane
DDR	DNA Damage Response
DDT	Dichlorodiphenyltrichloroethane
DEHP	Di(2-ethylhexyl) phthalate
DMA	Di-methylated Arsenic
DPCs	DNA–Protein Crosslinks
DRD	DNA Repair Deficiency
DRE	Dioxin-Responsive Element
DSBs	Double-Strand Breaks
ECHA	European Chemicals Agency
EDCs	Endocrine-Disrupting Chemicals
ELSI	Ethical, Legal, and Social Implications
ENU	Ethyl Nitrosourea
EPA	Environmental Protection Agency
EPFRs	Environmentally Persistent-Free Radicals
ETU	Ethylene Thiourea
EWASs	Epigenome-Wide Association Studies
ExAC	Exome Aggregation Consortium
FAO	Food and Agriculture Organization
FFF	Field-Flow Fractionation
FISH	Fluorescence In Situ Hybridization
FMOs	Flavin-containing Monooxygenases
GC	Gas Chromatography
GLDH	Glutamate Dehydrogenase
GnomAD	Genome Aggregation Database
GST	Glutathione-S-transferase
GWAS	Genome-Wide Association Studies
GWEIS	Genome-Wide Environment Interaction Studies
HAAs	Heterocyclic Aromatic Amines
HBCD	Hexabromocyclododecane
HBM4EU	Human Biomonitoring for Europe
HDAC	Histone Deacetylase
HD-PE	High-Density Polyethylene
HMGB1	High-Mobility Group Protein B1
HPLC	High-Performance Liquid Chromatography
HRR	Homologous Recombination Repair
HSP90	Heat Shock Protein 90
IARC	International Agency for Research on Cancer
IPM	Integrated Pest Management
LDL	Low-Density Lipoprotein
LMICs	Low- and Middle-Income Countries
LPO	Lipid Peroxidation
MBD	Methyl-Binding Domain
MCSFR	Macrophage Colony-Stimulating Factor Receptor
MDA	Malondialdehyde

mEH	Microsomal Epoxide Hydrolase
MGE	Mobile Genetic Elements
MMA	Monomethylated Arsenic
MMR	Mismatch Repair
MNU	N-methyl-N-nitrosourea
MPO	Myeloperoxidase
MRLs	Maximum Residue Limits
NATs	N-Acetyltransferases
NCAP	National Clean Air Programme
NER	Nucleotide Excision Repair
NOEC	No Observable Effect Concentration
NQO1	NAD(P)H: Quinone Oxidoreductase 1
NSCLC	Non-small Cell Lung Cancer
NTA	Nanoparticle Tracking Analysis
NTDs	Neural Tube Defects
OCPs	Organochlorine Pesticides
OCT	Octocrylene
PAH	Polycyclic Aromatic Hydrocarbons
PBDE	Polybrominated Diphenyl Ethers
PBPK	Physiologically Based Pharmacokinetic
PCB	Polychlorinated Biphenyl
PCPs	Personal Care Products
PCT	Polychlorinated Terphenyls
PEC	Predicted Environmental Concentration
PETE	Polyethylene Terephthalate
PFAS	Per-and Polyfluoroalkyl Substances
PFOA	Perfluorooctanoic Acid
PRS	Polygenic Risk Score
PS	Polystyrene
PTE	Phosphotriester
PTSs	Persistent Toxic Substances
PVC	Polyvinyl Chloride
PXR	Pregnane X Receptor
RNS	Reactive Nitrogen Species
SAH	S-adenosylhomocysteine
SAICM	Strategic Approach to International Chemicals Management
SAM	S-adenosylmethionine
SMRT	Single-Molecule Real Time
SNP	Single Nucleotide Polymorphism
SOD	Superoxide Dismutase
SPR	Surface Plasmon Resonance
SRT	Slime Retention Time
SSBs	Single-Strand Breaks
STM	Scanning Tunneling Microscope
SULTs	Sulfotransferases

SV	Structural Variations
TBBPA	Tetrabromobisphenol A
TCDD	2,3,7,8-Tetrachlorodibenzo-p-dioxin
TEI	Transgenerational Epigenetic Inheritance
TEM	Transmission Electron Microscopy
TLRs	Toll-Like Receptors
TMZ	Temozolomide
TSCA	Toxic Substances Control Act
UGTs	UDP-Glucuronosyltransferases
UPLC	Ultra-performance Liquid Chromatography
VOCs	Volatile Organic Compounds
WGBS	Whole-Genome Bisulfite Sequencing
XME	Xenobiotic Metabolism Enzymes

Chapter 1
Understanding Environmental Toxicology: An Overview of the Impact of Xenobiotics on the Environment and Human Health

Shikha Goswami, Shiv Kumar Giri, Ritu Dahiya, and Arzoo Pannu

Abstract Synthetic chemicals are flooding our environment and bodies. Such substances called xenobiotics enter the chemosphere (air, water and soil) with impacts ranging from ecosystem disruption to human toxicity at stages of acute or chronic exposure. This accumulating body of xenobiotics in ecosystems has redefined the context for environmental toxicology, focused primarily on the complex relationships among these compounds, human health and ecosystem integrity. This chapter provides an integrated overview of their biochemical behaviour, slime retention time (SRT) and eco-toxicity correlation combining the prospects of molecular toxicology with pharmacological considerations and global policy implementations. With omics tools, high-throughput screening and artificial intelligence entering the mainstream in xenobiotic science, the traditional approach of 'see-then-predict' toxicology is being replaced by 'predict-and-understand'. AOP models allow molecular initiating events to be linked to population-level outcomes and therefore toxicologists can predict toxicities more accurately. Genomics, metabolomics, high throughput screening and artificial intelligence technologies have moved xenobiotic research from the descriptive era into predictive toxicology. Mapping molecular initiating events to population-level effects of exposure using Adverse Outcome Pathway (AOP) models now allows for the prediction of toxicity risks with an unprecedent level of precision. The chapter highlights the necessity to reconsider environmental toxicology under a system perspective, which also comprises bioinformatics, computational modelling

S. Goswami
Department of Pharmacology, School of Pharmaceutical Sciences, DPSRU, Delhi, India

S. K. Giri
Departmentof Biotechnology, Maharaja Agrasen University, Solan, Himachal Pradesh, India

R. Dahiya
Department of Pharmacology, Lingaya's Vidyapeeth, Faridabad, Haryana, India

A. Pannu (✉)
School of Pharmacy and Research, Dev Bhoomi Uttarakhand University, Dehradun, India
e-mail: arzoopannu@gmail.com

K. Priya and S. K. Giri (eds.), *Environmental Toxicology: Xenobiotics and Effects on Human Genetic Material*, https://doi.org/10.1007/978-3-032-28619-2_1

and sustainable chemistry for the generation of safer compounds and making predictions with complex mixtures in the twenty-first century. The EU REACH, USA TSCA and India CMSR 2020 all policy tools are critiqued as an example of preventive chemical governance. The chapter contends that intervention in toxicological xenobiosis is scientifically mandatory and a moral issue. Fusing innovation, policy and ethics, environmental toxicology could be a principle-based discipline focused on safeguarding the integrity of genomes and ecological diversity across generations.

Keywords Xenobiotics · Genotoxicity · Environmental toxicology · Green chemistry · Predictive toxicology

1 Introduction

Environmental toxicology is a cross-disciplinary field of environmental science concerned with the study of the harmful effects of various chemical, biological and physical agents on living organisms. Xenobiotics comprise a vast array of synthetic chemicals—pesticides, industrial solvents, drugs, plasticizers, heavy metals such as newly emerging contaminants like PFAS and microplastics that do not occur naturally in life forms (Walker et al., 2023). Accumulation of these chemicals in the environment as a result of industrialization, urbanization and non-sustainable agriculture provides an important threat to ecological integrity and genetic stability of humans.

The use of pesticides is over 3.7 million tonnes per year worldwide, according to the FAO in 2022, but in 2023 the UNEP announced that there are currently more than 350 thousand synthetic chemicals registered for production and use globally. The majority of the contaminants are persistent, bio accumulative and they undergo long-range transport with consequent presence in ecosystems distant from sources, such as the Arctic and deep-sea sediments (Scheringer et al., 2023). Their global distribution provides evidence of worldwide exposure to xenobiotics and the merits discussing not only their effect on ecosystems, but also on maintenance of the genetic code in human.

1.1 Xenobiotics and Human Genetic Material

At the cellular level, this could be a genotoxic effect of xenobiotics: directly or indirectly interacting with DNA, RNA or chromosomes, causing mutations, chromosomal aberrations and epigenetic changes (Zhang & Zhao, 2024). The processes are mediated by reactive oxygen species production, DNA adduct formations, interference with repair processes and changes in methylation and histone modification profiles. For instance, benzene, cadmium and organophosphate insecticides have been associated with chromosomal instability and abnormal gene expression in human lymphocytes (Ali et al., 2023). Besides that, endocrine disruptors such as bisphenol A and phthalates have the ability to epigenetically control the expression of genes and affect the inheritance of susceptibility to diseases in successive generations (Vandenberg et al., 2024).

A pioneering article in Nature Communications in 2024 showed that pollutants in the environment select for microbiomes that metabolize xenobiotics, with a secondary effect on the metabolic fate of the latter. This new knowledge implies a complex interplay between environmental microbiota and host genome stability with the corollary that microbial metabolism may on one hand be capable of detoxifying or metabolizing xenobiotics to their active form, and then influence their genotoxic effects. The microbiome-genome interrelationship has emerged as a new paradigm for experimental approaches for environmental toxicology—a hybrid area of classical chemical toxicology with genomics and systems biology.

1.2 Relevance to Human Genetic Integrity

The Global Burden and Environmental Persistent Organic Pollutants (POPs) like dioxins, polychlorinated biphenyls (PCBs) and select pesticides are a concern since they have long environment half-lives and the potential for anthropogenic transport and accumulation through food webs (Stockholm Convention, 2023). Methyl mercury synthesized with the process of aquatic microbial methylation is biomagnified with predator fish such that neurotoxicity and teratogenicity are induced in humans that consume seafood with this contaminant (Wang et al., 2024). Once again, PFAS compounds that are informally termed as 'forever chemicals' are not broken down and are correlated with oxidative DNA damage and telomere decrease with human epidemiological studies (Liu et al., 2024) (Fig. 1).

These findings show that occupational exposure is not the exclusive domain of genetic toxicity, but that the public at large is exposed through dietary, environmental and transfer by the mother. Furthermore, climate change also abets this issue by altering the environmental mobility and bioavailability of xenobiotics—e.g. increased flooding pulls up buried pollutants into water supplies and crops (Rahman et al., 2024).

1.3 Problem Statement and New Experimental Idea

Despite decades of research, the mechanisms underlying the connection between chronic low-dose xenobiotic exposure and heritable genomic alterations have remained poorly defined. Most classical toxicological models are designed to evaluate acute toxicity or cancer endpoints without taking into consideration subtle, yet enduring, molecular changes in DNA and epigenetic programming. The following chapter, therefore, provides a new research framework.

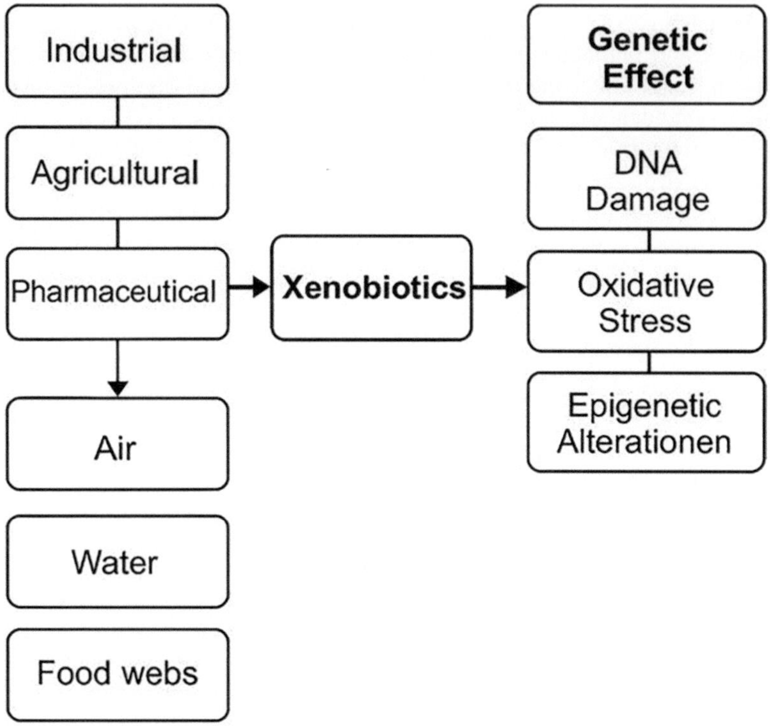

Fig. 1 Demonstrating the origin of xenobiotics in the environment (industrially, agriculturally, pharmacologically), their transport via air, water and food webs and their final interaction at cellular and genomic levels resulting in DNA damages, oxidative stress and epigenetic alterations

The aim was to investigate the interaction between xenobiotic exposure and microbiome-mediated metabolism in the maintenance of human genetic stability. This interdisciplinary area integrates environmental toxicology, genomics, microbiology, and systems biology approaches, which are essential for understanding the dynamic and multi-level mechanisms underlying xenobiotic-induced genetic effects.

This chapter herein lays the scientific basis for the study of how xenobiotics interact with the environment and biological entities to influence human genetic material. The chapter surveys the origin and fates of environmental xenobiotics, mechanisms of geno- and epino, microbiome-mediated modulation of toxicity, bioaccumulation and biomagnification in food webs, bio analytical techniques and biomarkers for tracking gene effects, regulatory and risk-assessment standards.

By combining latest findings, the chapter will emphasize novelties in challenges and experimental potential in the investigation of xenobiotic genotoxicity—above all under the dynamic world chemical climate.

2 Xenobiotics' Sources and Routes of Environmental Entry

Xenobiotics are a vast group of synthetic and natural chemicals that are not indigenous to biological systems. Their entry into the environment is due to anthropogenic activities with multiple dimensions such as industrialization, urbanization, intensive agriculture production and expansion in the pharmaceutical industry (Table 1). Persistence and bioactivity of the chemicals under soil, water and atmospheric exposures result in them being key environmental toxicants affecting human and ecological health (Gavrilescu et al., 2022a, 2022b). Identification of the main origin and transport routes of xenobiotics is indispensable for the definition of exposure routes and cumulative risk estimation.

2.1 Industrial and Pharmaceutical Sources

Industrial effluents are still the ultimate causes of xenobiotic contamination, with a substantial fraction due to petrochemicals, pigments, dyes, plastics and heavy metal-based manufacturing units (Sharma et al., 2020a, 2020b). Industrial pollutants such as persistent organic pollutants (POPs) under the head of polychlorinated biphenyls (PCBs) and dioxins, and industrial solvents such as benzene and toluene

Table 1 Principal classes of xenobiotics, their origins and their environmental half-lives

Class	Representative compounds	Primary sources	Approximate environmental half-life	References
Persistent organic pollutants (POPs)	PCBs, dioxins	Industrial discharge, waste incineration	10–50 years	Sharma et al. (2020a, 2020b)
Pesticides	DDT, glyphosate, atrazine	Agriculture, run off	2–15 years	Chakraborty et al. (2021)
Pharmaceuticals	Antibiotics, NSAIDs, hormones	Hospital and industrial effluents	Weeks to years	Kümmerer (2017)
heavy metals	Lead, cadmium, mercury	Mining, smelting, battery disposal	Indefinite (non-degradable)	Li et al. (2021a, 2021b, 2021c, 2021d)
Plasticizers & surfactants	Phthalates, alkylphenols	Urban waste, detergents, cosmetics	Months to years	Wang et al. (2023)

have been commonly found in surface and groundwaters within the vicinity of industrial corridors. Pharmaceutical and drug manufacturing units release active pharmaceutical ingredients (APIs) and metabolites into aqueous bodies that remain as micro-contaminants. Antibiotic residues in sewerage treatment facility effluents have been found to be above 10 μg/L in Asian and European nations (Kümmerer, 2017). The latter have a propensity to escape degradation processes and induce antimicrobial resistance (AMR) and environment imbalance.

2.2 Agricultural and Pesticide Residue

Agriculture is a key contributor to the xenobiotic loading with the utilization of pesticides, herbicides and fertilisers. Organophosphates, carbamates, neonicotinoids and glyphosate are a few of the most common agrochemicals with genotoxicity and neurotoxicity reported (Chakraborty et al., 2021). Their entry into water bodies with runoff leads to the accumulation of residues that suppress aquatic biodiversity and find their way into the human food chain. Studies have demonstrated notable levels of organochlorine pesticide residues in sediments along with fish tissues even several decades following the regulatory ban (Singh & Kumar, 2019). Excessive half-lives for the chemicals allow biomagnification, thereby increasing their toxicological impact throughout the trophic levels (Table 1).

2.3 Domestic and Urban Sources

Cities are a sourcc of origin of xenobiotic contamination due to transport emissions, construction materials, plasticizers, detergents and cosmetics. Polycyclic aromatic hydrocarbons (PAHs), phthalates and surfactants are typical urban xenobiotics with the potential to be endocrine disruptors (Wang et al., 2023). Wastewater treatment works are a sink and a source, with non-complete degradation leading to recirculation of micropollutants back into surface waters and percolation irrigation systems (Patel et al., 2020a, 2020b).

2.4 Environmental Pathways and Transport Mechanisms

After release, the xenobiotics also experience intricate physical, chemical and biological changes. Deposition, leaching and runoff transport them across compartments in the environment. Hydrophobic chemicals have the tendency to adsorb on sediments and particles, whereas hydrophilic chemicals become trapped in aqueous phases, also polluting the groundwater (Li et al., 2021a, 2021b, 2021c, 2021d). Other processes, such as biotransformation and volatilization, cause redistribution leading to their persistence in the environment. More seriously, the most of xenobiotics also possesses synergy: transformation products more similar or even more toxic than parent chemicals (Zhao et al., 2024a, 2024b, 2024c, 2024d).

The ongoing omnipresence of xenobiotic pollutants in each of the environmental media is testament to how difficult managing exposure and risk can be. Their resistance, linked with the spread of globalized industrial supply network, and the poorly equipped waste handling treatment units makes it tough to practice sustainable public health care. Their delivery and decay kinetics are pressing for predictive modelling and policy making with the intended aim of reducing their long-term effects on human genetics.

3 Toxicity Mechanism and Genomic Interaction

Adverse toxicological effect of xenobiotics is implemented through a chain of processes related by the cascade principle that result in genetic stability and cell homeostasis disturbances. Following absorption, xenobiotics are enzymatically activated (metabolically activated) to give a reactive intermediate capable of covalently reacting with cellular macromolecules such as DNA, proteins and lipids (Ghosh et al., 2019) (Table 2). Interactions can disrupt normal biochemistry, generate mutations and alter gene expression profiles. The extent of these activities is dependent on the physicochemical properties of the molecule, metabolism and duration of exposure.

Table 2 Representative classes of xenobiotics and their modes of genetic toxicity

Xenobiotic class	Representative compounds	Primary mechanism of toxicity	Genetic impact	References
Polycyclic aromatic hydrocarbons (PAHs)	Benzo pyrene	ROS generation, DNA adducts	Point mutations, strand breaks	Shen et al. (2018)
Heavy metals	Arsenic, cadmium, mercury	Oxidative stress, enzyme inhibition	Chromosomal aberrations	Rahman et al. (2023)
Endocrine disruptors	BPA, phthalates	Epigenetic modification, receptor binding	Gene silencing, transcriptional reprogramming	Dolinoy et al. (2016)
Pesticides	Organophosphates, glyphosate	Mitochondrial dysfunction, ROS	Genomic instability	Yadav et al. (2021)
Industrial solvents	Benzene, toluene	Metabolic activation, bone marrow toxicity	DNA crosslinking, leukemogenesis	Zhang et al. (2022)

3.1 Oxidative Stress and Reactive Oxygen Species (ROS)

Among the major routes adopted by xenobiotics in causing toxicity is oxidative stress induction. Most environmental pollutants, such as polycyclic aromatic hydrocarbons (PAHs), pesticides and heavy metals, induce the excessive production of reactive oxygen species (ROS) during the process of metabolic detoxification processes (Yadav et al., 2021). Under regular physiological states, antioxidant barriers like superoxide dismutase (SOD), catalase and glutathione peroxidase preserve ROS homeostasis. On the other hand, excessive exposure exceeds such barriers, resulting in oxidative cellular component damage, mostly DNA bases and mitochondrial membranes (Sies & Jones, 2020). Oxidative modifications such as 8-hydroxydeoxyguanosine (8-OHdG) formation are highly acknowledged biomarkers for genotoxicity induced by xenobiotics.

3.2 Pathways of DNA Adduct Generation and Genotoxicity

A number of xenobiotics, particularly electrophilic chemicals such as aflatoxins, benzo[a]pyrene and select nitrosamines, covalently bind with nucleophilic moieties in DNA and give rise to the formation of DNA adducts (Shen et al., 2018). Miscode events during replication can lead to point mutations, strand breaks or chromosomal aberrations. Ongoing accumulation of adducts disables DNA repair processes, including nucleotide excision repair (NER) and base excision repair (BER), and accelerates mutagenesis and carcinogenesis (Zhang et al., 2022). As a result, long exposure to benzene has been linked with chromosomal translocations and hematopoietic malignancies due to cumulative genotoxic stress. In addition, xenobiotic metabolism via the cytochrome P450 enzyme family can yield reactive metabolites that further contribute to DNA harm (Gavrilescu et al., 2022a, 2022b).

3.3 Epigenetic Modulation

Besides direct DNA damage, xenobiotics also acquire epigenetic effects that alter gene expression without involving a change in nucleotide sequencing. Exposures to heavy metals, endocrine-disrupting chemicals (EDCs) and pesticides have been shown to alter methylation of DNA, histone acetylation and microRNA expression (Greally & Jacobs, 2020). The changes may silence tumour suppressors or turn on carcinogens, and have long-term health effects passed down generations. Prenatal exposure to bisphenol A (BPA) and phthalates, for example, was found to be associated with differential gene methylation that influences neural and immune development (Dolinoy et al., 2016). These results imply that not only can xenobiotics act directly on individuals who are exposed but they also have the potential to trigger transgenerational epigenetic inheritance of perturbed expression trajectories for genes.

3.4 Cellular Signalling and Inflammatory Cascades

Xenobiotics also typically disrupt cellular activities including apoptosis, cell proliferation and inflammation at the levels of genetic and epigenetic polymorphisms. Heavy metals such as arsenic and cadmium activate pro-inflammatory cytokines (TNF-α, IL-6) and stress kinases (MAPK, NF-κB) by chronic inflammation reaction and injury cell (Rahman et al., 2023). Chronic inflammatory processes lead to oxidative stress and genomic instability causing a vicious self-perpetuating cycle of cellular injury. In addition, the disruption of endocrine and immune regulatory mechanisms that can be induced by EDCs has been causally associated with developmental and reproductive disorders in both animals and humans.

The underlying mechanisms of xenobiotics acting as toxicants are complicated, including oxidative stress, genotoxicity and epigenetic conflict; it causes irreversible damage to the genome. Against such a background, continuous environmental exposure and preference to biochemical over-accumulation compounds the health risk especially in vulnerable populations. Newly published studies highlight the importance of toxicogenomic analysis that can identify potential biomarkers of exposure and susceptibility before they fully manifest. This data is, in turn, essential for risk analysis and regulatory decision-making and the development of appropriate interventions to prevent gene damage by xenobiotic agents.

4 Human Health Implications and Epidemiological Documentation

Exposures to xenobiotics create among the most formidable environmental health threats of the twenty-first century. Human society remains continuously exposed to an inseparably blended concoction of industrial chemicals, pharmaceutical medicines, pesticides and microplastics by air, water, food and workplace exposure. WHO (2023a, 2023b, 2023c, 2023d) epidemiology studies forecast that chemical pollution leads to more than 9 million premature deaths annually, predominantly through non-communicable diseases (NCD) such as cancer, cardiovascular and neurodegenerative diseases. Xenobiotic toxicodynamic are complex, consisting of bioaccumulation, metabolic activation and disruption of molecular cascades of signalling maintaining human cellular and gene stability.

4.1 Genetic and Molecular Effects

At the molecular level, the xenobiotics like polycyclic aromatic hydrocarbons (PAHs), heavy metals (Pb, Cd, Hg), bisphenols, phthalates and organophosphates can cause DNA damage, oxidative lesions and chromosomal instability. The epidemiological biomonitoring has established the association of elevated urinary concentrations of the agents with an increased frequency of micronuclei, DNA strand breaks (Comet assay) and aberrantly expressed DNA-repair gene expression (XRCC1, OGG1 and p53) (Zhao et al., 2024a, 2024b, 2024c, 2024d). In addition, the chronic low-dose exposure can derange the epigenetic programming—both DNA methylation and histone acetylation—to impair gene expression patterns associated with carcinogenesis, metabolic homeostasis and immune response (Li et al., 2021a, 2021b, 2021c, 2021d).

4.2 Systemic Toxicities and Disease Associations

Longitudinal epidemiological findings reveal extensive correlations among exposure to xenobiotics and numerous systemic diseases (Table 3).

Respiratory and Cardiovascular: The repeated exposure of particulate-bound metals and polyaromatic hydrocarbons aggravates the oxidative stress, leading to the endothelial dysfunction and hypertension (Patel et al., 2020a, 2020b).

Endocrine and Reproductive: Endocrine-disrupting chemicals (EDCs) such as bisphenol A and dioxins inhibit hormonal receptor signalling, among other mechanisms, resulting in infertility, metabolic syndrome and premature pubertal onset (WHO, 2023a, 2023b, 2023c, 2023d).

Neurological: Neurotoxicity related to the use of certain OP and solvent vapours has been linked to Parkinsonian syndromes, developmental delay and dementia (Brack et al., 2019a, 2019b).

Carcinogenic Effects: A number of xenobiotics including arsenic, benzene and formaldehyde have been classified by the International Agency for Research on Cancer as human Group 1 carcinogens for their involvement in DNA adduct formation, chromosomal aberrations and mutational hotspots (UNEP, 2022a, 2022b).

4.3 Population-Level Epidemiology and Vulnerable Groups

Recent multicentre cohort studies (Al-Kandari et al., 2022; OECD, 2020a, 2020b, 2020c) observe an over proportioned risk among industrial workers, farm workers and people in low-income urban areas. Children, expectant females and the elderly show increased susceptibility by virtue of poor detoxification mechanisms and bioaccumulation in adipose tissues. The Global Human Biomonitoring Initiative (2022)

Table 3 Overview of xenobiotic classes, biomarkers and associated human toxicities

Xenobiotic class	Primary source	Documented human effect	Key biomarker/study evidence	Xenobiotic class
Heavy metals (Pb, Cd, Hg)	Industrial effluents, batteries	Neurotoxicity, nephrotoxicity, anaemia	↑ Blood Pb; WHO (2023a, 2023b, 2023c, 2023d)	Heavy metals (Pb, Cd, Hg)
Polycyclic aromatic hydrocarbons (PAHs)	Combustion, diesel exhaust	Lung and skin cancer	DNA adducts (8-OHdG); Brack et al. (2019a, 2019b)	Polycyclic aromatic hydrocarbons (PAHs)
Organophosphate pesticides	Agriculture	Neurodegeneration, developmental delay	↓ AChE activity; UNEP (2022a, 2022b)	Organophosphate pesticides
Bisphenols and phthalates	Plastics, consumer goods	Endocrine disruption, reproductive toxicity	Altered hormone levels; Li et al. (2021a, 2021b, 2021c, 2021d)	Bisphenols and phthalates
Volatile organic compounds (benzene, formaldehyde)	Solvents, industry	Hematological malignancies	Chromosomal aberrations; Zhao et al. (2024a, 2024b, 2024c, 2024d)	Volatile organic compounds (benzene, formaldehyde)

documents the presence of discernible levels of xenobiotics in >92% of adults and in cord blood of neonates, underlining the risk of transgenerational transmission.

Epidemiological modelling also points to geographical hotspots of exposure to pollution, where industrial effluent release and informal waste recycling are least regulated, i.e. in South and Southeast Asia. Both case–control studies in Delhi-NCR (Sharma et al., 2020a, 2020b) showed higher biomarkers of oxidative stress and altered antioxidant enzyme activity (SOD, CAT, GSH) in populations residing in the vicinities of industrial concentration zones.

4.4 Risk Assessment and Public Health Policy

The risk-assessment framework has incorporated the integration of 'omics' biomarkers as well as systems toxicology in the prediction of the health outcomes of xenobiotics. Global surveillance harmonizes biomarkers like 8-hydroxy-2′-deoxyguanosine (8-OHdG) and malondialdehyde (MDA) (WHO, 2023a, 2023b, 2023c, 2023d). Global policy frameworks like UNEP's Strategic Approach to International Chemicals Management (SAICM) and OECD's Global Chemicals Outlook II (2020c) uphold the refrain of precautionary measures, safe chemical replacement.

Despite these advances, enforcement in low-income countries is still fragmented, which pinpoints the need for the overall risk governance that merges environmental monitoring, clinical toxicology and genomics-based surveillance.

Cumulative epidemiological data support that prolonged exposure to xenobiotics is an underappreciated but increasing contributor to human morbidity. The molecular indicators and global data are singing in harmony: chemical safety can no longer be distinguished from genomic integrity. Effective prevention involves transdisciplinary collaboration of toxicologists, geneticists and policy makers to transform scientific knowledge into actionable public health information.

5 Environmental Distribution and Modelling of Risk

Comprehension of the environment distribution of xenobiotics provides essential insight for the prediction of human and ecological risk. Xenobiotic chemicals are subject to a sequence of transport, transformation and partitioning procedures when released into air, soil or water matrices. These comprise the process of sorption, volatilization, biotransformation, bioaccumulation and trophic transfer, altogether delineating the environment persistence as well as exposure potential (Abraham & D'Souza 2023). Recent UNEP (2022a, 2022b) reports project that over 350 000 synthetic chemicals are registered for international use, of which over 60% are insufficient in the environment fate data, indicating an extensive gap in the risk assessment models.

5.1 Environmental Compartments and Pathways

5.1.1 Atmospheric Transport

Volatile organic compounds (VOCs), halogenated hydrocarbons and particle-bound polycyclic aromatic hydrocarbons (PAHs) are disseminated by atmospheric winds and long-range transboundary movement. In the atmosphere, they are exposed to photochemical oxidation and form secondary air toxics in the form of aldehydes and ozone (OECD, 2020a, 2020b, 2020c). Redistribution by rain deposition and dust resuspension spread the xenobiotics to water and land environments, producing a world pollution cycle (UNEP, 2022a, 2022b).

5.1.2 Aquatic Systems

Rivers, lakes and seas serve as the final sinks for hundreds of xenobiotics such as pharmaceutical residues, cosmetics and pesticides. The wastewater treatment plants in the world remove less than 50% of micropollutants, and most of the metabolites are biologically active (Patel et al., 2020a, 2020b). Persistent organic pollutants (POPs) are highly hydrophobic, allowing them to bioaccumulate in aquatic biota and biomagnify along food chains (Brack et al., 2019a, 2019b). Coastal sediments serve as the secondary sources by resuspension during storms or dredging.

5.1.3 Land and Soil Distribution

Industrial effluents, agricultural chemical application and municipal waste leachates deposit xenobiotics in soils where they encounter organic matter as well as inorganic fractions. Hydrophobic chemicals and heavy metals intensely bond with humic matter, resulting in half-lives spanning a few decades or centuries (Li et al., 2021a, 2021b, 2021c, 2021d). However, climatic conditions, pH, as well as the structure of microbial assemblages, widely impact the kinetics of degradation. Crop-to-soil transfer remains the prime pathway of human exposure by the dietary assimilation of polluted crops (Sharma et al., 2020a, 2020b).

5.2 Modelling Environmental Fate

Risk modelling offers a numerical platform to forecast the concentration dynamics of xenobiotics among compartments. Multimedia fugacity models, including Level III Mackay models, are broadly used to model distribution under non-equilibrium situations (Gavrilescu et al., 2022a, 2022b). These models combine the parameters of partition coefficients (Kow, Koc), the degradation rate constants and advection fluxes. Recent methods involve the use of machine-learning algorithms and Bayesian probabilistic models to enhance accuracy, particularly for intricate mixtures and transformation products (Zhao et al., 2024a, 2024b, 2024c, 2024d). Remote sensing using satellites and geospatial cartographies also enable near-real-time observation of the dispersion of the pollutant, as the case of the European Copernicus Atmospheric Monitoring Service.

5.3 Bioaccumulation and Trophic Transfer

Bioaccumulation and bioconcentration factors (BCF > 2000) are among the best indicators of potential ecotoxicity and human risk. Xenobiotics that are highly bio accumulative, such as polychlorinated biphenyls (PCBs) and per- and polyfluoroalkyl substances (PFAS), accumulate in substantial numbers because they are resistant to metabolic elimination. There are registered PFAS levels in the Arctic polar bears and in remote bird species, as indicators of the world distillation phenomena (Al-Kandari et al., 2022).

The biomagnification index (BMF) measures the movement of pollutants through the food chain, with the top carnivores and humans at the top having the highest aggregate dose. Inclusion of biomonitoring information of freshwater food webs in predictive models provides improved characterization of the risk of dietary exposure.

5.4 Risk Characterization and Exposure Assessment

Environmental risk assessment (ERA) involves the determination of Predicted Environmental Concentration (PEC) and Predicted No-Effect Concentration (PNEC) ratios. PEC/PNEC > 1 indicates potential for ecological risk. The current ERA software, as adopted by the European Chemicals Agency (ECHA) and US EPA, flags the majority of industrial xenobiotics as being of medium to high concern due to persistence and endocrine activity (WHO, 2023a, 2023b, 2023c, 2023d).

In order to integrate human health perspectives, the Aggregate Exposure Pathway (AEP) model links fate in the environment to measures of internal dose. In conjunction with physiologically based pharmacokinetic (PBPK) modelling, these models can narrow the gap between external exposure and biomarker response (OECD, 2020a, 2020b, 2020c).

5.5 Predictive Toxicology Developments and Data Integration

Recent advances in computational toxicology, especially the application of Quantitative Structure–Activity Relationships (QSARs) and adverse outcome pathways

(AOPs), allow the in silico prediction of the toxicity of untested chemicals. The OECD QSAR Toolbox (v 5.0) adds molecular descriptors that are appropriate for persistence, bioaccumulation and genotoxicity. Combination of the biomarkers based on the field of omics (transcriptomics, metabolomics) with exposure modelling constitutes the foundation of systems toxicology, providing the high-resolution in silico prediction of real-world risk scenarios (Abraham & D'Souza, 2023; WHO, 2023a, 2023b, 2023c, 2023d).

5.6 Policy Implications and Sustainable Mitigation

The communication of environment modelling outputs to inform usable policy remains an imperative. Global regulations such as the UN 2030 Agenda for Sustainable Development (Goal 12) advocate good chemical practice and reduction of pollution. Model-based identification of hotspots to help prioritize intervention has now been incorporated into local policies like India's National Clean Air Programme (NCAP) and Integrated Waste Management Rules (UNEP, 2022a, 2022b).

Novel environmentally friendly remediation technologies, including green chemistry approaches, bioremediation using microbial consortia, and advanced oxidation processes (AOPs), are being developed to effectively degrade refractory environmental contaminants. Online, shared-data portals like IPCHEM (European Commission) and India's ENVIS to promote cross-sectorial access ease the availability of data for predictive risk models. Distribution and risk modelling, however, show the complexity in cycling of xenobiotics within ecosystems and recirculation back into humans. Cutting-edge predictive systems are merging computational modelling, biomonitoring and policy interfaces to further define an integrative approach to the management of chemical risks. Nevertheless, voids in knowledge of the information poor countries do motivate building capacity in analytical toxicology, model validation and environmental governance. The goal is the incorporation of scientific modelling with preventative interventions, protecting human genetic integrity and restoring ecosystem vigour.

6 Molecular Biomarkers and Assessment of Heritable Effects

The measurement of xenobiotic-promoted genetic damage is mostly based on are molecular biomarkers responsive to early biological modifications preceding the clinical presentation of diseases. These biomarkers are quantifiable indexes of exposure, effect and susceptibility (WHO, 2023a, 2023b, 2023c, 2023d). Their use in epidemiological work has the potential to place external chemical exposure in relation with genomic instability, oxidative stress and changes in gene expression.

Modern toxicogenomic is a synthesis between classic cytogenetic tests and genome wide ‘omics technologies, providing for the first time a deeply profound visualization of xenobiotic–genomic gene interaction (Zhao et al., 2024a, 2024b, 2024c, 2024d).

6.1 Biomarkers of Exposure

Biomarkers of exposure are the earliest of evidence of environmental toxicology, reflecting the internal chemical dose, not the external concentration. Analytical advances of LC–MS/MS and HR-GC–MS permit the detection of trace doses of xenobiotics and their metabolites in the blood, urine, hair and expired air (Abraham & D'Souza, 2023).

Major examples are:

- Organophosphate metabolites of pesticide exposure (Dialkyl phosphates).
- Urinary bisphenol A and phthalate esters as exposure biomarkers to plasticizers.
- Blood cadmium and lead levels related to metal toxicity and haematological abnormalities (Li et al., 2021a, 2021b, 2021c, 2021d).

New investigations have also indicated the function of the biotransformation enzymes, including cytochrome P450 enzymes of the *CYP1A1, CYP2E1* type, as active biomarkers of metabolic activation of the xenobiotics.

6.2 Biomarkers of Effect

Biomarkers of effect express the biochemical or molecular changes caused by exposure to xenobiotics. These include:

- 8-hydroxy-2′-deoxyguanosine (8-OHdG)—a marker of oxidative DNA damage.
- Malondialdehyde (MDA) and protein carbonyls—indices of lipid and protein oxidation.

Micronucleus frequency and comet assay tail moment—measures of chromosomal breakage and single-strand DNA lesions, respectively (Sharma et al, 2020a, 2020b). For chronic exposure conditions, these biomarkers have dose-dependent increase even in the sub-toxic ranges, vindicating their use in early hazard identification (UNEP, 2022a, 2022b).

6.3 Genetics and Epigenetic Biomarkers

6.3.1 DNA Damage and Repair Markers

Lesions produced by genotoxic xenobiotics are removed by the base excision repair and nucleotide excision repair mechanisms. Altered expression of the *XRCC1, OGG1, APE1* and p53 genes is molecular signatures of damaged genomic integrity (Gavrilescu et al., 2022a, 2022b).

With molecular tests such as quantitative PCR, western blot and RNA-Seq, it is possible to detect transcriptional dysregulation within exposed populations.

6.3.2 Epigenetic Modifications

Epigenetic biomarkers of non-mutagenic but heritable alterations induced by xenobiotics have given us understanding of the same. Global DNA hypomethylation, histone acetylation and miRNA dysregulation have been noted for exposure to heavy metals and persistent organic pollutants (Patel et al., 2020a, 2020b). Such adaptations shift gene silencing and stress-response networks, spanning exposure through long-lasting phenotypic outcomes such as cancer and metabolic disorders.

6.4 Omics-Based Biomarkers

The latest advances in omics technologies, including transcriptomics, proteomics, metabolomics, and epigenomics, have significantly enhanced the understanding of system-wide molecular alterations. Integrated multi-omics analyses enable the identification of key disrupted pathways, such as;

- Nrf2/ARE-mediated antioxidant defence activation,
- NF-κB-driven inflammatory response and
- MAPK and the p53 signal transduction of genotoxic stress (Brack et al., 2019a, 2019b).

Adverse Outcome Pathway paradigm of the OECD (2020a, 2020b, 2020c) connects these molecular signatures to organism effects so toxicity predictive modelling becomes possible. Machine learning technologies are increasingly applied for the purpose of classifying exposure profiles based on omics data, applying precision toxicology.

6.5 Biomarkers of Susceptibility

Individual-to-individual variability of the xenobiotic response is influenced by genetic polymorphisms of detoxification and repair enzymes. *CYP450* mutations, *GSTT1, GSTM1* deletions and polymorphisms of the SOD2 are associated with increased susceptibility to oxidation damage and carcinogenesis (Li et al., 2021a, 2021b, 2021c, 2021d).

Population-based genetic screening could therefore be used to illuminate personalized thresholds of exposure and permit better risk stratification among vulnerable populations like children, prenatal individuals and occupational workers.

6.6 Integrative Biomonitoring and Risk Correlation

Current biomonitoring schemes integrate exposure, effect and susceptibility biomarkers into extensive data sets amenable to modelling for correlating risks.

Studies, e.g. the Global Human Biomonitoring Initiative (2022) and EU HBM4EU Project, harmonize multi-country study protocols. Integrative data thus generated are applied towards the derivation of biological limit values (BLVs) and human biomonitoring guidance values (HBM-GVs) for the purpose of the regulatory response (European Commission, 2023a, 2023b, 2023c).

Statistical systems like Structural Equation Modelling (SEM) and Bayesian inference now permit causal connection of molecular changes and endpoints of disease and link laboratory-based toxicology to outcomes of public health.

Studies of molecular biomarkers are the confluence of exposure science and molecular genetics and allow for a mechanistic explanation of the action of xenobiotics upon the human genome. From classical genotoxicity assays to present omics-based analytics, biomarkers cover the entire continuum of evidence of the type of exposure and functional impairment. Prospects include data integration driven by artificial intelligence, cheap handheld biosensors and multi-omics integration for mass surveillance. Ultimately, the definitive measurement of molecular biomarkers becomes the key to realizing predictive, preventive and personalized toxicology, and crucial for the prevention of human genetic material degradation, particularly through the current rise of environmental complexity.

7 Preventive Strategies and Policy Integration

The rising prevalence of xenobiotic contamination calls for a switch from reactive toxicological assessment to proactive prevention and integration within policy. Current environmental management also focuses on the ‘One Health’ approach, which links ecosystem, human and animal health by reducing the chemical burden

(WHO, 2023a, 2023b, 2023c, 2023d). Accordingly, preventive strategies now incorporate green chemistry, life-cycle risk analysis, biomonitoring and evidence-based regulation for minimum genetic and ecological disruption (UNEP, 2022a, 2022b).

7.1 Source Reduction and Green Chemistry

Source-level elimination of hazardous compounds is the most sustainable form of toxicological prevention. Green chemistry principles advocate designing safer chemicals, the use of renewable feedstocks and the use of benign solvents. For instance, the use of ionic liquids or supercritical CO_2 instead of chlorinated solvents has reduced the generation of mutagenic wastes in industrial processes (Anastas et al., 2021).

Emerging 'safe-by-design' frameworks begin to assess molecular reactivity, biodegradability and bioaccumulation potential before commercialization (OECD, 2020a, 2020b, 2020c). Pharmaceutical and pesticide industries are taking up computational QSAR (Quantitative Structure-Activity Relationship) modelling in order to predict genotoxic potential and prevent the release of DNA-reactive intermediates into the environment.

7.2 Cleaner Production and Circular Economy

Preventive toxicology also extends to cleaner production and circular economy models that minimize the leakage of xenobiotics during manufacture, use and disposal. Extended producer responsibility and chemical leasing promote accountability throughout supply chains (European Commission, 2023a, 2023b, 2023c). Technological interventions include:

- Zero-liquid-discharge effluent systems in pharmaceutical industries,
- Photocatalytic and enzymatic degradation of persistent organics and
- Recovery of heavy metals through bioremediation or electrochemical processes (Patel et al., 2022a, 2022b).

Such innovations reduce environmental load and are in line with the Sustainable Development Goals related to water quality, responsible production and climate action (respectively, SDG 6, 12 and 13).

7.3 Exposure Monitoring and Early-Warning Systems

Integration of biomarker-based monitoring with geospatial surveillance can indeed provide early warnings for xenobiotic hotspots. Real-time sensors for VOCs, nitrates and heavy metals, in concert with satellite-derived pollution indices, enable rapid interventions to take place (Zhao et al., 2024a, 2024b, 2024c, 2024d).

Periodic biomonitoring programmes in occupational groups and sensitive populations can be helpful in the identification of subclinical genetic damage in public health contexts, thus allowing early regulatory response. Mobile diagnostic kits for measuring 8-OHdG or micronucleus frequency are being tried in India and the EU for on-site screening of genotoxicity (Sharma & Ghosh, 2021).

7.4 Risk Communication and Community Awareness

Preventive strategies succeed only when translated into public literacy and behavioural change. Transparent communication of environmental risks—through citizen science platforms, mobile apps and public databases (e.g. ToxCast, REACH)—empowers communities to make informed choices (Gavrilescu, 2022).

Educational programmes incorporating environmental genetics into the school curriculum strengthen grassroots awareness about hereditary impacts of pollution. Community-based risk education has been shown to reduce pesticide misuse by more than 30% in agrarian districts (Li et al., 2021a, 2021b, 2021c, 2021d).

7.5 Regulatory and Policy Frameworks

In general, environmental toxicology is controlled under a mosaic of multilateral and national regulations globally. Some key frameworks include:

- Basel, Rotterdam and Stockholm Conventions (for hazardous waste and POPs),
- EU REACH Regulation (Registration, Evaluation, Authorisation and Restriction of Chemicals),
- India's Chemical Management & Safety Rules 2020 and
- US Toxic Substances Control Act (TSCA) amendments of 2016.

Integration of human biomonitoring data into regulatory decisions has enhanced risk quantification. Today, the EEA (2023) requires biomarker evidence in setting exposure limits; this example is gradually being followed by Asian regulatory bodies.

7.6 Policy Harmonization and International Collaboration

Xenobiotic threats are transboundary; therefore, global policy coherence is essential. Initiatives such as Strategic Approach to International Chemicals Management (SAICM) and Global Alliance on Health and Pollution (GAHP) promote data sharing and capacity building (UNEP, 2022a, 2022b).

India's inclusion in the OECD Mutual Acceptance of Data scheme has facilitated cross-border acceptance of toxicological testing results. Collaboration between ICMR, CPCB and MoEFCC further enhances national surveillance capacity and fosters research–policy synergy.

7.7 Ethical, Socioeconomic and Genetic Considerations

Genetic biomonitoring needs ethical oversight. Policy needs to strike an appropriate balance between the public health and privacy of consented genetic data. The formulation of WHO Genomic Data Protection Guidelines (2024) puts the experiments using biomarkers as compelled to not just human rights but also evidence-based governance. From an economic perspective, preventive toxicology is cost-effective: For instance, WHO has estimated that for every USD spent on preventing of pollution, a return up to 30 times the investment will be realized (WHO). This expresses the double health and economic case for embedding prevention into policy.

At the forefront of environmental toxicology is prevention, requiring a fusion of mechanistic insight and pragmatic governance. Combined with sound regulatory systems, green chemistry, biomonitoring and public engagement are all good strategies for society to use to 'make sure that this trusts situation stays minimized' with regard to both exposure of foreign chemicals and genetic risk. The way to that future is through converging policy and science, AI-enabled surveillance and international alignment so environmental safety and genomic fidelity become mutually reinforcing.

8 Future Directions and Research Innovations

The future of environmental toxicology entails anticipatory precision-based discipline permeated with molecular clarity, digital efficiency and planetary policy harmonization. With the increasing complexity of mixtures, traditional toxicology has become inadequate for testing xenobiotic mixtures. It is future research that thus resides in mechanistic modelling, multi-omics integration and AI-assisted risk prediction following sustainable innovation in material design and environmental monitoring (OECD, 2023).

8.1 Predictive and Computational Toxicology

Now, computational modelling serves as the backbone of next-generation toxicological assessment. The prediction of genotoxic potential before synthesis can be done using QSARs, molecular docking and machine-learning algorithms by Zhao et al. (2024a, 2024b, 2024c, 2024d).

Integrating deep-learning neural networks with toxicogenomic databases, such as Tox21 or Pub Chem Bio Assay, enables automated screening of thousands of chemicals. Predictive toxicology is hence accelerating the regulatory evaluation while reducing dependence on animal testing in accordance with the 3Rs principle (Replacement, Reduction, Refinement) and the EU directives (European Commission, 2023a, 2023b, 2023c).

8.2 Multi-omics and Systems Biology Integration

Convergence of genomics, transcriptomics, proteomics, metabolomics and epigenomics is now driving a new understanding in xenobiotic action. Multi-omics pipelines give holistic views on molecular perturbation networks, key biomarkers and regulatory nodes by Brack et al. (2022).

Future research is directed towards the establishment of systems toxicology databases that correlate exposure signatures with human disease outcomes. For instance, network analyses of xenobiotic-responsive genes such as Nrf2, p53 and AHR revealed a cross-talk among the oxidative, inflammatory and apoptotic pathways (Li et al., 2021a, 2021b, 2021c, 2021d). Such integrated datasets will underpin the expansion of AOPs and allow the computational extrapolation from in vitro to in vivo effects.

8.3 Human Organoid and Organ-on-Chip Technologies

New models developed for human relevance bridge the translational gap that exists from laboratory testing into clinical reality. Organoid and microfluidic organ-on-chip systems replicate physiological microenvironments, enabling dynamic assessment of xenobiotic metabolism, DNA repair and immune crosstalk (Sharma & Ghosh, 2022).

Validation of liver- and kidney-on-chip models for testing pharmaceuticals and agrochemicals will provide real-time biomarkers of genotoxic stress. Integration with biosensors and microelectrode arrays allows the continuous monitoring of cellular oxidative states, enabling ethical and reproducible toxicology research.

8.4 Artificial Intelligence and Big-Data Toxicology

The future of environmental toxicology lies in AI-driven analytics that will turn heterogeneous omics, environmental and clinical datasets into actionable insights.

Knowledge graphs, Bayesian inference models and reinforcement learning algorithms are now applied to identify unknown exposure–effect relationships (Patel et al., 2022a, 2022b).

AI-powered early warning systems can merge meteorological data with data of industrial emissions to predict pollution spikes, thus enabling public health interventions. Biomonitoring platforms will link wearable sensors to environmental databases and allow for near real-time mapping of genotoxic exposure risk.

8.5 Nanotoxicology and Emerging Contaminants

Fast growth in nanotechnology and new-generation materials opens new, unexplored toxicological frontiers. ENPs, micro- and nanoplastics and new flame retardants have unique surface reactivity and bio-persistence. Their detailed, long-term genotoxicity, transport across barriers and epigenetic modulation remain to be determined by future studies (Gavrilescu et al., 2023).

Advanced characterization tools—such as atomic-force microscopy, synchrotron spectroscopy and single-cell sequencing—will elucidate nano-bio interactions, guiding safe-by-design synthesis of next-generation materials.

8.6 Environmental DNA (eDNA) and Metagenomic Surveillance

Application of environmental DNA and metagenomics is a major recent innovation in ecological toxicology. eDNA sampling from air, soil and water enables detection of microbial and genetic perturbations caused by xenobiotic contamination (UNEP, 2022a, 2022b).

Community-level genetic responses provide a way of inferring ecosystem health and resilience. Integration of CRISPR-based detection systems with eDNA analytics is expected to enable real-time bio surveillance of genotoxic pollutants in complex habitats.

8.7 Policy-Science Synergy and Capacity Building

Cross cutting policy integration will also be critical for the way forward. Such harmonized global frameworks—bridging WHO one-health agenda, UNEP global chemicals outlook and OECD AOP platforms—will be needed to allow data sharing and consistent risk assessment (WHO, 2024). Supporting South-South partnerships, regional toxico genomics centres and open-source exposure databases can strengthen readiness of developing nations to respond to chemical threats without compromising genomic health.

8.8 *Ethical and Societal Dimensions*

Once genomics and AI infiltrate toxicology, ethical oversight will be needed to ensure transparency, privacy of data and equitable sharing of the benefits. Creation of bioethics committees for both environmental genomics and public participation models will be important. Furthermore, citizen science crowdsourcing networks for collection of biomarker data can democratize exposure mapping and help build public trust in science risk communication within a society. The next 10 years will be a period of rebirth for environmental toxicology, converging together molecular precision, digital intelligence and sustainability ethics. New frontiers in our ability to predict and prevent xenobiotic-induced genetic injury will undoubtedly be achieved with emerging technologies, including organ-on-chip testing and AI-based exposomics. In the end, it is only by globally integrated ethical and evidence-based toxicology that protection of human genomic integrity and planetary health will be ensured against accelerating anthropogenic change.

9 Policy Frameworks, Ethical Considerations and Global Governance of Xenobiotics

The control and steering policy for xenobiotics has evolved from fragmented national acts to globally coordinated regulations, which together converge environmental safety, protection of human health and permanent accessibility. New regulations such as EU REACH (Registration, Evaluation, Authorization and Restriction of Chemicals), USA TSCA (Toxic Substances Control Act) or India's Draft CMSR (Chemical Management and Safety Rules) 2020 have paved the way for an emerging 'no data, no market' policy. To do so, testing has to be done in toxicology, ecotoxicology and genotoxic of every chemical substance prior their commercial release (European Commission, 2023a, 2023b, 2023c; MoEFCC, 2020). These frameworks also call for cautious risk assessment, extended producer responsibility (EPR) and transparency along the lifecycle of a substance—from an idea to its waste form.

9.1 Global Governance and Harmonization

The Strategic Approach to International Chemicals Management (SAICM) under UNEP and WHO has placed chemical safety in the general framework of sustainable development (UNEP, 2022a, 2022b). Chemical safety is specifically mentioned in several of the targets of the 2030 Agenda for Sustainable Development, notably under SDG 3.9 and under SDG 12.4, thereby reaffirming the collective objective to prevent deaths and illnesses from chemical exposure as well as pollution. The OECD Test Guidelines Programme provides a key driver in the standardization of toxicological test methods with sorted conflicts and animal use reduced through validated in vitro and (Q)SAR alternative methodologies (OECD, 2023).

9.2 Ethical Dimensions and Data Transparency

Ethical issues in environmental toxicology encompass not just the welfare of members of experimental species, and even vulnerable human populations more easily exposed to xenobiotics, but also native populations and ecosystems. Contemporary methods applied 3Rs (Replacement, Reduction, Refinement) in toxicological testing (Russell & Burch, 2021). And there are also open-data efforts, like the OECD eChem Portal and US EPA Comp Tox Chemicals Dashboard that promote transparency by making chemical hazard data available to the public at no cost. These platforms democratize environmental health knowledge and enable predictive modelling and AI-based hazard assessment (Patel et al., 2022a, 2022b).

9.3 The Ethics of Emerging Technologies

Emerging technologies such as AI-based predictive toxicology, the engineering of nanomaterials and gene editing technologies create new policy implications (including implications for ethics of responsibility; algorithmic bias and unintended ecological impacts (Brack et al., 2022; Gavrilescu et al., 2023). For instance, models developed using machine learning techniques can perpetuate systemic bias when trained on a database of incomplete or outdated toxicity data that leads to the misclassification of chemical hazards. Likewise, potential deleterious health effects of nanomaterial delivery/removal are also expected such as prolonged bioaccumulation and intergenerational genotoxicity. Therefore, an ethical governance would request algorithmic transparency, stakeholder consent and post-market setting surveillance.

9.4 Equity, Justice and Environmental Policy

The convergence of toxicology and social justice (environmental justice) recognizes that the distribution of xenobiotic exposure does not unfold uniformly across different socioeconomic strata (WHO, 2023a, 2023b, 2023c, 2023d). Industrial zoning close to poor or underprivileged communities increases the chance of a chronic exposure. Inclusion of equity metrics into environmental risk assessment is a mechanism for policies to target interventions in communities most impacted by pollution-associated genotoxicity. Furthermore, there is a network of international conventions such as the Basel, Rotterdam and Stockholm Conventions that also regulate trade in hazardous waste, pesticide usage and POPs (persistent organic pollutants) and harmonize chemical safety with human rights systems (UNEP, 2022a, 2022b).

9.5 Future Policy Outlook

The new 'Global Framework on Chemicals—Beyond 2020' aims to expand governance from chemical safety towards circular-economy-oriented chemicals design, encompassing green chemistry and life-cycle thinking (Anastas et al., 2021). Ethical preparation for the future will increasingly depend on toxicology as a field of study, of policy-making, ethical reflection and civil-society action. In India, the progress made under initiatives of CPCB and MoEFCC of establishing a National Chemicals Management Authority reflects an intuitional acceptance to genetic integrity (and ecosystem) integrity from xenobiotic challenge.

10 Conclusion and Key Recommendations

The contemporary environmental toxicology has evolved into an interdisciplinary border which involves chemistry, genomics, ecology and data science for decoding the intricate relationships between xenobiotics and life. In fact, an accumulating evidence base stresses that exposure to human toxicants and waste-by-products (including pesticides, heavy metals, industrial solvents, pharmaceuticals and microplastics) triggers molecular/genotoxic cascades resulting in a loss of human/ecological resilience (Gavrilescu et al., 2023; WHO, 2023a, 2023b, 2023c, 2023d). The identification of xenobiotics as a cause for genomic instability, endocrinologic disruption and modification of epigenomes has established their importance as factors in planetary health.

10.1 Holistic Approach to Xenobiotic Hazards

In various sections, this chapter has shown that the genotoxic effects of xenobiotics are multifaceted and are influenced by chemical structure, persistence in the environment, bioavailability and synergistic actions. Xenobiotic residues in air, water or even human biological matrices are also cloudy and according to biomonitoring studies they have shown significant association with biomarkers of effect like oxidative stress and DNA strand breaks (Singh et al., 2022; WHO, 2023a, 2023b, 2023c, 2023d). This is the type of evidence that stresses the inappropriateness of single-compound toxicity tests—and, finally, the necessity of mixture toxicology concepts that resemble nature on-site (Brack et al., 2022).

The integration of several omics like the transcriptome, the metabolome and epigenome has modified predictive capabilities in toxicology. These systems are mechanistically connecting molecular perturbations to phenotypes that might predict early warning for disease and environmental harm (Anastas et al., 2021; OECD, 2023).

10.2 Future of Research and Innovation

Environmental toxicology in the twenty-first century: Challenges and opportunities that could derive from the digital revolution and precision toxicity. Artificial Intelligence, machine learning and high throughput screening technologies will take over traditional assays with data-enabled models predicting genotoxicity up to subcellular level even in the absence of active chemicals (Patel et al., 2022a, 2022b). The creation of adverse outcome pathway frameworks will deliver a coherent mechanistic road map from molecular initiating events to population-level outcomes.

Equally crucial is the emergence of 'green' chemistry principles in molecular design for low toxicity or lack of environmental persistence. Sustainable materials and greener routes to public interest materials will serve as the foundations of circular chemical economies resulting in xenobiotic load levels similar to, for example, Anastas and Erythropel (2021). But that progress must be undergirded by ethics: The tech leap needs to promote equity, justice and transparency in environmental decision-making.

10.3 Policy and Governance Priorities

At the policy level, planetary governance must advance from reacting to acting proactively, embedding early warnings and cumulative exposure tracking within national frameworks (MoEFCC, 2020; UNEP, 2022a, 2022b). Revision of the current international conventions (Stockholm, Rotterdam and Basel Convention) might contribute to 6 regulating [trans-border transport of hazardous substance]. Concurrently, the emerging chemical regulation in India being coordinated through MoEFCC and CPCB will provide an opportune time to include genetic toxicology in environmental risk assessment to develop National Chemical Safety Database (NCSD) for public health surveillance.

This will necessitate a parallel development in the oversight of ethics to mandate open-access data-sharing, algorithmic accountability and post-market biomonitoring for new compounds and nanomaterials. The interactional model among academy, regulators and industry that facilitates the translation of xenobiotic research to policy outcomes is vital.

10.4 Key Recommendations

- Adopt Systems Toxicology: Combine omics, modelling and AI to predict complex xenobiotic interactions and genotoxic effects.
- Foster the research and production of more benign products through toxicity prediction tools and life cycle assessment.
- Regional Biomonitoring Programmes: Establish programmes for trends monitoring of xenobiotic exposure in at-risk populations.
- To Improve Ethical and Transparency Framework: The open databases of toxicity data will prevent scientific bias, and the validation of machine learning rule sets should be a mandatory requirement.
- Foster Global Collaboration: Harmonize country toxicology strategies with those of the international community (e.g. UNEP Beyond 2020 Framework; WHO Health and Pollution Agenda).
- Education and Understanding: Create an awareness programme by integrating the environmental toxicology curricula into tertiary education so as to train future sustainable scientists.

Final Perspective: The xenobiotic toxic dynamics would be a human ethics responsibility for all species' protection. And in the future of environmental toxicology it will be a predictable, preventive and personalized intervention on chemical risk. Interdisciplinary work—inclusive of pharmacology, data science, environment policy and ethics—is set to drive the emerging frontier in a toxic-free & genomically sustainable globe.

References

Abraham, J., & D'Souza, R. (2023). Integrated waste management approaches for sustainable environmental protection in developing countries. *Waste Management, 168,* 45–58.

Ali, H., Khan, E., & Ilahi, I. (2023). Environmental chemistry and ecotoxicology of emerging xenobiotic contaminants. *Chemosphere, 318*, 137923.

Al-Kandari, H., Rana, S., & Al-Ghafari, A. (2022). Global distribution and persistence of environmental contaminants: A meta-analysis. *Nature Sustainability, 5*, 784–796.
Anastas, P. T., & Erythropel, H. C. (2021). Green chemistry: Principles and practice in the 21st century. *Nature Reviews Chemistry, 5*(5), 301–316.
Anastas, P. T., Zimmerman, J. B., & Erythropel, H. C. (2021). Green chemistry: Principles and practice in the 21st century. *Nature Reviews Chemistry, 5*(5), 301–316. https://doi.org/10.1038/s41570-021-00284-3
Brack, W., Alygizakis, N., & Schüürmann, G. (2019a). Integrating omics data in environmental toxicology: Pathway analysis and systems-level interpretation. *Nature Reviews Genetics, 20*(12), 768–784.
Brack, W., et al. (2019b). Towards a holistic understanding of contaminants of emerging concern in aquatic systems. *Science of the Total Environment, 678*, 267–282.
Brack, W., Alygizakis, N., & Schüürmann, G. (2022). Integrating omics and adverse outcome pathways for next-generation toxicology. *Environmental Toxicology and Chemistry, 41*(8), 1664–1682.
Chakraborty, P., Adhikary, S., Bhattacharya, S., Dutta, S., & Chatterjee, S. (2021). Pharmaceuticals and personal care products as emerging environmental contaminants: Prevalence, toxicity, and remedial approaches. *ACS Chemical Health & Safety, 28*(6), 387–399.
Dolinoy, D. C., et al. (2016). Environmental epigenomics in human health and disease. *Annual Review of Pharmacology and Toxicology, 56*, 81–99.
European Commission. (2023a). Chemicals strategy for sustainability: Towards a toxic-free environment. Publications Office of the European Union.
European Commission. (2023b). Human biomonitoring for Europe (HBM4EU): Final technical report. Publications Office of the European Union.
European Commission. (2023c). IPCHEM: The information platform for chemical monitoring—Data integration report 2023. European Environment Agency.
European Environment Agency (EEA). (2023). Human biomonitoring data integration in chemical risk assessment. EEA Technical Report.
Gavrilescu, M. (2022). Environmental governance and communication in toxicology: Engaging citizens for sustainable pollution control. *Environmental Science and Pollution Research, 29*(11), 15542–15558.
Gavrilescu, M., Cristea, V., & Kumar, P. (2022a). DNA repair gene expression profiling in environmental genotoxicity studies. *Environmental Pollution, 310*, Article 119824.
Gavrilescu, M., et al. (2022b). Emerging contaminants in the environment: Challenges and perspectives. *Environmental Engineering Science, 39*(4), 355–374.
Gavrilescu, M., Kumar, P., & Cristea, V. (2023). Nanomaterials and environmental safety: Challenges for genotoxicity testing and regulation. *Journal of Hazardous Materials, 451*, 131–149.
Ghosh, R., Banerjee, S., & Singh, A. (2019). Mechanistic insight into xenobiotic-induced genotoxicity. *Toxicology Mechanisms and Methods, 29*(2), 97–108.
Global Human Biomonitoring Initiative. (2022). Human biomonitoring data summary report 2022. European Environment Agency.
Greally, J. M., & Jacobs, M. N. (2020). Inheritance of environmental exposures through epigenetic mechanisms. *Nature Reviews Endocrinology, 16*(9), 512–526.
Kümmerer, K. (2017). *Pharmaceuticals in the environment: Sources, fate, effects, and risks*. Springer Nature.
Li, Y., Hu, W., & Zhang, X. (2021a). Community-based interventions in pesticide risk communication and exposure reduction. *Environmental Health Perspectives, 129*(9), 96003.
Li, Y., Zhang, X., & Wu, J. (2021b). Human biomonitoring of heavy metals: Exposure biomarkers, susceptibility genes, and health implications. *Environmental Research, 196*, Article 110987.
Li, X., Zhang, H., & Zhou, Q. (2021c). Environmental behavior and risk assessment of heavy metals. *Science of the Total Environment, 765*, 144–158.

Li, Y., Zhang, X., & Wu, J. (2021d). Multi-omics integration for exposome research: Applications to environmental genotoxicants. *Environmental Research, 196*, Article 110987.

Liu, H., Zhang, T., & Sun, Q. (2024). Advances in biomonitoring and toxicogenomics of xenobiotic pollutants. *Ecotoxicology and Environmental Safety, 276*, 116287.

Ministry of Environment, Forest and Climate Change (MoEFCC). (2016). *Solid Waste Management Rules, 2016*. Government of India, New Delhi.

Ministry of Environment, Forest and Climate Change (MoEFCC). (2020). *Draft chemical (management and safety) rules*. Government of India.

Ministry of Environment, Forest and Climate Change (MoEFCC). (2023). *National clean air programme (NCAP): Progress report and implementation framework*. Government of India, New Delhi.

OECD. (2020a). Guidance on developing and assessing adverse outcome pathways (AOPs). Organisation for Economic Co-operation and Development.

OECD. (2020b). Guidance on key considerations for the safe-by-design approach in chemicals and materials. Organisation for Economic Co-operation and Development.

OECD. (2020c). Global chemicals outlook II—From legacies to innovative solutions. Organisation for Economic Co-operation and Development.

OECD (2023). Guidance on digital and computational approaches in chemical safety assessment. OECD.

Patel, D. M., Rao, P., & Mehta, N. (2020a). Epigenetic signatures of environmental pollutants: DNA methylation and miRNA perturbations in human exposure cohorts. *Environmental Epigenetics, 6*(1), dvaa004.

Patel, R., Singh, J., & Mehta, S. (2020b). Micropollutant removal in wastewater treatment plants. *Water Research Advances, 5*, 100–112.

Patel, R., Mehta, D., & Singh, P. (2022a). Circular economy innovations for chemical pollution mitigation: Integrating cleaner production with toxicology. *Journal of Cleaner Production, 370*, Article 133492.

Patel, R., Mehta, D., & Singh, P. (2022b). Artificial intelligence and big-data approaches for environmental health surveillance: Applications in pollutant forecasting and exposure mapping. *Journal of Cleaner Production, 370*, Article 133492.

Rahman, S., et al. (2023). Heavy metal toxicity: Inflammatory mechanisms and health impacts. *Environmental Pollution, 323*, 121–138.

Rahman, M. M., Islam, M. S., & Karim, M. R. (2024). Environmental and human health risks of persistent xenobiotic compounds: Current perspectives and future challenges. *Environmental Research, 245,* 118102.

Russell, W. M. S., & Burch, R. L. (2021). *The principles of humane experimental technique*. Methuen.

Scheringer, M., Schulz, R., MacLeod, M., & Cousins, I. T. (2023). Stories of global chemical pollution: Will we ever understand environmental persistence? *Environmental Science & Technology, 57*(4), 1452–1460.

Sharma, S., Gupta, P., & Khurana, R. (2020a). Micronucleus and comet assay biomarkers for chronic pesticide exposure: A comparative field study. *Toxicology Reports, 7*, 1101–1110.

Sharma, P., Gupta, N., & Singh, S. (2020b). Industrial xenobiotics: Environmental persistence and toxicity. *Toxicology Letters, 334*, 45–57.

Sharma, N., & Ghosh, A. (2021). Portable biosensing technologies for rapid genotoxicity detection in field environments. *Sensors and Actuators B: Chemical, 340*, Article 129957.

Sharma, N., & Ghosh, A. (2022). Organoids and organ-on-chip platforms for human-relevant genotoxicity testing. *Advanced Drug Delivery Reviews, 183*, 114–128.

Shen, J., Zhao, C., & Liu, Y. (2018). DNA adducts and mutagenesis in environmental toxicology. *Mutation Research, 828*, 1–13.

Sies, H., & Jones, D. P. (2020). Reactive oxygen species and cellular redox homeostasis. *Nature Reviews Molecular Cell Biology, 21*(7), 363–383.

Singh, A., & Kumar, V. (2019). Xenobiotics, oxidative stress, and antioxidant defense mechanisms in human health. *Drug and Chemical Toxicology, 42*(4), 377–387.

Singh, R., Tiwari, D., & Bhattacharya, P. (2022). Molecular biomarkers of xenobiotic-induced DNA damage: Advances and translational relevance. *Mutation Research/genetic Toxicology and Environmental Mutagenesis, 874*, Article 503458.

Stockholm Convention. (2023). *Report of the conference of the parties to the stockholm convention on persistent organic pollutants*. United Nations Environment Programme (UNEP), Geneva, Switzerland.

UNEP. (2022a). Global chemicals outlook III: Towards a resilient and healthy planet. United Nations Environment Programme.

UNEP. (2022b). Global chemicals and waste outlook 2022: Towards a pollution-free planet. United Nations Environment Programme.

Vandenberg, L. N., Nadal, A., & Zoeller, R. T. (2024). Endocrine-disrupting chemicals and public health protection strategies. *Nature Reviews Endocrinology, 20*(2), 85–101.

Walker, T. R., Blais, M., & Xavier, J. C. (2023). Emerging concerns associated with xenobiotic pollutants and environmental sustainability. *Environmental Pollution, 325*, 121456.

Wang, L., Chen, Y., & Zhao, P. (2023). Urban pollutants and endocrine disruption: A global review. *Environmental Pollution, 324*, 121–139.

Wang, L., Chen, X., & Li, Y. (2024). Xenobiotic-induced oxidative stress and inflammatory signaling in environmental toxicology. *Science of the Total Environment, 912*, 168742.

WHO. (2023a). Human biomonitoring and environmental health surveillance: Methodological framework. World Health Organization.

WHO. (2023b). One health joint plan of action (2022–2026): Preventing and mitigating health risks from environmental contaminants. World Health Organization.

WHO. (2023c). Global status report on health and pollution 2023. World Health Organization.

WHO. (2023d). State of the science on endocrine disrupting chemicals—Update 2023. World Health Organization.

WHO. (2024). Ethical guidelines for human genomic data and environmental health research. World Health Organization.

Yadav, R., Tripathi, A., & Prasad, R. (2021). Pesticide-induced oxidative stress and genotoxicity in mammals. *Environmental Toxicology and Pharmacology, 84*, Article 103601.

Zhang, H., Wu, J., & Li, X. (2022). Molecular basis of benzene toxicity and chromosomal damage. *Toxicology Reports, 9*, 51–64.

Zhang, Y., & Zhao, L. (2024). Xenobiotic exposure and oxidative stress-mediated toxicological pathways in humans. *Journal of Hazardous Materials, 462*, 132327.

Zhao, L., Qian, M., & Han, M. (2024a). Smart sensing and AI-based pollutant tracking for environmental health monitoring. *Nature Sustainability, 7*(2), 118–130.

Zhao, L., Qin, X., & Han, M. (2024b). Toxicogenomic insights into xenobiotic-induced oxidative stress and DNA damage response pathways. *Frontiers in Toxicology, 2*, 143–155.

Zhao, D., Chen, L., & Wang, Q. (2024c). Transformation products of xenobiotics and their environmental relevance. *Nature Sustainability, 7*, 654–669.

Zhao, L., Qian, M., & Han, M. (2024d). Environmental DNA (eDNA) and metagenomic surveillance in pollution monitoring: Prospects for real-time biosurveillance. *Nature Sustainability, 7*(2), 118–130.

Chapter 2
Xenobiotics: Sources, Fate, and Transport in the Human Body

Udaykumar Reddy, Linju Joy, Usha Padmanabhan, and Ashish Jain

Abstract Xenobiotics are undesirable and hazardous molecules that are not normally formed or expected to occur in the human body. They enter the body through environmental exposures. These xenobiotics, originating from a variety of sources, can be categorized as industrial pollutants, agrochemicals, food additives, pharmaceuticals, combustion by-products, household products, cosmetics, and naturally occurring phytochemicals and poisons. Their primary route of entry into the body is through ingestion of contaminated food and water, inhalation, or skin absorption. After introduction into the body, these xenobiotics are transported and accumulated in different body tissues, which is decided by their molecular makeup, solubility, and interactions with different tissues and plasma proteins. Lipid-soluble substances accumulate in adipose tissues, while hydrophilic substances predominantly exist in blood plasma and interstitial fluids. Xenobiotic metabolism is predominantly mediated by the liver, which uses enzymatic mechanisms to transform lipophilic chemicals into water-soluble compounds for easier excretion. Significant biotransformation also occurs in the gastrointestinal tract and other tissues. After these biotransformation reactions, the majority of xenobiotic compounds are excreted via urine, faeces, and respiratory routes. Inadequate excretion can lead to their accumulation, which can have negative health consequences such as oxidative stress, DNA damage, chronic inflammation, an unbalanced gut flora, and cancer. The magnitude of their impact may vary based on multiple factors, including dosage, age, polymorphism in metabolic enzymes, nutritional status, health status, and genetic predisposition. To

U. Reddy · L. Joy · A. Jain (✉)
Department of Microbiology, Smt. Chandibai Himathmal Mansukhani College, Ulhasnagar, Thane, Maharashtra, India
e-mail: ajcdri@gmail.com

U. Reddy
e-mail: uday4microbio@gmail.com

L. Joy
e-mail: linjujoy90@gmail.com

U. Reddy · U. Padmanabhan
Department of Cell Biology & Virology, Haffkine Institute for Training Research and Testing, Parel, Mumbai, India
e-mail: U.padmanabhan@haffkineinstitute.org

K. Priya and S. K. Giri (eds.), *Environmental Toxicology: Xenobiotics and Effects on Human Genetic Material*, https://doi.org/10.1007/978-3-032-28619-2_2

assess human risk, drug development, and to formulate anti-xenobiotic therapies, it is essential to understand their exposure, absorption, metabolism, and excretion. The proposed book chapter outlines the sources, transport modes, metabolism, accumulation, and excretion of xenobiotic substances, providing crucial information about their possible health risks.

Keywords Xenobiotics · Hazardous molecules · Inflammation · Transport modes · Biotransformation

1 Introduction

Progress in technology in the twentieth century permitted the production and widespread use of a wide range of chemical compounds aimed to improve human existence, including antibiotics, insecticides, dyes, personal care products (PCPs), and different additives. However, many of these chemicals are either missing in natural settings or present at much lower quantities than those produced by anthropogenic activity. The primary challenge arises from their different physicochemical properties, such as size, degree of ionization, solubility in water, and volatile nature, which impact their environmental behavior and persistence (Štefanac et al., 2021). Xenobiotics are chemical substances that are not produced by the body and are considered to be incompatible with standard biological functions. There are several factors, including dosage, exposure time, chemical constituents of the drug, and individual differences in hepatic metabolism and detoxification, that impact their physiological and toxicological effects (Shreenidhi et al., 2024). To reduce the harmful effects of xenobiotics and identify major environmental contaminants, many global and regional regulatory bodies have researched how these hazardous compounds affect humans and animals. They have also assessed pollutant levels in various ecosystems and developed sophisticated analytical instruments for the detection and surveillance of these harmful substances (Štefanac et al., 2021). The concentration of these dangerous xenobiotics in the environment may fluctuate from high to low, depending on their geographic distribution (Atashgahi et al., 2018). Xenobiotics, particularly those derived from pesticides, are a significant contributor to cancer globally. They lead to mutation and increase cancer risk, induce immune response, cause inflammation, disrupt hormonal balance, and alter crucial metabolic pathways (Kumar et al., 2022). Continuous or prolonged exposure to these xenobiotics has been linked to the development of autoimmune diseases, metabolic problems such as diabetes, and neurological diseases like Parkinson's disease (Sajid & Agrewala, 2019). Drugs and other xenobiotic substances can alter the composition and function of gut flora (Pant et al., 2023). The makeup and activity of gut microbes can be changed by drugs and other xenobiotic molecules. Furthermore, these drugs can cause epigenetic changes that increase the risk of disease later in life if they are exposed during crucial phases of development, such as embryonic development and the neonatal phase (Sobral et al., 2025). Human activities like the dumping of household trash, industrial discharges, and agricultural practices are the primary means by which xenobiotic chemicals enter the environment and ecosystems. These elements are susceptible to physical, chemical, and biological transformations that could make it impossible for them to continue in their initial configurations (Ortiz et al., 2022).

This chapter reviews the origins of xenobiotic substances, how they enter living organisms, how they interact with biological systems, how the human body processes them, and the potential health risks linked to their presence and transformation.

2 Sources of Xenobiotics

Xenobiotics are substances that are introduced into the environment mainly via human activity rather than their natural existence. These include synthetic medicines, fertilizers, plastics, dyes, pesticides, heavy metals, environmental pollutants, and other industrial chemicals (Dirbeba Dinka, 2018). The most extensively researched xenobiotic groups, such as pesticides, preservatives, plasticizers, personal care products, dyes, and pigments, are now found in a wide range of substances, including agricultural chemicals, pharmaceuticals, cosmetics, flavorings, fragrances, food additives, industrial chemicals, and environmental pollutants (Ortiz et al., 2022). These pollutants spread throughout the environment as a result of human activities, leading to environmental damage. They can be absorbed by plants or swallowed by aquatic living organisms, eventually affecting human health, thereby creating a continual and destructive cycle (Singh et al., 2024). Human exposure to xenobiotics is extensive, with people possibly meeting between 1 and three million such molecules over their lifetime. Although some of these substances may be harmless, while others are toxic, the body consistently recognizes them as foreign elements (Johnson et al., 2012). Industries that process oil and gas, thermal power industries, and transportation networks are among the main sources of modern toxic xenobiotics. These xenobiotics are also found in food products, where they appear as colorants, emulsifiers, stabilizers, thickeners, flavouring agents, and preservatives (Kucherenko et al., 2021). Metals such as lead are regarded as important environmental contaminants because of their toxicity, long-term persistence, ability to bioaccumulate, and detrimental consequences on human health (Tam & Wong, 2000), cigarette smoking, eating contaminated vegetables, and breathing in air contaminated by fuel combustion are the major ways that the general public is exposed to cadmium. Smokers are constantly exposed to cadmium since tobacco has a high cadmium level. Actually, 10% of the total cadmium in tobacco is usually found in cigarettes (Jaishankar et al., 2014; Kim et al., 2015). Over 200 chemicals generated from petroleum and about 10,000 polycyclic aromatic hydrocarbons are carcinogenic in nature. Among them, the EPA has selected 16 chemicals as priority pollutants (Minteer et al., 2007). Benzo[a]pyrene, a main representative of polycyclic aromatic hydrocarbons, has been regularly discovered in various environmental samples (Bukowska et al., 2022). Chemical dyes are extensively utilized in the clothing industry and are considered the largest source of dyes in industrial wastewater, with an estimated global discharge of around 280,000 tons per year. Many of these colorants are highly toxic and harmful, causing significant negative impacts on the ecosystems they enter (Gupta et al., 2022; Jin et al., 2007; Kuhad et al., 2004). One of the main issues is their physicochemical qualities, which include tiny molecular dimensions, ionization ability, ability to dissolve in water, nature of fatty acids, and volatile nature (de Oliveira et al., 2020). Through food, water, and the environment, some commodities, such as pesticides and insecticides, pose serious risks to people. Foodborne pathogens, preservatives, and artificial sweeteners pose significant health risks (Goel, 2021). Certain high-temperature cooking processes produce xenobiotic compounds found in meat and fish, such as nitrosamines and heterocyclic amines, which have been linked to carcinogenesis (Nogacka et al., 2019). Pollutants can enter the natural world either directly or indirectly through sources like hospital effluents. Furthermore, xenobiotics might be released either during the manufacturing process or as part of the finished product (Embrandiri et al., 2016). Pharmaceuticals and endocrine-disrupting chemicals are a structurally diverse category of organic pollutants found worldwide in wastewater, surface water, groundwater, estuarine ecosystems, and drinking water (Gerrity et al., 2011).

3 Entry of Xenobiotics

Every day, people are exposed to xenobiotics via skin contact, respiration, and consumption. Humans may consume these substances knowingly, through supplements or pharmaceuticals, or unknowingly, through contaminated food and water. Once ingested, the gut microbes can metabolize and modify xenobiotics both before and after absorption (Lindell et al., 2022). In most of the living organisms, consuming contaminated food serves as the main route of exposure (Calatayud Arroyo et al., 2021). Dietary intake and occupational activities are common routes of toxic exposure to pesticides (Fig. 1). Occupational exposure is primarily related to agricultural activities, although personnel in pesticide manufacture and formulation, road and railway construction workers, those working in greenhouses, forestry, and nurseries are also at significant risk of contact with neurotoxic pesticides (Caroline & Frederic, 2012). Polycyclic aromatic hydrocarbons are present in a wide range of foods—which includes oils, cereals, and various vegetables—when exposed to thermal cooking techniques such as baking, frying, and grilling, as well as during food processing (Nogacka et al., 2019). Evidence has demonstrated the relationship between various amines and polycyclic aromatic hydrocarbons that are principal genotoxic substances created during cooking and food preparation (Nogacka et al., 2019). The thalidomide disaster demonstrated that drugs are capable of crossing the placenta and potentially causing harmful effects on fetal development (Myllynen et al., 2007). Dietary intake represents the primary pathway through which humans are exposed to xenobiotics (Abdelsalam et al., 2020). When taken orally, xenobiotics enter the upper gastrointestinal system, and those that are absorbed are delivered to the liver via the hepatic portal vein, an event generally referred to as the first-pass effect (Abdelsalam et al., 2020). When substances are ingested, they may be absorbed in the small intestine and processed by the liver, or they may reach the colon, where the gut microbiota can further metabolize them, acting much like an additional detoxifying organ. Moreover, compounds that were absorbed earlier or entered the body through different exposure pathways can be excreted into the intestine through bile, allowing them to undergo additional interactions with the gut microbiota (Roberts et al., 2002; Zimmermann et al., 2019). Benzo[a]pyrene is a polycyclic aromatic hydrocarbon, mainly originates from the incomplete combustion of organic matter. Main sources are smoke generated from tobacco, wood, and grilled food smoke, along with vehicular and industrial exhaust. The body encounters Benzo[a]pyrene mainly through inhalation of smoke, ingestion of food, and dermal contact with coal tar products (Bukowska et al., 2022). Particulate matter can enter the human body through several mechanisms, each associated with distinct health risks. Inhalation is the primary route, directly impacting the lungs and allowing compounds to enter the bloodstream. Ingestion affects the digestive system, while exposure through the eyes or skin results in localized surface effects (Lang et al., 2025). The extent of xenobiotic absorption and distribution within the body is determined by its chemical properties, including polarity, molecular mass, and lipid solubility, as well as the functioning of biological barriers such as the acidic environment of the stomach and the blood-brain barrier (Wu et al., 2023). Tobacco smoke, a complex combination of hundreds of chemicals, is significantly associated with colorectal cancer (CRC), albeit the actual processes behind this relationship are complicated and multivariate (Aguayo et al., 2025; Jin et al., 2023). Heavy metal exposure can occur through polluted air, food, or water, as well as hazardous occupational environments. Key sources include environmental media (water, air, soil, dust), occupational activities, medicinal products, and diet (Goyer et al., 2004; Neal & Guilarte, 2013). Endogenous xenobiotics are chemicals generated by the human body, such as acids from bile, bilirubin, cholesterol, eicosanoids, among others steroid hormones and certain fatty acids. These compounds resemble environmental xenobiotics in structure and activity. Endogenous xenobiotics also include bacterial toxins, cytotoxins, and herbal toxins, which are created as part of an organism's defensive system (Mishra et al., 2019). Xenobiotics must pass through the intestinal epithelium, basement membrane, and capillary endothelium before entering the circulation. In mammals, xenobiotics are not absorbed via specialized transport mechanisms; instead, they utilize the same pathways as nutrients. These include active transport, pinocytosis, filtration through pores, lymphatic absorption, and passive diffusion (Chhabra, 1979).

Active Transport—Active transport mechanisms allow solutes, including xenobiotics, to enter the cell against their concentration gradient, through carrier proteins, using cellular energy stored in the form of ATP. This transport mechanism is designed for the cellular uptake of those nutrients present in lower concentrations. Xenobiotics hijack this mechanism for their transport, both uptake and excretion (George et al., 2017). **Pinocytosis**—Pinocytosis is non-specific mode of transport. In this mechanism cell membrane folds inward and forms vesicles that enclose extracellular fluid. Once within the cell, the contents of the vesicles are discharged into the cytoplasm

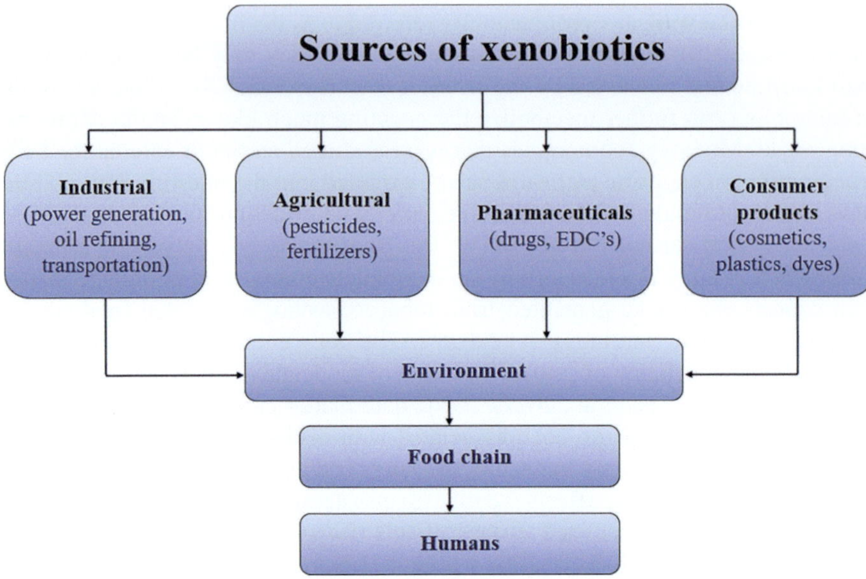

Fig. 1 Sources and route of xenobiotics entering the human system

(Ritter et al., 2021). Pinocytosis is one of the several transport mechanisms through which xenobiotics enter the cells. Xenobiotics mainly enter through active transport and passive diffusion, but larger molecules enter through pinocytosis. **Filtration through "Pores"**—This mechanism includes the transit of tiny, water-soluble xenobiotics through aqueous channels or fenestrations in capillary endothelium, such as those in the renal glomerulus. Filtration depends on the size and charge of molecules and plays a crucial role in the transit of xenobiotics from the blood to the renal tubules (Kaufman et al., 2025). **Lymphatic Absorption**—This pathway is critical for the transport of lipophilic xenobiotics, especially those ingested via the gastrointestinal system. After entering the intestinal cells, some of these chemicals are packaged into chylomicrons and transferred via the lymphatic system. This pathway enables them to bypass the liver's initial round of processing and enter the circulation directly (Yáñez et al., 2011). **Passive Diffusion**—Passive diffusion is the major mode of transport for xenobiotics across cell membranes from high to low concentrations without the need of energy or transport proteins. Since their lipid solubility, xenobiotics can easily enter the cell, but their excretion is difficult through passive diffusion. Their detoxification and conversion in water-soluble form is necessary for their excretion through passive diffusion. It works well for tiny, nonpolar, and fat-soluble compounds and is one of the primary ways through which many xenobiotics, including medications and environmental toxins, enter the body (Xu et al., 2005; Yáñez et al., 2011).

4 Metabolism of Xenobiotics

In recent decades, the consumption of pharmaceutical products has risen significantly. With pharmacies more accessible and the prevalence of chronic diseases increasing, medications have become an essential component of daily existence. Pharmaceuticals' effects on human health have been thoroughly investigated, and current investigations are concentrating on their interactions with the human microbiota. Several organs, including the mouth, skin, stomach, and reproductive system, are inhabited by distinct microbial communities that may regulate the way drugs function in the human system (Turnbaugh et al., 2007). Understanding xenobiotic metabolism is essential for drug development, determining dosages, anticipating interactions, and identifying toxicities, such as liver damage and the formation of dangerous reactant intermediates (Ndlovu, 2024). Figure 2 depicts how multiple factors affect xenobiotic metabolism. The bucket represents the entire metabolic capacity of body, while the arrows represent different factors of xenobiotic metabolism. Variables such as dose, route, enzyme competition, hazardous chemicals, gut microbiome activity, and enzyme modulation affect xenobiotic detoxification rate, efficiency, and outcome, resulting in effective clearance or toxicity.

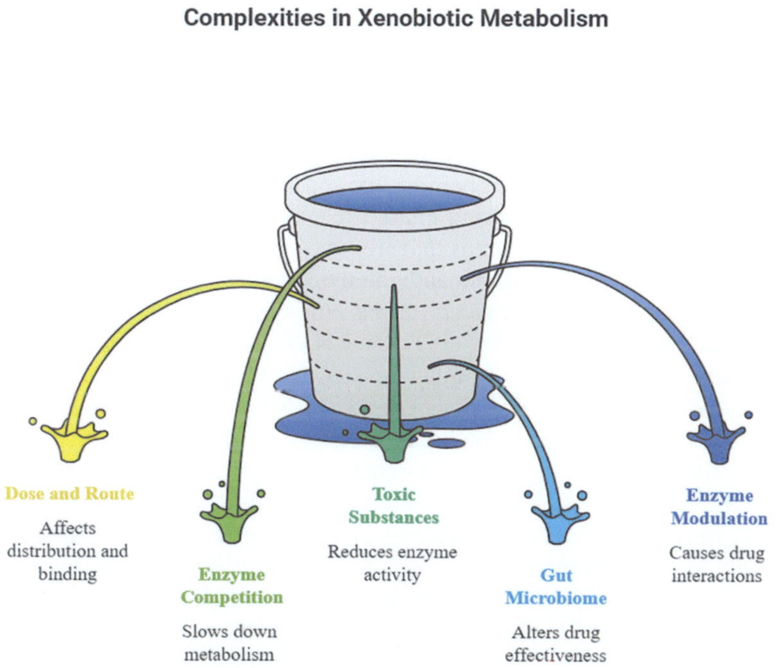

Fig. 2 Complexities in xenobiotic metabolism

Xenobiotic metabolism is dependent on several factors. Age is one of the most significant parameters since metabolism changes with aging. Infants are more susceptible because their enzymes for oxidative metabolism are not fully developed. The dosage of xenobiotics, their mode of administration, dispersion within the body, and interaction with biomolecules also have an effect on their metabolism. Xenobiotics may compete for enzyme active sites, leading to reduced enzyme activity and slower metabolism. In addition, xenobiotic substances can slow down enzyme activity, making metabolism even less efficient (Kucherenko et al., 2021). The major enzyme system responsible for drug metabolism is Cytochrome P450 monooxygenases, which account for more than 75% of all drug-metabolizing processes. The pharmacokinetics of medications and xenobiotics, and their pharmacodynamic effects, are influenced by the rate at which P450 enzymes metabolize them (Tracy et al., 2016). The human gut microbiome also plays a vital role in the metabolism of a wide range of drugs. They modify their chemical structures, thus influencing both the efficacy and possible toxicity of medication. Investigations have shown that gut microbiota can modify the effects of over 50 drugs, including antibiotics and chemotherapeutic treatments (Koppel et al. 2017). The interactions between drugs and microorganisms can either induce or reduce the effectiveness of the drug, and in some cases, they may increase its toxicity. The branch of science that deals with such outcomes is known as toxico microbiomes (Abdelsalam et al., 2020). Metabolism of xenobiotics increases their water solubility, allowing them to be excreted from the body. When xenobiotics are ingested orally, they pass through the upper digestive system and enter the liver via the hepatic portal vein. Cytochrome P450 enzymes in the liver modify both endogenous and foreign chemical substances (Clarke et al., 2019; Michalopoulos, 2007). Several enzymes participate in the metabolism of xenobiotic substances. These enzymes help change and break down chemicals that are not normally found in the body. During Phase I reactions, cytochrome P450 enzymes, which are heme-containing proteins that are very good at recognizing and breaking down specific substances, predominantly do oxidation and help break down a wide range of medications and chemicals in the environment. Phase II reactions use enzymes such as UGTs, SULTs, GSTs, and NATs to attach chemical groups to the modified compounds. This conjugation step increases their solubility in water, which makes it easier to eliminate from the body (Ndlovu, 2024). Another instance of nitro reductive metabolism of organ halogens is observed with clonazepam, a drug commonly prescribed to treat seizures and panic disorder. Studies have reported that intestinal tract microbes in rats transform it into 7-aminoclonazepam (Elmer & Remmel, 1984). Advancements in xenobiotic metabolism research have facilitated the development of non-invasive urinary biomarkers. In rats, the cytotoxic lipid peroxidation product 4-Hydroxynonenal was metabolized into several urinary mercapturic acids, with 1,4-dihydroxynonane mercapturic acid being the most abundant (3.5% of the dose). Within 48 hours, ~25% of 1,4-dihydroxynonane mercapturic acid-derived radioactivity was excreted in urine, 18% in faeces, and only a minimal fraction remained covalently bound in the liver or other organs (Van Vugt-Lussenburg et al., 2022). Xenobiotic metabolism is primarily the enzyme-mediated conversion of readily ingested lipophilic substances into hydrophilic forms that may be eliminated from the body. Although the liver is the principal location of these processes, analogous biotransformation occurs in the gastrointestinal tract and other organs (Carmody & Turnbaugh, 2014). There are three distinct stages of liver metabolism:

Phase I—Activation: The substance is oxidized, reduced, or hydrolyzed to introduce or expose functional groups. **Phase II—Conjugation**: Polar groups are incorporated into the molecule, improving its water solubility and aiding excretion. **Phase III—Transport**: The resultant metabolites are excreted from the cells without any additional modification (Kumar Jain et al., 2005). UDP-glucuronosyltransferases (UGTs) are Phase II enzymes that are crucial during detoxification reactions by converting lipid-soluble substances to more water-soluble β-D-glucuronides for excretion. These enzymes act on Phase I metabolites and substances containing hydroxyl, carboxyl, amine, or thiol groups. Their activity is essential for the easy excretion of these xenobiotics. UGTs are found mostly in endothelial cells and astrocytes, and protect the brain from toxic xenobiotics by forming the blood-brain barrier. They also regulate endogenous compounds such as steroids and dopamine (Ouzzine et al., 2014). After entering the circulation, xenobiotic distribution is determined by parameters such as tissue blood flow, membrane permeability, and active transport. Small lipophilic compounds can easily traverse cell membranes and accumulate in adipose tissues, thus extending their life in the body. Some medications can accumulate in tissues due to pH variations, binding to biological components, or storage in lipid-rich regions (Iyevhobu et al., 2025). Under normal environments, flavor compounds are generally metabolized by the TCA cycle and converted into non-toxic products. For example, alanine (FEMA 3818) is broken down by oxidative deamination into an α-keto acid, which may then be further oxidized into CO_2 and water or utilized to create glucose through gluconeogenesis or ketogenesis. In mitochondria, L-glutamic acid is deaminated by L-glutamate dehydrogenase, resulting in NH_4^+ that can be utilized in the urea cycle and other anabolic pathways or expelled as urea. The remaining carbon skeletons eventually join the citric acid cycle, as alanine is converted to acetyl-CoA (Smith et al., 2018).

Gut microbes impact numerous gastrointestinal medicines, frequently activating prodrugs. For example, intestinal bacteria metabolize sulfasalazine into active 5-ASA and sulphapyridine, whereas subsequent microbial conversion to inert N-acetyl-5-ASA might decrease medication efficacy (Van Hogezand et al., 1992). N-acetyl-5-ASA can also discourage anaerobic bacteria, including *Clostridium difficile*, possibly changing the gut microbiota balance (Deloménie et al., 2001). *Pseudomonas* sp. ST1 may convert paracetamol, a commonly used anti-inflammatory medication, to 4-aminophenol, a hazardous metabolite. ST1 removed 0.77 g/L of acetaminophen in 3 days and 0.65 g/L of p-aminophenol in 2 days. P-aminophenol is toxic to the kidneys, mutagenic, and capable of damaging DNA (Żur et al., 2018). Gut organisms can activate medications such as sulindac, which is an anti-inflammatory drug, and loperamide (an anti-diarrheal) through a reduction process (Koppel et al., 2017). Xenobiotics can affect medication metabolism by activating or inhibiting enzymes, particularly cytochrome P450s such as CYP3A4. Such modulation can alter drug processing, induce drug-drug interactions, and result in unpredictable drug responses

(Croom, 2023). Cytochrome P-450 enzymes are principally oxidative in action, converting lipophilic and halogenated medicines like PCDDs and diclofenac into more water-soluble forms. However, the breakdown and elimination of halogenated xenobiotics by gut microbes, primarily through hydrolytic and reductive processes, is still poorly understood. A knowledge gap exists between the breakdown of specific xenobiotics and the responsible microorganism, gene, and enzyme in the complex gut habitat (Atashgahi et al., 2018).

5 Fate of Xenobiotics

The Industrial Revolution's breakthroughs in science and technology led to the inappropriate use of resources, leading to a lot of harm to the natural environment (Sikandar et al., 2013). Microorganisms can break down xenobiotics; however, it depends on the chemical structure of the substance and the type of microorganisms that are present. Enzymes, such as liver cytochrome P450, change these foreign compounds so that they can be removed by urine, breath, sweat, and faeces. During this process, xenobiotics are broken down into carbon dioxide and water. Each step of the pathway is catalyzed by specific enzymes (Embrandiri et al., 2016).

Xenobiotics mainly enter the aquatic environments through industrial and domestic waste discharge, stormwater, and atmospheric deposition. Since conventional treatment facilities are not designed to specifically remove or detoxify these compounds, they persist in sewer systems. Many of these compounds also show resistance to the natural biodegradation process and bioaccumulate in the tissues of aquatic organisms and enter the food chain, ultimately leading to human exposure (Štefanac et al., 2021). Xenobiotics may leach into groundwater and contaminate surface water bodies, including ponds, lakes, rivers, etc., affecting aquatic ecosystems. These compounds bioaccumulate in aquatic animals via absorption through the body or gills and diet. Their metabolism causes oxidative stress in these animals, disrupting their homeostasis by producing excessive reactive oxygen species and other toxic intermediates, hence compromising their antioxidant defences and overall health. Finally, it leads to cellular damage, an impaired immune system, organ dysfunction, and even death (Miglani et al., 2022). The improper use of human and veterinary pharmaceuticals, along with inadequate disposal of pharmaceutical waste, significantly contributes to environmental pollution. Continuous presence of xenobiotics in the aquatic system creates a positive pressure and is responsible for producing drug-resistant strains of bacteria, posing a major threat to aquatic animals and human health. Although these substances rarely induce acute toxicity in aquatic organisms. There is a knowledge gap in understanding and establishing a relation between their continuous exposure and long-term effects. Their multigenerational effects are poorly understood, even though many species experience continuous exposure throughout their lifespans (Fent et al., 2006).

Physical parameters of soil, including temperature, pH, and moisture, are critical factors influencing the fate and behavior of xenobiotics in soils. Due to their structural similarities to humic substances, many xenobiotics and their metabolites can bind to soil organic matter, which reduces their bioavailability and possibility of leaching into groundwater and plant absorption. This binding also facilitates natural humification processes and reduces soil pollution. In addition to microbial biotransformation by soil microbiome, inorganic minerals and soil metals act as catalysts that interact with xenobiotics and play a big role in their detoxification and transformation (Hayat et al., 2010). Incomplete degradation of xenobiotics can result in the formation of hazardous intermediates that may cause more harm to the environment due to their enhanced stability, toxicity, and mobility. Soil microorganisms degrade these substances; however, their metabolites may accumulate over time. Due to a lack of suitable microbial species in contaminated soils and a lack of microbial enzymes to break them down into harmless by-products, some xenobiotics exhibit high persistence and decompose slowly and posing risks to both human health and ecological systems (Borpatragohain et al., 2019). Persistent xenobiotics in soil exert multiple effects. These compounds may be absorbed by plants and accumulate in lipid-rich tissues, edible parts, and transported to animals and humans by consumption of contaminated edible parts, increasing exponentially in successive levels of the food chain and ultimately disrupting ecosystem equilibrium. Some xenobiotics target beneficial soil bacteria and interfere with biogeochemical cycles (Wang et al., 2022). Pesticide xenobiotics such as DDT, PCBs, chlordane, PAHs, and dioxins present significant risks to soil ecosystems, plants, animals, and humans. The bioavailability and distribution of these compounds in soil depend on their chemical properties, site-specific conditions, and the presence of soil organic matter (Miglani et al., 2022). Octocrylene (OCT), a common sunscreen agent, persists in aquatic environments and can form toxic chlorination by-products (Table 1). Under typical chlorination, 15–20% of OCT is transformed into 11 by-products, six reported for the first time. Most by-products inhibited algal growth by 20–50%, showing higher toxicity than OCT itself and posing ecological risks (Medici et al., 2022). A large population around the world chew areca nuts, exposing themselves to alkaloids like arecoline, arecaidine, guvacine, and guvacoline. This habit, the fourth most common reason which is linked to oral cancer in parts of Asia due to nitrosamine formation, while the metabolism and effects of arecoline and arecaidine remain poorly understood (Patterson et al., 2010). Pesticides and other major xenobiotic compounds pose serious threat to the plants, animals, and human population. Substances such as dioxins, DDT, polychlorinated biphenyls, chlordane, polycyclic aromatic hydrocarbons along with nitroaromatic compounds represent significant contaminants in the soils of industrialized regions. The chemical make-up of these xenobiotics, together with environmental conditions, influences their bioavailability and dispersion in soil ecosystems, where soil organic matter plays a critical regulatory function (Miglani et al., 2022). A substantial amount of microplastics in the surrounding environment causes their buildup within living organisms. Since organisms are unable to efficiently break down and excrete microplastics, they accumulate and disrupt normal metabolic activities. They are responsible for obstruction in the digestive tract, sensation of false satiation due to an indigestible mass in the stomach, malnutrition, energy depletion, and interfere with essential functions such as compromised immune, endocrine function and other metabolic processes. (Gall & Thompson, 2015).

6 Health Hazards of Xenobiotics

A large number of xenobiotic compounds are lipophilic, making them difficult to eliminate from the body and causing their accumulation over a lifetime. Around the world, pollutants like polychlorinated biphenyls and perfluorooctanoic acid are known to persist in the environment. Long-term exposure to these chemicals can cause serious respiratory problems, such as pulmonary tuberculosis, pulmonary fibrosis, and asthma. These illnesses are common among coal miners and people exposed to asbestos at work. Few studies have raised concern over the use of nanomaterials, which are commonly incorporated into certain medicines, and have been reported to act as toxic xenobiotics (Ramaka et al., 2020). Living organisms are constantly and unavoidably exposed to a variety of xenobiotics in their daily lives, which are present in the environment and daily products. Intentional exposures of xenobiotics occur with pharmaceuticals, cosmetics, and food additives; unintentionally, these compounds enter the body through contaminated food, water, air, pesticide residues, industrial chemicals, and environmental pollutants present in the surroundings (King et al., 2000). Humans are exposed to xenobiotics daily through multiple routes, such as dermal contact, inhalation, and ingestion. After entering the body, these compounds can be changed by the gut microbiota, either before they are absorbed or after they reach the bloodstream (Koppel et al., 2017). If xenobiotics are not broken down or removed from the body, they can build up and lead to long-term disease and inflammation. Those that are not absorbed in the small intestine move to the large intestine, where gut microbes break them into smaller parts. The by-products are then either filtered out by the kidneys or passed in faeces, helping the body keep its metabolic processes going (Abdelsalam et al., 2020). Some xenobiotics, such as stimulants and vasoconstrictors, can affect the sympathetic nervous system and can increase blood pressure and heart rate. This may lead to heart palpitations, chest pain, and, in severe cases, heart failure. Exposure to certain xenobiotics, like allergens found in foods, plants, or insect stings, can also cause sudden allergic reactions. Hives, skin irritation, itching, and redness are common signs of these reactions. In rare cases, xenobiotic allergies can lead to anaphylaxis. This is a life-threatening condition that brings on sudden trouble breathing, swelling of the face and throat, and a rapid drop in blood pressure. It needs immediate medical attention (Chatterjee Ranjeeta P. et al., 2024). Xenobiotics, mainly pharmaceutical drugs, can interfere with drug metabolism pathways, which can lead to adverse drug reactions by making drugs more toxic or less effective. Such interactions commonly occur through the inhibition or activation of cytochrome P450 enzymes, which play an important role in drug biotransformation (Zanger & Schwab, 2013)Bisphenols, ethanolamine, triclosan (TCS), phthalates, and parabens are often found in personal care products and have been reported to disrupt endocrine function (Khalid & Abdollahi, 2021).

Table 1 Xenobiotic compounds and their biotransformation metabolic products

Xenobiotic compound	Type/source	Degradation or transformation product(s)	Mechanism/process	Environmental or biological effect	Reference
Octocrylene (OCT)	Cosmetic (sunscreen agent)	11 chlorinated by-products (6 novel); more toxic than OCT	Chlorination under aquatic conditions	Inhibits algal growth (20–50%), potential ecological risk	(Medici et al., 2022)
Areca alkaloids (Arecoline, Arecaidine, Guvacine, Guvacoline)	Natural xenobiotics (areca nut chewing)	Nitrosamine derivatives	Nitrosation during metabolism	Linked to oral cancer; major public health concern in Asia	(Patterson et al., 2010)
Persistent organic pollutants (POPs)—*DDT, PCBs, chlordane, PAHs, dioxins*	Pesticides/industrial pollutants	Slow degradation; formation of hazardous intermediates	Incomplete microbial mineralization; binding with SOM	Bioaccumulation in soil, plants, food chain; ecological and human toxicity	(Borpatragohain et al., 2019; Miglani et al., 2022; Wang et al., 2022)
Microplastics	Synthetic polymer contaminants	Non-degradable; accumulate within organisms	Physical accumulation, not biodegradation	Disrupts metabolism, feeding, and digestion in aquatic and soil fauna	(Gall & Thompson, 2015)
General xenobiotics (organic pollutants)	Wastewater, stormwater runoff	Carbon dioxide and water (upon complete oxidation)	Enzymatic degradation (e.g., oxygenases, cytochrome P450)	Detoxification and energy release during microbial metabolism	(Embrandiri et al., 2016; Sikandar et al., 2013)

Endocrine-disrupting compounds (EDCs) have been linked to cancer, metabolic disorders, diabetes, obesity, and infertility. When triclosan (TCS) is exposed to UV light or sunlight, it breaks down and produces reactive oxygen species (ROS) and free radicals. These substances can cause lipid peroxidation, stress in the endoplasmic reticulum, cell cycle arrest, skin cell damage, and changes in the proteins involved in apoptosis (Dubey et al., 2019). Prolonged exposure to triclosan (TCS) from personal care items can also damage liver and kidney function (Khalid & Abdollahi, 2021). UV filters like oxybenzone and octinoxate help protect the skin from harmful ultraviolet radiation. However, they can also interfere with hormone regulation. Long-term exposure to these xenobiotics has been linked to reproductive and developmental problems and a higher risk of some cancers. Over time, endocrine-disrupting chemicals in cosmetics may also affect thyroid function and disrupt other metabolic processes (Li et al., 2017; Liu et al., 2021). Many cosmetic products contain harmful metals like lead, cadmium, and mercury. These substances can be dangerous for both the environment and human health. Individuals may be exposed to them through skin absorption, inhalation, or ingestion (Issac & Kandasubramanian, 2021; Juliano & Magrini, 2017). Numerous studies have shown that the use of cosmetics may result in neurotoxicity, genetic mutations, and reproductive disorders (Amasa et al., 2012). Exposure to these compounds has also been associated with endocrine disorders and different types of cancer. For example, endocrine disruptors are linked to breast cancer due to their interference with estrogen regulation in women. Additionally, these compounds are associated with immune system impairment, which increases susceptibility to a range of health conditions (Adegoke et al., 2021). Cadmium exposure from cosmetics has been associated with lung damage and renal complications. Extended exposure to cadmium has also been linked to cancer and hormonal imbalance (Charkiewicz et al., 2023; Miglani et al., 2022). Arsenic functions as a carcinogen regardless of the route of exposure, whether oral or nasal. Long-term exposure to arsenic can raise the risk of cancers in the skin, lungs, bladder, and liver. It has also been connected to skin damage, breathing problems, and heart disease (Speer et al., 2023).

7 Conclusion

Xenobiotics are compounds that are foreign to the biological systems of living organisms, often resistant to breakdown, accumulate in tissues, and are potentially harmful. The main sources of xenobiotics are industrial activities, agricultural practices, domestic waste, pharmaceuticals, and cosmetic products. Common xenobiotic compounds include pesticides, fertilizers, industrial chemicals, heavy metals, dyes, environmental pollutants, pharmaceuticals, and food additives. They enter the body mainly through diet, water, inhalation, or dermal contact. After entering the body, they are transported to the liver and processed through enzymes such as cytochrome P450. During their detoxification, these compounds are first altered by oxidation, reduction, hydrolysis, and by the addition or removal of functional groups. Polar groups are attached to increase their water solubility, and efflux pumps move them out of the cell, finally excreted via urine, bile, or sweat. Rarely, during their biotransformation, highly reactive intermediates are formed that are more harmful than the original molecule. Exposure to xenobiotics can lead to several health problems, such as acute toxicity, hormonal imbalances, chronic diseases, tissue damage, cancer, and immune system disorders. Specific molecular pathways of xenobiotic-induced toxicity and their health consequences are not clearly understood. Thorough and interdisciplinary research is needed to better understand how different xenobiotic compounds are metabolized and build up in the body, as well as their effects on health. Advanced analytical methods, molecular toxicology, and omics-based research can improve our understanding of xenobiotic interactions with biological systems. Additionally, studying microbial degradation and developing long-term detoxification strategies are essential to reduce their impact on health and the environment. In summary, xenobiotics are a critical area in environmental and medicinal research. Continuous and in-depth research is essential to understand their metabolism, assess their health hazards, and explore their complex nature. Future xenobiotic research should be focused on environmental clean-up, epigenetic impact, biomonitoring of xenobiotics in biospecimens, developing better risk assessment methods, and developing safer chemical substitutes for environmental sustainability.

References

Abdelsalam, N. A., Ramadan, A. T., ElRakaiby, M. T., & Aziz, R. K. (2020). Toxicomicrobiomics: The human microbiome vs. pharmaceutical, dietary, and environmental xenobiotics. *Frontiers in pharmacology, 11*). Frontiers Media S.A. https://doi.org/10.3389/fphar.2020.00390.

Adegoke, E. O., Rahman, M. S., Park, Y. J., Kim, Y. J., & Pang, M. G. (2021). Endocrine-disrupting chemicals and infectious diseases: From endocrine disruption to immunosuppression. *International Journal of Molecular Sciences, 22*(8). MDPI AG. https://doi.org/10.3390/ijms22083939.

Aguayo, F., Tapia, J. C., Calaf, G. M., Muñoz, J. P., Osorio, J. C., Guzmán-Venegas, M., Moreno-León, C., Levican, J., & Andrade-Madrigal, C. (2025). The role of xenobiotics and Anelloviruses in colorectal cancer: Mechanisms and perspectives. *International Journal of Molecular Sciences, 26*(9). Multidisciplinary Digital Publishing Institute (MDPI). https://doi.org/10.3390/ijms26094354.

Amasa, W., Santiago, D., Mekonen, S., & Ambelu, A. (2012). Are cosmetics used in developing countries safe? Use and dermal irritation of body care products in Jimma town, southwestern Ethiopia. *Journal of Toxicology, 2012*. https://doi.org/10.1155/2012/204830

Atashgahi, S., Shetty, S. A., Smidt, H., & de Vos, W. M. (2018). Flux, impact, and fate of halogenated xenobiotic compounds in the gut. *Frontiers in physiology, 9*. Frontiers Media S.A. https://doi.org/10.3389/fphys.2018.00888.

Borpatragohain, B., Sahoo, S., & Rai, A. (2019). An overview on the impact, interaction and fate of xenobiotic-soil organic matter complexes. *International Journal of Chemical Studies*. https://www.researchgate.net/publication/340949690

Bukowska, B., Mokra, K., & Michałowicz, J. (2022). Benzo[a]pyrene—Environmental occurrence, human exposure, and mechanisms of toxicity. *International Journal of Molecular Sciences, 23*(11). MDPI. https://doi.org/10.3390/ijms23116348.

Calatayud Arroyo, M., García Barrera, T., Callejón Leblic, B., Arias Borrego, A., & Collado, M. C. (2021). A review of the impact of xenobiotics from dietary sources on infant health: Early life exposures and the role of the microbiota. *Environmental Pollution, 269*. Elsevier Ltd. https://doi.org/10.1016/j.envpol.2020.115994.

Carmody, R. N., & Turnbaugh, P. J. (2014). Host-microbial interactions in the metabolism of therapeutic and diet-derived xenobiotics. *Journal of Clinical Investigation, 124*(10), 4173–4181). American Society for Clinical Investigation. https://doi.org/10.1172/JCI72335.

Caroline, M.-D., & Frederic, D. (2012). Nervous system disorders induced by occupational and environmental toxic exposure. *Open Journal of Preventive Medicine, 02*(03), 272–278. https://doi.org/10.4236/ojpm.2012.23039

Charkiewicz, A. E., Omeljaniuk, W. J., Nowak, K., Garley, M., & Nikliński, J. (2023). Cadmium toxicity and health effects—A brief summary. *Molecules, 28*(18). Multidisciplinary Digital Publishing Institute (MDPI). https://doi.org/10.3390/molecules28186620.

Chatterjee, R. P., Mane, V. K., & Seth, P. N. (2024). Effect Of xenobiotics on human health. *Futuristic Trends in Agriculture Engineering & Food Sciences, 3*.

Chhabra, R. S. (1979). Intestinal absorption and metabolism of xenobiotics. *Environmental Health Perspectives, 33*, 61–69. https://doi.org/10.1289/ehp.793361

Clarke, G., Sandhu, K. V., Griffin, B. T., Dinan, T. G., Cryan, J. F., & Hyland, N. P. (2019). Gut reactions: Breaking down xenobiotic–microbiome interactions. *Pharmacological Reviews, 71*(2), 198–224. https://doi.org/10.1124/pr.118.015768

Croom, E. (2023). The effects of xenobiotics on metabolism: A comprehensive review. *Journal of Clinical & Experimental Pharmacology, 13*, 1000373–1000374. https://doi.org/10.35248/2161-1459.23.13.373

de Oliveira, M., Frihling, B. E. F., Velasques, J., Filho, F. J. C. M., Cavalheri, P. S., & Migliolo, L. (2020). Pharmaceuticals residues and xenobiotics contaminants: Occurrence, analytical techniques and sustainable alternatives for wastewater treatment. *Science of the Total environment, 705*. Elsevier B.V. https://doi.org/10.1016/j.scitotenv.2019.135568.

Deloménie, C., Fouix, S., Longuemaux, S., Brahimi, N., Bizet, C., Picard, B., Denamur, E., & Dupret, J. M. (2001). Identification and functional characterization of arylamine N-acetyltransferases in eubacteria: Evidence for highly selective acetylation of 5-aminosalicylic acid. *Journal of Bacteriology, 183*(11), 3417–3427. https://doi.org/10.1128/JB.183.11.3417-3427.2001

Dirbeba Dinka, D. (2018). Environmental xenobiotics and their adverse health impacts-a general review. *Journal of Environment Pollution and Human Health, 6*(3), 77–88. https://doi.org/10.12691/jephh-6-3-1

Dubey, D., Srivastav, A. K., Singh, J., Chopra, D., Qureshi, S., Kushwaha, H. N., Singh, N., & Ray, R. S. (2019). Photoexcited triclosan induced DNA damage and oxidative stress via p38 MAP kinase signaling involving type I radicals under sunlight/UVB exposure. *Ecotoxicology and Environmental Safety, 174*, 270–282. https://doi.org/10.1016/j.ecoenv.2019.02.065

Elmer, G. W., & Remmel, R. P. (1984). Role of the intestinal microflora in clonazepam metabolism in the rat. *Xenobiotica, 14*(11).

Embrandiri, A., Katheem Kiyasudeen, S., Rupani, P. F., & Ibrahim, M. H. (2016) Environmental xenobiotics and its effects on natural ecosystem. *Plant responses to xenobiotics*, 1–18). Springer Singapore. https://doi.org/10.1007/978-981-10-2860-1_1

Fent, K., Weston, A. A., & Caminada, D. (2006). Ecotoxicology of human pharmaceuticals. *Aquatic toxicology, 76*(2), 122–159). Elsevier. https://doi.org/10.1016/j.aquatox.2005.09.009.

Gall, S. C., & Thompson, R. C. (2015). The impact of debris on marine life. *Marine Pollution Bulletin, 92*(1–2), 170–179. https://doi.org/10.1016/j.marpolbul.2014.12.041

George, B., You, D., Joy, M. S., & Aleksunes, L. M. (2017). Xenobiotic transporters and kidney injury. In *Advanced drug delivery reviews, 116*, 73–91). Elsevier B.V. https://doi.org/10.1016/j.addr.2017.01.005.

Gerrity, D. W., Benotti, M. J., Reckhow, D. A., & Snyder, S. A. (2011). Pharmaceuticals and endocrine-disrupting compounds in drinking water. In X. Baoshan, S. Nicola, & H. M. Pan

(Eds.), *Biophysico-chemical processes of anthropogenic organic compounds in environmental systems* (pp. 233–249). John Wiley & Sons, Inc..

Goel, A. (2021). Short review on xenobiotics and human health. *International Journal of Nutrition & Lifestyle, 1*(3) https://www.researchgate.net/publication/358424045

Goyer, R., Golub, M., Choudhury, H., Hughes, M., Kenyon, E., & Stifelman, M. (2004). *Issue paper on the human health effects of metals*. Environmental Protection Agency.

Gupta, S. K., Rachna, Singh, B., Mungray, A. K., Bharti, R., Nema, A. K., Pant, K. K., & Mulla, S. I. (2022). Bioelectrochemical technologies for removal of xenobiotics from wastewater. *Sustainable Energy Technologies and Assessments, 49*. https://doi.org/10.1016/j.seta.2021.101652.

Hayat, M. T., Xu, J., Ding, N., & Mahmood, T. (2010). Dynamic behavior of persistent organic pollutants in soil and their interaction with organic matter. In *Molecular environmental soil science at the interfaces in the earth's critical zone* (pp. 217–222). Springer. https://doi.org/10.1007/978-3-642-05297-2_65

Issac, M. N., & Kandasubramanian, B. (2021). Effect of microplastics in water and aquatic systems. *Environmental Science and Pollution Research*. https://doi.org/10.1007/s11356-021-13184-2/Published

Iyevhobu, K. O., Okparaku, S. O., Obohwemu, K. O., Ken-Iyevhobu, B. A., Asibor, E., Oseni, D. I., Momodu, K. O., Nnamudi, P., Abayomi, G., Chauhan, R., Sharma, S., & Akabuokwu, K. C. (2025). Xenobiotic distribution: A comprehensive review and toxicological implications. *JORMA International Journal of Health and Social Sciences, 2*(2), 58–66. https://doi.org/10.64074/99wreq81

Jaishankar, M., Tseten, T., Anbalagan, N., Mathew, B. B., & Beeregowda, K. N. (2014). Toxicity, mechanism and health effects of some heavy metals. *Interdisciplinary toxicology, 7*(2), 60–72. Slovak Toxicology Society. https://doi.org/10.2478/intox-2014-0009.

Jin, X. C., Liu, G. Q., Xu, Z. H., & Tao, W. Y. (2007). Decolorization of a dye industry effluent by aspergillus fumigatus XC6. *Applied Microbiology and Biotechnology, 74*(1), 239–243. https://doi.org/10.1007/s00253-006-0658-1

Jin, Y., Chi, J., LoMonaco, K., Boon, A., & Gu, H. (2023). Recent review on selected xenobiotics and their impacts on gut microbiome and metabolome. In *TrAC—Trends in analytical chemistry* (Vol. 166). Elsevier B.V. https://doi.org/10.1016/j.trac.2023.117155

Johnson, C. H., Patterson, A. D., Idle, J. R., & Gonzalez, F. J. (2012). Xenobiotic metabolomics: Major impact on the metabolome. *Annual Review of Pharmacology and Toxicology, 52*, 37–56. https://doi.org/10.1146/annurev-pharmtox-010611-134748

Juliano, C., & Magrini, G. A. (2017). Cosmetic ingredients as emerging pollutants of environmental and health concern. A mini-review. *Cosmetics, 4*(2). MDPI AG. https://doi.org/10.3390/cosmetics4020011.

Kaufman, D. P., Basit, H., & Knohl, S. J. (2025). *Physiology, Glomerular Filtration Rate*.

Khalid, M., & Abdollahi, M. (2021). Environmental distribution of personal care products and their effects on human health. *Iranian Journal of Pharmaceutical Research, 20*(1), 216–253. https://doi.org/10.22037/ijpr.2021.114891.15088

Kim, N. H., Hyun, Y. Y., Lee, K. B., Chang, Y., Rhu, S., Oh, K. H., & Ahn, C. (2015). Environmental heavy metal exposure and chronic kidney disease in the general population. *Journal of Korean Medical Science, 30*(3), 272–277. https://doi.org/10.3346/jkms.2015.30.3.272

King, C. D., Rios, G. R., Green, M. D., & Tephly, T. R. (2000). UDP-Glucuronosyltransferases. *Current Drug Metabolism, 1*.

Koppel, N., Rekdal, V. M., & Balskus, E. P. (2017). Chemical transformation of xenobiotics by the human gut microbiota. *Science, 356*(6344), 1246–1257). American Association for the Advancement of Science. https://doi.org/10.1126/science.aag2770.

Kucherenko, S. V., Ovcharenko, A. M., & Pushenko, S. L. (2021). Xenobiotics: A threat to the health of living organisms. *E3S Web of Conferences, 285*. https://doi.org/10.1051/e3sconf/202128503006

Kuhad, R. C., Sood, N., Tripathi, K. K., Singh, A., & Ward, O. P. (2004). Developments in microbial methods for the treatment of dye effluents. *International Journal of Nutrition & Lifestyle, Elsevier Inc., 56*.

Kumar Jain, R., Kapur, M., Labana, S., Lal, B., Sarma, P. M., Bhattacharya, D., & Thakur, I. S. (2005). Microbial diversity: Application of micro-organisms for the biodegradation of xenobiotics. *Current Science, 89*(1).

Kumar, V., Yadav, C. S., & Banerjee, B. D. (2022). Xeno-estrogenic pesticides and the risk of related human cancers. *Journal of Xenobiotics, 12*(4), 344–355). Multidisciplinary Digital Publishing Institute (MDPI). https://doi.org/10.3390/jox12040024.

Lang, T., Lipp, A. M., & Wechselberger, C. (2025). Xenobiotic toxicants and particulate matter: Effects, mechanisms, impacts on human health, and mitigation strategies. *Journal of Xenobiotics, 15*(4). Multidisciplinary Digital Publishing Institute (MDPI). https://doi.org/10.3390/jox15040131.

Li, X., Gao, Y., Wang, J., Ji, G., Lu, Y., Yang, D., Shen, H., Dong, Q., Pan, L., Xiao, H., & Zhu, B. (2017). Exposure to environmental endocrine disruptors and human health. *Journal of Public Health and Emergency, 1*, 8–8. https://doi.org/10.21037/jphe.2016.12.09

Lindell, A. E., Zimmermann-Kogadeeva, M., & Patil, K. R. (2022). Multimodal interactions of drugs, natural compounds and pollutants with the gut microbiota. *Nature Reviews Microbiology*, 20(7), 431–443). Nature Research. https://doi.org/10.1038/s41579-022-00681-5

Liu, G., Cai, W., Liu, H., Jiang, H., Bi, Y., & Wang, H. (2021). The association of bisphenol a and phthalates with risk of breast cancer: A meta-analysis. In *International Journal of Environmental Research and Public Health, 18*(5), 1–15). MDPI AG. https://doi.org/10.3390/ijerph18052375.

Medici, A., Saviano, L., Siciliano, A., Libralato, G., Guida, M., Previtera, L., Di Fabio, G., & Zarrelli, A. (2022). Octocrylene: From sunscreens to the degradation pathway during chlorination processes: Formation of byproducts and their Ecotoxicity assessment. *Molecules, 27*(16). https://doi.org/10.3390/molecules27165286

Michalopoulos, G. K. (2007). Liver regeneration. *Journal of Cellular Physiology, 213*(2), 286–300. https://doi.org/10.1002/jcp.21172

Miglani, R., Parveen, N., Kumar, A., Ansari, M. A., Khanna, S., Rawat, G., Panda, A. K., Bisht, S. S., Upadhyay, J., & Ansari, M. N. (2022). Degradation of xenobiotic pollutants: An environmentally sustainable approach. *Metabolites, 12*(9). MDPI. https://doi.org/10.3390/metabo12090818.

Minteer, S. D., Liaw, B. Y., & Cooney, M. J. (2007). Enzyme-based biofuel cells. *Current opinion in biotechnology, 18*(3), 228–234. https://doi.org/10.1016/j.copbio.2007.03.007

Mishra, V. K., Singh, G., & Shukla, R. (2019). Impact of xenobiotics under a changing climate scenario. In *Climate change and agricultural ecosystems: Current challenges and adaptation* (pp. 133–151). Elsevier. https://doi.org/10.1016/B978-0-12-816483-9.00006-2

Myllynen, P., Pasanen, M., & Vähäkangas, K. (2007). The fate and effects of xenobiotics in human placenta. *Expert Opinion on Drug Metabolism and Toxicology, 3*(3), 331–346. https://doi.org/10.1517/17425255.3.3.331

Ndlovu, H. (2024). Stem cell research and regenerative medicine xenobiotic metabolism: Understanding biotransformation pathways, enzymatic mechanisms, and multifaceted impacts. *Cell Research and Regenerative Medicine, 7*(2), 185–186. https://doi.org/10.37532/SRRM.2024.7(2).185-186

Neal, A. P., & Guilarte, T. R. (2013). Mechanisms of lead and manganese neurotoxicity. *Toxicology research, 2*(2), 99–114). Royal Society of Chemistry. https://doi.org/10.1039/c2tx20064c.

Nogacka, A., Gómez-Martín, M., Suárez, A., González-Bernardo, O., de los Reyes-Gavilán, C., & González, S. (2019). Xenobiotics formed during food processing: Their relation with the intestinal microbiota and colorectal cancer. *International Journal of Molecular Sciences, 20*(8), 2051. https://doi.org/10.3390/ijms20082051

Ortiz, P., Torres-Sánchez, A., López-Moreno, A., Cerk, K., Ruiz-Moreno, Á., Monteoliva-Sánchez, M., Ampatzoglou, A., Aguilera, M., & Gruszecka-Kosowska, A. (2022). Impact of cumulative environmental and dietary xenobiotics on human microbiota: Risk assessment for one health. *Journal of Xenobiotics, 12*(1), 56–63. https://doi.org/10.3390/jox12010006

Ouzzine, M., Gulberti, S., Ramalanjaona, N., Magdalou, J., & Fournel-Gigleux, S. (2014). The UDP-glucuronosyltransferases of the blood-brain barrier: Their role in drug metabolism and detoxication. *Frontiers in Cellular Neuroscience, 8*. https://doi.org/10.3389/fncel.2014.00349

Pant, A., Maiti, T. K., Mahajan, D., & Das, B. (2023). Human gut microbiota and drug metabolism. *Microbial Ecology,* 86(1), 97–111). Springer. https://doi.org/10.1007/s00248-022-02081-x

Patterson, A. D., Gonzalez, F. J., & Idle, J. R. (2010). Xenobiotic metabolism: A view through the metabolometer. *Chemical Research in Toxicology, 23*(5), 851–860. https://doi.org/10.1021/tx100020p

Ramaka, S., Murthy, S. V., & Rao, A. R. (2020). Xenobiotics in health and disease: The two sides of a coin: A clinician's perspective. *Open Access Journal of Toxicology, 4*(4). https://doi.org/10.19080/oajt.2020.04.555641

Ritter, M., Bresgen, N., & Kerschbaum, H. H. (2021). From pinocytosis to Methuosis—Fluid consumption as a risk factor for cell death. In *Frontiers in cell and developmental biology* (Vol. 9). https://doi.org/10.3389/fcell.2021.651982

Roberts, M. S., Magnusson, B. M., Burczynski, F. J., & Weiss, M. (2002). Enterohepatic circulation: Physiological, pharmacokinetic and clinical implications. *Clinical Pharmacokinetics, 41*(10), 751–790. https://doi.org/10.2165/00003088-200241100-00005

Sajid, M., & Agrewala, J. N. (2019). Low prevalence of anti-xenobiotic antibodies among the occupationally exposed individuals is associated with a high risk of cancer. *Cancer Medicine, 8*(1), 246–260. https://doi.org/10.1002/cam4.1773

Shreenidhi, K. S., Vigneshwar, R., Sai Rahul, S., Jothi Murugan, S., Maline, M., Roy, S., & Geetha, V. (2024). A study on toxicity mechanisms of xenobiotics using network pharmacology approach. *Current Trends In Biotechnology And Pharmacy, 18*(4s), 138–151. https://doi.org/10.5530/ctbp.2024.4s.11

Sikandar, A., Shehzadi, K., Arshad, Q., & Munir, K. (2013). Phytoremediation: An analytical technique for the assessment of biodegradation of organic xenobiotic pollutants: A review. *International Journal of Science and Research, 4*. www.ijsr.net

Singh, V. K., Gunasekaran, P., Kumari, M., Krishnan, D., & Ramachandran, V. K. (2024). Animal sourced biopolymer for mitigating xenobiotics and hazardous materials. *Physical Sciences Reviews, 9*(4), 1731–1746. https://doi.org/10.1515/psr-2022-0221

Smith, R. L., Cohen, S. M., Fukushima, S., Gooderham, N. J., Hecht, S. S., Guengerich, F. P., Rietjens, I. M. C. M., Bastaki, M., Harman, C. L., McGowen, M. M., & Taylor, S. V. (2018). The safety evaluation of food flavouring substances: The role of metabolic studies. *Toxicology research*, 7(4), 618–646). Royal Society of Chemistry. https://doi.org/10.1039/C7TX00254H.

Sobral, A. F., Cunha, A., Costa, I., Silva-Carvalho, M., Silva, R., & Barbosa, D. J. (2025). Environmental xenobiotics and epigenetic modifications: Implications for human health and disease. *Journal of Xenobiotics, 15*(4). Multidisciplinary Digital Publishing Institute (MDPI). https://doi.org/10.3390/jox15040118.

Speer, R. M., Zhou, X., Volk, L. B., Liu, K. J., & Hudson, L. G. (2023). Arsenic and cancer: Evidence and mechanisms. *Advances in pharmacology 96*, 151–202). Academic Press https://doi.org/10.1016/bs.apha.2022.08.001

Štefanac, T., Grgas, D., & Dragičević, T. L. (2021). Xenobiotics—Division and methods of detection: A review. *Journal of xenobiotics*, 11(4), 130–141). MDPI. https://doi.org/10.3390/jox11040009

Tam, N. F., & Wong, Y. S. (2000). Spatial variation of heavy metals in surface sediments of Hong Kong mangrove swamps. *Environmental Pollution (Barking, Essex: 1987), 110*(2), 195–205. https://doi.org/10.1016/s0269-7491(99)00310-3

Tracy, T. S., Chaudhry, A. S., Prasad, B., Thummel, K. E., Schuetz, E. G., Zhong, X. B., Tien, Y. C., Jeong, H., Pan, X., Shireman, L. M., Tay-Sontheimer, J., & Lin, Y. S. (2016). Interindividual variability in cytochrome P450-mediated drug metabolism. *Drug Metabolism and Disposition, 44*(3), 343–351. https://doi.org/10.1124/dmd.115.067900

Turnbaugh, P. J., Ley, R. E., Hamady, M., Fraser-Liggett, C. M., Knight, R., & Gordon, J. I. (2007). The human microbiome project: Exploring the microbial part of ourselves in a changing world. *Nature*, 449(7164), 804–810. Nature Publishing Group. https://doi.org/10.1038/nature06244.

Van Hogezand, R. A., Kennis, H. M., Van Schaik, A., Koopman, J. P., Van Hees, P. A. M., & Van Tongeren, J. H. M. (1992). Bacterial acetylation of 5-aminosalicylic acid in faecal suspensions cultured under aerobic and anaerobic conditions. *European Journal of Clinical Pharmacology, 43*.

Van Vugt-Lussenburg, B. M. A., Capinha, L., Reinen, J., Rooseboom, M., Kranendonk, M., Onderwater, R. C. A., & Jennings, P. (2022). Commandeuring xenobiotic metabolism: Advances in understanding xenobiotic metabolism. *Chemical research in toxicology,* 35(7), 1184–1201. American Chemical Society. https://doi.org/10.1021/acs.chemrestox.2c00067.

Wang, X., Sial, M. U., Bashir, M. A., Bilal, M., Raza, Q.-U.-A., Ali Raza, H. M., Rehim, A., & Geng, Y. (2022). Pesticides xenobiotics in soil ecosystem and their remediation approaches. *Sustainability, 14*(6), 3353. https://doi.org/10.3390/su14063353

Wu, D., Chen, Q., Chen, X., Han, F., Chen, Z., & Wang, Y. (2023). The blood–brain barrier: Structure, regulation and drug delivery. *Signal Transduction and Targeted Therapy, 8*(1), 217. https://doi.org/10.1038/s41392-023-01481-w

Xu, C., Yong-Tao Li, C., & Tony Kong, A.-N. (2005). Induction of phase I, II and III drug metabolism/transport by xenobiotics. *Archives of Pharmacal Research, 28*(3) http://apr.psk.or.kr

Yáñez, J. A., Wang, S. W. J., Knemeyer, I. W., Wirth, M. A., & Alton, K. B. (2011). Intestinal lymphatic transport for drug delivery. *Advanced Drug Delivery Reviews, 63*(10–11), 923–942. https://doi.org/10.1016/j.addr.2011.05.019

Zanger, U. M., & Schwab, M. (2013). Cytochrome P450 enzymes in drug metabolism: Regulation of gene expression, enzyme activities, and impact of genetic variation. *Pharmacology and Therapeutics, 138*(1), 103–141. https://doi.org/10.1016/j.pharmthera.2012.12.007

Zimmermann, M., Zimmermann-Kogadeeva, M., Wegmann, R., & Goodman, A. L. (2019). Separating host and microbiome contributions to drug pharmacokinetics and toxicity. *Science, 363*(6427). https://doi.org/10.1126/science.aat9931

Żur, J., Piński, A., Marchlewicz, A., Hupert-Kocurek, K., Wojcieszyńska, D., & Guzik, U. (2018). Organic micropollutants paracetamol and ibuprofen—Toxicity, biodegradation, and genetic background of their utilization by bacteria. *Environmental science and pollution research,* 25(22), 21498–21524). Springer Verlag. https://doi.org/10.1007/s11356-018-2517-x.

Chapter 3
New Concerns in Environmental Toxicology: Nanoparticles Microplastics and Other Innovative Xenobiotics

Krishnan Anandhan, Kumaraswamy Sharavanabhavan Janani, Ahmed Mustafa, Mathan Ramesh, and Thirunavukkarasu Muralisankar

Abstract Environmental toxicology faces challenges posed by the incidence of emerging contaminants in the environment. Nanoparticles (NPs), Microplastics (MPs), Persistent Organic Pollutants (POPs), Plastic additives, Pharmaceutical and Personal Care Products (PPCPs), and Brominated Flame Retardants (BFRs) pose interconnected multi-level toxicity to ecosystems and human health. NPs are intentionally engineered carriers with unique size and physicochemical properties for the biomedical and industrial applications. They penetrate into the various levels of the biotic and abiotic compartments, enter into the food chain, and pose long-term health risks, producing toxicity through oxidative stress, cellular dysfunction, apoptosis, and DNA damage. MPs are distributed into environmental matrices and are exposed to human tissues through ingestion, inhalation, and dermal contact, and are evidently found in the lungs, sputum, bloodstream, and feces. Their accumulation in organisms, alone and along with other emerging contaminants, triggers oxidative stress, inflammatory issues, major organ dysfunctions, as well as cellular organelles and genes related to human physiological systems. POPs are the contaminants with low water solubility and a higher lipophilic nature; their higher aquatic and atmospheric transfer, as well as bioaccumulation in fatty tissues, leads to widespread distribution across different regions of the globe from their origin. BFRs are chemicals added to consumer products to improve fire resistance. Persistence and lower degradation rates of BFRs in the environment, contaminating ecosystems and causing health implication to biota, including humans. Constant disposal of PPCPs in the environment and their partial persistence affect ecological processes, promote anti-drug resistance in pathogenic microbes, and cause physiological changes in the biota, including humans. Hence, this chapter attempts to explore and distinguish the crucial current knowledge on the origin, sources, transfer, exposure routes, and ecological and public health implications of these xenobiotics, to improve their mitigation measures, and

K. Anandhan · K. S. Janani · M. Ramesh · T. Muralisankar (✉)
Department of Zoology, Bharathiar University, Coimbatore, Tamil Nadu, India
e-mail: tsmuralisankar@buc.edu.in

A. Mustafa
Director of Life Sciences Resource Center, Purdue University Fort Wayne, Fort Wayne, IN, USA

K. Priya and S. K. Giri (eds.), *Environmental Toxicology: Xenobiotics and Effects on Human Genetic Material*, https://doi.org/10.1007/978-3-032-28619-2_3

to establish regulatory frameworks to conserve the environmental health and human well-being.

Keywords Pesticides · POPs · PPCPs · Environmental health · Human health risk

1 Introduction

Xenobiotics (XBs) are organic and inorganic substances produced from natural and synthetic sources that can be found in environmental matrices and biota (Patel & Sen, 2013; Zachou et al., 2021). Their accumulation and fate in the environment have ecotoxicological implications for both biotic and abiotic components of ecosystems (Ramade, 2018). There are two fundamental classifications of the XBs based on their origin: natural XBs, which are the toxins produced by biota, including microbes, flora, fauna, and other species; and synthetic XBs, which are chemical components produced by human activities. Synthetic XBs are human-made chemical components, such as industrial chemicals, pesticides, pharmaceuticals, and synthetic polymers, which exhibit their anthropogenic properties when exposed to the environment, unlike natural XBs (Bharti, 2024). However, XBs are considered harmless chemicals until they exceed their optimum concentrations due to their persistence in the environment (Štefanac et al., 2021).

XBs exist in the environment in various physical forms, including gaseous, dust, and liquid forms. Additionally, based on their functions, such as active and passive XBs, the active functions of the XBs involve the ability to carry and transfer pesticides and dyes among different ecosystems, which causes toxicity to organisms, whereas the passive functions of the XBs are the unintentional presence without any bioactive role in the organism, causing toxicity to organisms (McLachlan, 1993; Pelkonen et al., 2008; Manzetti et al., 2014).

Direct discharges, such as the effluents from the pharmaceutical, petrochemical, agrochemical, paper, and miscellaneous industries, carry XBs into the environment. In contrast, the indirect discharge of XBs acts as a secondary source after consumer use, also entering the environment (Donner et al., 2010). Discharges from the disastrous events at the chemical storage and manufacturing places also contribute as an acute point-source of XBs (Singh et al., 2023). The nature of these XBs poses a greater threat to the environment, even at lower concentrations, than typical contaminants. This new era of contaminants has emerged as a result of industrial development and anthropogenic activities (Pereira, 2015; Štefanac et al., 2021), and these XBs pose emerging toxicological challenges. Therefore, further research on the emerging XBs, like nanoparticles (NPs), microplastics (MPs), pharmaceuticals and personal care products (PPCPs), pesticides, persistent organic pollutants (POPs) as well as brominated flame retardants (BFRs), is crucial to rectify the existing research gaps on XBs to develop sustainable and eco-friendly strategies to remediate them in the environment.

NPs are ultrafine, engineered particles with unique physicochemical and mobility properties, used for various applications (Kwon et al., 2020). When these properties are altered, they raise ecotoxicological risks to organisms in various ecosystems of the biosphere (Joudeh & Linke, 2022). Environmental transformation of NPs and their ability to penetrate biological membranes cause oxidative stress in biological systems, raising public health concerns and presenting challenges to mitigate their ecological impacts (Abbas et al., 2020). NPs interactions with other hazardous substances, as well as their synergistic relationship with pathogenic microbial communities, are poorly understood (Sathyamoorthy et al., 2025).

Plastics are synthetic materials that carry various applications due to their demand. Post-consumer plastic waste generated from the urban conditions is fragmented by environmental factors into size-reduced Microplastics (MPs) (Dris et al., 2018). MPs are classified as an emerging contaminant, ranging from 5 mm to 1 μm. They became omnipresent in every ecosystem around the globe, particularly in aquatic ecosystems (Chauhan & Sharma, 2024). Continuous weathering processes of plastics ultimately transform them into new contaminants, such as nanoplastics, which can easily penetrate cellular membranes. The environmental persistence of plastics, which enables them to carry toxic chemicals, additives, and pathogens, poses significant environmental and human health risks (Atugoda et al., 2022).

PPCPs, pesticides, herbicides, biocides, and POPs, as well as BFRs, also disrupt the ecological balance; the inadequate exploration of these contaminants poses a significant emerging threat to the environment (Encarnação et al., 2019; Sánchez-Resino et al., 2023; Segev et al., 2009). Therefore, the present chapter will provide an overview of the current state of knowledge regarding the prevalence, fate, transport, accumulation, and toxicity of XBs in the biotic and abiotic compartments of the environment. Conclusively, this chapter elucidates the interdisciplinary approaches to tackle the ecotoxicological risks by addressing the challenges posed by XBs to ecosystems and humans. This chapter will serve as a significant resource for researchers and policymakers to enhance their efforts in improving environmental sustainability.

2 Emerging Xenobiotics: Nanoparticles and Their Ecological Concerns

2.1 Nanoparticles—An Introduction

Materials with particle sizes ranging from 1 to 100 nm are considered NPs (Eker et al., 2024). NPs are dominating the entire pharmaceutical, cosmetic, electronic, energy, and environmental industries due to their unique properties, such as a higher surface area-to-volume ratio (Khan et al., 2019). NPs take up several forms, which include quantum dots, nanorings, nanobelts, nanowires, and nanorods that are widely used in electronics and energy sectors; liposomes, dendrimers, nanospheres, nanocapsules, and fullerenes that are often used for drug delivery in biomedical industries; nanopores and nanotubes, which are used in the process of catalysis (Mirsasaani et al., 2019). NPs are simple but structurally complex, comprising surface, core, and shell layers (Khan et al., 2019). The core is the largest and center portion of NPs, and carries the fundamental features of the NPs, often the produced NPs are only referred by their core material. The middle shell layer usually chemically differs from the core in all aspects, it holds the stability, biocompatibility, prevention of aggregation, and oxidation reactivity of the core material (Khan et al., 2019). The outermost surface layer of the NPs, which facilitates functionalization by attaching specific chemical groups, such as metal ions, surfactants, polymers, and small molecules (Ealia & Saravanakumar, 2017). The scientific exploration of NPs began in the 19th century, when the fundamental understanding of NPs was studied using colloidal gold by Michael Faraday. Later in the 20th century, study on NPs evolved into a proper discipline, followed by the development of several microscopy tools, such as the atomic force microscope (AFM) and the scanning tunneling microscope (STM), which facilitated visualization, design, and fabrication (Jeevanandam et al., 2018).

2.2 *Types of NPs and Their Applications:*

NPs can be broadly classified into various classes based on their origin, structure, dimensions, and pore size (Mekuye & Abera, 2023). Based on their origin, NPs can be natural or artificial; natural NPs include spider silk, ocean spray, viruses, and proteins, among others. Artificial NPs are manufactured at the nanoscale for their industrial purposes (Tadesse & Hailemariam, 2025). Based on structural composition, NPs are classified into metal-based, metal oxide, semiconductor, ceramic; lipid-based, carbon-based, and composite-based NPs (Khan et al., 2019). Metal-based NPs are made of pure metals such as Ag, Zn, Cu, Au, and Fe with thermal, catalytic, electrical, and optical properties which make them an apt option for usage in antimicrobial coating, biosensing, catalysis, and drug delivery (Luo et al., 2024). Metal oxide NPs are made of metal oxides like TiO_2, ZnO, SiO_2, and Al_2O_3, and these possess UV absorption and photocatalytic properties and are often used in cosmetics (Wu, 2023).

Semiconductor NPs use materials like Si, ZnS, and CdS, which possess both metallic and insulating properties, often used in solar panels, LEDs, and photo detectors (Ma et al., 2024). Ceramic NPs were made up of inorganic non-metallic solids such as oxides and phosphates, possessing heat resistance, chemical inertness, and hardness, which make them applicable for catalysis, material coating, imaging, and drug delivery (Thomas et al., 2015). Lipid-based NPs are composed of a large lipid core in a solid form, and these are biodegradable and biocompatible, often used in cancer therapy, vaccines, and genes transportations (Chauhan et al., 2025; Thi et al., 2021). Carbon-based NPs exhibit higher mechanical strength, electrical conductivity, and stability; hence, these NPs are being widely used in electronics and energy storage (Saleh et al., 2024). Composite NPs are multifunctional NPs that exhibit synergistic properties from the combined presence of two or more NPs and bulk materials; these NPs demonstrate enhanced mechanical strength and stability, which facilitates their use in the aerospace, automotive, packaging, and biomedical industries (Bhat et al., 2021).

According to the number of dimensions, NPs can be of four types, namely zero-dimensional (0D), 1D, 2D, and 3D. In 0D NPs, all three dimensions, length, width, and height, are confined to nanoscale. In 1D NPs, two dimensions are in nanoscale. In 2D NPs, only one dimension, which is often the thickness, lies within the nanoscale. In case of 3D, all the dimensions are above the nanoscale but they are made of building blocks that occur in size of nanoscale (Joudeh & Linke, 2022). In accordance with the material pore size, nanomaterials are classified into microporous, mesoporous, and macroporous. The pore size indirectly relates to the size of the molecule, interaction, and diffusion properties, which aid in determining the applications of synthesized NPs (Mekuye & Abera, 2023). Microporous nanomaterials possess pores with the diameters < 2 nm; mesoporous materials have comparatively larger pores of 2–50 nm diameters; macroporous materials have pore diameters > 50 nm (Mekuye & Abera, 2023).

2.3 *Exposure and Toxicity of NPs*

Intentionally or unintentionally released NPs undergo a complex transformation, including dissolution, oxidation, aggregation, and surface modifications in the environment (Pulido-Reyes et al., 2017; Zahra et al., 2022). NPs are commonly released through industrial effluents, exhaust gases, accidental spills, and usage phases (Saravanakumar et al., 2022). Additionally, the release of cosmetics, sunscreens, paints, textile dyes, and pharmaceuticals is another common way in which NPs enter the ecosystem (Zahra et al., 2022). The inefficient removal of nanosized particles from wastewater treatment plants is a key reason for NPs seeping into the environment (Yu et al., 2022). Additionally, the entry of agricultural runoff containing pesticides into water bodies, soil, and groundwater can also contribute to the transfer of NPs into aquatic bodies (Mishra & Sundaram, 2023). Conclusively, the atmospheric deposition of NPs of vehicle exhaust and combustion leads to airborne NPs pollution (Kumar et al., 2013). Hence, over time, these NPs dissolve, absorb, aggregate, and bioaccumulate in food webs, causing long-term and invasive ecological risks (Wang & Liu, 2022). With this lifecycle, NPs do not cease to exist; instead, NPs upon prolonged exposure to the environment, react with organic matter and modify their surface chemistry, forming "eco-coronas" (Cao et al., 2022).

The detection of NPs from the environment is a significant challenge; the first barrier is the collection of samples, as their nanometer size often causes them to escape during collection. For airborne NPs, collection requires highly efficient filters, impactors, and electrostatic precipitators. For soil and water collection, inert containers are used (Lee et al., 2023). Then the samples undergo specific procedures to obtain the concentrated nanoparticle-rich fraction (Azari et al., 2023). These fractions are then subjected to many analytical techniques which includes single-particle ICP-MS is used for detecting individual metal or metal-oxide NPs, they reveal the size distribution and number concentrations (Fernández-Trujillo et al., 2024). Transmission electron microscopy (TEM)/scanning electron microscopy (SEM) is used for imaging to analyze the surface morphology, shape, size, crystal structure, and elemental composition of NPs. The DLS and nanoparticle tracking analysis (NTA), used to measure the hydrodynamic diameters and particle concentrations, respectively (Fernández-Trujillo et al., 2024). Field-flow fractionation (FFF), coupled with ICP-MS or DLS, is used to separate particles based on their size (Wang et al., 2024a, 2024b). Spectroscopic tools, including UV–Vis, fluorescence, Raman, FT-IR, XPS, and XRD, are commonly used for this purpose (Fernández-Trujillo et al., 2024).

Since NPs release toxic effects at low concentrations, biological assays and bioindicators based on NPs detection become more effective before transitioning to higher-end analytical techniques (Khan et al., 2019). Molecular and biomarker assays: oxidative stress biomarkers which include activities of superoxide dismutase (SOD), catalase (CAT), and glutathione peroxidase (GPx), and the level of lipid peroxidation (LPO) (Nie et al., 2023), followed by micronuclei and DNA fragmentation, become reliable genotoxic indicators (Rim et al., 2013). Heat shock proteins and metal stress genes like metallothionein and glutathione s-transferase (GST) are important indicators of even a subtle NPs exposure (Nie et al., 2023). Invertebrate assays: Earthworms (e.g., *Eisenia fetida*), daphnids, and benthic invertebrates accumulate NPs, and their responses (survival, reproduction, and behavioral assays) can reveal the effect of bioavailable NPs fractions; reduced growth, chlorophyll degradation, or reactive oxygen species (ROS)-induced stress in algae like *Chlorella* sp. and *Scenedesmus* sp., serves as a early warning indicator; changes in soil or aquatic microbial communities (e.g., shifts in species abundance), declines in dehydrogenase, urease or phosphatase activity, or altered nitrification rates can signal NPs stress (Khanna et al., 2021; Qiao et al., 2022; Saxena et al., 2021).

2.4 Environmental and Health Implications of NPs

NPs and other engineered nanomaterials have become an integral part of all technologies (Eker et al., 2024). Their minute size, higher surface area, and remarkable functionality have made NPs inseparable from humankind; however, the same characteristics also make their effects and behavior in the environment unpredictable upon release (Das et al., 2025). Recent reviews suggest that titanium dioxide (TiO_2), silver (Ag), and zinc oxide (ZnO) NPs are the most prevalent types (Das et al., 2025). The Ag, TiO_2, CuO, and ZnO are metal-based and metal-oxide NPs that are reported to produce higher genotoxic and cytotoxic effects; they release ions that in turn produce ROS leading to oxidative stress, cell membrane disruption, inflammation, DNA damage, chromosomal mutations, mitotic aberrations, glutathione depletion, lipid damage, and cell dysfunction in organisms, particularly under chronic exposure.

They also interfere with detoxification process and metabolic enzymes by disguising and affecting cellular homeostasis (Das et al., 2025).

Humans are exposed to NPs through various pathways, including inhalation, skin absorption, and ingestion of contaminated water and food (Tran et al., 2024). There is a high possibility that NPs get inhaled into the body through polluted air and deposit in the tissues of lungs causing fibrosis, bronchitis, asthma, inflammation, and chronic obstructive pulmonary disease (COPD). Statistics suggest that metal-oxide NPs and nanotubes of carbon are the dominant type of NPs causing respiratory diseases in a wide range of population; they are also reported to reduce the capacity of lung clearance that results in increased lung infection (Portugal et al., 2024). Increased NPs accumulation in the bloodstream implies increased oxidative stress, which in turns to elevates blood pressure, promoting atherosclerosis, heart attacks, and strokes. The ultrafine industrial particles and carbon-based NPs are the most common causes of NP-mediated cardiovascular diseases (Khan et al., 2019).

NPs showcase a remarkable ability to bioaccumulate and transfer across the trophic levels. For instance, cerium oxide, one of the less common NPs, is also reported to accumulate in roots and leaves of the plant Silene flos-cuculi (Caryophyllaceae), and this supports the fact that NPs bioaccumulate in the plant and contribute to trophic transfer (Lizzi et al., 2021). Similarly, quantum dots exhibit higher trophic magnification, with a maximum of fivefold magnification between species (Rocha et al., 2017). Similarly, metal-oxide NPs exposed on algal cells contribute to cellular damage and metal ion release (Soares & Soares, 2021). The Ag and ZnO NPs are reported to be highly toxic to microalgae such as *Chlorella vulgaris*, zooplankton like *Daphnia magna*, and fish, including *Danio rerio,* at very low concentrations, primarily because of ROS generation and ion release (Khoshnamvand et al., 2020; Mahjoubian et al., 2023a; Suman et al., 2015). Experimental studies on zebrafish have demonstrated NPs-induced hepatotoxicity and neurotoxicity (Mahjoubian et al., 2023a, 2023b).

Human exposure studies reveal that there is an increased cancer risk noticed in people who are occupationally exposed to TiO_2 NPs (Le et al., 2018). TiO_2-NPs inhalation causes pulmonary inflammation. In animals, it leads to tumors and affects intestinal bacterial community by disrupting its structure at a very low dose of 1 mg/kg concentration (Li & Tang, 2024). Prolonged exposure often reduces the abundance of genes responsible for nitrogen transformations (Simonin et al., 2016). The alarming effects of NPs on ecosystems are found to be due to the mixture of NPs rather than isolated exposures, which demands an understanding of how the combined toxicity of NPs affects the ecosystem (Forest, 2021). A combined study involving a mixture of NPs demonstrates that interactions between different NPs can produce cumulative, amplified, or counteractive effects, depending on their ratios and physicochemical characteristics (Li et al., 2020).

For instance, the TiO_2 NPs were found to reduce Cd^{2+}-induced growth inhibition in *Chlamydomonas reinhardtii,* showcasing antagonistic effects; similarly, TiO_2 NPs enhance Cd^{2+} accumulation in *Cyprinus carpio* by 2.5-fold, representing synergistic effects; Al_2O_3 NPs showed an antagonistic effect when combined with Cu^{2+} by suppressing the toxicity of Cu^{2+} in *Scenedesmus obliquus* (Li et al., 2020). Certain

NPs are also found to cross the blood–brain barrier (BBB), which is both beneficial and alarming. This property facilitates targeted drug delivery, on the positive note, while they are disadvantageous. In a study by Tang et al. (2010), an in vitro BBB model formed of rat endothelial cells and astrocytes revealed that AgNPs cross the BBB after 4 h of exposure, accumulating inside brain microvessel endothelial cells, while silver microparticles did not. Similarly, an in vivo rat study demonstrated that AgNPs injected subcutaneously traversed the BBB and migrated into the brain, causing neural degradation and necrosis upon prolonged exposure (Tang et al., 2008). In another similar study, where the AuNPs were injected intraperitoneally, spontaneous penetration was demonstrated, and the NPs were found in brain tissues after perfusion (Sela et al., 2015). In a study on the toxicity of shape and size-dependent TiO_2 NPs in rats, it was found that the smaller nanoparticles penetrated the BBB-like endothelial cell layer more efficiently and faster than the larger ones (Liu et al., 2017).

AgNPs crossed the blood–brain barrier and caused hippocampal oxidative stress with impaired neurosteroid balance and delayed excretion (Dziendzikowska et al., 2022). Likewise, a study using AgNPs of 8 nm and TiO_2 NPs of sizes 6 nm and 35 nm caused decreased BBB integrity and disrupted tight junctions (Chen et al., 2016). A study by Liu et al. (2019) reveals that TiO_2 NPs can suppress the toxicity of Ag NPs in soil ecosystems, emphasizing that mixtures of NPs behave differently than when they are standalone. This finding provides a future outlook for studying real-world combinations of NPs to suppress their toxicity. This interactive behavior is provided by a study showing that TiO_2 NPs in irrigation water mitigated the harmful impacts of aged Ag NPs on soil microorganisms, *Arabidopsis thaliana*, and the earthworm *Eisenia fetida* by reducing Ag^+ bioavailability and oxidative stress (Liu et al., 2019).

However, the NPs' risk assessment remains a tedious process, as conventional models for their assessment are not nanosize-specific and do not capture the aggregation and corona formation of NPs. Laboratory-based exposure studies often use doses that are several-fold higher than the environmental concentration, typically for only an acute or short span of time, and fail to accurately analyze long-term, low-dose exposure studies. Recent advancements, including single-particle spectroscopy, omics-based biomarkers, and electron microscopy, have helped to bridge several research gaps of environmentally realistic NPs exposure, with characterization, omics translation, and NPs' regulatory risk assessment (El-Kalliny et al., 2023; Lead et al., 2018). Mitigation strategies require synthesis modifications and surface coatings to reduce reactivity, designing biodegradable or dissolvable nanomaterials, and incorporating biosynthesis procedures. Maintaining a balance between technological advancement and environmental protection is a need of the hour.

3 MPs: Global Xenobiotic Concern

3.1 Plastics: A Source of an Emerging Contaminant

Plastics are synthetic and semi-synthetic fossil-based products manufactured through the polymerization and condensation processes. Plastics are widely known for their durability, versatility, and cost-effectiveness in all their applications. The modern polymer age began in 1907 with the invention of the first synthetic resin, "Bakelite," by Leo Hendrik Baekeland. Bakelite was primarily used in electrical and household applications. The invention of Bakelite sparked interest in synthetic plastics, marking the beginning of the industrial evolution of plastics, which utilized petrochemical feed stocks such as benzene, ethylene, and propylene for mass production (Spitz, 1988). Currently, the polymer industry serves as a backbone for every industrial development, driven by the demand for plastic products that facilitate cost-effective efficiencies. The production trend is expected to increase to 70% by 2040 (OECD, 2024). In 2024, over 387 million tons of fossil-based plastics were produced, whereas only 36.5 million metric tons of consumer plastic waste were recycled, and over 55.8% of the worldwide plastic production was contributed by Asian countries (Plastic Overshoot Day, 2024).

Due to the higher demand for plastic-based products, the generation of post-consumer plastic waste may exceed the global capacity of waste management sectors, along with a neglect of the implications of global plastic waste generation (Plastic Overshoot Day, 2024). The excess macroplastic waste was disposed of in the environment, where it is fragmented by environmental factors, such as tidal actions (Barnes et al., 2009; Fei et al., 2023; Lechthaler et al., 2020). It leads to the emergence of a widespread contaminant known as Microplastics (MPs), which are plastic particles less than 5 mm in size, increasing the threat levels to global ecosystems (Thompson et al., 2004). MPs were discovered in the dynamic ecosystems from Mount Everest to the Abyssal plains of the ocean (Ramadoss et al., 2025; Wang et al., 2025). MPs have also been found in primitive organisms (marine sponges, Cnidarians, and Protozoans), and in human bloodstream, testes, and ovaries (Devereux et al., 2021; Fallon & Freeman, 2021; Montano et al., 2025; Zhang et al., 2022; Zhao et al., 2023). Therefore, MPs were recognized as an emerging contaminant that is omnipresent, persistent, and pervasive, and their multifaceted toxicity and ability to carry other XBs and pathogenic microbes pose potential indirect toxicological risks to the biota of ecosystems (Stapleton & Hai, 2023).

3.2 Sources, Types, and Properties of MPs

MPs classified into two different types as primary and secondary MPs based on their origin. Primary MPs are synthetic plastic particles manufactured by industries and used in personal care products, discharged from urban runoff, city dust, synthetic textiles, vehicle tires, road markings, and through other sources MPs enter the environments (Boucher & Friot, 2017; Song et al., 2024). The secondary MPs originated from the discharged macroplastic wastes due to their weathering by physical, chemical, and biological processes in the environment (Horton et al., 2017). Secondary MPs mostly originates from household waste discharges, marine fishing gears, construction plastic materials, and abrasion of disposed plastic materials (Anandhan et al., 2024; Cole et al., 2011; Dris et al., 2016) MPs were identified in the environment based on their morphometric properties, including shape, size, and color. MPs exhibit a variety of shapes in the environment, including fibers, fragments, spheres, films, and foams (Gan et al., 2023).

The size of MPs is classified as from the 0.1 μm to 5 mm, based on the size they are categorized as large MPs (1– < 5 mm), small MPs (5– < 10 μm), smaller MPs (10– < 50 μm), intermediate MPs (50– < 100 μm), fine MPs (100– < 500 μm), and outstanding MPs (500– < 1000 μm) (ISO, 2023). Majorly, MPs were recorded from the environmental matrices with different colors like blue, black, transparent, red, white, green, yellow, and brown; hence, the color is one of the prominent distinguishing criteria for the detection of MPs in environments (Anandhan et al., 2022; Fox et al., 2024; Gholizadeh et al., 2023). Apart from these properties, the synthetic origin of MPs was confirmed by their polymer nature. MPs were distinguished as different polymers based on their source: Polyethylene Terephthalate (PETE), High-Density Polyethylene (HD-PE), Low-Density Polyethylene, Polypropylene (PP), Polystyrene (PS), Polyvinyl chloride (PVC), and other polymers such as nylon, polycarbonates, and bioplastics (Björkner, 2011; Anandhan et al., 2025). Also, MPs can also acquire the chemical properties of other XBs when they interact in the environment, such as POPs, heavy metals, and pesticides (Cao et al., 2021).

3.3 Exposure of MPs to the Environment with the Abiotic and Biotic Components

There are multiple entry routes for MPs enters to the environment. MPs primarily enter the aquatic bodies, through the discharges of wastewater treatment plants and direct urban sewage and rainwater runoff. The wastewater effluents primarily include the post-consumer plastic wastes such as personal care products, synthetic textile fibers, and other household and industrial wastes (Iyare et al., 2020; Yaseen et al., 2022; Acarer, 2023). Water column is the primary source of the MPs contamination in sediments, dense polymers like PVC and other polymers can sink into the sediment column through vertical distribution (Quintana et al., 2025), whereas, irrespective of their density, MPs sink and become trapped in the sediment's thorough biofouling processes (Raza et al., 2022; Silva et al., 2022). MPs in the sediments remain more than the MPs present in the water columns, and also where the higher MPs and biotic interactions increase (Liu & Li, 2025; Quintana et al., 2025). Soil ecosystems, such as agricultural plains, are primarily receiving MPs through sewage sludge applications, along with the use of plastic films in agricultural lands (Long et al., 2023; Ramage et al., 2025). Additionally, MPs are also received through atmospheric depositions and synthetic fibers containing composts used in soil conditions (Hee et al., 2023; Raza et al., 2022). Airborne MPs contamination compromises the air quality of the atmospheric conditions, which originates from the plastic abrasions, construction dust, and synthetic textile fibers (Long et al., 2022; Wright et al., 2020). Typical wind conditions can transport the MPs, and which contributes to their cross-environmental distribution into the soil and water ecosystems (Hee et al., 2023; Ryan et al., 2023; Weber & Bigalke, 2025).

After the transfer of MPs and plastic wastes, they were subjected to the increased fragmentation due to the breakdown of polymer chains (Sutkar et al., 2023). UV radiation is the major abiotic factor that fragments plastics into much smaller size. As well, mechanical abrasion and wind processes contribute to the fragmentation of MPs (Andrady et al., 2022; Cheng et al., 2021). Mechanical and oxidative fragmentation of the MPs is a serious concern that alters the surface morphology of the MPs in the environment. The increase in fragmentation of MPs will increase the absorption properties of the MPs on other XBs, and it means higher chances of ecotoxicological risks to ecosystems (Li et al., 2025; Sutkar et al., 2023). Alteration of the functional groups, increased roughness of the surface, and hydrophobicity of the MPs create two different fragmentation pathways: particle bulk breakup caused by cracks and their penetrations, and degradation of surface layers, releasing micro- and nanoplastic fragments (Pfohl et al., 2022; Wang et al., 2025). These abiotic fragmentation processes were significantly modulated by environmental factors, including salinity, temperature, and oxygen availability in water, soil, and atmospheric conditions (Cai et al., 2023; Khoshmanesh et al., 2024; Liu et al., 2022a, 2022b).

MPs also serve as a potential vector due to their absorption properties; they can absorb (POPs) and heavy metals (Li et al., 2023a, 2023b; Liu et al., 2022a, 2022b). The xenobiotic nature of the MPs prevails when the MPs were transferred from water, air, and soil conditions to biota (Jia et al., 2024). The MPs were reported to cause various physical entanglement and physiological toxicity in the biota of the environment (Chang et al., 2022; Wang et al., 2021), and their bioavailability across all trophic levels of the environments through their direct and indirect ingestion/exposure from the abiotic columns like water, air, and soil conditions (Qaiser et al., 2023; Yang et al., 2025a, 2025b, 2025c). Ingestion of MPs from invertebrates to higher trophic level species is dependent on MPs' abundance and the feeding ecology of the organisms (Nelms, 2018; Gouin, 2020).

The ingestion of MPs is mostly retained in the gastrointestinal tracts of organisms and creates oxidative stress when enters into the physiological process of the organisms. MPs and the associated chemicals alter the integrity of the cell membranes and elevate intercellular ROS production, thereby damaging mitochondrial membranes and DNA (Jeong et al., 2017; Kadac-Czapska et al., 2024). The cellular imbalance of antioxidant enzymes and ROS release results in the altered metabolic activities, endocrine disruption, hepatic stress, and cellular necrosis. Additionally, prolonged exposure to MPs compromises an organism's growth, reproduction, and survival (Kadac-Czapska et al., 2024; Yang et al., 2025a, 2025b, 2025c).

The increased surface of the MPs habitats rapid developments of biofilms on the surface of MPs, a microhabitat called "plastisphere" which harbor colonies of different microbial species including pathogenic microbes from the water and sediments (Ahmad et al., 2025; Fortin et al., 2025). Plastisphere enable anaerobic processes, such as denitrification and dissimilatory nitrate reduction to ammonium, in low oxygen concentrations. It influences the dynamics of nitrogen cycling, facilitating the movement of nitrogen from aquatic ecosystems (Fortin et al., 2025). Moreover, the extracellular enzymes from the microbes such as Polyethylene terephthalate hydrolase (PETase), cutinases, laccases, esterases, lipases, lipid peroxidases, and manganese peroxidases, target the amorphous regions of plastic polymers and increase their crystallinity causing the plastic polymers to degrade. (Cai et al., 2023). These enzymatic degradations begin with microbial adherence, which leads to biofilm formation, followed by depolymerization, and the subsequent mineralization of monomers for microbial metabolism (Anand et al., 2023; Temporiti et al., 2022). Moreover, the other microbial enzymes involved in the further degradation of byproducts derived from the primary degradation of MPs (Bhatia et al., 2018). Microbial consortia of different bacterial and fungal species, notably strains of *Vibrio* spp., *Pseudomonas* spp., *Bacillus* spp., *Ideonella sakaiensis*, and *Rhodococcus* spp., have the ability to completely degrade MPs (Allemann et al., 2024; Salinas et al., 2023). However, the degradation rates of MPs vary due to the influences of environmental conditions, polymer nature and associated chemicals, the longer time scales of the prominent polymer degradation rates of PP > 30 years, PE 100–1000 years and PET > 400 years and these rates were influenced by the climatic conditions, temperature, and the availability of oxygen and microbial communities (Ahmad et al., 2025).

3.4 Human Health Implications and Mitigation Measures on MPs Contamination

Ubiquitous nature of MPs in the environment, they enter the human body through air, water, and food. In recent years, research on MPs has revealed the accumulation of MPs within the organs of the human body, which evident the possibility of potential health risks to humans (Nouri, Massahi & Hossini, 2025). MPs enter the human body through three primary routes: inhalation, dermal contact and ingestion. Ingestion is a dominant pathway of annual exposure to 39,000–520,000 MPs through bottled water. Daily ingestion of MPs up to a million by humans was reported, through indoor and outdoor exposure, 21 MPs/day are entered into the human body through inhalation (Heo et al., 2024; Nawab et al., 2024). MPs were found in the drinking water, table salts, honey, fruits, and vegetables and also in food products in plastic containers (Al-Mansoori et al., 2026), MPs exposed through the food materials, settled in the semen and follicular fluids of human, also < 150 μm-sized MPs can penetrate the intestinal barrier from digestive system and ends in bloodstream from where they will be translocated into the other organs of the body (Enyoh et al., 2020; Rodrigues et al.,

2024). Lesser than 25 μm-sized MPs from the contaminated water, soil, and cosmetic products, synthetic textile fibers, can enter the body through dermal contact (Ageel et al., 2022; Enyoh et al., 2020). After entering the body, the MPs were transferred to different organs and systems of the human body. The detection of MPs in humans is interlinked with their cardiac health, and the risk is associated with up to 4.5-fold higher chances of stroke, cardiac arrest and lethal effects in the physiology of humans (Mahalingiah et al., 2025; Ponnana et al., 2025).

Likewise, the accumulation of MPs in brain tissues raises serious concerns about the health of the neurological system. Studies revealed the MPs' penetration into the blood–brain barrier and their deposition on the myelin sheath of the human brain (Jambeck & Walker-Franklin, 2023; Nihart et al., 2025). A study on Sprague–Dawley rats exposed to MPs showed an increase in neuroinflammation, decreased neuronal signaling, cognitive impairments, and an increase in neuronal death in the brain (Kim et al., 2025). Dementia patients were shown to have higher MPs accumulation in their cerebrovascular system and immune cells, which might increase phosphorylation and tau protein cleavage, resulting in Alzheimer's disease, which also links the possibilities of risk of dementia and stroke (Gaspar et al., 2025; Huang et al., 2025; Nihart et al., 2025). Inhalation exposure to MPs sized < 2.5 μm can cause respiratory system damage (Bardawil et al., 2026). There are 12 different types of MPs, including PE, PP, PET, PS, Polyethersulfone (PES), Polyurethane (PUR), Polytetrafluoroethylane (PTFE), Polyacrylonitrile (PAN), Poly (methyl methacrylate) (PMMA), Styrene Ethylene Butylene Styrene (SEBS), Thermoplastic Elastomer (TPE), and Resin, have been found in the human lungs (Jenner et al., 2022). Pulmonary immune cells are infected by the MPs, which disrupts tissue homeostasis. Chronic exposure of MPs in the respiratory system reduces the phagocytic activity of macrophages in lung cells and causes lung inflammation and oxidative stress. Furthermore, inhaled MPs are disseminated to vital organs, including the brain, liver, spleen, colon, and kidneys (Soloff et al., 2025). The size-reduced MPs penetrate into the alveoli of the lungs, affecting gas exchange and leading to asthma, cancer, and other pulmonary diseases (Chartres et al., 2024; Saha & Saha, 2024).

The primary exposure route of MPs into humans is ingestion, which allows them to reach the digestive system. The accumulation of MPs in the gastrointestinal tract causes structural changes and cracks in the intestinal walls, as well as inflammation in the intestinal tissues, which affects the functions of the microvilli. MPs alter mucus production in the gastrointestinal tract, which enhances the likelihood of toxins entering the bloodstream and being carried to different organs through central nervous system and gut axis connections (Sofield et al., 2024). Gut microbiome alterations caused by the accumulation of MPs have significant impacts, including inflammatory bowel disease and gastrointestinal disorders, which result from microbiome imbalance, with a reduction in Bacteroidetes and an increase in pathogenic microbes (Fournier et al., 2023; Tamargo et al., 2022).

MPs also target the immune system by generating ROS upon cellular uptake, which activates the inflammatory signaling pathways, leading to the production of danger-associated molecular patterns, which activate the toll-like receptors and promote the pro-inflammatory cytokines (Yang et al., 2022). Consistent activation of immunity by MPs' exposure and inflammation in the vital organs of the body leads to a reduction in anti-inflammatory macrophages in the gut (Cai et al., 2024; Soloff et al., 2025). Autoimmune diseases like sclerosis and rheumatoid arthritis were linked to the presence of MPs and their interactions with the immune cells of the brain (Nihart et al., 2025). Reproductive and developmental health of humans was also compromised due to the accumulation of MPs in the reproductive organs and their hormones, which threaten the fertility of humans (Montano et al., 2025). MP-exposed farm animals were found to have an increased accumulation of MPs in uterine tissues and ovaries, irregular hormonal profiles, a reduced number of follicles, and a decrease in ovarian weight, also the follicular fluid is observed to have an average of 120 MPs, which affects the formation and maturation of oocytes (Yang et al., 2023a, 2023b; Zurub et al., 2024). In the reproductive system of human males, there are testicular inflammations, impairments in spermatogenesis, a reduced quality and count of sperm, and altered male hormone levels.

Major plastic polymers, such as PS, PP, PC, PE, PVC, PTFE, PET, and ABS-based MPs, were observed to accumulate in the semen and urine samples of male subjects, with a notable decrease in quality and count (Wang et al., 2024a, 2024b; Zhang et al., 2024). The early developmental implications of the MPs, such as invading the placental tissue, umbilical cord blood, and amniotic fluid, increase the risk of miscarriage or premature growth of the fetus (Sun et al., 2024a, 2024b; Zhu et al., 2024). The major organs, such as the liver and kidneys, were also intoxicated by the MPs. The exposure of MPs in the liver causes metabolic disturbances (Goodman et al., 2022; Zhang et al., 2025), whereas in the kidney, the toxicity of MPs causes inflammation through tubular damage and podocyte injuries, as well as oxidative stress and the activation of anti-inflammatory pathways, which ultimately lead to total organ failure, and also the MPs' exposure is expected to increase the levels of creatinine and nitrogen in urine discharges (Goodman et al., 2022; Jiang et al., 2025; Meng et al., 2022). When MPs enter the cells of human tissues, they cause damage to the systems through various mechanisms, including inflammation, oxidative stress, genotoxicity, and apoptosis. Primarily, MPs lead to increased ROS production, which overwhelms the antioxidant defense system, resulting in protein oxidation, lipid and DNA damage. Through these ROS-mediated pathways, the results of programmed cell death induction events (Osman et al., 2023) are observed. The activation of inflammatory pathways triggers the possibility of cancer and other diseases. When MPs penetrate the nuclear membrane of cells, the DNA breaks, also defecting the DNA repair mechanisms, which contributes to mutation and cancer-associated risks (Joseph et al., 2025; Zhang et al., 2025).

4 New Age of Xenobiotics and Their Human Implications

Currently, the pre-existing products or chemicals are attaining the status of innovative XBs, with their persistence and bioaccumulation rates playing a significant role in their xenobiotic nature. Apart from MPs and NPs, POPs and PFAS, PPCPs, pesticides, plasticizers, and BFRs were also categorized as emerging XBs (Bu & Ma, 2025; Disner & Tareq, 2025). These emerging XBs share similar properties with major XBs, including lower degradation rates, higher persistence, and greater accumulation in the environmental matrix. They are acutely toxic, cause chronic effects at the lowest concentrations, and easily penetrate and accumulate in the food web of the ecosystem, causing chronic health risks to biota, including humans (Allen et al., 2007; Das et al., 2024).

4.1 Persistent Organic Pollutants (POPs)

POPs are carbon-based organic chemicals that resist natural degradation processes and remain in the environment for several years, traveling far from their sources via air, water, or other environmental factors (Pariatamby & Kee, 2016). Most POPs contain chlorine or fluorine atoms, which provide them with higher resistance against photolytic, chemical, and biological degradation, facilitating their accumulation in the food chain and posing significant health risks to the environment and biota (Negrete-Bolagay et al., 2021). Following industrial adaptation after World War II, POPs became a widespread contaminant detected in most dynamic ecosystems, including the Arctic and Antarctic regions, from their source of origin (Ashraf, 2015; Guo et al., 2019). POPs can be categorized into organochlorine pesticides, industrial chemicals, and the byproducts from other chemical compounds (Akhtar et al., 2021). Organochlorine pesticides are the major category of POPs, which include dichlorodiphenyltrichloroethane (DDT), aldrin, dieldrin, endrin, chlordane, heptachlor, and toxaphene. These were mainly used as pesticides, herbicides, and insecticides (Kumari, Swamy & Singh, 2021).

The industrial chemicals, including polychlorinated biphenyls (PCBs), polybrominated diphenyl ethers (PDBEs), perfluoroalkyl substances (PFAS), and perfluorooctanoic acid (PFOA), were used in electronics, coolants, heat exchange fluids, and additives in other products (Gaines & Gaines, 2023). PFAS and PFOA were considered "forever chemicals" in the environment, primarily used in firefighting foams, textiles, and electronics (Brunn et al., 2023). Byproducts such as dioxins and furans are substances with no commercial use, and, due to their high persistence in the environment, they also harm the environment (Mathew et al., 2025). All these major categories of POPs are sourced from agricultural pesticides and insecticides. After that, the industries also act as a primary source, especially the unintentional byproducts that are produced deliberately from the industrial chemical manufacturing sectors (Sharma et al., 2014; Aravind kumar, 2022), whereas the environmental factors like climatic changes, ice melting, and permafrost patterns also serve as a secondary source of POPs release into the environment (Butler & Wild, 2023).

POPs express their persistence, mobility, lipophilicity, and low solubility properties in water, which aid them in binding to sediments and organic matter. This occurs when they are transported by the circulation patterns of ocean water and the atmosphere to remote regions, such as the Arctic, Antarctic, and deep oceanic sediments. This long-range transportation bioaccumulates POPs into living communities despite contact with polluted conditions (Muir & de Wit, 2010; Wang et al., 2015). The transfer of POPs into biotic life forms occurs through the phenomenon of bioaccumulation, resulting in biomagnification (Armitage & Gobas, 2007; Ren et al., 2017). The biota of ecosystems is significantly affected by the accumulation of POPs, which cause disruptions to reproductive, neurological, and immune systems (Akhtar et al., 2021). A study revealed that prolonged exposure to POPs, such as dioxins and PCBs, reduced androgen-dependent growth in the cartilage and muscle of the larynx in male African clawed frogs (*Xenopus laevis*), leading to a reduction in the ability to make courtship calls to females (Aulsebrook et al., 2020).

Research reported by Ritscher et al. (2018) that PFAS exposure in polar bears disrupts their reproductive and endocrine systems, resulting in reduced reproductive capacity. Similarly, the Penguin tissues, feathers, and eggs were found to have accumulations of PCBs, PBDEs, and PFAS, which can cause potential reproductive defects (Okuthe et al., 2025). Ritscher et al. (2018) reported that prolonged exposures to PFAS, PCBs, and dioxins affected the immune system, making individuals vulnerable to infections and causing increased infections in Sea otters (*Enhydra lutris*), Bottlenose dolphins (*Tursiops truncatus*), wild birds, and also diabetes, congenital disabilities, endocrine disruption, immune system issues, and reproductive failures in humans (Guo et al., 2019; Rokni et al., 2023). These adverse health risks are mediated through neurological impairment, endocrine disruption, carcinogenesis, and immune suppression (Sadiq et al., 2025).

A study by Li et al. (2023a, 2023b) examined all the environmental implications related to POPs. Soils affected with POPs pollution are found to have reduced productivity, disrupted food chains, causing species decline and imbalances. PCBs: a group I carcinogen as per International Agency for Research on Cancer (IARC) is a commonly used chemical in electrical transformers, plasticizers, and hydraulic fluids, and is reported to be the primary cause for liver cancer on prolonged exposure, delayed development and immune dysfunction. The primary threat is that they exhibit higher bioaccumulation, leading to greater concentrations in higher organisms, such as seals, humans, and polar bears (Megson et al., 2022). PBDEs is a widely used compound in textiles, plastics, flame retardants, electrical and electronic components. They are also known to cause thyroid hormone dysfunction, delayed neural development, and immune suppression, particularly in fish and birds. They persist in sediments and accumulate in the tissues of marine mammals (Cowell et al., 2018). Similarly, Polychlorinated Terphenyls, a widely used lubricant and component of flame retardants, studies reveal that it is linked to hepatotoxicity, hypertrophy, and immunosuppression in laboratory animals such as rat, mice, bird and fish (Jensen & Jørgensen, 1983). Industrial byproducts such as dioxins, Hexachlorobenzene (HCB), and furans are commonly found in combustion byproducts. They also show extreme persistence and create health implications in animals and humans (Marinković et al., 2010). PFAS are synthetic fluorinated compounds commonly used in firefighting foams, cookware (Teflon), textiles, and food packaging, exhibiting water mobility, which proves to be the most alarming cause of pollution and health impacts (De Silva et al., 2021).

4.2 Pesticides

Pesticides are chemicals that impact the growth and metabolism of specific organisms, occurring in several forms, including insecticides, herbicides, and fungicides (Rajmohan et al., 2020). Organochlorines, organophosphates, carbamate esters, pyrethroids, acetamides, triazoles, triazines, and neonicotinoids are commonly found in insecticides; among these, organochlorine contamination is particularly alarming in the environment (Wang et al., 2022). Some of the commonly used classes of pesticides include dichlorodiphenyldichloroethylene (DDE), dichlorodiphenyltrichloroethane (DDT), hexachlorocyclohexane (HCH), chlordane, and dichlorodiphenyldichloroethane (DDD) (Pascal-Lorber & Laurent, 2011). These pesticides exhibit far greater environmental persistence than imagined; among them, a group of 12 chemicals such as, Aldrin, Chlordane, DDT, Dieldrin, Endrin, Heptachlor, Hexachlorobenzene (HCB), Mirex, Polychlorinated Biphenyls (PCBs), Dioxins, Furans, and Toxaphene, which are known as "Dirty Dozen," has been identified, of which eight are organochlorines (Wang et al., 2022).

The use of DDT, due to its high persistence, was banned a long time ago; however, its persistence is still evident in higher concentrations in adipose tissues of animals and humans, which impacts the mammalian endocrine system and causes bird eggshell thinning (Wang et al., 2022). Similarly, Atrazine, a popular herbicide used for weed control, its presence has been detected in water channels, leading to developmental disorders in amphibians and endocrine issues in mammals (Rohr & McCoy, 2010). Paraquat is a another important herbicide used for weed control, and its chronic exposure causes pulmonary fibrosis as a result of ROS production, LPO, inflammation, apoptosis or ferroptosis in lung tissue and neurons and Parkinson-like neurodegeneration (Subbiah & Tiwari, 2021; Yang et al., 2023a, 2023b). Malathion is a similar organophosphate insecticide which causes acute cholinergic toxicity. Malathion is bioactivated to malaoxon, which in turn inhibits acetylcholinesterase (AChE), that it causes neurotoxicity in developmental stages. (Salama et al., 2015). Likewise, diazinon—a organophosphate insecticide, which causes classic organophosphate poisoning that affects the nervous system with the signs like muscarinic signs that effects on organs that are controlled by parasympathetic nervous system with symptoms like excess salivation, diarrhea, sweating, and slow heart rate. Also, nicotinic signs with symptoms like weakness, paralysis and muscle twitching and aging (Boyda et al., 2021).

Similarly, mancozeb is a dithiocarbamate fungicide; on breakdown it produces a metabolite called ethylene thiourea (ETU), causing insufficient thyroid hormone production by blocking the synthesis of thyroid peroxidase and a protein called sodium/iodide symporter (NIS), in turn causing reduced brain development and fetal growth (Runkle et al., 2017; Skalny et al., 2021). Likewise, Captan, a phthalimide fungicide and a human carcinogen, these occupational exposures causes gastrointestinal irritation and systemic toxicity (Reuber, 1989). Hence, effective mitigation of these toxic pesticides is required to prevent the related human health implications, for which integrated pest management (IPM) strategies will be the most applicable approach that reduces the dependency on chemical pesticides by involving biological, cultural, and mechanical control methods (Chaudhary et al., 2024). An exploration into biopesticides and the use of non-chemicals would be a safer alternative to chemical pesticides. Additionally, implementing stricter regulatory frameworks and safety standards to limit the use of harmful pesticides is recommended (Wend et al., 2024).

4.3 Plastic Additives

Plastic additives are added chemical substances to synthetic polymers during plastic manufacture to alter their properties, such as enhanced durability, performance, appearance, and marketing purposes (Ambrogi et al., 2017). These can also be added in various physical forms, such as liquids, pellets, and powders, to enhance the polymers used for packaging, construction, electronics, automotive parts, and other consumer-based applications (Coleman, 2017). Plastic products are protected by stabilizers (benzotriazoles, benzophenone derivatives, calcium–zinc stabilizers, lead-based and other heavy-metal stabilizers, organotin compounds, hindered phenols, and phosphites) from degradation caused by heat, UV radiation, thermomechanical stress, and peroxidation processes. (Real, 2023). Plasticizers are another category used to improve the flexibility and processability of plastics, reducing brittleness and making them easier to use based on their applications. Phthalates are the most used plasticizer in PVC polymer-based products and soft plastics (Ambrogi et al., 2017). BFRs are used as plasticizers to reduce the flammability and by improving their fire resistance. These include the polybrominated diphenyl ethers (PBDEs) and tetrabromobisphenol A (TBBPA) (Coleman, 2017). Coloring agents, such as TiO_2 and FeO_2 for inorganic pigments, phthalocyanines and anthraquinones, were used to add different colors to improve their applications and marketing purposes (Randhawa, 2024). For the stiffness and strength of plastics, fillers such as mica, talc, kaolin, clay, and calcium carbonate, as well as reinforcements like glass and carbon fibers, were also used (Hawley, 1996).

These additives leach from plastic products throughout their lifecycle, from manufacturing to degradation. The leaching of plastic additives can vary based on their chemical composition and the type of plastic polymers used. Benzophenone-12 has the highest leaching rate (8.79%), and the PE-MPs leach a higher number of additives (2.3%) (Hamilton et al., 2022; Hasegawa et al., 2022). MPs are serving as a vector of these plastic additives into the environment and the biota along the food chain associated with the ecosystems. The prey- and predator-mediated ingestion of MPs carries a higher number of plastic additives than the direct ingestion of MPs from the water column at lower trophic levels (Hamilton et al., 2022). The human health implications of exposure to plastic additives through food, water, inhalation of contaminated dust and air, and dermal contact include endocrine disruption, cancer, obesity, reproductive and teratogenic disorders (Campanale et al., 2020; Flaws et al., 2020). Among the various plastic additives, bisphenols, phthalates, and per- and polyfluoroalkyl substances (PFAS) are considered the most concerning groups in terms of human health risks due to their presence in consumer-use plastics (Ramos et al., 2023; Yashwanth et al., 2025). Bisphenol is a common plastic additive with an annual global production of three billion tons, and it is used in epoxy resin and polycarbonate-based plastic products, such as food storage containers, water bottles, sports equipment, aluminum cans, and lining layers (Wang et al., 2024a, 2024b). The leaching of BPA from these plastic materials, particularly when consuming food and beverages, is a significant source of human exposure to this chemical. Hence, BPA is considered a major endocrine disruptor which causes reproductive toxicity (Hahladakis et al., 2023). In males, BPA causes testicular dysgenesis syndrome, reduced sperm quality, testosterone synthesis, and anogenital distance, expressing feminizing effects and cancer risks. In females, follicle loss, oocyte immaturities, and defects on the hypothalamic-pituitary-ovarian axis occur (Rich et al., 2016; Matuszczak, 2019).

Phthalates were used in PVC-based plastic products, consumer goods, and cosmetics. Di(2-ethylhexyl) phthalate (DEHP) is a widely used plastic additive from the phthalate group (Rodgers et al., 2014). Phthalates can cause irregularities in puberty and fertility syndromes in both males and females, leading to follicle developmental stimulations and inhibitions, as well as oocyte DNA damage and premature ovarian follicle development (Hlisníková et al., 2020; Pan et al., 2023; Priya et al., 2021). Increased exposure to DEHP carries the risks of miscarriage and other pregnancy-related issues, and permanent infertility in women is associated with phthalate metabolite exposures (Abdo et al., 2023; Dutta et al., 2020). Over 4,000 groups are in the class of PFAS, which are considered "forever chemicals" (Criswell et al., 2024). They were used to enhance resistance to degradation and to transform plastics into persistent with thermal and chemical stability. These were used as surfactants, flame retardants, additives, and lubricants (Longendyke et al., 2022; Sun et al., 2024a, 2024b). PFAS were detected in drinking water, food and human blood; the dietary intake of PFAS serves as a dominant source of human exposure. Contaminated meat, dairy products, and fish can introduce PFAS into the human diet (Domingo & Nadal, 2017; Jha et al., 2021; Torres & De-la-Torre, 2022). Likewise, BPA and Phthalates, as well as PFAS, were also associated with reproductive, developmental, metabolic, and endocrine disorders, including carcinogenic risks (Haverinen et al., 2021; Xie, Peng & Xiao, 2025).

4.4 Brominated Flame Retardants (BFRs)

BFRs are synthetic chemicals added to textiles, electronics, and plastics to enhance fire resistance thereby ensuring that consumers and products meet safety standards

(Fent et al., 2020). It includes polybrominated diphenyl ethers (PBDEs), polybrominated biphenyls (PBBs), hexabromocyclododecane (HBCD), and tetrabromobisphenol A (TBBPA) (Gorga et al., 2013; Wang & Dai, 2022). Their slower degradation and bioaccumulation properties support their higher persistence in the environments. The extensive presence of BFRs in the air, water, soil, and biota, including humans, reflects their lipophilicity, persistence, and environmental ubiquity (Thuy et al., 2025; Xiong et al., 2019) The leaching, adsorption, and desorption of BFRs is attributed to temperature changes in the environment, which promote their migration in aquatic environments (Enyoh et al., 2024; Watanabe & Sakai, 2003). Humans were exposed to BFRs through inhalation of BFR-contaminated air, dust, water, and food, as well as dermal contact. Notably, from the dairy milk and fish diet, it was estimated that they contained 50 ng/day of PBDEs, which were then passed on to their consumers (Kim et al., 2025; Marquès et al., 2022). BFRs were absorbed into the serum, placenta, liver, breast milk, fat storage and then distributed to the various tissues of the human body (Daso et al., 2010; Gouesse & Plante, 2022). MPs containing BFRs, which are added as additives in plastics enter the environment's food chain, where the trophic transfer of BFRs occurs (Chen et al., 2022; Sun et al., 2019; Yang et al., 2025a, 2025b, 2025c).The prey contaminated with MPs transfers higher concentrations of BFRs than the MPs ingested by fish from the water column (Hasegawa & Nakaoka, 2021; Herrera et al., 2023).

Human health implications caused by BFRs' toxicity include disruptions to endocrine, liver, and kidney functions, as well as neurological, reproductive, developmental, and immunological effects (Jarosiewicz & Krześlak, 2024; Wikoff & Birnbaum, 2011). Higher concentrations of the PBDEs lead to morphological impairments in the head of human infants. PBDEs also cause chronic defects in motor activities, behavioral patterns, and alter brain functions (Gibson et al., 2018; Herbstman et al., 2010; Linares et al., 2015). Dose-dependent neurological toxicity of PBDEs encompasses oxidative stress, calcium and thyroid homeostasis, as well as effects on human developmental and reproductive systems (Xue et al., 2023). Also, disruptions in the GABAergic, glutamatergic, and dopaminergic neurotransmitter systems lead to cell death in the brain, resulting in locomotor deficits characteristic of Parkinson's disease (Fonnum & Mariussen, 2009; García-Suastegui et al., 2025). Individual and combined exposure to PBDE47, 183, and 209 causes ventilatory dysfunctions in the lungs, and the influences of BFRs decrease pulmonary functions by inhibiting the vitality of bronchial epithelial cells, increasing cell death, DNA damage, inflammation, and oxidative stress (Koike et al., 2016; Mao et al., 2024). The long-term exposure of TBBPA analogues produces significant hepato, neuro, and reproductive toxicity (Lai et al., 2015; Wu et al., 2021).

4.5 Pharmaceuticals and Personal Care Products (PPCPs)

PPCPs are an emerging water contaminant encompassing a diverse group of pharmaceutical drugs and personal care products, including toothpastes, deodorants, insect repellents, cosmetics and cleaning products (Dey et al., 2019; Chakraborty et al., 2023). After human use of PPCPs, they are excreted and disposed of through sewage effluents, entering the terrestrial and aquatic ecosystems. When effluents and sewage sludges contain PPCPs from urban sectors and veterinary pharmaceuticals from aquaculture sources, which are used as fertilizers and manure for agricultural and livestock land applications, their transfer to terrestrial environments occurs through aquatic ecosystems (Keerthanan et al., 2021; Meyer et al., 2019; Ziylan-Yavas et al., 2022). Manufacturing facilities and ineffective treatment strategies for removing PPCPs from treatment plants also serve as a primary source of contamination (Reyes et al., 2021; Zaidi et al., 2025). Unlike the other XBs, PPCPs are present in the environment at lower concentrations, and they can persist for months to years when their disposal into the ecosystems is consistent, which promotes the exposure of PPCPs to biota (Beutel et al., 2021; Richardson, 2009).

Pharmaceuticals pose public health issues when they remain in their active states, even at trace concentrations, which allows the development of drug resistance by microbes (Kumar & Pal, 2018; Serwecińska, 2020). Additionally, PPCPs are capable of disrupting the endocrine system, as well as the growth, development, and reproduction of humans and other biota (Gonsioroski et al., 2020; Tijani et al., 2015). Chronic exposure to antibiotic pharmaceuticals promotes the progressive expression of antimicrobial resistance genes in biota along the food chain, leading to toxicity and chronic health risks for consumers of aquatic and terrestrial animals and plants (Kumar et al., 2020; Sun et al., 2020).

5 Conclusion

The emerging concerns on NPs, MPs, and other innovative XBs changed the course of ecotoxicological research. The chapter explored the types, origin, potential sources, transfer, fate, with environmental and public health concerns of emerging XBs based on current research trends. Their origin, sources, types, transfer, and toxic nature pose a challenge to current environmental management strategies. The ecological impact of XBs alters the nature of environmental matrix. It influences the biological food chain, from lower to higher trophic levels of ecosystems. Consequently, their interactions with the biology and physiology of organisms have been observed to have adverse health implications, ranging from behavioral patterns to the molecular level. The diverse impacts are increasing concern for the development of methods to detect, characterize, and understand their toxicological mechanisms, as well as to regulate and remediate their impacts on the environment. However, the research gaps in improving detection, characterization, and their ecological transfer require new perspectives through interdisciplinary research. Furthermore, the regulatory frameworks need to be transformed to tackle the dynamic nature of the emerging XBs in the environment. Futuristic outlooks should focus on predicting the fate of XBs and implementing sustainable and eco-friendly measures to remediate them from contaminated environments and their biotic systems, thereby restricting their access to human physiology. Finding eco-friendly and sustainable alternatives to replace the XBs and their applications through multidisciplinary collaborative efforts with the industries will pave the way for building harmony between humans and the environment.

References

Abbas, Q., Yousaf, B., Ullah, H., Ali, M. U., Ok, Y. S., & Rinklebe, J. (2020). Environmental transformation and nano-toxicity of engineered nano-particles (ENPs) in aquatic and terrestrial organisms. *Critical Reviews in Environmental Science and Technology, 50*(23), 2523–2581. https://doi.org/10.1080/10643389.2019.1705721

Abdo, N., Al-Khalaileh, H., Alajlouni, M., Hamadneh, J., & Alajlouni, A. M. (2023). Screening for phthalates biomarkers and its potential role in infertility outcomes in Jordan. *Journal of Exposure Science & Environmental Epidemiology, 33*(2), 273–282. https://doi.org/10.1038/s41370-022-00517-7

Acarer, S. (2023). Microplastics in wastewater treatment plants: Sources, properties, removal efficiency, removal mechanisms, and interactions with pollutants. *Water Science and Technology, 87*(3), 685–710. https://doi.org/10.2166/WST.2023.022

Ageel, H. K., Harrad, S., & Abdallah, M. A. E. (2022). Occurrence, human exposure, and risk of microplastics in the indoor environment. *Environmental Science: Processes & Impacts, 24*(1), 17–31. https://doi.org/10.1039/D1EM00301A

Ahmad, O. A., Jamal, M. T., Almalki, H. S., Alzahrani, A. H., Alatawi, A. S., & Haque, M. F. (2025). Microplastic pollution in the marine environment: Sources, impacts, and degradation. *Journal of Advanced Veterinary and Animal Research, 12*(1), 260. https://doi.org/10.5455/JAVAR.2025.L893

Akhtar, A. B. T., Naseem, S., Yasar, A., & Naseem, Z. (2021). Persistent organic pollutants (POPs): Sources, types, impacts, and their remediation. 213–246. https://doi.org/10.1007/978-981-15-5499-5_8

Allemann, M. N., Tessman, M., Reindel, J., Scofield, G. B., Evans, P., Pomeroy, R. S., Burkart, M. D., Mayfield, S. P., & Simkovsky, R. (2024). Rapid biodegradation of microplastics generated from bio-based thermoplastic polyurethane. *Scientific Reports, 14*(1), 1–10. https://doi.org/10.1038/S41598-024-56492-6

Allen, J. G., McClean, M. D., Stapleton, H. M., Nelson, J. W., & Webster, T. F. (2007). Personal exposure to polybrominated diphenyl ethers (PBDEs) in residential indoor air. *Environmental Science and Technology, 41*(13), 4574–4579. https://doi.org/10.1021/ES0703170

Al-Mansoori, M., Harrad, S., & Abdallah, M. A. E. (2026). Microplastics in commercial food products from the United Kingdom market: Implications for human exposure. *Environmental Research, 288*, 123178. https://doi.org/10.1016/j.envres.2025.123178

Ambrogi, V., Carfagna, C., Cerruti, P., & Marturano, V. (2017). Additives in polymers. In *Modification of polymer properties* (pp. 87–108). https://doi.org/10.1016/B978-0-323-44353-1.00004-X

Anand, U., Dey, S., Bontempi, E., Ducoli, S., Vethaak, A. D., Dey, A., & Federici, S. (2023). Biotechnological methods to remove microplastics: A review. *Environmental Chemistry Letters, 21*(3), 1787. https://doi.org/10.1007/S10311-022-01552-4

Anandhan, K., Tharini, K., Thangal, S. H., Yogeshwaran, A., & Muralisankar, T. (2022). Occurrence of microplastics in the gastrointestinal tracts of edible fishes from South Indian Rivers. *Bulletin of Environmental Contamination and Toxicology, 109*(6), 1023–1028. https://doi.org/10.1007/s00128-022-03595-3

Anandhan, K., Thangal, S. H., Yogeshwaran, A., Kaaran, S., Ajith Kumar, T. T., & Muralisankar, T. (2024). Microplastics contamination in the edible fish Mozambique tilapia (*Oreochromis mossambicus*) from the Selvampathy wetland of Coimbatore, Tamil Nadu, India. *Bulletin of Environmental Contamination and Toxicology, 112*(1). https://doi.org/10.1007/s00128-023-03839-w

Anandhan, K., Muralisankar, T., Yogeshwaran, A., Thangal, S. H., & Janani, K. S. (2025). Microplastics contamination in water and sediments: Seasonal and spatial distributions from the urban wetlands of Coimbatore, Tamilnadu, India. *Water, Air, and Soil Pollution, 236*(8). https://doi.org/10.1007/s11270-025-08117-5

Andrady, A. L., Barnes, P. W., Bornman, J. F., Gouin, T., Madronich, S., White, C. C., Zepp, R. G., & Jansen, M. A. (2022). Oxidation and fragmentation of plastics in a changing environment; from UV-radiation to biological degradation. *The Science of the Total Environment, 851*(Pt 2), Article 158022. https://doi.org/10.1016/J.scitotenv.2022.158022

Aravind kumar, J., Krithiga, T., Sathish, S., Renita, A. A., Prabu, D., Lokesh, S., Geetha, R., Namasivayam, S. K. R., & Sillanpaa, M. (2022). Persistent organic pollutants in water resources: Fate, occurrence, characterization and risk analysis. *Science of The Total Environment, 831*, 154808. https://doi.org/10.1016/J.SCITOTENV.2022.154808

Armitage, J. M., & Gobas, F. A. P. C. (2007). A terrestrial food-chain bioaccumulation model for POPs. *Environmental Science and Technology, 41*(11), 4019–4025. https://doi.org/10.1021/ES0700597

Ashraf, M. A. (2015). Persistent organic pollutants (POPs): A global issue, a global challenge. *Environmental Science and Pollution Research, 24*(5), 4223–4227. https://doi.org/10.1007/S11356-015-5225-9

Atugoda, T., Piyumali, H., Wijesekara, H., Sonne, C., Lam, S. S., Mahatantila, K., & Vithanage, M. (2022). Nanoplastic occurrence, transformation and toxicity: A review. *Environmental Chemistry Letters, 21*(1), 363–381. https://doi.org/10.1007/S10311-022-01479-W

Aulsebrook, L. C., Bertram, M. G., Martin, J. M., Aulsebrook, A. E., Brodin, T., Evans, J. P., Hall, M. D., O'Bryan, M. K., Pask, A. J., Tyler, C. R., & Wong, B. B. (2020). Reproduction in a polluted world: Implications for wildlife. *Reproduction, 160*(2), R13–R23. https://doi.org/10.1530/REP-20-0154

Azari, A., Vanoirbeek, J. A., Van Belleghem, F., Vleeschouwers, B., Hoet, P. H., & Ghosh, M. (2023). Sampling strategies and analytical techniques for assessment of airborne micro and nano plastics. *Environment International, 174*, Article 107885. https://doi.org/10.1016/j.envint.2023.107885

Bardawil, C. E., Dobbins, J., Lankford, S., Soloff, A. C., & Dhupar, R. (2026). Micro- and nanoplastics and pulmonary health: The current state of research. *Microplastics, 5*(1), 29. https://doi.org/10.3390/microplastics5010029

Barnes, D. K. A., Galgani, F., Thompson, R. C., & Barlaz, M. (2009). Accumulation and fragmentation of plastic debris in global environments. *Philosophical Transactions of the Royal Society b: Biological Sciences, 364*(1526), 1985–1998. https://doi.org/10.1098/RSTB.2008.0205

Beutel, M. W., Harmon, T. C., Novotny, T. E., Mock, J., Gilmore, M. E., Hart, S. C., Traina, S., Duttagupta, S., Brooks, A., Jerde, C. L., & Hoh, E. (2021). A review of environmental pollution from the use and disposal of cigarettes and electronic cigarettes: Contaminants, sources, and impacts. *Sustainability, 13*(23), 12994. https://doi.org/10.3390/SU132312994

Bharti, S. (2024). Xenobiotic remediation for sustainable urban development and ecosystem resilience. *Water, Air, & Soil Pollution, 235*(9), 1–40. https://doi.org/10.1007/S11270-024-07382-0

Bhat, A., Budholiya, S., Aravind Raj, S., Sultan, M. T. H., Hui, D., Md Shah, A. U., & Safri, S. N. A. (2021). Review on nanocomposites based on aerospace applications. *Nanotechnology Reviews, 10*(1), 237–253. https://doi.org/10.1515/ntrev-2021-0018

Bhatia, S. K., Bhatia, R. K., Choi, Y. K., Kan, E., Kim, Y. G., & Yang, Y. H. (2018). Biotechnological potential of microbial consortia and future perspectives. *Critical Reviews in Biotechnology, 38*(8), 1209–1229. https://doi.org/10.1080/07388551.2018.1471445

Björkner, B., Frick-Engfeldt, M., Pontén, A., & Zimerson, E. (2011). Plastic materials. In *Contact dermatitis* (5th ed., pp. 695–728). https://doi.org/10.1007/978-3-642-03827-3_37

Boucher, J., & Friot, D. (2017). International union for conservation of nature a global evaluation of sources primary microplastics in the oceans.

Boyda, J., Hawkey, A. B., Holloway, Z. R., Trevisan, R., Di Giulio, R. T., & Levin, E. D. (2021). The organophosphate insecticide diazinon and aging: Neurobehavioral and mitochondrial effects in zebrafish exposed as embryos or during aging. *Neurotoxicology and Teratology, 87*, Article 107011. https://doi.org/10.1016/j.ntt.2021.107011

Brunn, H., Arnold, G., Körner, W., Rippen, G., Steinhäuser, K. G., & Valentin, I. (2023). PFAS: Forever chemicals—persistent, bioaccumulative and mobile. Reviewing the status and the need for their phase out and remediation of contaminated sites. *Environmental Sciences Europe, 35*(1), 1–50. https://doi.org/10.1186/S12302-023-00721-8

Bu, Q., & Ma, Y. (2025). Emerging pollutants in the environment: occurrence, fate, risk assessment and degradation methods. *Toxics, 13*(7), 521. https://doi.org/10.3390/TOXICS13070521

Butler, T., & Wild, O. (2023). Hemispheric air pollution. In *Handbook of air quality and climate change* (pp. 1–29). https://doi.org/10.1007/978-981-15-2527-8_12-1

Cai, Z., Li, M., Zhu, Z., Wang, X., Huang, Y., Li, T., Gong, H., & Yan, M. (2023). Biological degradation of plastics and microplastics: A recent perspective on associated mechanisms and influencing factors. *Microorganisms, 11*(7), 1661. https://doi.org/10.3390/microorganisms11071661

Cai, B., Andino, F. D. J., McGrath, J. L., Romanick, S. S., & Robert, J. (2024). Ingestion of polyethylene terephthalate microplastic water contaminants by Xenopus laevis tadpoles negatively affects their resistance to ranavirus infection and antiviral immunity. *Environmental Pollution, 356*. https://doi.org/10.1016/j.envpol.2024.124340

Campanale, C., Massarelli, C., Savino, I., Locaputo, V., & Uricchio, V. F. (2020). A detailed review study on potential effects of microplastics and additives of concern on human health. *International Journal of Environmental Research and Public Health, 17*(4), 1212. https://doi.org/10.3390/ijerph17041212

Cao, Y., Zhao, M., Ma, X., Song, Y., Zuo, S., Li, H., & Deng, W. (2021). A critical review on the interactions of microplastics with heavy metals: Mechanism and their combined effect on organisms and humans. *Science of the Total Environment, 788*, Article 147620. https://doi.org/10.1016/j.scitotenv.2021.147620

Cao, J., Yang, Q., Jiang, J., Dalu, T., Kadushkin, A., Singh, J., Fakhrullin, R., Wang, F., Cai, X., & Li, R. (2022). Coronas of micro/nano plastics: A key determinant in their risk assessments. *Particle and Fibre Toxicology, 19*(1), 55. https://doi.org/10.1186/S12989-022-00492-9

Chakraborty, A., Adhikary, S., Bhattacharya, S., Dutta, S., Chatterjee, S., Banerjee, D., Ganguly, A., & Rajak, P. (2023). Pharmaceuticals and personal care products as emerging environmental contaminants: Prevalence, toxicity, and remedial approaches. *ACS Chemical Health & Safety, 30*(6), 362–388. https://doi.org/10.1021/acs.chas.3c00071

Chang, X., Fang, Y., Wang, Y., Wang, F., Shang, L., & Zhong, R. (2022). Microplastic pollution in soils, plants, and animals: A review of distributions, effects and potential mechanisms. *Science of the Total Environment, 850*, Article 157857. https://doi.org/10.1016/j.scitotenv.2022.157857

Chartres, N., Cooper, C. B., Bland, G., Pelch, K. E., Gandhi, S. A., BakenRa, A., & Woodruff, T. J. (2024). Effects of microplastic exposure on human digestive, reproductive, and respiratory health: A rapid systematic review. *Environmental Science & Technology, 58*(52), 22843–22864. https://doi.org/10.1021/acs.est.3c09524

Chaudhary, V., Kumar, M., Chauhan, C., Sirohi, U., Srivastav, A. L., & Rani, L. (2024). Strategies for mitigation of pesticides from the environment through alternative approaches: A review of recent developments and future prospects. *Journal of Environmental Management, 354*, Article 120326. https://doi.org/10.1016/j.jenvman.2024.120326

Chauhan, A. S., Chand, P., & Parashar, T. (2025). Lipid-based nanoparticles: Strategy for targeted cancer therapy. *BIO Integration, 6*(1), 994. https://doi.org/10.15212/bioi-2024-0107

Chen, I. C., Hsiao, I. L., Lin, H. C., Wu, C. H., Chuang, C. Y., & Huang, Y. J. (2016). Influence of silver and titanium dioxide nanoparticles on in vitro blood-brain barrier permeability. *Environmental Toxicology and Pharmacology, 47*, 108–118. https://doi.org/10.1016/j.etap.2016.09.009

Chen, Y., Chen, Q., Zhang, Q., Zuo, C., & Shi, H. (2022). An overview of chemical additives on (micro)plastic fibers: Occurrence, release, and health risks. *Reviews of Environmental Contamination and Toxicology, 260*(1), 1–25. https://doi.org/10.1007/S44169-022-00023-9

Cheng, F., Zhang, T., Liu, Y., Zhang, Y., & Qu, J. (2021). Non-negligible effects of UV irradiation on transformation and environmental risks of microplastics in the water environment. *Journal of Xenobiotics, 12*(1), 1. https://doi.org/10.3390/JOX12010001

Cole, M., Lindeque, P., Halsband, C., & Galloway, T. S. (2011). Microplastics as contaminants in the marine environment: A review. *Marine Pollution Bulletin, 62*(12), 2588–2597. https://doi.org/10.1016/j.marpolbul.2011.09.025

Coleman, E. A. (2017). Plastics additives. In *Applied plastics engineering handbook: Processing, materials, and applications* (2nd ed., pp. 489–500). https://doi.org/10.1016/B978-0-323-39040-8.00021-3

Cowell, W. J., Margolis, A., Rauh, V. A., Sjödin, A., Jones, R., Wang, Y., Garcia, W., Perera, F., Wang, S., & Herbstman, J. B. (2018). Associations between prenatal and childhood PBDE exposure and early adolescent visual, verbal and working memory. *Environment International, 118*, 9–16. https://doi.org/10.1016/j.envint.2018.05.004

Criswell, R., Fleisch, A. F., & Ducatman, A. (2024). Clinical implications of new drinking water regulation for 'forever chemicals.' *JAMA, 332*(10), 785–786. https://doi.org/10.1001/JAMA.2024.12705

Das, S., Parida, V. K., Tiwary, C. S., Gupta, A. K., & Chowdhury, S. (2024). Emerging contaminants in the aquatic environment: Fate, occurrence, impacts, and toxicity. *ACS Symposium Series, 1475*, 1–32. https://doi.org/10.1021/BK-2024-1475.CH001

Das, D., Ingti, B., Paul, P., Jamatia, J. P., Ete, T., Khan, T., & Kalita, J. (2025). Soil microbial dynamics in response to the impact of nanoparticles on agricultural implications. *Discover Soil, 2*(1), 58. https://doi.org/10.1007/s44378-025-00087-8

Daso, A. P., Fatoki, O. S., Odendaal, J. P., & Okonkwo, J. O. (2010). A review on sources of brominated flame retardants and routes of human exposure with emphasis on polybrominated diphenyl ethers. 239–254. https://doi.org/10.1139/A10-010

De Silva, A. O., Armitage, J. M., Bruton, T. A., Dassuncao, C., Heiger-Bernays, W., Hu, X. C., Kärrman, A., Kelly, B., Ng, C., Robuck, A., & Sun, M. (2021). PFAS exposure pathways for humans and wildlife: A synthesis of current knowledge and key gaps in understanding. *Environmental Toxicology and Chemistry, 40*(3), 631–657. https://doi.org/10.1002/etc.4935

Devereux, R., Hartl, M. G., Bell, M., & Capper, A. (2021). The abundance of microplastics in cnidaria and ctenophora in the North Sea. *Marine Pollution Bulletin, 173*, Article 112992. https://doi.org/10.1016/J.MARPOLBUL.2021.112992

Dey, S., Bano, F., & Malik, A. (2019). Pharmaceuticals and personal care product (PPCP) contamination—a global discharge inventory. In *Pharmaceuticals and personal care products: Waste management and treatment technology emerging contaminants and micro pollutants* (pp. 1–26). https://doi.org/10.1016/B978-0-12-816189-0.00001-9

Disner, G. R., & Tareq, S. M. (2025). Editorial: Emerging water contaminants in developing countries: Detection, monitoring, and impact of xenobiotics. *Frontiers in Water, 7*, 1584752. https://doi.org/10.3389/FRWA.2025.1584752/BIBTEX

Domingo, J. L., & Nadal, M. (2017). Per- and Polyfluoroalkyl Substances (PFASs) in food and human dietary intake: A review of the recent scientific literature. *Journal of Agricultural and Food Chemistry, 65*(3), 533–543. https://doi.org/10.1021/ACS.JAFC.6B04683

Donner, E., Eriksson, E., Holten-Lützhøft, H. C., Scholes, L., Revitt, M., & Ledin, A. (2010). Identifying and classifying the sources and uses of xenobiotics in urban environments. In *Xenobiotics in the urban water cycle: Mass flows, environmental processes, mitigation and treatment strategies* (pp. 27–50). https://doi.org/10.1007/978-90-481-3509-7_2

Dris, R., Gasperi, J., Saad, M., Mirande, C., & Tassin, B. (2016). Synthetic fibers in atmospheric fallout: A source of microplastics in the environment? *Marine Pollution Bulletin, 104*(1–2), 290–293. https://doi.org/10.1016/J.MARPOLBUL.2016.01.006

Dris, R., Imhof, H. K., Löder, M. G., Gasperi, J., Laforsch, C., & Tassin, B. (2018) Microplastic contamination in freshwater systems: Methodological challenges, occurrence and sources. In *Microplastic contamination in aquatic environments: An emerging matter of environmental urgency* (pp. 51–93). https://doi.org/10.1016/B978-0-12-813747-5.00003-5

Dutta, S., Haggerty, D. K., Rappolee, D. A., & Ruden, D. M. (2020). Phthalate exposure and long-term epigenomic consequences: A review. *Frontiers in Genetics, 11*, Article 499019. https://doi.org/10.3389/FGENE.2020.00405

Dziendzikowska, K., Wilczak, J., Grodzicki, W., Gromadzka-Ostrowska, J., Węsierska, M., & Kruszewski, M. (2022). Coating-dependent neurotoxicity of silver nanoparticles—an in vivo study on hippocampal oxidative stress and neurosteroids. *International Journal of Molecular Sciences, 23*(3), 1365. https://doi.org/10.3390/ijms23031365

Ealia, S. A. M., & Saravanakumar, M. P. (2017, November). A review on the classification, characterisation, synthesis of nanoparticles and their application. In *IOP conference series: materials science and engineering* (vol. 263(3), p. 032019). IOP Publishing. https://doi.org/10.1088/1757-899X/263/3/032019

Eker, F., Duman, H., Akdaşçi, E., Bolat, E., Sarıtaş, S., Karav, S., & Witkowska, A. M. (2024). A comprehensive review of nanoparticles: From classification to application and toxicity. *Molecules, 29*(15), 3482. https://doi.org/10.3390/molecules29153482

El-Kalliny, A. S., Abdel-Wahed, M. S., El-Zahhar, A. A., Hamza, I. A., & Gad-Allah, T. A. (2023). Nanomaterials: A review of emerging contaminants with potential health or environmental impact. *Discover Nano, 18*(1), 68. https://doi.org/10.1186/s11671-023-03787-8

Encarnação, T., Pais, A. A., Campos, M. G., & Burrows, H. D. (2019). Endocrine disrupting chemicals: Impact on human health, wildlife and the environment. *Science Progress, 102*(1), 3. https://doi.org/10.1177/0036850419826802

Enyoh, C. E., Shafea, L., Verla, A. W., Verla, E. N., Qingyue, W., Chowdhury, T., & Paredes, M. (2020). Microplastics exposure routes and toxicity studies to ecosystems: An overview. *Environmental Analysis, Health and Toxicology, 35*(1), Article e2020004. https://doi.org/10.5620/eaht.e2020004

Enyoh, C. E., Maduka, T. O., Rana, M. S., Osigwe, S. C., Ihenetu, S. C., & Wang, Q. (2024). Chemicals from brominated flame retardants: Analytical methods, occurrence, transport and risks. *Applied Sciences, 14*(17), 7892. https://doi.org/10.3390/APP14177892

Fallon, B. R., & Freeman, C. J. (2021). Plastics in Porifera: The occurrence of potential microplastics in marine sponges and seawater from Bocas del Toro, Panamá. *PeerJ, 9*, Article e11638. https://doi.org/10.7717/PEERJ.11638/SUPP-5

Fei, X., He, H., Pi, X., Lu, X., Chen, Q., Ma, J., Wang, Y., Fang, M., Wu, C., & Feng, S. (2023). The distribution, behavior, and release of macro-and micro-size plastic wastes in solid waste disposal sites. *Critical Reviews in Environmental Science and Technology, 53*(3), 366–389. https://doi.org/10.1080/10643389.2022.2054649

Fent, K. W., LaGuardia, M., Luellen, D., McCormick, S., Mayer, A., Chen, I. C., Kerber, S., Smith, D., & Horn, G. P. (2020). Flame retardants, dioxins, and furans in air and on firefighters' protective ensembles during controlled residential firefighting. *Environment International, 140*, Article 105756. https://doi.org/10.1016/J.ENVINT.2020.105756

Fernández-Trujillo, S., Jiménez-Moreno, M., Rodríguez-Fariñas, N., & Rodríguez Martín-Doimeadios, R. C. (2024). Critical evaluation of the potential of ICP-MS-based systems in toxicological studies of metallic nanoparticles. *Analytical and Bioanalytical Chemistry, 416*(11), 2657–2676. https://doi.org/10.1007/s00216-024-05181-4

Flaws, J., Damdimopoulou, P., Patisaul, H. B., Gore, A., Raetzman, L., & Vandenberg, L. N. (2020). *Plastics, EDCs and health.* Endocrine Society. https://www.endocrine.org/-/media/endocrine/files/topics/edc_guide_2020_v1_6bhqen.pdf

Fonnum, F., & Mariussen, E. (2009). Mechanisms involved in the neurotoxic effects of environmental toxicants such as polychlorinated biphenyls and brominated flame retardants. *Journal of Neurochemistry, 111*(6), 1327–1347. https://doi.org/10.1111/J.1471-4159.2009.06427.X

Forest, V. (2021). Combined effects of nanoparticles and other environmental contaminants on human health-an issue often overlooked. *NanoImpact, 23*, Article 100344. https://doi.org/10.1016/j.impact.2021.100344

Fortin, S. G., Uhlig, K., Hale, R. C., & Song, B. (2025). Microplastic biofilms as potential hotspots for plastic biodegradation and nitrogen cycling: A metagenomic perspective. *FEMS Microbiology Ecology, 101*(5), 35. https://doi.org/10.1093/FEMSEC/FIAF035

Fournier, E., Leveque, M., Ruiz, P., Ratel, J., Durif, C., Chalancon, S., Amiard, F., Edely, M., Bezirard, V., Gaultier, E., & Lamas, B. (2023). Microplastics: What happens in the human digestive tract? First evidences in adults using in vitro gut models. *Journal of Hazardous Materials, 442*. https://doi.org/10.1016/j.jhazmat.2022.130010

Fox, S., Stefánsson, H., Peternell, M., Zlotskiy, E., Ásbjörnsson, E. J., Sturkell, E., Wanner, P., & Konrad-Schmolke, M. (2024). Physical characteristics of microplastic particles and potential for global atmospheric transport: A meta-analysis. *Environmental Pollution, 342*, Article 122938. https://doi.org/10.1016/J.ENVPOL.2023.122938

Gaines, T. L. G., & Gaines, L. T. C.G. (2023). Historical and current usage of per- and polyfluoroalkyl substances (PFAS): A literature review. *American Journal of Industrial Medicine, 66*(5), 353–378. https://doi.org/10.1002/AJIM.23362

Gan, Q., Cui, J., & Jin, B. (2023). Environmental microplastics: Classification, sources, fates, and effects on plants. *Chemosphere, 313*, Article 137559. https://doi.org/10.1016/J.CHEMOSPHERE.2022.137559

García-Suastegui, W. A., Navarro-Mabarak, C., Silva-Adaya, D., Dolores-Raymundo, H. G., Alvarez-Gonzalez, M. Y., León-Olea, M., & Ramos-Chávez, L. A. (2025). Neurotransmitter systems affected by PBDE exposure: Insights from in vivo and in vitro neurotoxicity studies. *Toxics, 13*(4), 316. https://doi.org/10.3390/TOXICS13040316

Gaspar, L., Bartman, S., Tobias-Wallingford, H., Coppotelli, G., & Ross, J. M. (2025). Short-term exposure to polystyrene microplastics alters cognition, immune, and metabolic markers in an apolipoprotein E (APOE) genotype and sex-dependent manner. *Environmental Research Communications, 7*(8), Article 085012. https://doi.org/10.1088/2515-7620/ADF8AE

Gholizadeh, M., Shadi, A., Abadi, A., Nemati, M., Senapathi, V., & Karthikeyan, S. (2023). Abundance and characteristics of microplastic in some commercial species from the Persian Gulf, Iran. *Journal of Environmental Management, 344*, Article 118386. https://doi.org/10.1016/J.JENVMAN.2023.118386

Gibson, E. A., Siegel, E. L., Eniola, F., Herbstman, J. B., & Factor-Litvak, P. (2018). Effects of polybrominated diphenyl ethers on child cognitive, behavioral, and motor development. *International Journal of Environmental Research and Public Health, 15*(8), 1636. https://doi.org/10.3390/IJERPH15081636

Gonsioroski, A., Mourikes, V. E., & Flaws, J. A. (2020). Endocrine disruptors in water and their effects on the reproductive system. *International Journal of Molecular Sciences, 21*(6), 1929. https://doi.org/10.3390/IJMS21061929

Goodman, K. E., Hua, T., & Sang, Q. X. A. (2022). Effects of polystyrene microplastics on human kidney and liver cell morphology, cellular proliferation, and metabolism. *ACS Omega, 7*(38), 34136–34153. https://doi.org/10.1021/ACSOMEGA.2C03453

Gorga, M., Martínez, E., Ginebreda, A., Eljarrat, E., & Barceló, D. (2013). Determination of PBDEs, HBB, PBEB, DBDPE, HBCD, TBBPA and related compounds in sewage sludge from Catalonia (Spain). *Science of the Total Environment, 444*, 51–59. https://doi.org/10.1016/J.SCITOTENV.2012.11.066

Gouesse, R. J., & Plante, I. (2022). Environmental exposure to brominated flame retardants: Unraveling endocrine and mammary gland effects that may increase disease risk. *Toxicological Sciences, 186*(2), 190–207. https://doi.org/10.1093/TOXSCI/KFAC006

Gouin, T. (2020). Toward an improved understanding of the ingestion and trophic transfer of microplastic particles: Critical review and implications for future research. *Environmental Toxicology and Chemistry, 39*(6), 1119. https://doi.org/10.1002/ETC.4718

Guo, W., Pan, B., Sakkiah, S., Yavas, G., Ge, W., Zou, W., Tong, W., & Hong, H. (2019). Persistent organic pollutants in food: Contamination sources, health effects and detection methods. *International Journal of Environmental Research and Public Health, 16*(22), 4361. https://doi.org/10.3390/ijerph16224361

Hahladakis, J. N., Iacovidou, E., & Gerassimidou, S. (2023). An overview of the occurrence, fate, and human risks of the bisphenol-A present in plastic materials, components, and products. *Integrated Environmental Assessment and Management, 19*(1), 45–62. https://doi.org/10.1002/IEAM.4611;PAGEGROUP:STRING:PUBLICATION

Hamilton, B. M., Baak, J. E., Vorkamp, K., Hammer, S., Granberg, M., Herzke, D., & Provencher, J. F. (2022). Plastics as a carrier of chemical additives to the Arctic: Possibilities for strategic monitoring across the circumpolar North. *9*(2), 284–296. https://doi.org/10.1139/AS-2021-0055

Hasegawa, T., & Nakaoka, M. (2021). Trophic transfer of microplastics from mysids to fish greatly exceeds direct ingestion from the water column. *Environmental Pollution, 273*, Article 116468. https://doi.org/10.1016/J.ENVPOL.2021.116468

Hasegawa, T., Mizukawa, K., Yeo, B. G., Sekioka, T., Takada, H., Nakaoka, M. (2022). Trophic transfer of microplastics enhances plastic additive accumulation in fish. Preprint from bioRxiv. https://doi.org/10.1101/2021.03.09.434685

Haverinen, E., Fernandez, M. F., Mustieles, V., Tolonen, H. (2021). Metabolic syndrome and endocrine disrupting chemicals: An overview of exposure and health effects. *International Journal of Environmental Research and Public Health, 18*(24), 13047. https://doi.org/10.3390/IJERPH182413047

Hawley, G. (1996). Plastics. In *Industrial minerals and their uses* (pp. 353–400). https://doi.org/10.1016/B978-081551408-4.50009-0

Hee, Y. Y., Hanif, N. M., Weston, K., Latif, M. T., Suratman, S., Rusli, M. U., & Mayes, A. G. (2023). Atmospheric microplastic transport and deposition to urban and pristine tropical locations in Southeast Asia. *Science of the Total Environment, 902*. https://doi.org/10.1016/j.scitotenv.2023.166153

Heo, S. J., Moon, N., & Kim, J. H. (2024). A systematic review and quality assessment of estimated daily intake of microplastics through food. *Reviews on Environmental Health, 40*(2), 371–392. https://doi.org/10.1515/REVEH-2024-0111

Herbstman, J. B., Sjödin, A., Kurzon, M., Lederman, S. A., Jones, R. S., Rauh, V., Needham, L. L., Tang, D., Niedzwiecki, M., Wang, R. Y., & Perera, F. (2010). Prenatal exposure to PBDEs and neurodevelopment. *Environmental Health Perspectives, 118*(5), 712–719. https://doi.org/10.1289/EHP.0901340

Herrera, A., Acosta-Dacal, A., Pérez-Luzardo, O., Martínez, I., Rapp, J., Reinold, S., Montesdeoca-Esponda, S., Montero, D., & Gomez, M. (2023). Trophic transfer of DDE, BP-3 and chlorpyrifos from microplastics to tissues in *Dicentrarchus labrax*. *Science of the Total Environment, 882*, Article 163295.

Hlisníková, H., Petrovičová, I., Kolena, B., Šidlovská, M., & irotkin, A. (2020). Effects and mechanisms of phthalates' action on reproductive processes and reproductive health: A literature review. *International Journal of Environmental Research and Public Health, 17*(18), 6811. https://doi.org/10.3390/IJERPH17186811

Horton, A. A., Walton, A., Spurgeon, D. J., Lahive, E., Svendsen, C. (2017). Microplastics in freshwater and terrestrial environments: Evaluating the current understanding to identify the knowledge gaps and future research priorities. *Science of the Total Environment, 586*, 127–141. https://doi.org/10.1016/J.SCITOTENV.2017.01.190

Huang, H., Hou, J., Li, M., Wei, F., Liao, Y., & Xi, B. (2025). Microplastics in the bloodstream can induce cerebral thrombosis by causing cell obstruction and lead to neurobehavioral abnormalities. *Science Advances, 11*(4). https://doi.org/10.1126/SCIADV.ADR8243

ISO. (2023). Principles for the analysis of microplastics present in the environment. Retrieved October 16, 2025, from https://www.iso.org/standard/78033.html#lifecycle

Iyare, P. U., Ouki, S. K., & Bond, T. (2020). Microplastics removal in wastewater treatment plants: A critical review. *Environmental Science: Water Research & Technology, 6*(10), 2664–2675. https://doi.org/10.1039/D0EW00397B

Jambeck, J. R., & Walker-Franklin, I. (2023). The impacts of plastics' life cycle. *One Earth, 6*(6), 600–606. https://doi.org/10.1016/j.oneear.2023.05.015

Jarosiewicz, M., & Krześlak, A. (2024). Epigenetic implications of common brominated flame retardants (PBDEs and TBBPA): Understanding the health risks of BFRs exposure. *Chemosphere, 361*, Article 142488. https://doi.org/10.1016/J.CHEMOSPHERE.2024.142488

Jeevanandam, J., Barhoum, A., Chan, Y. S., Dufresne, A., & Danquah, M. K. (2018). Review on nanoparticles and nanostructured materials: History, sources, toxicity and regulations. *Beilstein Journal of Nanotechnology, 9*(1), 1050–1074. https://doi.org/10.3762/bjnano.9.98

Jenner, L. C., Rotchell, J. M., Bennett, R. T., Cowen, M., Tentzeris, V., & Sadofsky, L. R. (2022). Detection of microplastics in human lung tissue using μFTIR spectroscopy. *Science of the Total Environment, 831*, Article 154907. https://doi.org/10.1016/J.SCITOTENV.2022.154907

Jensen, A. A., & Jørgensen, K. F. (1983). Polychlorinated terphenyls (PCTs) use, levels and biological effects. *Science of the Total Environment, 27*(2–3), 231–250. https://doi.org/10.1016/0048-9697(83)90156-0

Jeong, C. B., Kang, H. M., Lee, M. C., Kim, D. H., Han, J., Hwang, D. S., Souissi, S., Lee, S. J., Shin, K. H., Park, H. G., & Lee, J. S. (2017). Adverse effects of microplastics and oxidative stress-induced MAPK/Nrf2 pathway-mediated defense mechanisms in the marine copepod *Paracyclopina nana. Scientific Reports, 7*, 41323. https://doi.org/10.1038/SREP41323

Jha, G., Kankarla, V., McLennon, E., Pal, S., Sihi, D., Dari, B., Diaz, D., & Nocco, M. (2021). Per- and polyfluoroalkyl substances (PFAS) in integrated crop-livestock systems: Environmental exposure and human health risks. *International Journal of Environmental Research and Public Health, 18*(23), 12550. https://doi.org/10.3390/IJERPH182312550

Jia, J., Liu, Q., Zhao, E., Li, X., Xiong, X., & Wu, C. (2024). Biofilm formation on microplastics and interactions with antibiotics, antibiotic resistance genes and pathogens in aquatic environment. *Eco-Environment & Health, 3*(4), 516. https://doi.org/10.1016/J.EEHL.2024.05.003

Jiang, N., Zheng, X., Zhang, N., & Cao, Y. (2025). The detrimental effects of microplastic exposure on kidney function. *Frontiers in Medicine, 12*, 1620733. https://doi.org/10.3389/FMED.2025.1620733

Joseph, M. M., Nair, J. B., & Joseph, A. M. (2025). Microscopic menace: Exploring the link between microplastics and cancer pathogenesis. *Environmental Science: Processes & Impacts, 27*(7), 1768–1795. https://doi.org/10.1039/D5EM00232J

Joudeh, N., & Linke, D. (2022). Nanoparticle classification, physicochemical properties, characterization, and applications: A comprehensive review for biologists. *Journal of Nanobiotechnology, 20*(1), 262. https://doi.org/10.1186/s12951-022-01477-8

Kadac-Czapska, K., Ośko, J., Knez, E., & Grembecka, M. (2024). Microplastics and oxidative stress—Current problems and prospects. *Antioxidants, 13*(5), 579. https://doi.org/10.3390/ANTIOX13050579

Keerthanan, S., Jayasinghe, C., Biswas, J. K., & Vithanage, M. (2021). Pharmaceutical and Personal Care Products (PPCPs) in the environment: Plant uptake, translocation, bioaccumulation, and human health risks. *Critical Reviews in Environmental Science and Technology, 51*(12), 1221–1258. https://doi.org/10.1080/10643389.2020.1753634

Khan, I., Saeed, K., & Khan, I. (2019). Nanoparticles: Properties, applications and toxicities. *Arabian Journal of Chemistry, 12*(7), 908–931. https://doi.org/10.1016/j.arabjc.2017.05.011

Khanna, K., Kohli, S. K., Handa, N., Kaur, H., Ohri, P., Bhardwaj, R., Yousaf, B., Rinklebe, J., & Ahmad, P. (2021). Enthralling the impact of engineered nanoparticles on soil microbiome: A concentric approach towards environmental risks and cogitation. *Ecotoxicology and Environmental Safety, 222*, Article 112459. https://doi.org/10.1016/j.ecoenv.2021.112459

Khoshmanesh, M., Sanati, A.M., Shahcheragh, S., Farjadfard, S., Bonyadi, Z., & Ramavandi, B. (2024). Recent advances in dyes uptake by microplastics in aquatic environments: Influencing factors and ecotoxicological behaviors. *Arabian Journal of Chemistry, 17*(5), Article 105737. https://doi.org/10.1016/J.ARABJC.2024.105737

Khoshnamvand, M., Hao, Z., Fadare, O. O., Hanachi, P., Chen, Y., & Liu, J. (2020). Toxicity of biosynthesized silver nanoparticles to aquatic organisms of different trophic levels. *Chemosphere, 258*, Article 127346. https://doi.org/10.1016/j.chemosphere.2020.127346

Kim, D. Y., Park, M. K., Yang, H. W., Woo, S. Y., Jung, H. H., Son, D. S., Choi, B. Y., & Suh, S. W. (2025). Effects of microplastic accumulation on neuronal death after global cerebral ischemia. *Cells, 14*(4), 241. https://doi.org/10.3390/CELLS14040241

Koike, E., Yanagisawa, R., & Takano, H. (2016). Brominated flame retardants, hexabromocyclododecane and tetrabromobisphenol A, affect proinflammatory protein expression in human bronchial epithelial cells via disruption of intracellular signaling. *Toxicology in Vitro, 32*, 212–219. https://doi.org/10.1016/J.TIV.2015.12.013

Kumar, A., & Pal, D. (2018). Antibiotic resistance and wastewater: Correlation, impact and critical human health challenges. *Journal of Environmental Chemical Engineering, 6*(1), 52–58. https://doi.org/10.1016/J.JECE.2017.11.059

Kumar, P., Pirjola, L., Ketzel, M., & Harrison, R. M. (2013). Nanoparticle emissions from 11 non-vehicle exhaust sources—A review. *Atmospheric Environment, 67*, 252–277. https://doi.org/10.1016/J.ATMOSENV.2012.11.011

Kumar, S. B., Arnipalli, S. R., & Ziouzenkova, O. (2020). Antibiotics in food chain: The consequences for antibiotic resistance. *Antibiotics, 9*(10), 688. https://doi.org/10.3390/ANTIBIOTICS9100688

Kumari, K., Swamy, S., & Singh, A. (2021). Global monitoring plan on persistent organic pollutants (POPs). *Persistent Organic Pollutants*, 227–252. https://doi.org/10.1201/9781003046806-9

Kwon, H. S., Ryu, M. H., & Carlsten, C. (2020). Ultrafine particles: Unique physicochemical properties relevant to health and disease. *Experimental & Molecular Medicine, 52*(3), 318–328. https://doi.org/10.1038/s12276-020-0405-1

Lai, D. Y., Kacew, S., & Dekant, W. (2015). Tetrabromobisphenol A (TBBPA): Possible modes of action of toxicity and carcinogenicity in rodents. *Food and Chemical Toxicology, 80*, 206–214. https://doi.org/10.1016/J.FCT.2015.03.023

Le, H. Q., Tomenson, J. A., Warheit, D. B., Fryzek, J. P., Golden, A. P., & Ellis, E. D. (2018). A review and meta-analysis of occupational titanium dioxide exposure and lung cancer mortality. *Journal of Occupational and Environmental Medicine, 60*(7), e356–e367. https://doi.org/10.1097/jom.0000000000001314

Lead, J. R., Batley, G. E., Alvarez, P. J., Croteau, M. N., Handy, R. D., McLaughlin, M. J., Judy, J. D., & Schirmer, K. (2018). Nanomaterials in the environment: Behavior, fate, bioavailability, and effects—an updated review. *Environmental Toxicology and Chemistry, 37*(8), 2029–2063. https://doi.org/10.1002/etc.4147

Lechthaler, S., Waldschläger, K., Stauch, G., & Schüttrumpf, H. (2020). The way of macroplastic through the environment. *Environments, 7*(10), 73. https://doi.org/10.3390/environments7100073

Lee, J., Hong, S., An, S. A., & Khim, J. S. (2023). Methodological advances and future directions of microalgal bioassays for evaluation of potential toxicity in environmental samples: A review. *Environment International, 173*, Article 107869. https://doi.org/10.1016/j.envint.2023.107869

Li, C., & Tang, M. (2024). The toxicological effects of nano titanium dioxide on target organs and mechanisms of toxicity. *Journal of Applied Toxicology, 44*(2), 152–164. https://doi.org/10.1002/jat.4534

Li, M., Liu, W., & Slaveykova, V. I. (2020). Effects of mixtures of engineered nanoparticles and metallic pollutants on aquatic organisms. *Environments, 7*(4), 27. https://doi.org/10.3390/environments7040027

Li, L., Chen, C., Li, D., Breivik, K., Abbasi, G., & Li, Y. F. (2023a). What do we know about the production and release of persistent organic pollutants in the global environment? *Environmental Science: Advances, 2*(1), 55–68. https://doi.org/10.1039/D2VA00145D

Li, W., Zu, B., Yang, Q., Guo, J., & Li, J. (2023b). Sources, distribution, and environmental effects of microplastics: A systematic review. *RSC Advances, 13*(23), 15566. https://doi.org/10.1039/D3RA02169F

Li, M., Ma, W., Fang, J. K. H., Mo, J., Li, L., Pan, M., Li, R., Zeng, X., & Lai, K. P. (2025). A review on the combined toxicological effects of microplastics and their attached pollutants. *Emerging Contaminants, 11*(2), Article 100486. https://doi.org/10.1016/J.EMCON.2025.100486

Linares, V., Bellés, M., & Domingo, J. L. (2015). Human exposure to PBDE and critical evaluation of health hazards. *Archives of Toxicology, 89*(3), 335–356. https://doi.org/10.1007/S00204-015-1457-1/TABLES/3

Liu, J., & Li, C. (2025). Impact of microplastics on aquatic ecosystems. *Water, 17*(14), 2124. https://doi.org/10.3390/W17142124

Liu, X., Sui, B., & Sun, J. (2017). Size-and shape-dependent effects of titanium dioxide nanoparticles on the permeabilization of the blood–brain barrier. *Journal of Materials Chemistry B, 5*(48), 9558–9570. https://doi.org/10.1039/C7TB01314K

Liu, J., Williams, P. C., Goodson, B. M., Geisler-Lee, J., Fakharifar, M., & Gemeinhardt, M. E. (2019). TiO_2 nanoparticles in irrigation water mitigate impacts of aged Ag nanoparticles on soil microorganisms, *Arabidopsis thaliana* plants, and *Eisenia fetida* earthworms. *Environmental Research, 172*, 202–215. https://doi.org/10.1016/J.ENVRES.2019.02.010

Liu, Q., Wu, H., Chen, J., Guo, B., Zhao, X., Lin, H., Li, W., Zhao, X., Lv, S., & Huang, C. (2022a). Adsorption mechanism of trace heavy metals on microplastics and simulating their effect on microalgae in river. *Environmental Research, 214*, Article 113777. https://doi.org/10.1016/J.ENVRES.2022.113777

Liu, S., Junaid, M., Liao, H., Liu, X., Wu, Y., & Wang, J. (2022b). Eco-corona formation and associated ecotoxicological impacts of nanoplastics in the environment. *Science of the Total Environment, 836*, Article 155703. https://doi.org/10.1016/j.scitotenv.2022.155703

Lizzi, D., Mattiello, A., Piani, B., Gava, E., Fellet, G., & Marchiol, L. (2021). Single and repeated applications of cerium oxide nanoparticles differently affect the growth and biomass accumulation of silene floscuculi L.(Caryophyllaceae). *Nanomaterials*, 11(1), 229. https://doi.org/10.3390/nano11010229

Long, X., Fu, T. M., Yang, X., Tang, Y., Zheng, Y., Zhu, L., Shen, H., Ye, J., Wang, C., Wang, T., & Li, B. (2022). Efficient atmospheric transport of microplastics over Asia and adjacent oceans. *Environmental Science & Technology, 56*(10), 6243–6252. https://doi.org/10.1021/ACS.EST.1C07825

Long, B., Li, F., Wang, K., Huang, Y., Yang, Y., Xie, D. (2023). Impact of plastic film mulching on microplastic in farmland soils in Guangdong province, China. *Heliyon, 9*(6), Article e16587. https://doi.org/10.1016/J.HELIYON.2023.E16587

Longendyke, G. K., Katel, S., & Wang, Y. (2022). PFAS fate and destruction mechanisms during thermal treatment: A comprehensive review. *Environmental Science: Processes & Impacts, 24*(2), 196–208. https://doi.org/10.1039/D1EM00465D

Luo, L., Huang, W., Zhang, J., Yu, Y., & Sun, T. (2024). Metal-based nanoparticles as antimicrobial agents: A review. *ACS Applied Nano Materials, 7*(3), 2529–2545. https://doi.org/10.1021/acsanm.3c05615

Ma, T., Xue, N., Muhammad, A., Fang, G., Yan, J., Chen, R., Sun, J., & Sun, X. (2024). Recent progress in photodetectors: From materials to structures and applications. *Micromachines, 15*(10), 1249. https://doi.org/10.3390/mi15101249

Mahalingiah, S., Nadeau, K. C., & Christiani, D. C. (2025). Microplastics and human health. JAMA [Preprint]. https://doi.org/10.1001/JAMA.2025.14718

Mahjoubian, M., Naeemi, A. S., Moradi-Shoeili, Z., Tyler, C. R., & Mansouri, B. (2023a). Toxicity of silver nanoparticles in the presence of zinc oxide nanoparticles differs for acute and chronic exposures in zebrafish. *Archives of Environmental Contamination and Toxicology, 84*(1), 1–17. https://doi.org/10.1007/s00244-022-00965-0

Mahjoubian, M., Naeemi, A. S., Moradi-Shoeili, Z., Tyler, C. R., & Mansouri, B. (2023b). Oxidative stress, genotoxic effects, and other damages caused by chronic exposure to silver nanoparticles (Ag NPs) and zinc oxide nanoparticles (ZnO NPs), and their mixtures in zebrafish (Danio rerio). *Toxicology and Applied Pharmacology, 472*, Article 116569. https://doi.org/10.1016/j.taap.2023.116569

Manzetti, S., Van Der Spoel, E. R., & Van Der Spoel, D. (2014). Chemical properties, environmental fate, and degradation of seven classes of pollutants. *Chemical Research in Toxicology, 27*(5), 713–737. https://doi.org/10.1021/TX500014W

Mao, H., Lin, T., Huang, S., Xie, Z., Jin, S., Shen, X., Jin, Y., & Ding, Y. (2024). The impact of brominated flame retardants (BFRs) on pulmonary function in US adults: A cross-sectional study based on NHANES (2007–2012). *Scientific Reports, 14*(1), 1–11. https://doi.org/10.1038/s41598-024-57302-9

Marinkovic, N., Pasalic, D., Ferencak, G., Grskovic, B., & Rukavina, A. S. (2010). Dioxins and human toxicity. *Arhiv Za Higijenu Rada i Toksikologiju, 61*(4), 445. https://doi.org/10.2478/10004-1254-61-2010-2024

Marquès, M., Nadal, M., & Domingo, J. L. (2022). Human exposure to polybrominated diphenyl ethers (PBDEs) through the diet: An update of the scientific literature. *Food and Chemical Toxicology, 167*, Article 113322. https://doi.org/10.1016/J.FCT.2022.113322

Mathew, N., Somanathan, A., Tirpude, A., Pillai, A. M., Mondal, P., & Arfin, T. (2025). Dioxins and their impact: A review of toxicity, persistence, and novel remediation strategies. *Analytical Methods, 17*(8), 1698–1748. https://doi.org/10.1039/D4AY01767F

Matuszczak, E., Komarowska, M. D., Debek, W., & Hermanowicz, A. (2019). The impact of bisphenol A on fertility, reproductive system, and development: A review of the literature. *International Journal of Endocrinology, 2019*(1), 4068717. https://doi.org/10.1155/2019/4068717

McLachlan, J. A. (1993). Functional toxicology: A new approach to detect biologically active xenobiotics. *Environmental Health Perspectives, 101*(5), 386–387. https://doi.org/10.1289/EHP.93101386;WGROUP:STRING:PUBLICATION

Megson, D., Brown, T., Jones, G. R., Robson, M., Johnson, G. W., Tiktak, G. P., Sandau, C. D., & Reiner, E. J. (2022). Polychlorinated biphenyl (PCB) concentrations and profiles in marine mammals from the North Atlantic Ocean. *Chemosphere, 288*, Article 132639. https://doi.org/10.1016/j.chemosphere.2021.132639

Mekuye, B., & Abera, B. (2023). Nanomaterials: An overview of synthesis, classification, characterization, and applications. *Nano Select, 4*(8), 486–501. https://doi.org/10.1002/nano.202300038

Meng, X., Zhang, J., Wang, W., Gonzalez-Gil, G., Vrouwenvelder, J. S., & Li, Z. (2022). Effects of nano- and microplastics on kidney: Physicochemical properties, bioaccumulation, oxidative stress and immunoreaction. *Chemosphere, 288*, Article 132631. https://doi.org/10.1016/J.CHEMOSPHERE.2021.132631

Meyer, M. F., Powers, S. M., & Hampton, S. E. (2019). An evidence synthesis of pharmaceuticals and personal care products (PPCPs) in the environment: Imbalances among compounds, sewage treatment techniques, and ecosystem types. *Environmental Science & Technology, 53*(22), 12961–12973. https://doi.org/10.1021/ACS.EST.9B02966

Mirsasaani, S. S., Hemati, M., Dehkord, E. S., Yazdi, G. T., & Poshtiri, D. A. (2019). Nanotechnology and nanobiomaterials in dentistry. In *Nanobiomaterials in clinical dentistry* (pp. 19–37). Elsevier. https://doi.org/10.1016/B978-0-12-815886-9.00002-4

Mishra, S., & Sundaram, B. (2023). Fate, transport, and toxicity of nanoparticles: An emerging pollutant on biotic factors. *Process Safety and Environmental Protection, 174*, 595–607. https://doi.org/10.1016/j.psep.2023.04.037

Montano, L., Raimondo, S., Piscopo, M., Ricciardi, M., Guglielmino, A., Chamayou, S., Gentile, R., Gentile, M., Rapisarda, P., Conti, G. O., & Ferrante, M. (2025). First evidence of microplastics in human ovarian follicular fluid: An emerging threat to female fertility. *Ecotoxicology and Environmental Safety, 291*. https://doi.org/10.1016/j.ecoenv.2025.117868

Muir, D. C. G., & de Wit, C. A. (2010). Trends of legacy and new persistent organic pollutants in the circumpolar arctic: Overview, conclusions, and recommendations. *Science of the Total Environment, 408*(15), 3044–3051. https://doi.org/10.1016/J.SCITOTENV.2009.11.032

Nawab, A., Ahmad, M., Khan, M. T., Nafees, M., Khan, I., & Ihsanullah, I. (2024). Human exposure to microplastics: A review on exposure routes and public health impacts. *Journal of Hazardous Materials Advances, 16*, Article 100487. https://doi.org/10.1016/J.HAZADV.2024.100487

Negrete-Bolagay, D., Zamora-Ledezma, C., Chuya-Sumba, C., De Sousa, F. B., Whitehead, D., Alexis, F., & Guerrero, V. H. (2021). Persistent organic pollutants: The trade-off between potential risks and sustainable remediation methods. *Journal of Environmental Management, 300*, Article 113737. https://doi.org/10.1016/j.jenvman.2021.113737

Nelms, S. E., Galloway, T. S., Godley, B. J., Jarvis, D. S., & Lindeque, P. K. (2018). Investigating microplastic trophic transfer in marine top predators. *Environmental Pollution (Barking, Essex: 1987), 238*, 999–1007. https://doi.org/10.1016/J.ENVPOL.2018.02.016

Nie, P., Zhao, Y., & Xu, H. (2023). Synthesis, applications, toxicity and toxicity mechanisms of silver nanoparticles: A review. *Ecotoxicology and Environmental Safety, 253*, Article 114636. https://doi.org/10.1016/j.ecoenv.2023.114636

Nihart, A. J., Garcia, M. A., El Hayek, E., Liu, R., Olewine, M., Kingston, J. D., Castillo, E. F., Gullapalli, R. R., Howard, T., Bleske, B., & Scott, J. (2025). Bioaccumulation of microplastics in decedent human brains. *Nature Medicine, 31*(4), 1114–1119. https://doi.org/10.1038/S41591-024-03453-1

Nouri, M., Massahi, T., & Hossini, H. (2025). Microplastics in human body: a narrative on routes of exposure to contamination and potential health effects. *Environmental Pollutants and Bioavailability, 37*(1). https://doi.org/10.1080/26395940.2025.2538535

OECD. (2024). Policy scenarios for eliminating plastic pollution by 2040. https://doi.org/10.1787/76400890-EN

Okuthe, G. E., Dube, E., & Mafunda, P. S. (2025). Effects of pharmaceuticals and endocrine-disrupting chemicals on reproductive biology of aquatic fauna: penguins as sentinel species. *Journal of Xenobiotics, 15*(4), 110. https://doi.org/10.3390/jox15040110

Osman, A. I., Hosny, M., Eltaweil, A. S., Omar, S., Elgarahy, A. M., Farghali, M., Yap, P. S., Wu, Y. S., Nagandran, S., Batumalaie, K., & Gopinath, S. C. (2023). Microplastic sources, formation, toxicity and remediation: A review. *Environmental Chemistry Letters, 21*(4), 1. https://doi.org/10.1007/S10311-023-01593-3

Pan, J., Liu, P., Yu, X., Zhang, Z., & Liu, J. (2023). The adverse role of endocrine disrupting chemicals in the reproductive system. *Frontiers in Endocrinology, 14*, 1324993. https://doi.org/10.3389/FENDO.2023.1324993

Pariatamby, A., & Kee, Y. L. (2016). Persistent organic pollutants management and remediation *Procedia Environmental Sciences, 31*, 842–848. https://doi.org/10.1016/j.proenv.2016.02.093

Pascal-Lorber, S., & Laurent, F. (2011). Phytoremediation techniques for pesticide contaminations. In *Alternative farming systems, biotechnology, drought stress and ecological fertilisation* (pp. 77–105). https://doi.org/10.1007/978-94-007-0186-1_4

Patel, D., & Sen, D. J. (2013). Xenobiotics: An essential precursor for living system. *American Journal of Advanced Drug Delivery, 1*, 262–270.

Pelkonen, O., Kapitulnik, J., Gundert-Remy, U., Boobis, A., & Stockis, A. (2008). Local kinetics and dynamics of xenobiotics. *Critical Reviews in Toxicology, 38*(8), 697–720. https://doi.org/10.1080/10408440802194931

Pereira, L. C., de Souza, A. O., Bernardes, M. F. F., Pazin, M., Tasso, M. J., Pereira, P. H., & Dorta, D. J. (2015). A perspective on the potential risks of emerging contaminants to human and environmental health. *Environmental Science and Pollution Research, 22*(18), 13800–13823. https://doi.org/10.1007/S11356-015-4896-6

Pfohl, P., Wagner, M., Meyer, L., Domercq, P., Praetorius, A., Huffer, T., Hofmann, T., & Wohlleben, W. (2022). Environmental degradation of microplastics: How to measure fragmentation rates to secondary micro- and nanoplastic fragments and dissociation into dissolved organics. *Environmental Science & Technology, 56*(16), 11323–11334. https://doi.org/10.1021/ACS.EST.2C01228

Plastic Overshoot Day—Report. (2024). EA-Earth Action, 2024. https://plasticovershoot.earth/wp-content/uploads/2024/04/EA_POD_report_2024.pdf

Ponnana, S. R., Rajagopalan, T., Zhang, T., Chen, Z., Sirasapalli, S. K., Deo, S. (2025). Microplastic concentration, social, and environmental features and their association with chronic disease prevalence: An analysis across U.S. Census Tracts. *Journal of the American College of Cardiology, 85*(12), 374. https://doi.org/10.1016/S0735-1097(25)00858-7

Portugal, J., Bedia, C., Amato, F., Juárez-Facio, A. T., Stamatiou, R., Lazou, A., Campiglio, C. E., Elihn, K., & Piña, B. (2024). Toxicity of airborne nanoparticles: Facts and challenges. *Environment International, 190*, 108889. https://doi.org/10.1016/j.envint.2024.108889

Priya, K., Setty, M., Babu, U. V., & Pai, K. S. R. (2021). Implications of environmental toxicants on ovarian follicles: How it can adversely affect the female fertility? *Environmental Science and Pollution Research, 28*(48), 67925–67939. https://doi.org/10.1007/S11356-021-16489-4

Pulido-Reyes, G., Leganes, F., Fernández-Piñas, F., & Rosal, R. (2017). Bio-nano interface and environment: A critical review. *Environmental Toxicology and Chemistry, 36*(12), 3181–3193. https://doi.org/10.1002/etc.3924

Qaiser, Z., Aqeel, M., Sarfraz, W., Rizvi, Z. F., Noman, A., Naeem, S., & Khalid, N. (2023). Microplastics in wastewaters and their potential effects on aquatic and terrestrial biota. *Case Studies in Chemical and Environmental Engineering, 8*, Article 100536. https://doi.org/10.1016/J.CSCEE.2023.100536

Qiao, R., Mortimer, M., Richter, J., Rani-Borges, B., Yu, Z., Heinlaan, M., Lin, S., & Ivask, A. (2022). Hazard of polystyrene micro-and nanospheres to selected aquatic and terrestrial organisms. *Science of the Total Environment, 853*, Article 158560. https://doi.org/10.1016/j.scitotenv.2022.158560

Quintana, R., Manzano-Medina, S., Pérez-López, L., Oyón-Sanz, A., González-Fernández, D., González-Gordillo, J. I., Marti, E., Echevarría, F., & Morales-Caselles, C. (2025). Vertical distribution and composition of plastics in coastal areas of the Gulf of Cádiz: Insights into transport dynamics. *Environmental Science & Technology, 59*(33), 17760. https://doi.org/10.1021/ACS.EST.5C03513

Rajmohan, K. S., Chandrasekaran, R., & Varjani, S. (2020). A review on occurrence of pesticides in environment and current technologies for their remediation and management. *Indian Journal of Microbiology, 60*(2), 125–138. https://doi.org/10.1007/s12088-019-00841-x

Ramade, F. (2018). The pollution of the hydrosphere by global contaminants and its effects on aquatic ecosystems. *Aquatic Ecotoxicology*, 151–184. https://doi.org/10.1201/9781351069854-10

Ramadoss, D., Biju, A., Rathore, C., Saha, M., Kolandhasamy, P., Palogi, C., John, J., & Behera, A. K. (2025). The first report on emerged microplastics in deep-sea sediment: Insights from the Central Indian Ocean Basin. *Marine Pollution Bulletin, 211*, 117435. https://doi.org/10.1016/j.marpolbul.2024.117435

Ramage, S. J., Coull, M., Cooper, P., Campbell, C. D., Prabhu, R., Yates, K., Dawson, L. A., Devalla, S., & Pagaling, E. (2025). Microplastics in agricultural soils following sewage sludge applications: Evidence from a 25-year study. *Chemosphere, 376*, Article 144277. https://doi.org/10.1016/J.CHEMOSPHERE.2025.144277

Ramos, T. M., Christensen, T. B., Bour, A., Almroth, B. C., Kristensen, D. M., Selck, H., & Syberg, K., (2023). A not so circular healthcare economy: A review of challenges with plastic associated chemicals. *TrAC Trends in Analytical Chemistry, 166*, Article 117191. https://doi.org/10.1016/J.TRAC.2023.117191

Randhawa, K. S. (2024). Synthesis, properties, and environmental impact of hybrid pigments. *The Scientific World Journal, 2024*(1), 2773950. https://doi.org/10.1155/TSWJ/2773950

Raza, M., Lee, J. Y., & Cha, J. (2022). Microplastics in soil and freshwater: Understanding sources, distribution, potential impacts, and regulations for management. *Science Progress, 105*(3), 00368504221126676. https://doi.org/10.1177/00368504221126676

Real, L. E. P. (2023). Degradation and stabilization of polymers. In *Weathering of polymers and plastic materials* (pp. 1–33). https://doi.org/10.1007/978-3-031-33285-2_1

Ren, J., Wang, X., Wang, C., Gong, P., Wang, X., & Yao, T. (2017). Biomagnification of persistent organic pollutants along a high-altitude aquatic food chain in the Tibetan Plateau: Processes and mechanisms. *Environmental Pollution, 220*, 636–643. https://doi.org/10.1016/J.ENVPOL.2016.10.019

Reuber, M. D. (1989). Carcinogenicity of captan. *Journal of Environmental Pathology, Toxicology and Oncology: Official Organ of the International Society for Environmental Toxicology and Cancer*, *9*(2), 127–143. https://pubmed.ncbi.nlm.nih.gov/2732908/

Reyes, N. J. D., Geronimo, F. K. F., Yano, K. A. V., Guerra, H. B., & Kim, L. H. (2021). Pharmaceutical and personal care products in different matrices: Occurrence, pathways, and treatment processes. *Water*, *13*(9), 1159. https://doi.org/10.3390/W13091159

Rich, A. L., Phipps, L. M., Tiwari, S., Rudraraju, H., & Dokpesi, P. O. (2016). The increasing prevalence in intersex variation from toxicological dysregulation in fetal reproductive tissue differentiation and development by endocrine-disrupting chemicals. *Environmental Health Insights, 10*. https://doi.org/10.4137/EHI.S39825

Richardson, S. D. (2009). Water analysis: Emerging contaminants and current issues. *Analytical Chemistry, 81*(12), 4645–4677. https://doi.org/10.1021/AC9008012

Rim, K. T., Song, S. W., & Kim, H. Y. (2013). Oxidative DNA damage from nanoparticle exposure and its application to workers' health: a literature review. *Review Safety and Health at Work, 4*(4), 177–186. https://doi.org/10.1016/j.shaw.2013.07.006

Ritscher, A., Wang, Z., Scheringer, M., Boucher, J. M., Ahrens, L., Berger, U., Bintein, S., Bopp, S. K., Borg, D., Buser, A. M., & Cousins, I. (2018). Zürich statement on future actions on per- and polyfluoroalkyl substances (PFASs). *Environmental Health Perspectives, 126*(8). https://doi.org/10.1289/EHP4158

Rocha, T. L., Mestre, N. C., Sabóia-Morais, S. M. T., & Bebianno, M. J. (2017). Environmental behaviour and ecotoxicity of quantum dots at various trophic levels: A review. *Environment International, 98*, 1–17. https://doi.org/10.1016/j.envint.2016.09.021

Rodgers, K. M., Rudel, R. A., & Just, A. C. (2014). Phthalates in food packaging, consumer products, and indoor environments. In *Molecular and integrative toxicology* (pp. 31–59). https://doi.org/10.1007/978-1-4471-6500-2_2

Rodrigues, W. J., R, M. K. H., & J, S. K. (2024). Microplastics and nanoplastics: Environmental sources, human exposure pathways, and potential health impacts. *Revista Electronica de Veterinaria, 25*(1S), 537–543. https://doi.org/10.69980/redvet.V25I1S.750

Rohr, J. R., & McCoy, K. A. (2010). A qualitative meta-analysis reveals consistent effects of atrazine on freshwater fish and amphibians. *Environmental Health Perspectives, 118*(1), 20–32. https://doi.org/10.1289/ehp.0901164

Rokni, L., Rezaei, M., Rafieizonooz, M., Khankhajeh, E., Mohammadi, A. A., & Rezania, S. (2023). Effect of persistent organic pollutants on human health in South Korea: A review of the reported diseases. *Sustainability, 15*(14), 10851. https://doi.org/10.3390/su151410851

Runkle, J., Flocks, J., Economos, J., & Dunlop, A. L. (2017). A systematic review of Mancozeb as a reproductive and developmental hazard. *Environment International, 99*, 29–42. https://doi.org/10.1016/j.envint.2016.11.006

Ryan, A. C., Allen, D., Allen, S., Maselli, V., LeBlanc, A., Kelleher, L., Krause, S., Walker, T. R., & Cohen, M. (2023). Transport and deposition of ocean-sourced microplastic particles by a North Atlantic hurricane. *Communications Earth and Environment, 4*(1), 1–10. https://doi.org/10.1038/s43247-023-01115-7

Sadiq, F. K., Sadiq, A. A., Matsika, T. A., Momoh, B. A. (2025). Sustainable remediation of persistent organic Pollutants: A review on Recent innovative technologies. *Current Research in Biotechnology*, 100293. https://doi.org/10.1016/j.crbiot.2025.100293

Saha, S. C., & Saha, G. (2024). Effect of microplastics deposition on human lung airways: A review with computational benefits and challenges. *Heliyon, 10*(2), Article e24355. https://doi.org/10.1016/J.HELIYON.2024.E24355

Salama, M., Lotfy, A., Fathy, K., Makar, M., El-Emam, M., El-Gamal, A., El-Gamal, M., Badawy, A., Mohamed, W. M., & Sobh, M. (2015). Developmental neurotoxic effects of Malathion on 3D neurosphere system. *Applied & Translational Genomics, 7*, 13. https://doi.org/10.1016/j.atg.2015.07.001

Saleh, M., Gul, A., Nasir, A., Moses, T. O., Nural, Y., & Yabalak, E. (2024). Comprehensive review of Carbon-based nanostructures: Properties, synthesis, characterization, and cross-disciplinary applications. *Journal of Industrial and Engineering Chemistry*. https://doi.org/10.1016/j.jiec.2024.11.052

Salinas, J., Carpena, V., Martínez-Gallardo, M. R., Segado, M., Estrella-González, M. J., Toribio, A. J., Jurado, M. M., López-González, J. A., Suárez-Estrella, F., & López, M. J. (2023). Development of plastic-degrading microbial consortia by induced selection in microcosms. *Frontiers in Microbiology, 14*, 1143769. https://doi.org/10.3389/fmicb.2023.1143769

Sánchez-Resino, E., Marquès, M., Gutiérrez-Martín, D., Restrepo-Montes, E., Martinez, M. A., Salas-Huetos, A., Babio, N., Salas-Salvado, J., Gil-Solsona, R., & Gago-Ferrero, P. (2023). Exploring the occurrence of organic contaminants in human semen through an innovative LC-HRMS-based methodology suitable for target and nontarget analysis. *Environmental Science & Technology, 57*(48), 19236–19252. https://doi.org/10.1021/acs.est.3C04347

Saravanakumar, K., De Silva, S., Santosh, S. S., Sathiyaseelan, A., Ganeshalingam, A., Jamla, M., Sankaranarayanan, A., Veeraraghavan, V. P., Mubarak Ali, D., Lee, J., & Thiripuranathar, G. (2022). Impact of industrial effluents on the environment and human health and their remediation using MOFs-based hybrid membrane filtration techniques. *Chemosphere, 307*, Article 135593. https://doi.org/10.1016/j.chemosphere.2022.135593

Sathyamoorthy, S., Venugopal, A., Lenin, M. J., Tirkey, S., & Sevanan, M. (2025). Nanomaterials and microbial compatibility: Synergistic and antagonistic mechanisms. *Nano-Bioremediation for Wastewater Treatment, 35*, 72. https://doi.org/10.1002/9781394271641

Saxena, P., Saharan, V., Baroliya, P. K., Gour, V. S., & Rai, M. K. (2021). Mechanism of nanotoxicity in Chlorella vulgaris exposed to zinc and iron oxide. *Toxicology Reports, 8*, 724–731. https://doi.org/10.1016/j.toxrep.2021.03.023

Segev, O., Kushmaro, A., & Brenner, A. (2009). Environmental impact of flame retardants (Persistence and Biodegradability). *International Journal of Environmental Research and Public Health, 6*(2), 478–491. https://doi.org/10.3390/IJERPH6020478

Sela, H., Cohen, H., Elia, P., Zach, R., Karpas, Z., & Zeiri, Y. (2015). Spontaneous penetration of gold nanoparticles through the blood brain barrier (BBB). *Journal of Nanobiotechnology, 13*(1), 71. https://doi.org/10.1186/s12951-015-0133-1

Serwecińska, L. (2020). Antimicrobials and antibiotic-resistant bacteria: A risk to the environment and to public health. *Water, 12*(12), 3313. https://doi.org/10.3390/W12123313

Sharma, B. M., Bharat, G. K., Tayal, S., Nizzetto, L., Čupr, P., & Larssen, T. (2014). Environment and human exposure to persistent organic pollutants (POPs) in India: A systematic review of recent and historical data. *Environment International, 66*, 48–64. https://doi.org/10.1016/j.envint.2014.01.022

Silva, C. J., Machado, A. L., Campos, D., Rodrigues, A. C., Silva, A. L. P., Soares, A. M., & Pestana, J. L. (2022). Microplastics in freshwater sediments: Effects on benthic invertebrate communities and ecosystem functioning assessed in artificial streams. *The Science of the Total Environment, 804*. https://doi.org/10.1016/j.scitotenv.2021.150118

Simonin, M., Richaume, A., Guyonnet, J. P., Dubost, A., Martins, J. M., & Pommier, T. (2016). Titanium dioxide nanoparticles strongly impact soil microbial function by affecting archaeal nitrifiers. *Scientific Reports, 6*(1), 33643. https://doi.org/10.1038/srep33643

Singh, M., Mishra, R. C., Shah, I., Wadhwa, V., & Mor, V. (2023). Xenobiotics: Sources, pathways, degradation, and risk associated with major emphasis on pharmaceutical compounds. In *Xenobiotics in urban ecosystems: Sources, distribution and health impacts* (pp. 87–106). https://doi.org/10.1007/978-3-031-35775-6_5

Skalny, A., Aschner, M., Paoliello, M., Santamaria, A., Nikitina, N., Rejniuk, V., Jiang, Y., Rocha, J., & Tinkov, A. (2021). Endocrine-disrupting activity of mancozeb. *Arhiv Za Farmaciju, 71*(6), 491. https://doi.org/10.5937/arhfarm71-34359

Soares, E. V., & Soares, H. M. (2021). Harmful effects of metal (loid) oxide nanoparticles. *Applied Microbiology and Biotechnology, 105*(4), 1379–1394. https://doi.org/10.1007/s00253-021-11124-1

Sofield, C. E., Anderton, R. S., & Gorecki, A. M. (2024). Mind over microplastics: Exploring microplastic-induced gut disruption and gut-brain-axis consequences. *Current Issues in Molecular Biology, 46*(5), 4186. https://doi.org/10.3390/CIMB46050256

Soloff, A. C., Zhao, X., Udoh, H., Bardawil, C., & Dhupar, R. (2025). Inhaled microplastics inhibit tissue maintenance functions of pulmonary macrophages. *211*(Abstracts), A3126–A3126. https://doi.org/10.1164/AJRCCM.2025.211.ABSTRACTS.A3126

Song, J., Wang, C., & Li, G. (2024). Defining primary and secondary microplastics: A connotation analysis. *ACS ES&T Water, 4*(6), 2330–2332. https://doi.org/10.1021/ACSESTWATER.4C00316

Spitz, P. H. (1988). Petrochemicals: the rise of an industry. Retrieved December 13, 2025, from https://cir.nii.ac.jp/crid/1971712334794861441

Stapleton, M. J., & Hai, F. I. (2023). Microplastics as an emerging contaminant of concern to our environment: A brief overview of the sources and implications. *Bioengineered, 14*(1), 2244754. https://doi.org/10.1080/21655979.2023.2244754

Štefanac, T., Grgas, D., & Dragičević, T. L. (2021). Xenobiotics—Division and methods of detection: A review. *Journal of Xenobiotics, 11*(4), 130–141. https://doi.org/10.3390/JOX11040009

Subbiah, R., & Tiwari, R. R. (2021). The herbicide paraquat-induced molecular mechanisms in the development of acute lung injury and lung fibrosis. *Critical Reviews in Toxicology, 51*(1), 36–64. https://doi.org/10.1080/10408444.2020.1864721

Suman, T. Y., Rajasree, S. R., & Kirubagaran, R. (2015). Evaluation of zinc oxide nanoparticles toxicity on marine algae Chlorella vulgaris through flow cytometric, cytotoxicity and oxidative stress analysis. *Ecotoxicology and Environmental Safety, 113*, 23–30. https://doi.org/10.1016/j.ecoenv.2014.11.015

Sun, B., Hu, Y., Cheng, H., & Tao, S. (2019). Releases of brominated flame retardants (BFRs) from microplastics in aqueous medium: Kinetics and molecular-size dependence of diffusion. *Water Research, 151*, 215–225. https://doi.org/10.1016/J.WATRES.2018.12.017

Sun, S., Korheina, D. K., Fu, H., & Ge, X. (2020). Chronic exposure to dietary antibiotics affects intestinal health and antibiotic resistance gene abundance in oriental river prawn (*Macrobrachium nipponense*), and provokes human health risk. *Science of the Total Environment, 720*, Article 137478. https://doi.org/10.1016/J.SCITOTENV.2020.137478

Sun, H., Su, X., Mao, J., Liu, Y., Li, G., & Du, Q. (2024a). Microplastics in maternal blood, fetal appendages, and umbilical vein blood. *Ecotoxicology and Environmental Safety, 287*, Article 117300. https://doi.org/10.1016/J.ECOENV.2024.117300

Sun, R., Alinezhad, A., Altarawneh, M., Ateia, M., Blotevogel, J., Mai, J., Naidu, R., Pignatello, J., Rappe, A., Zhang, X., & Xiao, F. (2024b). New insights into thermal degradation products of long-chain per-and polyfluoroalkyl substances (PFAS) and their mineralization enhancement using additives. *Environmental Science & Technology, 58*(50), 22417–22430. https://doi.org/10.1021/ACS.EST.4C05782

Sutkar, P. R., Gadewar, R. D., & Dhulap, V. P. (2023). Recent trends in degradation of microplastics in the environment: A state-of-the-art review. *Journal of Hazardous Materials Advances, 11*, Article 100343. https://doi.org/10.1016/J.HAZADV.2023.100343

Tadesse, S. H., & Hailemariam, T. T. (2025). A review on the classification, characterisation, synthesis of nanoparticles and their application. https://doi.org/10.11648/j.advances.20250602.15

Tamargo, A., Molinero, N., Reinosa, J. J., Alcolea-Rodriguez, V., Portela, R., Bañares, M. A., Fernández, J. F., & Moreno-Arribas, M. V. (2022). PET microplastics affect human gut microbiota communities during simulated gastrointestinal digestion, first evidence of plausible polymer biodegradation during human digestion. *Scientific Reports, 12*(1), 1–15. https://doi.org/10.1038/s41598-021-04489-w

Tang, J., Xiong, L., Wang, S., Wang, J., Liu, L., Li, J., Wan, Z., & Xi, T. (2008). Influence of silver nanoparticles on neurons and blood-brain barrier via subcutaneous injection in rats. *Applied Surface Science, 255*(2), 502–504. https://doi.org/10.1016/j.apsusc.2008.06.058

Tang, J., Xiong, L., Zhou, G., Wang, S., Wang, J., Liu, L., Li, J., Yuan, F., Lu, S., Wan, Z., & Chou, L. (2010). Silver nanoparticles crossing through and distribution in the blood-brain barrier in vitro. *Journal of Nanoscience and Nanotechnology, 10*(10), 6313–6317. https://doi.org/10.1166/jnn.2010.2625

Temporiti, M. E. E., Nicola, L., Nielsen, E., & Tosi, S. (2022). Fungal enzymes involved in plastics biodegradation. *Microorganisms, 10*(6), 1180. https://doi.org/10.3390/MICROORGANISMS10061180

Thi, T. T. H., Suys, E. J., Lee, J. S., Nguyen, D. H., Park, K. D., & Truong, N. P. (2021). Lipid-based nanoparticles in the clinic and clinical trials: From cancer nanomedicine to COVID-19 vaccines. *Vaccines, 9*(4), 359. https://doi.org/10.3390/vaccines9040359

Thomas, S., Kumar Mishra, P., & Talegaonkar, S. (2015). Ceramic nanoparticles: Fabrication methods and applications in drug delivery. *Current Pharmaceutical Design, 21*(42), 6165–6188. https://doi.org/10.2174/1381612821666151027153246

Thompson, R. C., Olsen, Y., Mitchell, R. P., Davis, A., Rowland, S. J., John, A. W., McGonigle, D., & Russell, A. E. (2004). Lost at sea: Where is all the plastic? *Science, 304*(5672), 838. https://doi.org/10.1126/SCIENCE.1094559

Thuy, T. L., Hoang, T. D., Hoang, V. H., & Nguyen, M. K. (2025). A review on flame retardants in soils: Occurrence, environmental impact, health risks, remediation strategies, and future perspectives. *Toxics, 13*(3), 228. https://doi.org/10.3390/TOXICS13030228

Tijani, J. O., Fatoba, O. O., Babajide, O. O., & Petrik, L. F. (2015). Pharmaceuticals, endocrine disruptors, personal care products, nanomaterials and perfluorinated pollutants: A review. *Environmental Chemistry Letters, 14*(1), 27–49. https://doi.org/10.1007/S10311-015-0537-Z

Torres, F. G., & De-la-Torre, G. E. (2022). Per- and polyfluoroalkyl substances (PFASs) in consumable species and food products. *Journal of Food Science and Technology, 60*(9), 2319–2336. https://doi.org/10.1007/S13197-022-05545-7

Tran, T. K., Nguyen, M. K., Lin, C., Hoang, T. D., Nguyen, T. C., Lone, A. M., Khedulkar, A. P., Gaballah, M. S., Singh, J., Chung, W. J., & Nguyen, D. D. (2024). Review on fate transport toxicity and health risk of nanoparticles in natural ecosystems: Emerging challenges in the modern age and solutions toward a sustainable environment. *Science of The Total Environment, 912*, 169331. https://doi.org/10.1016/j.scitotenv.2023.169331

Wang, J., & Dai, G. D. (2022). Comparative effects of brominated flame retardants BDE-209, TBBPA, and HBCD on neurotoxicity in mice. *Chemical Research in Toxicology, 35*(9), 1512–1518. https://doi.org/10.1021/ACS.CHEMRESTOX.2C00126

Wang, T., & Liu, W. (2022). Emerging investigator series: Metal nanoparticles in freshwater: Transformation, bioavailability and effects on invertebrates. *Environmental Science: Nano, 9*(7), 2237–2263. https://doi.org/10.1039/D2EN00052K

Wang, Z., Na, G., Ma, X., Ge, L., Lin, Z., & Yao, Z. (2015). Characterizing the distribution of selected PBDEs in soil, moss and reindeer dung at Ny-Ålesund of the Arctic. *Chemosphere, 137*, 9–13. https://doi.org/10.1016/J.CHEMOSPHERE.2015.04.030

Wang, L., Nabi, G., Yin, L., Wang, Y., Li, S., Hao, Z., & Li, D. (2021). Birds and plastic pollution: Recent advances. *Avian Research, 12*(1), 1–9. https://doi.org/10.1186/S40657-021-00293-2

Wang, X., Sial, M. U., Bashir, M. A., Bilal, M., Raza, Q. U. A., Ali Raza, H. M., Rehim, A., & Geng, Y. (2022). Pesticides xenobiotics in soil ecosystem and their remediation approaches. *Sustainability, 14*(6), 3353. https://doi.org/10.3390/su14063353

Wang, J., Chan, F. K. S., Johnson, M. F., Chan, H. K., Cui, Y., Chen, J., & Chen, W. Q. (2024a). Material cycles, environmental emissions, and ecological risks of bisphenol A (BPA) in China and implications for sustainable plastic management. *Environmental Science & Technology, 59*(3), 1631–1646. https://doi.org/10.1021/ACS.EST.4C09876

Wang, M., Wu, Y., Li, G., Xiong, Y., Zhang, Y., & Zhang, M. (2024b). The hidden threat: Unraveling the impact of microplastics on reproductive health. *Science of the Total Environment, 935*, Article 173177. https://doi.org/10.1016/J.SCITOTENV.2024.173177

Wang, T., Liu, D., Liu, R., Yuan, F., Ding, Y., Tao, J., Wang, Y., Yu, W., Fang, Y., & Li, B. (2025). Weathering process and characteristics of microplastics in coastal wetlands: A 24-month in situ study. *Environmental Science & Technology, 59*(15), 7667–7677. https://doi.org/10.1021/ACS.EST.4C12140

Watanabe, I., & Sakai, S. I. (2003). Environmental release and behavior of brominated flame retardants. *Environment International, 29*(6), 665–682. https://doi.org/10.1016/S0160-4120(03)00123-5

Weber, C. J., & Bigalke, M. (2025). Forest soils accumulate microplastics through atmospheric deposition. *Communications Earth and Environment, 6*(1), 1–9. https://doi.org/10.1038/S43247-025-02712-4

Wend, K., Zorrilla, L., Freimoser, F. M., & Gallet, A. (2024). Microbial pesticides–Challenges and future perspectives for testing and safety assessment with respect to human health. *Environmental Health, 23*(1), 49. https://doi.org/10.1186/s12940-024-01090-2

Wikoff, D. S., & Birnbaum, L. (2011). Human health effects of brominated flame retardants. pp. 19–53. https://doi.org/10.1007/698_2010_97

Wright, S. L., Ulke, J., Font, A., Chan, K. L. A., & Kelly, F. J. (2020). Atmospheric microplastic deposition in an urban environment and an evaluation of transport. *Environment International, 136*, Article 105411. https://doi.org/10.1016/J.ENVINT.2019.105411

Wu, H., Wang, J., Xiang, Y., Li, L., Qie, H., Ren, M., Lin, A., & Qi, F. (2021). Effects of tetrabromobisphenol A (TBBPA) on the reproductive health of male rodents: A systematic review and meta-analysis. *Science of the Total Environment, 781*, Article 146745. https://doi.org/10.1016/J.scitotenv.2021.146745

Wu, H. (2023). The application of metal oxide nanoparticles in sunscreen and their potential risks. *Applied and Computational Engineering, 24*, 210–215. https://doi.org/10.54254/2755-2721/24/20230710

Xie, Y., Peng, R., & Xiao, L. (2025). Environmental chemicals and female reproductive health: Unraveling mechanisms and societal impacts—A narrative review. *Clinical and Experimental Obstetrics and Gynecology, 52*(8), 39882. https://doi.org/10.31083/CEOG39882/PDF

Xiong, P., Yan, X., Zhu, Q., Qu, G., Shi, J., Liao, C., & Jiang, G. (2019). A review of environmental occurrence, fate, and toxicity of novel brominated flame retardants. *Environmental Science & Technology, 53*(23), 13551–13569. https://doi.org/10.1021/ACS.EST.9B03159

Xue, J., Xiao, Q., Zhang, M., Li, D., & Wang, X. (2023). Toxic effects and mechanisms of polybrominated diphenyl ethers. *International Journal of Molecular Sciences, 24*(17), 13487. https://doi.org/10.3390/ijms241713487

Yang, W., Jannatun, N., Zeng, Y., Liu, T., Zhang, G., Chen, C., & Li, Y. (2022). Impacts of microplastics on immunity. *Frontiers in Toxicology, 4*, Article 956885. https://doi.org/10.3389/FTOX.2022.956885

Yang, J., Kamstra, J., Legler, J., & Aardema, H. (2023a). The impact of microplastics on female reproduction and early life. *Animal Reproduction, 20*(2), Article e20230037. https://doi.org/10.1590/1984-3143-AR2023-0037

Yang, X., Xiao, P., & Shi, X. (2023b). Molecular mechanism of paraquat-induced ferroptosis leading to pulmonary fibrosis mediated by Keap1/Nrf2 signaling pathway. *Molecular Biology Reports, 50*(11), 9249–9261. https://doi.org/10.1007/s11033-023-08756-z

Yang, H., Niu, S., Guo, M., & Xue, Y. (2025a). A critical review of the ecotoxic effects of microplastics on aquatic, soil and atmospheric ecosystems and current research challenges. *Environmental Research, 274*, Article 121361. https://doi.org/10.1016/J.ENVRES.2025.121361

Yang, W., Wang, Z., Jiang, Y., Cui, S., Yang, M., Li, C., Li, Y. F., & Jia, H. (2025b). Bioaccumulation of novel brominated flame retardants in a marine food web: A comprehensive analysis of occurrence, trophic transfer, and interfering factors. *Science of the Total Environment, 962*, Article 178428. https://doi.org/10.1016/J.scitotenv.2025.178428

Yang, Y., Chen, Y., Gao, Y., Jin, Y., & Xu, F. (2025c). Mapping research frontiers in microplastics-induced oxidative stress: A bibliometric analysis (2010–2024). *Frontiers in Environmental Science, 13*, 1555341. https://doi.org/10.3389/FENVS.2025.1555341/BIBTEX

Yaseen, A., Assad, I., Sofi, M. S., Hashmi, M. Z., & Bhat, S.U. (2022). A global review of microplastics in wastewater treatment plants: Understanding their occurrence, fate and impact. *Environmental Research, 212*(Pt B). https://doi.org/10.1016/j.envres.2022.113258

Yashwanth, A., Huang, R., Iepure, M., Mu, M., Zhou, W., Kunadu, A., Carignan, C., Yegin, Y., Cho, D., Oh, J. K., & Taylor, M. T. (2025). Food packaging solutions in the post-per- and polyfluoroalkyl substances (PFAS) and microplastics era: A review of functions, materials, and bio-based alternatives. *Comprehensive Reviews in Food Science and Food Safety, 24*(1), Article e70079. https://doi.org/10.1111/1541-4337.70079

Yu, C., Kim, S., Jang, M., Park, C. M., & Yoon, Y. (2022). Occurrence and removal of engineered nanoparticles in drinking water treatment and wastewater treatment processes: A review. *Environmental Engineering Research, 27*(5). https://doi.org/10.4491/EER.2021.339

Zachou, K., Arvaniti, P., Lyberopoulou, A., & Dalekos, G. N. (2021). Impact of genetic and environmental factors on autoimmune hepatitis. *Journal of Translational Autoimmunity, 4*, Article 100125. https://doi.org/10.1016/j.jtauto.2021.100125

Zahra, Z., Habib, Z., Hyun, S., & Sajid, M. (2022). Nanowaste: Another future waste, its sources, release mechanism, and removal strategies in the environment. *Sustainability, 14*(4), 2041. https://doi.org/10.3390/su14042041

Zaidi, N., Mir, M. A., Chang, S. K., Abdelli, N., Hasnain, S. M., Ali Khan, M. A., & Andrews, K. (2025). Pharmaceuticals and personal care products as emerging contaminants: Environmental fate, detection, and mitigation strategies. *International Journal of Environmental Analytical Chemistry*. https://doi.org/10.1080/03067319.2025.2484456

Zhang, E., Kim, M., Rueda, L., Rochman, C., VanWormer, E., Moore, J., & Shapiro, K. (2022). Association of zoonotic protozoan parasites with microplastics in seawater and implications for human and wildlife health. *Scientific Reports, 12*(1), 1–11. https://doi.org/10.1038/S41598-022-10485-5

Zhang, X., Yu, C., Wang, P., & Yang, C. (2025). Microplastics and human health: Unraveling the toxicological pathways and implications for public health. *Frontiers in Public Health, 13*, 1567200. https://doi.org/10.3389/fpubh.2025.1567200

Zhang, C., Zhang, G., Sun, K., Ren, J., Zhou, J., Liu, X., Lin, F., Yang, H., Cao, J., Nie, L., & Zhang, P. (2024). Association of mixed exposure to microplastics with sperm dysfunction: A multi-site study in China. *eBioMedicine*, 108. https://doi.org/10.1016/j.ebiom.2024.105369

Zhao, Q., Zhu, L., Weng, J., Jin, Z., Cao, Y., Jiang, H., & Zhang, Z. (2023). Detection and characterization of microplastics in the human testis and semen. *Science of the Total Environment, 877*. https://doi.org/10.1016/j.scitotenv.2023.162713

Zhu, M., Li, X., Lin, W., Zeng, D., Yang, P., Ni, W., Chen, Z., Lin, B., Lai, L., Ouyang, Z., & Fan, J. (2024). Microplastic particles detected in fetal cord blood, placenta, and meconium: A pilot study of nine mother-infant pairs in South China. *Toxics, 12*(12), 850. https://doi.org/10.3390/toxics12120850

Ziylan-Yavas, A., Santos, D., Flores, E. M. M., & Ince, N. H. (2022). Pharmaceuticals and personal care products (PPCPs): Environmental and public health risks. *Environmental Progress and Sustainable Energy, 41*(4), Article e13821. https://doi.org/10.1002/ep.13821

Zurub, R. E., Cariaco, Y., Wade, M. G., & Bainbridge, S. A. (2024). Microplastics exposure: Implications for human fertility, pregnancy and child health. *Frontiers in Endocrinology, 14*, 1330396. https://doi.org/10.3389/FENDO.2023.1330396

Chapter 4
Genetic Damage: Mechanism and Consequences of Xenobiotic-Induced Mutation

Tanzeel Ahmed, Smrita Singh, and Mohammad Khalid Parvez

Abstract Xenobiotics are chemical substances that are extraneous to living organisms, including substances such as pesticides, and even microorganisms. Pesticides are utilized to kill, eradicate insects, pests, insects, weeds, or any other living thing. Pesticides are routinely employed in agriculture despite the harmful genetic effects of exposure. Their extensive use leads to the buildup of pesticides in food, soil, running water, etc., even though they are beneficial in boosting crop yields. Because of their delayed biodegradation and high exposure levels, pesticides find their way into the food chain. Occupational exposure to pesticides occurs mainly through the breath or through skin contact while preparation of the solutions to be sprayed or during spraying. Additionally, exposure to pesticides can also occur through consuming contaminated food. This book chapter explores adverse effects caused by pesticide, such as genetic damage and its associated mechanism, and consequences like neurotoxicity, endocrine disruption, carcinogenesis, reproductive development and immunological effects. The chapter also discusses the biological risk of pesticides and the potential for genetic damage, especially in humans, are dependent on a multigrazing process whose unknown effects are affected by independent or dependent characters such as age, smoking habits, contact periods, and food-related patterns. Many pesticides cause oxidative DNA damage due to generation of free radicals in non-target species and become susceptible to infections. The potential mechanism of pesticide-induced genotoxicity has been reported as oxidative stress in humans. DNA alteration is the primary risk factor and acts as an indicator in the affected organisms which identifies the genotoxic potential of xenobiotics, so the effective strategy for risk assessment can be followed. Several antioxidants have been reported to counteract the genotoxicity due to its neutralizing potential of reactive oxygen species (ROS). This book chapter throws light on the effectiveness of various antioxidants in preventing genotoxicity after exposure to pesticides.

T. Ahmed (✉) · S. Singh
School of Biotechnology, IFTM University, Moradabad, U.P, India
e-mail: ahmed.tanzeel@gmail.com

M. K. Parvez
Deparment of Pharmacognosy, School of Pharmacy, King Saud University, Riyadh, Saudi Arabia

K. Priya and S. K. Giri (eds.), *Environmental Toxicology: Xenobiotics and Effects on Human Genetic Material*, https://doi.org/10.1007/978-3-032-28619-2_4

Keywords Xenobiotics · Pesticides · DNA damage · Genotoxicity · Oxidative stress · Antioxidants

1 Introduction

Pesticides are a type of xenobiotics which are chemical agents that play a critical role in agriculture, public health, and vector control programs. The U.S. Environmental Protection Agency (EPA) defines a pesticide as "*any substance or mixture of substances intended for preventing, destroying, repelling, or mitigating any pest*" (EPA, 2020). Similarly, the Food and Agriculture Organization (FAO) emphasizes that pesticides include both synthetic and naturally derived compounds, as well as biological agents such as bacteria and viruses used in pest control (FAO, 2018).

More than 4 million tones of pesticides are globally used with highest use in Asia followed by the America and Europe (FAO, 2021). To maximize the crop yield, agriculture demands are increasing for the pesticide in developing countries (Sharma et al., 2019). The most important use of pesticides worldwide is that of herbicides followed by pesticides, insecticides and fungicides which show the prevalence of weed and pest management challenges in farming systems (Schreinemachers & Tipraqsa, 2012). Irrespective of regional differences in the usage rate, pesticides usage has become a universal factor in agricultural strengthening and food safety strategies.

The primary goal of using pesticides is to protect crops from insects, weeds, fungi, and other pests, on the other hand that may cause overwhelming loss in crop yield. It has been reported that in cases where pesticides are not used 35–42% annual global food losses occur due to pests (Oerke, 2006). The use of pesticides to reduce crop damage significantly contributes to food availability, stability of agricultural markets, and farmer income. This also plays a role in controlling vector-borne diseases in humans, such as malaria and dengue, thereby extending their benefits beyond agriculture. On the other hand, following are the risks associated with pesticide use.

- Environmental contamination
- Non-target organism toxicity
- Persistence in soil and water ecosystems

Pesticides get accumulated in the food chain easily at low concentrations also that lead to acute to chronic human exposure (Mostafalou & Abdollahi, 2017). Over reliance on pesticides lead to resistant pest species creating an increased chemical dependency. This indicates an alarming concern for the serious ecological and public health challenges despite the fact that pesticides help ensure food security. Pesticides exposure is connected with adverse health outcomes in humans. This exposure can occur through two ways:

- Acute exposure

- Chronic exposure

- Acute Exposure: This exposure is for a short period of time that results in poisoning, respiratory distress, skin irritation, and neurological symptoms, and even death (Ecobichon, 2001).
- Chronic Exposure: This exposure is long term that has subtle effects including endocrine disruption, carcinogenesis, neurotoxicity, reproductive impairment, and immune dysfunction (Kim et al., 2017).

The particular concern is the genotoxicity of pesticides where chromosomal aberrations, micronucleus formation, and oxidative DNA lesions, cause DNA damage and thereby increase cancer risk (Bolognesi, 2003). The most common is the occupational exposure among agriculture workers where the person is involved in pesticide application, manufacturing, and handling these chemicals. The general population is at risk due to dietary intake of pesticide residues in food and drinking water (Costa et al., 2014). Pesticides do have two facets of a coin where on one side it is necessary for securing food production and on the other hand it is harmful to health.

2 Classification of Pesticides

Pesticides can be classified based on target organisms, chemical nature, and mode of action.

1. **Based on Target Organisms**: Based on the target organism, pesticides are classified as follows:

S. No.	Pesticides	Action	Example
1	Insecticides	Targets insects and kill	Organophosphates, carbamates
2	Herbicides	Control or eliminate weeds	Glyphosate, atrazine
3	Fungicides	Suppress fungal pathogens	Mancozeb, carbendazim
4	Rodenticides	Kill rodents such as rats and mice	Warfarin
5	Nematicides	Eliminate nematodes affecting plant roots	Fluensulfone, aldicarb, ethoprop
6	Bactericides	Destroy harmful bacteria in crops	Copper hydroxide, copper sulphate, chlorothalonil
7	Molluscicides	Target snails and slugs	Metaldehyde and iron phosphate, copper sulphate
8	Acaricides	Control mites and ticks	Permethrin, chlorpyrifos, amitraz, abamectin

2. **Based on Chemical Composition**: Based on chemical compositions, pesticides are classified as follows:

S. No.	Pesticides	Action	Example
1	Organochlorines	Neurotoxicants, disrupt transmission of nerve impulse	DDT, aldrin
2	Organophosphates	Inhibits enzyme acetylcholinesterase (AchE), overaccumulation of neurotransmitter acetylcholine (Ach)	Parathion, malathion
3	Carbamates	Inhibit acetylcholinesterase (AchE)	Carbaryl, aldicarb
4	Pyrethroids	Neurotoxic, preventing the closure of voltage gated sodium channels in nerve cells membranes, leading to prolonged opening and continuous influx of sodium ions (Na+)	Permethrin, cypermethrin
5	Neonicotinoids	Neurotoxic binds to nicotinic acetylcholine receptors (nAchRs)	Imidacloprid
6	Biopesticides	Disrupt biochemical processes, interference with nervous system	Bacillus thuringiensis, neem extract

3. **Based on Mode of Action**: Based on mode of action, pesticides are classified as follows:

S. No.	Pesticides	Action	Example
1	Contact pesticides	Act on pests upon direct exposure	Pyrethroids
2	Systemic pesticides	Absorbed and transported through plant tissues, protecting the whole plant	Neonicotinoids
3	Fumigants	Volatile compounds acting in gaseous form	Methyl bromide
4	Stomach poisons	Effective when ingested by pests	Arsenical compounds

3 Routes of Human Exposure to Pesticides

There are three routes of human exposure to pesticides based on occupation, environment, and lifestyle. The degree of exposure depends on factors such as duration of contact, protective measures, pesticide formulation, and individual susceptibility. Following are the three primary routes of exposure:

- Occupational exposure
- Environmental exposure
- Dietary exposure

1. **Occupational Exposure**

Agricultural workers, pesticide sprayers, and pest control operators are at the greatest risk of this exposure. In this exposure, there are three routes through which pesticides enter into the human body.

- **Dermal absorption**: This is the most common route, in which pesticides get absorbed through the skin. This occurs when pesticides are handled without adequate protective equipment (Fenske et al., 2002).
- **Inhalation exposure**: This takes place due to inhaling pesticides present in aerosols, vapors, or dusts during spraying or in enclosed areas like greenhouses (Hernández et al., 2011).
- **Oral exposure**: This type of exposure occurs from hand-to-mouth contact while eating, drinking, or smoking during pesticide handling.

There are several studies in which it has been reported that farmers and applicators often show higher levels of pesticide metabolites in blood and urine compared to the general population, confirming occupational risk (Alavanja & Bonner, 2012a, 2012b). Lack of personal safety measures, poor training, and insecure institutions, especially in developing countries, increase this risk.

2. **Environmental Exposure**

The main cause of environmental exposure is the contamination of air, soil, and water bodies after the use of insecticides. These chemicals can be released from fields during spraying, can leak into groundwater, or can remain in the environment due to slow degradation.

- **Airborne exposure**: This is due to the evaporation and drift of pesticides, which can affect nearby residents, including children (Harnly et al., 2005).
- **Water contamination**: Long-term ingestion of low-level pesticides in drinking water through agricultural runoff (Stehle & Schulz, 2015).
- **Soil contamination**: This not only affects agricultural workers but also contributes to the bioaccumulation of the ecosystem.

3. **Dietary Exposure**

The most common way of contact of pesticides for the general population is through the dietary exposure. Pesticide residues remain on fruits, vegetables, grains, and animal products until they are properly washed, handled, or processed.

- WHO and FAO continuously monitor the maximum residue limits (MRLs), but the common problems have been observed in both the developing and industrial countries (WHO, 2020).
- Bioaccumulation in the food chain (e.g., organochlorines in fish and dairy) further increases long-term risks (Jensen et al., 2012).

4 Persistence and Bioaccumulation of Pesticides

The pesticide's persistence in soil, water, and air for extended periods after application is the critical concern. Its persistency is influenced by chemical structure, solubility, volatility, and environmental conditions such as temperature, pH, and microbial activity (Fenner et al., 2013). Compounds such as organochlorines (e.g., DDT, aldrin, dieldrin) have their long half-lives, that remain in soil for decades (Zhang et al., 2011). The persistent pesticides do not biodegrade for a long time and therefore leak in groundwater and surface water and contaminate them. This raises the risk of long-term exposure to humans and wildlife (Aktar et al., 2009).

Bioaccumulation and Biomagnifications

- The continuous accumulation of pesticides leads to bioaccumulation, the process by which pesticides accumulate in living organisms at a faster rate than their metabolism or excretion. Lipophilic pesticides, such as organochlorines and some pyrethroids, are accumulated in adipose tissue, liver, and breast milk (Li et al., 2016).
- The pesticide concentration increases at higher trophic levels through the food chain, and is known as biomagnification. For example, small aquatic organisms absorb pesticide residues through water, which are then consumed by fish, and then by humans. As a result, pesticide burdens on top predators and humans than to lower organisms (Walker et al., 2012).

Public Health Implications

Pesticides are widely used in agriculture and public health. The lipophilic nature and persistency of pesticides result in their accumulation in human tissues and fluids. The persistence and bioaccumulation of pesticides pose serious health risks (Kim et al., 2017). To reduce these risks, it is necessary to adopt safe alternatives like effective monitoring and stricter regulation of persistent pesticides.

Pesticide Residues in Human Tissues

There are several reported studies that have shown the pesticide residues in the human tissues like adipose tissue, liver and kidney, brain tissue, and breast milk.

Pesticide Metabolites in Human Fluids

Pesticide metabolites act as biomarkers for various insecticides, such as 3-phenoxybenzoic acid (3-PB) for pyrethroids and para-nitrophenol for organophosphates, the total daily intake which is thought to predict possible health effects including oxidative stress and chronic diseases. The biomarkers in different human fluids are shown in (Table 1).

Table 1 The biomarkers in different human fluids

S. No	Human fluids	Biomarkers
1	Blood/serum	p,p′-DDE and 3,5,6-trichloro-2-pyridinol
2	Urine	Dialkylphosphate (DAP), 3-PBA (3-phenoxybenzoic acid)
3	Semen and reproductive fluids	DDE, lindane
4	Hair and nails	Organochlorines and pyrethroids

5 Adverse Health Effects of Pesticide Exposure

The environment and public health are at risk due to the usage of pesticides. Nowadays, due to continuous use of large amount of pesticide, its components are present in our atmosphere and have contaminated our food, water, and soil, causing acute and long-term health risks. High doses of pesticides caused acute toxicity which happens due to inhalation consumption, or coming into contact with the skin or eyes. Chronic toxicity is caused by repeated exposure to pesticides. Different forms of toxicity, such as neurotoxicity, mutagenicity, carcinogenicity, teratogenicity, and endocrine disruption, are produced by pesticides (Ahmad et al., 2024).

In order to reduce the health risks associated with contemporary pesticides, there are need for safety considerations. This chapter examined an overview of pesticides with respect to genetic damage and its associated mechanism.

5.1 Neurotoxicity and Endocrine Disruption

Acetylcholinesterase is the enzyme that breaks down the neurotransmitter acetylcholine and many organophosphates and carbamate pesticides function by blocking this enzyme. These insecticides overstimulate cholinergic receptors in the neurological system by blocking AChE, which causes acetylcholine to build up. Symptoms including increased salivation may result from this (Colovic et al., 2013; Kumar et al., 2018). Organochlorines like Indane and endosulfan are among the pesticides that may interact with GABA receptors in the central nervous system. They have the power to alter inhibitory neurotransmission by altering GABA receptor activation. Increased lipid peroxidation and DNA damage in brain tissue, along with neuronal hyperexcitability, seizures, and other neurological symptoms, can result from the disruption of GABAergic signaling. Estradiol can cause anxiety disorders (Li & Graham, 2017), and dicofol can affect sensory, motor, or cognitive functions (Duffard, 1996; Seralini & Jungers, 2021). Certain pesticides, such as organophosphates and pyrethroids, have been shown to cause oxidative stress in the nervous system. Oxidative stress can accelerate the onset of neurodegenerative illnesses, alter cell function, and harm neurons (Gargouri et al., 2018; Pearson & Patel, 2016; Yang et al., 2020).

Pesticides are among the many substances that have been found to be endocrine disruptors. Due to their high ability to bind to estrogen or androgen receptors, endocrine disrupting chemicals (EDCs) primarily work by interfering with natural hormones. Specifically, EDCs can imitate the natural hormone's effect (agonist action) by binding to and activating a variety of hormone receptors (AR, ER, AhR, PXR, CAR, and ERR). Additionally, EDCs may attach to these receptors without causing them to become active. The receptors are blocked and their function is inhibited by this antagonist effect. Lastly, EDCs may also reduce the concentration of endogenous hormones by interfering with hormone production, transport, metabolism, and excretion. According to a study by Mnif et al. (2007), effects associated with endocrine disruption have been observed mostly in fish, birds, mammals, reptiles, and invertebrates. The majority of them impact reproductive function and are associated with exposure to organochlorine insecticides (OC). Endocrine disruptor pesticides have also been demonstrated to impair sexual and reproductive development in humans; the effects of these pesticides appear to be contingent on a number of variables, including age, gender, occupation, and food (Mnif et al., 2011).

5.2 Carcinogenesis

Worldwide, the most common cause of cancer is direct contact with chemicals. Exposure to pesticides is significantly linked to DNA damage and mutations that facilitate carcinogenesis via several pathways. Numerous pesticides produce reactive oxygen species (ROS), resulting in oxidative damage such as 8-oxo-deoxyguanosine, which induces mispairing and point mutations (Kaur & Kaur, 2018). Certain pesticides or their metabolites create DNA adducts and substantial cross-links, beyond the capacity of base excision repair (BER) and nucleotide excision repair (NER) mechanisms (Kaur & Kaur, 2024). Data from comet and cytogenetic assays indicate that pesticides cause single- and double-strand breaks, leading to chromosomal rearrangements (Varsha et al., 2020). Polymorphisms in DNA repair genes, including XRCC1, OGG1, XPF, XPG, and ERCC1, enhance vulnerability by diminishing repair efficacy (Siddique et al., 2017). Pesticide exposure induces epigenetic alterations that silence tumor suppressor genes and activate oncogenes, hence initiate cancer and its progression (Ferreira et al., 2022).

5.2.1 Breast Cancer

There is growing scientific evidence that exposure pesticides, is associated with a higher risk of breast cancer. Pesticides have been shown to increase malignancy and DNA mutation in susceptible people by interfering with estrogen receptors or causing damage to breast tissue's DNA. Because they can have estrogenic effects on mammary cells, a number of OCs have been linked to breast cancer on its own. Pesticide residues are detected in milk samples for example heptachlor epoxide, chlordane, dichlorodiphenyltrichloroethane (DDT), and heptachlor. In a nested case–control research, Cohn et al. (2007) examined samples of females in Oakland, California, and found that higher blood p,p′-DDT concentrations were associated with an increased risk of breast cancer. Exposure to pesticides has been demonstrated to promote apoptosis, genetic damage, and mutations that facilitate the development of breast cancer. This exposure induces mutations in DNA damage response (DDR) and homologous recombination repair (HRR) genes, such as BRCA1 and ATM, hence enhancing genomic instability (Scandolara et al., 2022). Pesticides may also modify epigenetic regulation and obstruct normal apoptotic pathways, facilitating the malignant transformation of breast epithelial cells (Kaur & Kaur, 2018). The interaction among oxidative stress, DNA repair deficiencies, and apoptotic dysregulation is fundamental to pesticide-related breast carcinogenesis.

5.2.2 Bladder and Colon Cancers

Numerous cases of bladder and colon cancer have been linked to heterocyclic aromatic amines. A prospective cohort study of pesticide applicators in the United States found that imidazoline herbicides, such as imazaquin and imazethapyr, had a role in bladder cancer (Koutros et al., 2016). Furthermore, exposure to pesticides was associated with an OR of 1.68 and a 95 percent confidence interval (CI) of 1.23–2.29% for bladder cancer in a case–control study of male agricultural laborers in Egypt, which included 881 controls and 953 cases (Ahmad et al., 2024). Chronic exposure to organophosphates and carbamates in colon cells induces ROS-mediated apoptosis and mitochondrial dysfunction, facilitating malignant transformation (Mostafalou & Abdollahi, 2017). In the bladder tissue, pesticide metabolites can create DNA adducts and obstruct normal apoptotic signaling, hence increasing cancer risk (Alavanja & Bonner, 2012a, 2012b).

5.2.3 Liver Cancer

The study conducted by Ahmad et al. (2024) reported that long-term exposure to a number of pesticides may increase the risk of liver cancer. Certain pesticides, including several fungicides and herbicides, have been demonstrated to possess genotoxic properties, meaning they may damage DNA and may result in malignant changes in liver cells. Among the specific pesticides that have been connected to liver cancer are organochlorine pesticides. Research has examined the connection between occupational exposure to particular pesticides and the risk of liver cancer. Although studies have linked occupational pesticide exposure to an increased risk of liver cancer, further research is required to establish a clear cause-and-effect relationship.

5.3 *Developmental and Reproductive Toxicity*

5.3.1 Fertility

The ability of a woman to conceive within a year is referred to as fertility, and it includes both male and female characteristics such as sperm quality, ova maturity etc. A connection is found between sperm sex aneuploidies and organophosphate metabolites, while another study found a connection between pesticide exposure and erectile dysfunction (Ahmad et al., 2024). Women who handled herbicides two years before attempting to conceive were more likely to experience infertility. Consuming fruits and vegetables with high levels of pesticide residues decreased the volume of total ejaculate, sperm count, and percentage of morphologically normal sperm in men attending a reproductive clinic. Sperm morphology was correlated with the intake of low-to-moderate pesticide residue. These findings show that exposure to agricultural pesticides through the food may be sufficient to change human spermatogenesis (Chiu et al., 2015). DNA damage in ovarian and sperm cells, raises the chance of mutation and reduces fertility. Pesticides cause oxidative stress in sperms, which can result in chromatin instability, DNA breakage, and mutations that can be passed on to the progeny (Meeker & Stapleton, 2010). Double-strand breaks and impaired DNA repair are observed in ovarian cells exposed to organophosphate and organochlorine pesticides, which leads to apoptosis and lower oocyte quality (Guillette & Edwards, 2005). Additionally, epigenetic changes in reproductive cells brought on by prolonged exposure can impact gene expression and reproductive health (Kaur & Kaur, 2018).

5.3.2 Birth Defects

Neural tube defects (NTDs) are the most common and severe congenital conditions in fetuses. One study in China reported that 119 women with pregnancies were afflicted by NTD following the exposure to organochlorine pesticides as compared to women with healthy newborns. Organochlorine pesticides (OCPs) concentrations in the umbilical cord tissue were employed as indicators of prenatal exposure in order to investigate the relationship between in utero exposure to OCPs and the risk of NTDs (Yin et al., 2021). Exposure to pesticides has been repeatedly linked to an increased risk of various complications, such as ocular abnormalities, limb diseases, orofacial clefts, and urogenital abnormalities as particular issues. The probability of any birth defect was likewise increased by parental pesticide exposure (Ahmad et al., 2024).

5.3.3 Fetal Death

Effects of pesticide exposure include fetal death, neonatal death, stillbirth, and natural abortion (El-Baz et al., 2023). Ahmad et al. (2024) summarized that nine out of eleven trials showed positive relationships with pesticide exposure, and the results were consistent across several study designs. The Ontario Farm Study's results indicated crucial periods when pesticide exposure is the most harmful. Preconception exposure was associated with early first-trimester abortions, but post-conception exposure was associated with late spontaneous abortions (Arbuckle et al., 2001). According to a Philippine study, farming households that used a lot of pesticides had a six-fold higher chance of spontaneous abortion than those that employed integrated pest management (Crisostomo & Molina, 2002).

5.4 Immunotoxic Consequences

When exposed to chemicals, such as pesticides, the immune system is more susceptible. Xenobiotics such as pesticides can cause, exacerbate, or initiate pathological immune processes. Gene alterations in immunoregulatory factors and modifications to immunological tolerance pathways are examples of immunotoxicity processes. Experimental and modest human studies have demonstrated that pesticides can negatively impact immune function (Ahmed et al., 2016; Banerjee et al., 2001). However, epidemiological evidence is still insufficient to establish definite links to immunotoxic risk. Epidemiological data on pesticide-induced immunotoxicity still heavily rely on in vitro and animal studies (Corsini et al., 2013).

6 The Genotoxic Mechanism

Nonetheless, a wealth of data indicates that a large number of pesticides have genotoxic effects, which means they harm the cells' genetic material and cause chromosomal abnormalities, mutations, and perhaps cancer. The production of free radicals and oxidative stress is one of the main processes behind pesticide-induced genotoxicity.

6.1 Oxidative Stress as a Genotoxic Mechanism

When the generation of reactive oxygen species (ROS) and reactive nitrogen species (RNS) surpasses the cell's antioxidant defense capabilities, oxidative stress results.

Pesticides can cause excessive ROS generation by interfering with cytochrome P450 enzyme activity, mitochondrial function, and cellular redox equilibrium (Banerjee et al., 2001).

Increased ROS levels damage DNA bases, break strands, and cause proteins to cross-link, which eventually results in chromosomal instability and mutations (Kaur & Kaur, 2018).

6.2 Pesticide-Induced Free Radical Generation

During metabolism, a lot of pesticides go through redox cycling, producing hydrogen peroxide (H_2O_2), hydroxyl radicals (•OH), and superoxide anions (O_2•−).

It has been demonstrated that pyrethroids, carbamates, and organophosphates induce oxidative DNA lesions like 8-hydroxy-2′-deoxyguanosine (8-OHdG) and lipid peroxidation (Pilger & Rüdiger, 2006).

These extremists lead to:

- Base changes, such as guanine oxidation.
- DNA breaking into single and double strands.
- Sister chromatid exchange and micronuclei production (Bolognesi & Morasso, 2000).

6.3 Disruption of Antioxidant Systems

According to López et al. (2007), pesticides weaken the antioxidant defenses of the cells, including reduced glutathione (GSH), glutathione peroxidase (GPx), catalase (CAT), and superoxide dismutase (SOD). By increasing DNA adducts and chromosomal abnormalities, this imbalance amplifies ROS-mediated genotoxicity.

6.4 Downstream Genetic Consequences

- Damage from oxidative DNA can result in
- tumor suppressor gene mutations (p53, BRCA1).
- Proto-oncogene activation (H-ras, K-ras, N-ras, HER2).
- Epigenetic changes, like changed methylation. According to Mostafalou and Abdollahi (2017), prolonged exposure is associated with a higher risk of cancer, neurological illnesses, and reproductive damage.
- DNA modification as a genotoxic biomarker

6.5 DNA Alteration as Genotoxicity Biomarkers

The ability of chemical, physical, or biological substances to harm the genetic material and perhaps cause mutations, chromosomal abnormalities, and cancer is known as genotoxicity. Assessing the dangers to occupational health and the environment requires the detection of such changes. Since they directly represent the chemical interactions between the toxicants and genetic material, DNA changes are regarded as trustworthy biomarkers of genotoxicity among the various endpoints (Albertini, 2000).

6.5.1 Types of DNA Alterations as Biomarkers

1. DNA Strands Break

Single-strand breaks (SSBs) and double-strand breaks (DSBs) can be caused by pesticides and other xenobiotics. Strand breaks in human lymphocytes and other cells are commonly measured using the Comet test, DNA fragmentation etc. (Ahmed et al., 2021).

2. DNA Adducts

It is created when DNA bases covalently bond to reactive pesticide metabolites. Mutagenesis results from DNA adduct interfering with transcription and replication. Various techniques like LC–MS/MS and ^{32}P-postlabeling are used to obtain detection (Phillips & Arlt, 2007).

3. Lesions in Oxidative DNA

Pesticide-generated reactive oxygen species (ROS) lead to oxidative alterations like 8-hydroxy-2′-deoxyguanosine (8-OHdG). Its mutagenic potential and genotoxic stress are closely linked to elevated 8-OHdG levels (Valavanidis et al., 2009).

4. Formation of Micronuclei

Acentric chromosomal fragments or entire chromosomes that are removed from daughter nuclei during cell division are found in the micronuclei. In human biomonitoring research, the micronucleus assay is a proven biomarker of chromosomal damage (Fenech, 2007).

5. Aberrations in the Chromosome

Genotoxic effects are reflected in structural alterations such translocations, deletions, or aneuploidy. Common cytogenetic tests include fluorescence in situ hybridization (FISH) and karyotyping (Gonzales et al., 2016).

6. Mutations in Genes

Tumor suppressor genes like p53 are susceptible to deletions, frameshifts, and point mutations brought on by pesticides. Toxicogenomic makes extensive use of PCR-based mutation identification and sequencing (Caba & Aubrecht, 2008).

6.5.2 Limitations of DNA Alterations

Following are the limitations of DNA alterations:

- Damage to DNA may be repaired, which lowers the stability of biomarkers.
- Results may be affected by inter-individual variability, including age, diet, lifestyle, and genetic background.
- For tests to be reproducible across trials, they must be standardized.
- It is generally acknowledged that oxidative stress is linked to a number of mechanisms that might accelerate DNA breaks and cause genomic instability. These include harmful lifestyle choices, smoking, body composition abnormalities, and a family history of cancer. Assessing the degree of genetic instability in smokers and non-smokers, as well as how the risk may vary based on lifestyle and other factors, was the goal.
- Various techniques have been employed to assess genetic instability or damage. Single-cell gel electrophoresis, DNA fragmentation etc. are used to assess genomic instability in peripheral blood mononuclear cells under the distinct oxidative stress scenarios (Ahmed et al., 2016, 2018, 2021; Locken Castilla et al., 2019).

7 Pesticides and Genetic Damage

7.1 Oxidative DNA Damage and Apoptosis

Pesticide labels also advise limiting exposure to these dangerous substances. The development of oxidative stress in diverse cell types and animal models appears to be a common impact of different pesticide kinds, according to experimental findings. Both reactive oxygen species (ROS) and reactive nitrogen species (RNS), which are linked to a number of illnesses such as cancer, inflammation, and cardiovascular and neurological conditions, are responsible for pesticide-induced oxidative stress. At least five distinct signaling cascades, including mitochondrial-induced apoptosis, can be triggered by ROS and RNS (Fig. 1). Exogenous antioxidants may help lessen or stop the harmful effects of pesticides (Sule et al., 2022).

8 Antioxidants and Defense Mechanisms

The creation of oxidative stress is one of the chemical processes via which pesticides impact living things. However, because of their capacity to eliminate free radicals, a number of plant-based dietary components with antioxidant qualities have recently been regarded as chemo-preventive agents. Such a food ingredient needs to be able

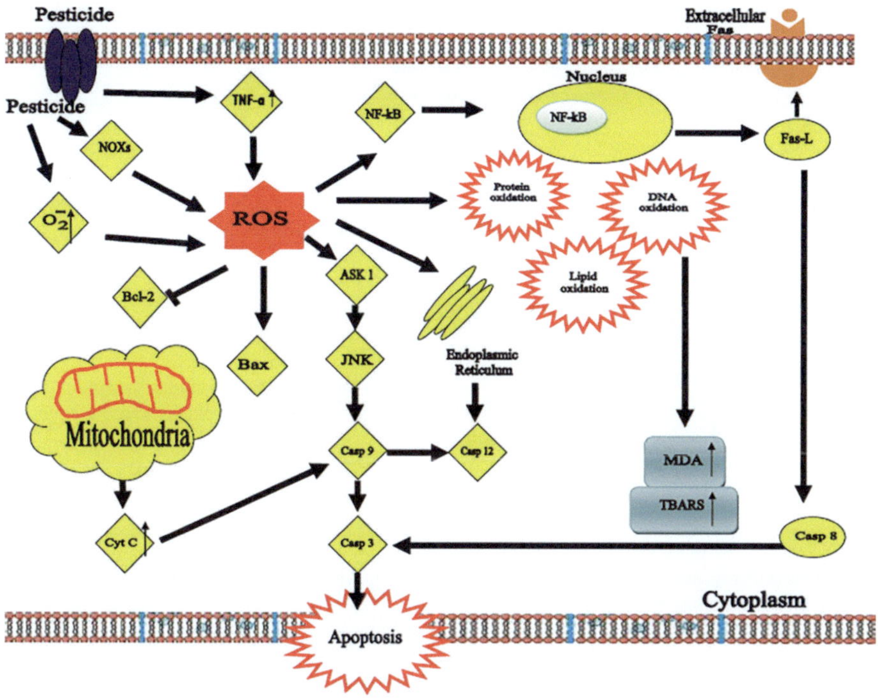

Fig. 1 Mechanism of action in DNA damage and apoptosis of cells caused by pesticides that overwhelm antioxidant defenses

to remove free radicals, be readily absorbed, and operate at the proper physiological level, among other requirements. Its primary purpose is to decrease ROS-induced cellular damage and preserve the redox balance. As a result, it ought to function in membrane domains and aqueous solutions. Certain plant hormones and phenolic chemicals have these characteristics. People have long used medicinal plants as a traditional means of treating a wide range of illnesses. There is still a need for strong and efficient analgesic medications, even with notable advancements in therapeutic development. In this regard, it has been extensively reported that a variety of substances originating from plants play a crucial role in the process of creating innovative methods to cure a range of illnesses. The pharmacological evaluation of many plants such as curcumin used in traditional medicinal systems has advanced significantly as a result of recent study on herbal plants or medicine (Ahmed et al., 2016). Alkaloids, flavonoids, terpenoids, and tannins are examples of secondary metabolites or compounds that are present in medicinal plants and that influence the plant's therapeutic efficacy, especially its antioxidant capacity. Secondary metabolites interact with proteins, nucleic acids, and other biomembranes and are active and volatile targets of cells. They are tiny molecular weight compounds with a wide range of biological properties. Nutraceuticals and functional meals are also made with medicinal herbs. These goods offer more health benefits than just basic

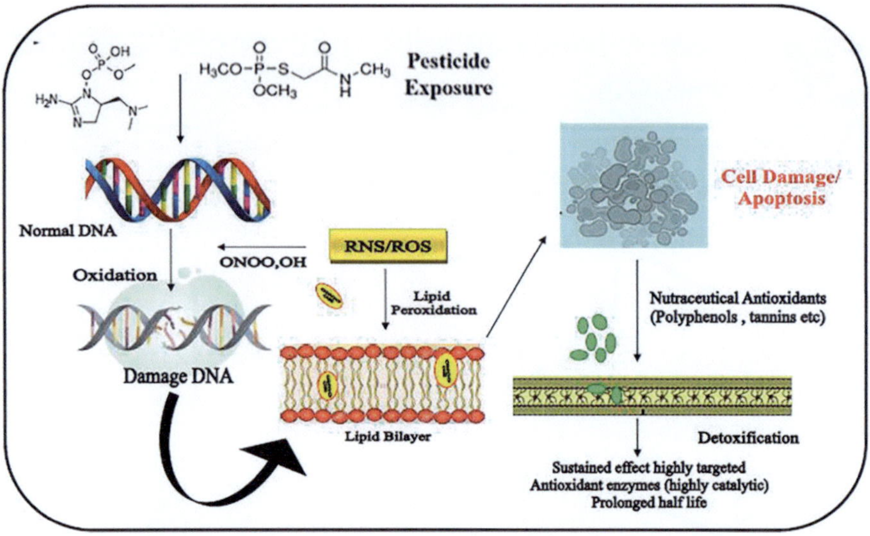

Fig. 2 Detoxifying, anti-inflammatory, and antioxidant qualities of natural compounds over coming negative effects of some pesticides

sustenance. They are made with specific plant extracts or bioactive compounds to support several aspects of health, including digestive wellness, cognitive function, and cardiovascular health. The possible protective effects of several nutraceuticals against pesticide-induced toxicities have been investigated. These naturally occurring substances that come from food sources may have detoxifying, anti-inflammatory, and antioxidant qualities that could lessen the negative effects of some pesticides (Sajad et al., 2024) (Fig. 2).

9 Conclusion

The mutations produced by xenobiotics pose a serious threat to genomic integrity as they can result in genetic instability, cancer, reproductive problems, and many other long-term health issues. The production of free radicals, oxidative stress, and DNA adducts are the main processes behind this damage, as they weaken DNA repair and increase the mutation rates. The urgent need for mitigation therapy is highlighted by the fact that these molecular changes result in real health concerns for the general public as well as for affected groups in the workplace.

Endogenous and dietary antioxidants are essential for preventing oxidative damage caused by xenobiotics. Their potential as therapeutic and preventative agents is highlighted by their ability to strengthen cellular defenses and neutralize free radicals. A comprehensive strategy for reducing genetic damage and maintaining long-term health may include incorporating a diet rich in antioxidants and investigating pharmaceutical therapies. Discovering new biomarkers for rapid detection of xenobiotic-induced mutations and investigating genetic susceptibility factors that put certain people at higher risk should be key goals of future research. Furthermore, developments in omics and molecular toxicology can help us understand the gene–environment relationship more deeply. Stricter laws governing xenobiotic exposure, better risk assessment strategies, and the promotion of sustainable alternatives with lower toxicity are crucial from a policy perspective. Ultimately, reducing the burden of xenobiological-induced damage and guaranteeing a better future will depend largely on aligning scientific research with public health regulations.

References

Ahmad, M. F., et al. (2024). Pesticides impacts on human health and the environment with their mechanisms of action and possible countermeasures. *Heliyon, 10*(7), Article e29128.

Ahmed, T., & Banerjee, B. D. (2021). Permethrin induced cytotoxicity of rat splenocytes: Protective effect of N-acetylcysteine. *Indian Journal of Experimental Biology, 59*, 11–18.

Ahmed, T., Goel, V., & Banerjee, B. D. (2018). Propoxur-induced oxidative DNA damage in human peripheral blood mononuclear cells: Protective effects of curcumin and α-tocopherol. *Drug and Chemical Toxicology, 41*(2), 128–134.

Ahmed, T., & Banerjee, B. D. (2016). HSP27 modulates survival signaling in endosulfan-exposed human peripheral blood mononuclear cells treated with curcumin. *Human & Experimental Toxicology, 35*(7), 695–704.

Aktar, W., Sengupta, D., & Chowdhury, A. (2009). Impact of pesticides use in agriculture: Their benefits and hazards. *Interdisciplinary Toxicology, 2*(1), 1–12.

Alavanja, M. C. R., & Bonner, M. R. (2012a). Occupational pesticide exposures and cancer risk: A review. *Journal of Toxicology and Environmental Health, Part B, 15*(4), 238–263.

Alavanja, M. C. R., & Bonner, M. R. (2012b). Pesticides and human cancers. *Cancer Investigation, 30*(2), 243–250.

Albertini, R. J. (2001). HPRT mutations in humans: Biomarkers for mechanistic studies. *Mutation Research, 489*, 1–16.

Arbuckle, T. E., Lin, Z., & Mery, L. S. (2001). An exploratory analysis of the effect of pesticide exposure on the risk of spontaneous abortion in an Ontario farm population. *Environmental Health Perspectives, 109*(8), 851–857.

Banerjee, B. D, Seth, V., & Ahmed, R. S. (2001) Pesticide-induced oxidative stress: perspectives and trends. *Reviews on Environmental Health, 16*(1), 1–40.

Bolognesi, C. (2003). Genotoxicity of pesticides: A review of human biomonitoring studies. *Mutation Research/Reviews in Mutation Research, 543*(3), 251–272.

Bolognesi, C., & Morasso, G. (2000). Genotoxicity of pesticides: Potential risk for consumers. *Trends in Food Science & Technology, 11*(4–5), 182–187.

Caba, E., & Aubrecht, J. (2008). *Genomic approaches for investigating mechanisms of genotoxicity* (pp.69–70).

Chiu, Y. H., et al. (2015). Fruit and vegetable intake and their pesticide residues in relation to semen quality among men from a fertility clinic. *Human Reproduction, 30*(6), 1342–1351.

Cohn, B. A., et al. (2007). DDT and breast cancer in young women: New data on the significance of age at exposure. *Environmental Health Perspectives, 115*(10), 1406–1414.

Colovic, M. B., et al. (2013). Acetylcholinesterase inhibitors: Pharmacology and toxicology. *Current Neuropharmacology, 11*(3), 315–335.

Corsini, E., et al. (2013). Pesticide induced immunotoxicity in humans: A comprehensive review of the existing evidence. *Toxicology, 307*, 123–135.
Costa, L. G., et al. (2014). Neurotoxicity of pesticides: A brief review. *Frontiers in Bioscience, 6*, 270–289.
Crisostomo, L., & Molina, V. V. (2002). Pregnancy outcomes among farming households of Nueva Ecija with conventional pesticide use versus integrated pest management. *International Journal of Occupational and Environmental Health, 8*(3), 232–242.
Ecobichon, D. J. (2001). Pesticide use in developing countries. *Toxicology, 160*(1–3), 27–33.
El-Baz, M. A., Amin, A. F., & Mohany, K. M. (2023). Exposure to pesticide components causes recurrent pregnancy loss by increasing placental oxidative stress and apoptosis: A case–control study. *Scientific Reports, 13*(1), 9147.
EPA. (2020). *What is a pesticide?* United States Environmental Protection Agency.
FAO. (2018). *International code of conduct on pesticide management*. Food and Agriculture Organization of the United Nations.
FAO. (2021). *FAOSTAT: Pesticides use data.* Food and Agriculture Organization of the United Nations.
Fenech, M. (2007). Cytokinesis-block micronucleus cytome assay. https://doi.org/10.1038/nprot.2007.77
Fenner, K., et al. (2013). Evaluating pesticide degradation in the environment: Blind spots and emerging opportunities. *Science, 341*(6147), 752–758.
Fenske, R. A., Lu, C., Barr, D., & Needham, L. (2002). Children's exposure to chlorpyrifos and parathion in an agricultural community in central Washington State. *Environmental Health Perspectives, 110*(5), 549–553.
Ferreira, J. D., et al. (2022). Somatic DNA damage response and homologous repair gene alterations in breast cancer patients with occupational exposure to pesticides. *Frontiers in Oncology, 12*, Article 892984.
Gargouri, B., et al. (2018). Pyrethroid bifenthrin induces oxidative stress, neuroinflammation, and neuronal damage, associated with cognitive and memory impairment in murine hippocampus. *Neurochemistry International, 120*, 121–133.
Gonzales, P. R., Carroll, A. J., & Korf, B. R. (2016). Overview of clinical cytogenetics. *Current Protocols in Human Genetics, 89*, 8.1.1–8.1.13.
Guillette, L. J., & Edwards, T. M. (2005). Exposure to environmental contaminants and reproductive dysfunction in wildlife and humans. *Reproductive Toxicology, 20*(2), 279–288.
Harnly, M. E., et al. (2005). Pesticides in dust from homes in an agricultural area. *Environmental Science & Technology, 39*(23), 8253–8261.
Hernández, A. F., Parrón, T., & Alarcón, R. (2011). Pesticides and asthma. *Current Opinion in Allergy and Clinical Immunology, 11*(2), 90–96.
Jensen, B. H., et al. (2012). Pesticide residues in fruits and vegetables from conventional and organic production. *Food Additives & Contaminants: Part B, 5*(3), 182–188.
Kaur, R., & Kaur, S. (2018). Occupational pesticide exposure, impaired DNA repair, and diseases. *International Journal of Occupational and Environmental Health, 24*(3–4), 207–214.
Kaur, R., & Kaur, S. (2024). Polymorphisms in DNA repair genes as biomarkers of susceptibility for pesticide-induced DNA damage among agricultural workers. *Mutation Research/genetic Toxicology and Environmental Mutagenesis, 895*, 504–512.
Kim, K. H., Kabir, E., & Jahan, S. A. (2017). Exposure to pesticides and the associated human health effects. *Science of the Total Environment, 575*, 525–535.
Koutros, S., et al. (2016). Occupational exposure to pesticides and bladder cancer risk. *International Journal of Epidemiology, 45*(3), 792–805.
Kumar, S. L., Margekar, P., & Margekar, V. (2018). Recent advances in management of organophosphate & carbamate poisoning. *Indian Journal of Medical Specialities, 9*(3), 154–159.
Li, S. H., & Graham, B. M. (2017). Why are women so vulnerable to anxiety, trauma-related and stress-related disorders? The potential role of sex hormones. *The Lancet Psychiatry, 4*(1), 73–82.

Li, Y. F., et al. (2016). The transport of persistent organic pollutants to the Arctic and their fate. *Science of the Total Environment, 342*(1–3), 5–86.

Locken-Castilla, A., et al. (2019). Smoking index, lifestyle factors, and genomic instability assessed by single-cell gel electrophoresis: A cross-sectional study in subjects from Yucatan, Mexico. *Clinical Epigenetics, 11*(1), 150.

López, O., et al. (2007). Changes in antioxidant enzymes in humans with long-term exposure to pesticides. *Toxicology Letters, 171*(3), 146–153.

Meeker, J. D., & Stapleton, H. M. (2010). House dust concentrations of organophosphate flame retardants in relation to hormone levels and semen quality parameters. *Environmental Health Perspectives, 118*(3), 318–323.

Mnif, W., Pillon, A., Balaguer, P., & Bartegi, A. (2007). Endocrine xenoestrogenic disrupters, molecular mechanisms and detection methods. *Thérapie, 62*, 369–386.

Mnif, W., et al. (2011). Effect of endocrine disruptor pesticides: A review. *International Journal of Environmental Research and Public Health, 8*(6), 2265–2303.

Mostafalou, S., & Abdollahi, M. (2017). Pesticides: An update of human exposure and toxicity. *Archives of Toxicology, 91*(2), 549–599.

Oerke, E. C. (2006). Crop losses to pests. *Journal of Agricultural Science, 144*(1), 31–43.

Pearson, J. N., & Patel, M. (2016). The role of oxidative stress in organophosphate and nerve agent toxicity. *Annals of the New York Academy of Sciences, 1378*(1), 17–24.

Phillips, D. H., & Arlt, V. M. (2007). The 32P-postlabeling assay for DNA adducts. *Nature Protocols, 2*(11), 2772–2781.

Pilger, A., & Rüdiger, H. W. (2006). 8-Hydroxy-2′-deoxyguanosine as a marker of oxidative DNA damage related to occupational and environmental exposures. *International Archives of Occupational and Environmental Health, 80*(1), 1–15.

Sajad, M., et al. (2024). Role of nutraceutical against exposure to pesticide residues: Power of bioactive compounds. *Frontiers in Nutrition, 11*, 1342881.

Scandolara, T. B. et al. (2022). Somatic DNA damage response and homologous repair gene alterations and its association with tumor variant burden in breast cancer patients with occupational exposure to pesticides. *Frontiers in Oncology*, 12:1–13

Schreinemachers, P., & Tipraqsa, P. (2012). Agricultural pesticides and land use intensification in high, middle and low income countries. *Food Policy, 37*(6), 616–626.

Seralini, G., & Jungers, G. (2021). Endocrine disruptors also function as nervous disruptors and can be renamed endocrine and nervous disruptors (ENDs). *Toxicology Reports, 8*, 1538–1557.

Sharma, A., et al. (2019). Worldwide pesticide usage and its impacts on ecosystem. *SN Applied Sciences, 1*(11), 1446.

Siddique, T., Banerjee, B. D., & Ray, M. R. (2017). GSTP1 and XRCC1 polymorphisms and DNA damage in agricultural workers exposed to pesticides. *Environmental Toxicology and Pharmacology, 54*, 217–224.

Stehle, S., & Schulz, R. (2015). Agricultural insecticides threaten surface waters at the global scale. *Proceedings of the National Academy of Sciences, 112*(18), 5750–5755.

Sule, R. O., et al. (2022). A common feature of pesticides: Oxidative stress—the role of oxidative stress in pesticide-induced toxicity. *Oxidative Medicine and Cellular Longevity, 2022*(1), 5563759. https://doi.org/10.1155/2022/5563759

Valavanidis, A., Vlachogianni, T., & Fiotakis, C. (2009). 8-hydroxy-2′-deoxyguanosine (8-OHdG): A critical biomarker of oxidative stress and carcinogenesis. *Journal of Environmental Science and Health. Part c, Environmental Carcinogenesis & Ecotoxicology Reviews, 27*(2), 120–139.

Varsha, H., Kaur, R., & Kaur, S. (2020). Modulation of DNA damage by XPF, XPG and ERCC1 gene polymorphisms in pesticide-exposed agricultural workers of Punjab, North-West India. *Mutation Research/genetic Toxicology and Environmental Mutagenesis, 861–862*, Article 503257.

Walker, C. H. et al. (2012). *Principles of ecotoxicology* (4th ed.). CRC Press.

WHO. (2020). *Pesticide residues in food: 2018 evaluations.* World Health Organization & FAO.

Yang, C., Lim, W., & Song, G. (2020). Mediation of oxidative stress toxicity induced by pyrethroid pesticides in fish. *Comparative Biochemistry and Physiology Part C: Toxicology & Pharmacology, 234*, Article 108758.

Yin, S., Zhang, Y., Zhang, X., & Wang, Y. (2021). Prenatal exposure to organochlorine pesticides is associated with increased risk for neural tube defects. *Science of the Total Environment, 784*, Article 147197.

Zhang, W., Jiang, F., & Ou, J. (2011). Global pesticide consumption and pollution: With China as a focus. *Proceedings of the International Academy of Ecology and Environmental Sciences, 1*(2), 125–144.

Chapter 5
DNA Adduct Mediated Carcinogenesis: The Consequences of Xenobiotic Bioactivation

Savita Bansal, Meenakshi Vachher, Taruna Arora, Archana Burman, Radhika Bakhshi, and Sandeep Yadav

Abstract Increasing concentrations of hazardous chemicals in the environment and diet has become a major health concern due to their association with genomic mutation and onset of cancer. Many chemicals or xenobiotics require metabolic activation by various enzymes which are classified into phase I and phase II. These metabolic reactions of xenobiotics generate electrophiles and a free radical that can interact covalently with DNA resulting in DNA adducts formation. These DNA lesions are genotoxic and have a potential to induce DNA damage. Some conjugation reactions generate reactive intermediates that can interact with DNA to form DNA adducts catalyzed by phase II enzymes involving acetylation or sulfation reactions of aromatic and heteroaromatic hydroxyl-amines and glutathione-S conjugation of haloalkanes. DNA adducts drive mutations in genes involved in cancer suppression, oncogenes, and apoptosis. DNA adducts formed by xenobiotics or their reactive metabolites can occur at various sites on the DNA, including ring nitrogen atoms, exocyclic nitrogen atoms, and oxygen atoms of the nucleobases. The formation and type of DNA adduct depends on the structure of reactive metabolites, nature of electrophiles, ability of the compound to intercalate within the DNA, and their interaction with nucleophilic sites on the DNA. The DNA adducts drive the mutations by site-specific mutagenesis and have the ability to interfere with the process of DNA transcription and replication. DNA adduct can also result in point mutations, deletions, or insertions by mispairing during DNA replication. Understanding the role of DNA adducts in cancer pathology can provide insights into both carcinogenesis and potential therapeutic interventions. Also, DNA adducts can be used as potential biomarkers to predict, prevent, and diagnose the diseases. This chapter will focus on the identification, formation, and measurement of DNA adducts and their association with the development and progression of cancers as well as described the impact of genotoxic chemicals in cancer risk.

S. Bansal · M. Vachher · T. Arora · A. Burman · S. Yadav (✉)
Department of Biochemistry, Institute of Home Economics, University of Delhi, New Delhi, India
e-mail: sandeep.yadav@ihe.du.ac.in

R. Bakhshi
Institute of Home Economics, University of Delhi, New Delhi, India

K. Priya and S. K. Giri (eds.), *Environmental Toxicology: Xenobiotics and Effects on Human Genetic Material*, https://doi.org/10.1007/978-3-032-28619-2_5

Keywords DNA adduct · Carcinogenesis · DNA adduct omics · Xenobiotics · Biomarkers

1 Introduction

Continuous exposure to the hazardous chemicals generated either from environmental sources, dietary sources, as well as to endogenously generated electrophiles can form covalent adducts with DNA (Hwa Yun et al., 2020; Wogan et al., 2004). The formation of DNA adducts represents one of the earliest and most decisive molecular events in chemical carcinogenesis, serving as a direct mechanistic link between exposure to genotoxic agents and the initiation of malignant transformation (Cao & Zhang, 2024). The term "DNA adductome" was given by Kanaly et al., in 2006 and refers to the entirety of DNA adducts present in a given biological target including cells, tissues, or whole organisms. Many chemicals or xenobiotics require metabolic activation by various enzymes which are classified into phase I and phase II (Guengerich, 2008; van Vugt-Lussenburg et al., 2022). These metabolic reactions of xenobiotics generate electrophiles and free radicals that have the ability to interact with DNA covalently to form DNA adducts (Chandimali et al., 2025; Rendic & Guengerich, 2012). These oxidized DNA lesions are genotoxic and have a potential to induce DNA damage. Some conjugation reactions generate reactive intermediates that can interact with DNA to form DNA adducts catalyzed by phase II enzymes such as acetylation or sulfation of aromatic and heteroaromatic hydroxyl-amines and glutathione-S conjugation of haloalkanes (Miller et al., 1991; Minchin et al., 1992; Thier et al., 1996). Fortunately, the majority of DNA adducts are efficiently eliminated by cellular DNA repair systems. However, a small proportion may evade repair and, during cell division, give rise to mutations that can drive cells toward a cancerous state (Sancar et al., 2004; Schärer, 2003). DNA adducts that drive the mutations in cancer-associated genes, including oncogenes such as H-Ras and K-Ras and the tumor suppressor gene p53, have drawn major attention (Greenman et al., 2007; Stratton et al., 2009). Such mutations are associated with carcinogenesis in experimental models and in human cancers due to environmental exposures (Fournier & Thomas, 1984; Loeb & Harris, 2008; Rebhan et al., 2020).

DNA adducts formed by xenobiotics or their reactive metabolites can be generated at various sites on the DNA, including ring nitrogen atoms, exocyclic nitrogen atoms, and oxygen atoms of the nucleobases. Also, certain chemical carcinogens are capable of forming adducts with the phosphate backbone of the DNA (Gaskell et al., 2007; Turesky & Le Marchand, 2011). The formation and type of DNA adducts depend on the structural properties of reactive metabolites, nature of electrophiles, intercalating properties of the compound within the DNA, and their interaction with nucleophilic sites on the DNA (Cavalieri et al., 2012; Shrivastav et al., 2010). The DNA adducts drive the mutations by site-specific mutagenesis and have the ability to interfere with the process of DNA transcription and replication (Cai et al., 2012). The causality of

DNA adducts in cancer development lies in their potential to cause genomic instability and interference with the transcription process and disruption of normal cellular functions. DNA adduct can also result in point mutations, deletions, or insertions by mispairing during DNA replication (Conney et al., 2001; Seo et al., 2005; Wei et al., 1994). In the last decade, association of DNA adducts with the development and progression of various types of cancer such as lung tumors, hepatocellular carcinoma, urothelial carcinomas, kidney cancer, breast cancer, and many more has been explored. Understanding the role of DNA adducts in cancer pathology can provide insights into both carcinogenesis and potential therapeutic interventions.

Numerous chemicals and lifestyle factors have been linked to the development of different types of cancer such as polycyclic aromatic hydrocarbons (PAHs) present in cigarette smoke are the potential risk for lung cancer; aromatic amines are linked to bladder cancer; aristolochic acids are associated with upper urothelial cancer; and aflatoxin B1, a mycotoxin is a well-established risk factor for liver cancer (Schmeiser et al., 2014; Shirabe et al., 2011; Smith et al., 2000; Xu et al., 2025). Therefore, DNA adducts can be used as potential biomarkers to predict, prevent, and diagnose the diseases and the identification and quantification of DNA adducts represent the essential initial steps in elucidating the role of genotoxic chemicals in cancer etiology.

DNA adduct omics encompasses the comprehensive, global analysis of DNA adducts. DNA adduct omics is primarily applied as a powerful screening tool for the detection of DNA adducts, providing critical evidence of exposure to specific carcinogens and their potential biological consequences (Balbo et al., 2014; Guo & Turesky, 2019). Advances in mass spectrometry, particularly hyphenated liquid chromatography mass spectrometry platforms, have now enabled the sensitive detection of diverse DNA adducts in human biospecimens, greatly enhancing the utility of DNA adduct analysis in molecular epidemiology (Carlsson & Törnqvist, 2017; Chang et al., 2018; Takeshita et al., 2019). DNA adducts are chemically diverse lesions whose biological consequences depend on their structure, genomic location, persistence, and interaction with DNA repair pathways. Advances in analytical chemistry, molecular biology, and high-resolution mass spectrometry have enabled the sensitive detection, structural characterization, and genome-wide mapping of these lesions, giving rise to the emerging field of DNA adduct omics. Although DNA adduct omics is a relatively recent addition to the omics landscape, it has benefited substantially from advances in mass spectrometric technologies, particularly tandem mass spectrometry (MS/MS) and multistage mass spectrometry (Hemeryck et al., 2015; Martella et al., 2023). This chapter provides a comprehensive overview of DNA adducts and their formation, classification and biological significance of adducts, their role in cancer development, and modern approaches for their detection.

1.1 Metabolism and Bioactivation of Xenobiotics

Xenobiotic compounds' metabolism primarily occurs in the liver which by converting their nonpolar nature to a more polar one makes them water soluble so that they are easily excreted through urine and/or bile. The metabolism is primarily composed of three phases namely, functionalization or activation, conjugation, and finally detoxification. The process is performed with the help of two sets of enzymes namely phase I and II which work separately or in conjunction depending on the nature of the xenobiotic (Omiecinski et al., 2011). During the preemptive activation step, certain functional groups like the hydroxyl, sulfhydryl, and amino of the xenobiotic compounds are either exposed or inserted. The molecule is either reduced, oxidized, or hydrolyzed so that a polar water-soluble moiety is achieved. This is done primarily through phase I enzymes like cytochrome P450 (CYP) monooxygenases family comprised of around 57 members and 18 families as observed in humans, flavin-containing monooxygenases (FMOs), and reductases that function through oxidation or reduction at both the activation and detoxification steps. Other enzymes like esterases and amidases also participate in the process by hydrolyzing and cleaving the molecule (Guengerich, 2008; Lewis, 2004).

The altered site created by the phase I enzymes is conjugated to a small polar molecule like glucuronic acid, sulfate, glutathione, amino acids like glycine, methyl and acetyl groups which makes the xenobiotic compound more water soluble and easily excretable with the help of phase II enzymes. The phase II enzymes include Glutathione S-Transferases (GSTs), UDP-Glucuronosyltransferases (UGTs), Sulfotransferases (SULTs), Acyl-CoA:Amino Acid N-Acyltransferases, and N-Acetyltransferases (NATs). Both the set of enzymes reflect significant genetic variation in humans and therefore are responsible for differential efficacy and toxicity for various xenobiotic compounds (Ingelman-Sundberg et al., 2018; Ring et al., 2017). Finally, these water-soluble conjugates are excreted out of the body through urine and/or bile with the help of energy-driven transporters which belong to the ATP-binding cassette (ABC) transporter family. These transporters tend to move the metabolites from several organs like kidney, liver, brain, placenta, blood brain barrier, and intestine wherein the compounds are excreted from the cytosol to the extracellular spaces, urine, or bile and detoxification occurs (How et al., 2025; Mao et al., 2018). Some of the key transporters include MDR1 (Multidrug resistance protein), MRP (Multidrug resistant associated protein), and BCRP (Breast cancer resistance protein) which play an important role in pumping out the metabolites produced by phase 1 and 2 enzymes (Poku & Iram, 2022; Wilkens, 2015). The three components together are associated to work as a Xenobiotic Triumvirate. The xenobiotics metabolism even though is working for their general detoxification by converting them as an inert, water soluble, excretable metabolite, tends to sometimes malfunction and the xenobiotics thus get bioactivated in the form of a chemically reactive intermediate which behaves like a toxin, carcinogen, or a teratogen. The malfunctioning is a consequence of failure of the protective role of both phase I and II sets of enzymes which are responsible for the activation and detoxification of xenobiotics. The phase 1 enzymes primarily CYP get involved in the trap step. CYP was originally supposed to attach a reactive functional group for easier phase II conjugation step, but now it tends to bioactivate and creates an extremely labile and electrophilic intermediate like epoxides, quinones, and free radicles which eagerly react with the cellular components. Some of the examples of bioactivation include BaP (Benzo[a] pyrene) converted to Bay-region epoxides by *CYP1A1*, acetaminophen converted to N-acetyl-p-benzoquinone imine (NAPQI) by *CYP2E1*, aflatoxin converted to aflatoxin B1-8,9-epoxide by *CYP1A2* and *CYP3A4* enzymes. Few of them bind to the DNA resulting in hepatic carcinogenesis while NAPQI causes liver necrosis by binding to hepatic proteins and reducing cellular glutathione (Guengerich, 2008; Hinson et al., 2010; Shiizaki et al., 2017). The phase II enzymes which normally detoxify the phase I byproducts by conjugation tend to fail in their scavenger role leading to bioactivation. This could be due to several reasons like lowered cellular glutathione levels as seen in the case of NAPQI, or reduced expression of these enzymes leading to insufficient activity and intermediate accumulation. These accumulated intermediates now tend to start searching the potential nucleophilic sites and thus initiate binding the critical cellular macromolecules like DNA, proteins, and lipids in a covalent manner hereby starting the process of toxicity in the organism. The net product could be the formation of DNA and protein adducts and lipid peroxides which are even more toxic than the xenobiotics thus paving the way of free radical generation, necrosis, membrane damage, and carcinogenesis (Hughes et al., 2021; Wang et al., 2021).

2 DNA Adducts and Their Formation

Formation of the DNA adduct is one of the deadliest outcomes of bioactivation as these molecules play a pivotal role as a carcinogen. CYP family enzymes, especially *CYP1A1*, *CYP1A2*, and *CYP1B1* bioactivate xenobiotics like polycyclic aromatic hydrocarbons and aflatoxins which are initially lipophilic and inert into reactive electrophile intermediates by the addition of oxygen moiety leading to their hydroxylation, dealkylation, or epoxidation. They actively look for electrons in the nucleophilic sites in the DNA structure, especially nitrogenous bases of DNA like guanine which are modified at N7 or O6 for the formation of covalent bonds finally leading to the formation of the DNA adduct. CYP is known to oxidize chemicals like polycyclic aromatic hydrocarbons (PAH) into diol epoxides which is a recognized carcinogen (Moorthy et al., 2015). The adducts formed are quite site specific as benzo[a] pyrene uniquely targets N2 guanine, while aromatic amines attach to guanine C5 position and aflatoxin binds to guanine N7 position (Shiizaki et al., 2017). Similarly, the phase II enzymes involved in conjugation even though they are performing a protective function sometimes led to de novo activation and formation of DNA reactive species especially when they are slow or polymorphic in nature. The N-hydroxylation of aromatic amines is further processed and highly reactive esters like N-acetoxy or N-sulfonyloxy derivatives are created by phase II enzymes like NATs or SULTs which dissociate as nitrenium ion that are actively involved in DNA adduct formation (Barnes et al., 2018; Hwa Yun et al., 2020; Shimada, 2006).

Single nucleotide polymorphism (SNP) and gene deletion are known to create the imbalance for the phase I and II enzymes leading to increased DNA adduct formation and carcinogenesis. Other key genetic factors include the presence of *GSTM1* null genotype which induces phase II enzyme deficiency (Coric et al., 2016), *CYP1A1* polymorphism responsible for hyper-induced phase I enzymes (Kiyohara et al., 2002), and slow NAT2 acetylators (Lee et al., 2010). Amongst all of them the highest risk combination genotype includes Fast *CYP1A1* + Null *GSTM1* as it induces rapid DNA adduct formation and the least risk genotype includes Normal *CYP1A1* + Present *GSTM1* as DNA adducts that are immediately nullified as they are formed (Autrup, 2000; Poirier, 2004).

3 Classification of DNA Adducts

DNA adducts represent the chemically and biologically diverse DNA lesions. These can be categorized in several ways based on their sources, kind of chemical bond formed, and modified nucleotide site.

3.1 On the Basis of Source

3.1.1 Endogenous DNA Adducts

These are formed as byproducts of normal metabolic processes including the formation of reactive oxygen species (ROS) as part of the metabolism, lipid peroxidation reactions or from natural metabolites like formaldehyde, acetaldehyde, bile acids, etc. (Swenberg et al., 2011). These accrue with aging and add to the associated increased frequency of mutations (Table 1). The most widespread endogenous DNA adduct includes the aldehydic lesions linked to apurinic/apyrimidinic (AP) sites that are generated following the loss of DNA bases as a consequence of oxidative stress and the chemical depurination of alkylated and normal bases. (Swenberg et al., 2011). Other endogenous adducts include oxidative stress generated adducts like 8 oxo-deoxyguanosine (8-oxo-dG); lipid peroxidation-derived adducts like malondialdehyde (MDA-dG) and 4-hydroxynonenal (4-HNE)-derived etheno adducts; endogenous formaldehyde adducts like N2-hydroxymethyl-dG and bile acid conjugated N-nitroso adducts like O6-carboxymethyl-dG and 3-ethane sulphonic acid-dC (Hwa Yun et al., 2020). These can serve as biomarkers for internal damage, aging, and cancer risk assessment. Endogenous carbonyl compounds including methylglyoxal and glyoxal generated as byproducts of several metabolic pathways lead to the formation of DNA adducts and advanced glycation end products (AGEs) that serve as key linkers of metabolic dysfunction to genomic instability (Lai et al., 2022).

Table 1 Represents a comprehensive framework of DNA adduct types and classification to understand their formation and biological relevance

Type of DNA adduct	Description	Key sources	Examples	Biological consequence	References
Alkyl adducts	Small alkyl (methyl/ethyl) groups are added to nucleophilic sites of DNA bases especially O-6 and N-7 sites of guanine	Endogenous S-adenosylmethionine (SAM) Exogenous Alkylating agents like dimethylnitrosamines, methyl methane sulfonate	O4-Methylthymine O6-Methylguanine (O6-MeG) N7-Methylguanine (N7-MeG)	G → A transitions Depurination (N7-MeG instability)	Lindahl (1993), Pegg (2011)
Bulky aromatic adducts (PAH derived)	PAHs upon metabolic activation (CYP450 enzymes) forms reactive diol epoxide intermediates that covalently bind N2 guanine	Cigarette smoke Vehicular smoke Industrial emissions	BPDE-dG adducts (benzo[a]pyrene) 7,8-Dihydroxy-9,10-epoxy-BaP adducts	Strong helix distortion Mutagenic hotspots in TP53 gene in smokers Potent carcinogen for lung cancer	Ewa and Danuta (2017), Phillips and Venitt (2012)
Aromatic amine and hetrocyclic amine adduct	Bioactivation by N-acetyltransferases and CYP enzymes form nitrenium ions reacting with C8-guanine	High-temperature cooking of meat Tobacco smoke (4-aminobiphenyl)	dG-C8-PhIP from cooked/grilled meat 4-ABP-DNA adducts	Frameshift and bases substitution mutations Correlated with bladder and colorectal cancers	Phillips and Venitt (2012)
Oxidative DNA adducts	ROS attack nucleobases primarily guanine; oxidation leads to altered base pairing	Reactive oxygen species (ROS) from normal metabolism Inflammation Ionizing radiation	8-oxo-7,8 Dihydroguanine (8-oxo-G) Thymine glycol FapyG	Leads to G → T transversions Halts DNA polymerase (thymine glycol) BER pathway activation	Cooke et al. (2003)

(continued)

Table 1 (continued)

Type of DNA adduct	Description	Key sources	Examples	Biological consequence	References
Aflatoxin B1 adducts	Metabolic epoxy activation (AFB1-exo-8,9-epoxide) reacts with N7 of Guanine	Grains and nuts contaminated with aflatoxin Aspergillus flavus fungus	Aflatoxin B1-N-7 guanine (AFB1-N7-G) AFB1-FaPy adducts	Leads G → T transversions Signature G → T mutation at TP53 gene 249 codons Potent hepatocarcinogen	Usman and Volpi (2018)
Lipid peroxidation-derived adducts	M1dG (pyrimidopurinone) HNE-dG (4-hydroxynonenal) Crotonaldehyde adducts	ROS induces membrane lipid peroxidation Chronic inflammation	Aldehydes (MDA, 4-HNE) form Michael adducts or Schiff bases with DNA bases	Highly mutagenic Leads to G → T and G → T mutations Associated with cancers caused due to inflammation	Swenberg et al. (2011)
Crosslink adducts (DNA–DNA/ DNA–protein adducts)	Intra and interstrand crosslinks are formed Mitomycin C crosslinks Aldehyde induced DNA –protein crosslinks (DPCs)	Chemotherapeutic drugs (e.g. Cisplatin) Formaldehyde exposure	Cisplatin 1,2-d(GpG), 1,2-d(ApG) and 1,3-(GpTpG) links Pt-G N6-HOMe-dAdo, N2-HOMe-dGuo, N4-HOMe-dCyd DPCs	Replication is halted Apoptosis of rapidly dividing cells Platinum drug toxicity	Tretyakova et al. (2015)
UV-induced DNA adducts	Photons induce covalent bonding between adjacent pyrimidines (T-T, C-T dimers)	UV-B and UV-C radiation exposure	Cyclobutane pyrimidine dimers (CPDs) 6-4 Pyrimidine-pyrimidone Photoproducts (6-4PPs)	C → T transitions at dipyrimidine sites Prominent in skin cancers (melanoma and non-melanoma)	Ning and Knox (1999), Schreier et al. (2007)

(continued)

Table 1 (continued)

Type of DNA adduct	Description	Key sources	Examples	Biological consequence	References
Endogenous aldehyde derived	Aldehydes react with nucleophilic DNA bases forming Schiff base intermediates	Metabolism of alcohol (acetaldehyde) Endogenous aldehyde generation	Formaldehyde-dG adducts Acetaldehyde-dG adducts	Mutagenic and carcinogenic, especially alcohol linked esophageal and oral cancer	Swenberg et al. (2011)
Sugar-phosphate backbone adducts/strand break precursor lesions	Loss of purine bases creates abasic sites, sugar oxidation breaks backbone	Oxidative stress Alkylation damage (N-7 alkylguanine instability)	AP sites after depurination Covalent sugar oxidation products	Precursor to strand breaks Leads to genomic instability	Tretyakova et al. (2015)
Epigenetic adducts and methylated adducts	DNA methyltransferases	Enzymatic hyper or hypo methylation of cytosine residues mostly in the promoter regions of various genes (CpGislands)	5-Methylcytosine 5-Hydroxymethyl cytosine	Heritable altered gene expression Mutation hotspot upon deamination of 5mC → T	Chappell et al. (2016)

3.1.2 Exogenous DNA Adducts

These are formed due to the exposure of DNA to environmental pollutants, tobacco smoke, chemicals, radiation, and diet (Phillips & Venitt, 2012). Most exogenous compounds need metabolic activation via phase I enzymes to electrophilic intermediates which can react with DNA bases. These act as early biomarkers for exposure and for early cancer risk assessment. Environmental carcinogens include Polycyclic Aromatic hydrocarbons (PAHs) like benzo[a]pyrene whose metabolic activation produces benzo[a]pyrene diol epoxide (BPDE), which largely binds to the N2 position of guanine forming bulky and highly reactive mutagenic adducts (Phillips, 2005). Charring or grilling of food leads to the production of Heterocyclic aromatic amines (HAAs) and PAHs that can form DNA adducts (Ewa & Danuta, 2017). Alcohol consumption contributes to DNA adduct and DNA protein cross-links via its toxic metabolite acetaldehyde with the most prominent being N2-ethyl-2′-deoxyguanosine (N2-ethyl-dG) (Guidolin et al., 2021).

Contaminated food grains and nuts have aflatoxins which are potent mycotoxins produced by Aspergillus species. AflatoxinB1 is metabolically transformed in the liver by *CYP450* to a highly reactive intermediate, aflatoxin B1-8,9-epoxide which forms covalent adducts primarily at the N7 guanine generating the AFB1-N7-dG adduct (Fasullo, 2019). These adducts can further lead to depurination or rearrange to more stable formamidopyrimidine (AFB1-FAPy) adducts causing characteristic GC to TA transversions correlated with hepatocellular carcinoma (Poirier, 2012).

3.2 On the Basis of Type of Chemical Bond Formed

3.2.1 Covalent Adducts

These adducts involve formation of a stable covalent bond between electrophilic compounds and nucleophilic sites on DNA bases. These include:

(A) **Monoadducts**: These are formed when DNA damage results in chemical modification of a single nucleotide base of the DNA. Such single bond adducts can halt DNA replication or transcription and sometimes be transformed into potentially more harmful interstrand cross-links (ICLs)/intrastrand crosslinks (ISLs) which can lead to impaired cellular function and even cancer (Souliotis et al., 2003). For example alkylating agents like N-methyl-N-nitrosourea (MNU) and temozolomide (TMZ) form O6-methylguanine which is a highly cytotoxic and mutagenic adduct that might result in mispairing during DNA replication (*Drinking Water and Health, Volume 9*, 1989).

(B) **Exocyclic Adducts**: These are harmful chemical moieties produced when reactive aldehydes often from lipid peroxidation or tobacco smoke produce cyclic structures (etheno adducts) attached to the exocyclic amino group of DNA bases. Examples include 1,N-6-ethenoadenine (εA), 3,N4-ethenocytosine

(ƐC), and 1,N2-ethenoguanine (ƐG). They disrupt normal DNA function having miscoding potential driving mutations, genomic instability, and carcinogenesis. They serve as important biomarkers for oxidative stress and mutagen exposure (Mishina et al., 2005).

(C) **Intrastrand Cross-links (ISLs)**: These are formed when a mutagen binds to adjacent bases present on the same DNA strand. Cisplatin-induced GG intrastrand cross-links distort DNA helix and block DNA replication (Wang & Lippard, 2005).

(D) Interstrand Cross links (ICLs): These cross link nucleo-bases on the opposite DNA strands, thus preventing strand separation. These are highly cytotoxic adducts and require intricate repair pathways including homologous recombination. DNA ICL forming agents include Nitrogen mustard, Mitomycin A and Platinum compounds (Muniandy et al., 2010).

(E) DNA–Protein crosslinks (DPCs): Few reactive aldehydes like formaldehyde and acetaldehyde, ionizing radiation, and chemotherapeutic drugs form covalent linkages among DNA proteins including repair enzymes and histones. Such bulky adducts lead to the disruption of chromatin structure and transcription and result in genomic instability, cancer, and aging (Torrecilla et al., 2024; Tretyakova et al., 2015).

3.3 On the Basis of Modified Nucleotide Site

3.3.1 Guanine Adducts

Guanine is the most commonly modified DNA base as it possesses high nucleophilicity. Reactive chemicals like carcinogens (aflatoxins), environment pollutants or chemotherapeutic drugs covalently bind guanine and lead to the formation of such adducts including N7-dG adducts that are frequent but chemically unstable and generally result in depurination. O6-dG adduct is highly mutagenic and results in GC $\rightarrow$ AT transitions. C8-dG adducts induce Z-DNA formation affecting gene expression and might lead to cancer establishment (Boysen et al., 2009; Pegg, 2011; Vongsutilers & Gannett, 2018).

3.3.2 Adenine Adducts

These are formed when a reactive chemical directly reacts with adenine of DNA. For example, N6-dA adduct formed from tumorigenic dibenzo[a,l]pyrene is highly genotoxic (Kropachev et al., 2013). N3-dAadducts are formed upon exposure to styrene and result in DNA damage (Mikeš et al., 2010).

3.3.3 Cytosine Adducts

These are modified cytosine bases in the DNA which are formed upon exposure to various pollutants, chemicals, or metabolic by products. Deamination of cytosine produces uracil or thymine adducts which can mispair with guanine during DNA replication driving mutations. MDA reacts with cytosine forming N4-(3-oxopropenyl) cytosine (M1-dC) adducts which are highly mutagenic. Oxidative stress creates lesions like 8-hydroxyguanine which can alter methylation of nearby cytosines thus influencing gene expression (Hall et al., 1996).

3.3.4 Thymine Adducts

Thymine is the most commonly modified base upon exposure to ultra-violet radiation forming cyclobutane thymine dimers (2 + 2) or (6–4) pyrimidine-pyrimidone photoproducts that lead to the distortion of DNA and replication errors. Also, covalent cross links among thymine and amino acids particularly lysine or tyrosine result in the formation of DNA–protein cross links (DPCs) upon UV exposure (Charlton et al., 1999; Schreier et al., 2007).

3.3.5 Sugar-Phosphate Backbone Adducts

These are formed when certain environmental carcinogens bind covalently to the sugar or phosphate groups of DNA/RNA backbone. These include phosphate and sugar adducts. Phosphate adducts are formed when an alkylating carcinogen like ethyl nitrosourea (ENU) reacts with oxygen atom of the phosphodiester linkage forming a phosphotriester (PTE). These might lead to conformational changes in the DNA as they point into the major groove of the DNA helix causing structural distortions which interfere with normal DNA replication (Jones et al., 2010). Ionizing radiation or oxidative stress leads to the generation of reactive radicals (C5-sugar radical) in the pentose of DNA backbone. This can result in immediate DNA breaks or formation of more complex stable adducts that can disrupt the DNA helix (Rivero et al., 1988; Yoshikura & Taguchi, 1979).

4 DNA Adducts and Cancer

The DNA adducts formation represents one of the primary key initial molecular events in chemical carcinogenesis. They also serve as important biomarkers of carcinogen exposure to genotoxic agents like tobacco smoke, environment pollutants, and chemotherapeutic drugs as well as assess cancer risk (Phillips, 2005). The carcinogenic nature of DNA adducts largely depends upon their chemical structure, stability, and precise location in the genome. Bulky DNA adducts like benzo[a]pyrene diol epoxide (BPDE) and polycyclic aromatic hydrocarbon (PAH) adducts distort the helical structure and thus affect DNA replication and transcription. Nucleotide excision repair (NER) system can repair such adducts. However, if not corrected by the NER system such lesions can result in point mutations frequently related with tobacco smoke-associated lung cancer incidence (Poirier, 2016). Patterns of p53 G to T transversions were found to represent the key mutagenic signature of tobacco smoke-induced DNA damage in lung cancer (Hainaut, 2001). Besides, bulky DNA adducts have been significantly associated with stomach, colon, and bladder cancer (Poirier, 2004). Similarly in populations exposed to contaminated food, aflatoxin B1 generates covalent adducts at N7-dG resulting in alterations in the p53 tumor suppressor gene, a characteristic of hepatocellular carcinoma (Kensler et al., 2011). DNA adducts generated endogenously also play a significant role in carcinogenesis. ROS produced as a byproduct of mitochondrial respiration or during inflammation leads to the production of oxidative DNA lesions like 8-oxo-dG. This lesion can mispair with adenine during DNA replication leading to GC → TA transversions in oncogenes and tumor suppressor genes and thus genomic instability (Cooke et al., 2003; Poetsch, 2020). Malondialdehyde (MDA) and acetaldehyde which are lipid peroxidation products generate exocyclic DNA adducts which are highly cytotoxic and contribute to genomic instability and carcinogenesis (Mishina et al., 2005). MDA serves a common marker for oxidative stress and is implicated in carcinogenesis. Significantly higher plasma MDA levels are found in lung, breast, ocular, and oral cancer patients. MDA levels are associated with increased tumor growth, aggressiveness, and poor prognosis (Mohideen et al., 2021).

Cellular DNA repair capacity greatly influences the persistence of DNA adducts. DNA repair deficiency (DRD) including nucleotide excision repair (NER), mismatch repair (MMR), etc. result in DNA repair syndromes exhibiting multisystemic indications and oncogenic susceptibility (Sharma et al., 2020). Also, certain chemotherapeutic drugs like cisplatin can induce secondary cancers primarily leukemia or myelodysplastic syndromes and certain solid cancers, years after treatment. The drug induces DNA adducts leading to various mutations in hematopoietic stem cells which can later manifest as secondary malignancies demanding ongoing surveillance in survivors (Chovanec et al., 2017). DNA adducts can be employed as molecular dosimeters to ascertain the exposure to various carcinogens. They serve to measure the actual internal exposure by incorporating complex factors comprising uptake, metabolism, and repair, thus forming a direct association between the external dose and effect, imperative for better cancer risk assessment by providing a more precise measure of exposure (Hwa Yun et al., 2020). Furthermore the sensitivity of DNA adduct measurements has increased due to developments in mass spectrometry and immunological detection methods, reinforcing their use in cancer risk assessment and preventative measures (Chao et al., 2024). In summary, DNA adduct production connects genetic insults, endogenous metabolism, and environmental exposure to the onset of cancer. Understanding the causes of carcinogenesis and creating successful preventive and therapeutic approaches requires an in depth understanding of the types, origins, and biological effects of DNA adducts.

5 Detection of DNA Adducts

The accurate DNA adducts detection and quantification is critical for understanding exposure, risk assessment, and their role in mutagenesis and carcinogenesis (Brown, 2012; Phillips, 2005). Since DNA adducts are rare as in 1 adduct per 106–1010 nucleotides, very sensitive, specific, and reliable analytical approaches are required to detect DNA adducts. Over time, a number of sophisticated and traditional analytical techniques have been developed for the detection of DNA adducts, ranging from radioisotope-based approaches to high-resolution spectrometry and next generation technologies (Table 2). Traditional methods use only microgram of DNA and have a high sensitivity for detecting very low frequencies of DNA adducts, but they are unable to describe the co-migrating endogenous interfering I-compounds (Phillips, 2005). One of the oldest and most popular approach for detecting DNA adducts is ^{32}P postlabeling, which involves enzymatically digesting the DNA, enriching it for adducted nucleotides and then radiolabeling the same (Phillips, 2005; Reddy & Randerath, 1986). Immunoassays are another traditional approach that offers further tools for tissue-specific adduct localization. Using the antibodies, immunoassays (Immunoblot ELISA and Immunohistochemistry) identify the particular adducts. These approaches are useful for viewing tissue dispersed adducts, but they rely only on variable antibody specificity and they typically have lower quantitative precision than other techniques (Santella, 1999).

As the technology advanced, the DNA adduct analysis was transformed by the liquid chromatography-tandem mass spectrometry (LC-MS/MS) approach, which uses standards labeled with isotopes to permit both structural characterization and quantification (Boysen & Hecht, 2003; Tretyakova et al., 2013). As high resolution MS approaches allow for untargeted DNA adduct omics; they also aid in the simultaneous detection of many known and novel adducts across the genomes in the biological samples (Kanaly et al., 2006; Rappaport et al., 2012). Recent developments in MS techniques have significantly enhanced the sensitivity and chemical coverage by enabling the identification of DNA alterations with minimum DNA inputs (Gerdemann et al., 2025). After administration of radiolabeled xenobiotics, accelerator mass spectrometry (AMS) offers an additional approach for highly sensitive ultra-trace measurement of DNA adducts at attomole levels (Coldwell et al., 2008). As it requires labeled compounds, AMS is more suited for experimental applications than epidemiological ones, despite its high sensitivity. By combining enzymatic or chemical identification of DNA adducts, next generation sequencing approaches such as XR-seq, cisplatin-seq, CPD-seq, and nanopore-seq enable genome-wide and single-molecule adduct mapping (Hu et al., 2019).

These approaches provide previously unheard-of spatial resolution, making it possible to locate genomic "hotspots" where mutational signatures and adducts coexist. Additionally, single molecule adduct detection in lengthy DNA strands may become possible with the development of nanopore sequencing, which can directly detect changed bases in the native DNA. These approaches demonstrate a range

Table 2 The principles, applications, advantages, and disadvantages of various DNA adduct detection approaches

Sl. No.	Name of method	Principle	Applications	Strengths	Limitations	References
1	^{32}P-Postlabelling	DNA digested enzymatically into nucleotides or nucleosides Adducted nucleotides are labelled at 5′-position using [γ-^{32}P]ATP and polynucleotide kinase Labeled adducts separated and quantify by measuring the radioactive signal	Environment monitoring	Small DNA inputs, high sensitivity	Provide poor structural information, potential for artifacts or interference by ‘I’ compounds	Phillips (2005), Reddy and Randerath (1986)
2	Targeted (LC-MS/MS)	DNA separated and hydrolyzed to produce changed 2′-deoxynucleosides or nucleobases Further separated and examined using LC-MS Targeted techniques use legitimate adduct standards, while untargeted uses “adductomics”	Mechanistic and exposome studies	Provides structural ID, quantification with isotope standards, simultaneously detect multiple adducts, discover novel adducts	Requires more DNA and complex instruments, less sensitive, loss or degrade some adducts during sample preparation, structural isomer can complicate interpretation	Boysen and Hecht (2003), Kanaly et al. (2006), Rappaport et al. (2012), Tretyakova et al. (2013)

(continued)

Table 2 (continued)

Sl. No.	Name of method	Principle	Applications	Strengths	Limitations	References
3	Immunoassays	Monoclonal or polyclonal antibodies bind selectively to particular DNA adducts antibody recognition of specific adducts Binding detected by immunoassay readouts like ELISA, immunohistochemistry and immunoblot	Useful especially when comparison between samples or spatial localization (in tumor tissue) is required	Tissue localization is possible tissue localization, relatively easy and straight forward	Required highly specific and well characterized, non-specific binding and cross reactivity, limited to known adducts	Santella (1999)
4	AMS (Accelerator MS)	Metabolic update and adduct formation of radiolabeled version of xenobiotic like ^{14}C labeled DNA extracted to use AMS to counts the individual ^{14}C atoms for quantifying the low levels of ^{14}C in DNA	Adriamycin-DNA adduct detection in cells, pharmacokinetics and carcinogenesis	High sensitivity at attomole, attomole sensitivity, track adduct persistence	Required isotopically labeled compounds, not always application in human s ethically, not feasible for all exposures	Coldwell et al. (2008)
5	Sequencing-based (XR-seq or nanopore)	Lesion recognition/repair coupled with NGS	Mechanistic studies and link to mutational signatures	Genome-wide localization of adducts	Expensive methods and specific biases	Hu et al. (2019)

of capabilities, including ultra-sensitive detection (^{32}P-postlabeling, AMS, structural identification (LC-MS/MS), and spatial mapping (sequencing-based methods). The choice of methods depends on the availability of standards, biological question, and the required resolution. DNA adducts are expected to play a major role in cancer risk classification, early detection, and chemoprevention measures as detection techniques become more accurate and comprehensive.

6 DNA Adducts as Potential Predictive Biomarkers

DNA adducts are increasingly considered as predictive biomarkers of cancer risk and are commonly acknowledged as direct signs of internal exposure to genotoxic chemicals. They are important because they are an early molecular event that connect exposure to carcinogens to later malignant transformation (Feng et al., 2006; Phillips, 2005). DNA adducts provide an integrated measure of the biologically effective dose by representing not only external exposure but also metabolic activation, detoxification processes, and DNA repair capacity. They are positioned at the crucial interface between environmental and endogenous carcinogens and cancer start. Because of this, they are very useful and appealing biomarkers for determining a person's vulnerability and estimating their risk of developing cancer (Perera, 1996).

DNA adducts are among the first biochemical effects of exposure to endogenous and environmental genotoxins that can be measured, and their persistence has been repeatedly associated with an increased risk of cancer in a variety of organs. Hepatocellular, lung, bladder breast, colorectal, and renal cancers are among the cancers that are more common in people with persistently high adduct loads, according to numerous research (Phillips, 2005). These lesions serve as sensitive biomarkers of early genotoxic stress and shed light on tissue-specific carcinogenic pathways since they develop prior to the development of overt mutations or tumors. Exposure to aflatoxin B1 (AFB1) is an example of the direct link between adduct production and cancer risk. The accumulation of AFB3-N3-guanine adducts closely correlates with the frequency of hepatocellular cancer in high-exposure areas such as Oidong, China, indicating the mycotoxin's ability to drive mutational events (Poirier, 1997; Siantar et al., 2008). Similarly, bulky dG adducts produced by tobacco smoke constituents like benzo[a]pyrene (BaP) are abundant in the lung tissues of smokers and correlates with TP53 mutational hotspots, demonstrating a mechanistic connection between exposure, DNA damage, and selective gene susceptibility (Boysen & Hecht, 2003).

Additional layers of risk are contributed by lifestyle and occupational exposures. While greater aromatic DNA adducts in breast tissue have been linked in population studies to increased tumor risk, aromatic amine adducts in the urothelium of workers in the dye industry are highly correlated with the incidence of bladder cancer (Gammon et al., 2004; Hernandez-Castillo et al., 2020). Not all carcinogenic adducts come from outside sources; internal metabolic processes can cause DNA damage patterns linked to kidney and colorectal malignancies. This is further demonstrated by endogenously formed lesions such 8-oxo-dG, which result from oxidative stress

(Phillips, 2005). Adduct loads are modulated by interindividual variability in the metabolic activity, detoxification pathways, and DNA repair capacity, according to a key finding from epidemiological and mechanistic studies. Certain people may be genetically predisposed to adduct accumulation due to variation in cytochrome P450 enzymes, glutathione-S-transferases, and repair proteins (Feng et al., 2006). This heterogeneity highlights the usefulness of DNA adduct measurement as a method for forecasting cancer before irreversible genetic changes occur and for identifying chemically sensitive subpopulations. Large-scale cohort studies and meta-analyses support DNA adducts predictive usefulness. The relative risk of lung and bladder malignancies is greatly increased by elevated levels by bulky adducts, especially in smokers and occupational exposure groups (Peluso et al., 2005). The idea that DNA adducts' profiling can function as a cross-cancer early warning indication of carcinogenic processes is supported by parallel findings in breast and liver cancers, where aromatic and AFB3-related adducts, respectively, demonstrate strong exposure-risk associations (Groopman et al., 2008). This collection of data highlights the fact that DNA adducts are active participants in carcinogenesis rather than just passive indicators. Persistent adducts shed light on the early stages of tumor development and identify susceptible populations where preventative measures may have the biggest effects by causing genomic instability, changing important regulatory genes like p53, H-ras, and K-ras, and forming mutational landscapes (Feng et al., 2006).

Despite their strong potential, a number of issues need to be resolved despite their great potential as they represent recent exposures and are impacted by DNA repair dynamics, their levels temporal fluctuation and may change significantly over time. Another difficulty is tissue specificity, which makes risk assessment more difficult because adducts' loads evaluated in surrogate tissues, like blood leukocytes, might not correctly reflect those in the target organs like liver or lung (Rundle, 2006). Further, greater application and cross-study comparability are hampered by the lack of standardized analytical techniques among labs. Lastly because the quantitative relationship between adduct load and cancer risk is not well established and biologically significant limits have not been established, the biological interpretation of DNA adduct levels is still unclear.

7 Future Directions in DNA Adduct Research

DNA adduct formation has become a major health concern and there is a need to explore on genome-chemical interactions, multi-omics approaches, DNA adducts biomarkers, high-resolution analytical technologies, and high-resolution mapping strategies to better understand the origins, diversity, and biological consequences of DNA modifications. DNA is constantly challenged by chemical modification and spontaneous loss or modification of its bases. Linking adduct formation to mutational signatures through whole-genome sequencing and computational modeling will be helpful to clarify the causal relationships between environmental exposures, endogenous metabolites, and carcinogenesis (Alexandrov et al., 2013; Phillips, 2018;

Poetsch, 2020). A major future priority lies in linking adduct formation to functional consequences such as transcriptional dysregulation, replication stress, chromatin remodeling, and mutation signatures leading to carcinogenesis (Cheng et al., 2022; La Barbera et al., 2022). Moreover, there is growing interest in connecting DNA adducts with epigenomic alterations, including perturbations in chromatin architecture and interference with transcription factor binding, which may contribute to disease phenotypes beyond their classical mutagenic effects (Barrett et al., 2017; Mangerich et al., 2013).

DNA adducts serve as biomarkers of carcinogen exposure, and their quantification is especially valuable in predicting the cancer risk assessment. They are also crucial tools in understanding the association of tumorigenic effects of environmental, dietary, and industrial carcinogens. DNA adduct omics involves the screening DNA adducts in the cellular genome or biological samples. Although a wide range of experimental methodologies has been developed for the analysis of DNA adducts in biological samples, there is a need to develop a ultra-sensitive mass spectrometry and single-molecule sequencing platforms capable of detecting extremely low-abundance adducts with greater structural specificity which enable the comprehensive adductome profiling across tissues and disease states (Cao & Zhang, 2024; Kanaly et al., 2006; Tretyakova et al., 2015). Emerging mass-spectrometry platforms such as ultra-high-resolution Orbitrap and ion-mobility MS now enable the detection of extremely low-abundance adducts, providing unprecedented sensitivity for mapping the adductome across different tissues, developmental stages, and disease states (Chang et al., 2018). Complementing these advances, single-molecule sequencing technologies including nanopore and single-molecule real-time (SMRT) sequencing are beginning to allow direct readout of bulky or structurally unique DNA lesions without the need for chemical derivatization, thus offering the potential for genome-wide, base-specific localization of adducts with structural annotations (Dorey & Howorka, 2024; Eid et al., 2009; Gilpatrick et al., 2020). Another important ultrasensitive approach for DNA adduct detection is the ^{32}P-Postlabeling, which is used to identify DNA adducts formed through the direct covalent binding of carcinogens and mutagens to DNA bases, as well as other DNA lesions arising from base modifications induced by endogenous or exogenous agents, including oxidative damage (Phillips & Arlt, 2020).

Future studies also consider the dynamics of the DNA repair system during which cells tolerate, misrepair, or eliminate specific adduct types, and how these processes differ among individuals or tumor subtypes. Robust DNA repair system protects genomic integrity by removing lesions or tolerating damage to ensure overall cellular survival. Disruption of this finely tuned balance can destabilize cellular metabolic homeostasis, as observed in many cancers, where deregulation of DNA repair pathways leads to genomic instability. Elucidating the mechanisms of DNA damage and the corresponding repair or tolerance pathways provides critical insight into the molecular basis of genotoxicity, thereby laying the foundation for a deeper understanding of chemical-genome interactions and their roles in disease development and progression (Chatterjee & Walker, 2017; Marteijn et al., 2014). Genetic variability in xenobiotic-metabolizing enzymes (XMEs) influences an individual's capacity to process and eliminate xenobiotics and, consequently, their susceptibility to diseases. Such differences are likely driven by polymorphisms in detoxification enzymes, DNA repair pathways, and metabolic activation processes, and may help explain the wide variability in the susceptibility at the individual level to environmental carcinogens (Bolufer et al., 2006; Crocco et al., 2019).. There is also growing interest in developing real-time, non-invasive biomarkers of DNA adduct stress which involve the circulating modified nucleosides or lesion-specific repair products to support early detection of exposure-related pathologies preventing the development and progression of cancer, also helpful to monitor therapy-induced genotoxicity. Cooke et al. analyzed urine samples to detect a broad spectrum of both expected and previously unrecognized 2′-deoxyribonucleoside DNA adducts and explained their clinical relevance (Cooke et al., 2018; Vodicka et al., 2023). Beyond serving as biomarkers of exposure and genotoxic effects, DNA adducts also provide the understanding of biological consequences and potential risk associated with chemical exposure. Therefore, any compound capable of forming DNA adducts, even at low concentration, should be considered to possess carcinogenic and mutagenic potential.

Finally, the relevance of DNA adducts as predictive biomarkers is being actively explored in personalized medicine and cancer risk assessment, necessitating standardized pipelines for adduct quantification, validation in large cohorts, and integration with exposure and clinical outcome data (Phillips & Arlt, 2020). Moreover, future work aims to assess the combined effects of complex chemical mixtures, lifestyle factors, and endogenous electrophiles, to understand the DNA adduct formation and associated damage. Collectively, these future directions aim to transform DNA adduct research from a lesion-centric discipline into an integrated framework for predicting genomic instability, cancer risk, and therapeutic response.

8 Conclusion

DNA adduct formation represents a fundamental paradigm through which environmental, dietary, occupational, and endogenous chemical exposures affect the genomic homeostasis and increase the risk of cancer development. The xenobiotics metabolizing enzymes can paradoxically generate highly reactive intermediates during xenobiotics biotransformation that covalently modify DNA, producing structurally diverse adducts with mutagenic and cytotoxic potential. When such lesions persist due to overwhelming exposure, their bioactivation, or compromised DNA repair capacity, they can initiate and promote malignant transformation by inducing mutations, transcription dysregulation, altering gene expression, and disrupting chromatin structure. Both endogenous and exogenous DNA adducts contribute to the overall burden of genomic damage, linking metabolism, inflammation, oxidative stress, and environmental exposure to cancer risk. Inter-individual variability in xenobiotic-metabolizing enzymes, DNA repair pathways, and cellular defense mechanisms further modulates susceptibility, emphasizing the importance of genetic and exposomic context in carcinogenesis.

Technological advances in mass spectrometry and sequencing-based lesion mapping make possible to detect and interpret DNA adducts with high sensitivity and resolution. These developments in DNA adduct omics provide predictive biomarkers with significant translational potential in cancer prevention, early detection, and risk stratification. Integrating DNA adduct profiling with genomics, epigenomics, transcriptomics, and clinical data will be essential to fully elucidate chemical-genome interactions and to develop precision-based strategies for cancer prevention and therapy. In conclusion, understanding DNA adduct-mediated carcinogenesis provides critical insight into how xenobiotic activation influences cancer risk and its progression, while also aiding in the prediction, prevention, and mitigation of chemically induced cancers, ultimately contributing to improved public health outcomes and personalized approaches to disease management. Future studies are needed to elucidate the combined influences of complex chemical mixtures, lifestyle factors, and endogenous electrophiles to better elucidate DNA adduct formation and the resulting genomic damage.

References

Alexandrov, L. B., Nik-Zainal, S., Wedge, D. C., Aparicio, S. A. J. R., Behjati, S., Biankin, A. V., Bignell, G. R., Bolli, N., Borg, A., Børresen-Dale, A.-L., Boyault, S., Burkhardt, B., Butler, A. P., Caldas, C., Davies, H. R., Desmedt, C., Eils, R., Eyfjörd, J. E., Foekens, J. A., … Stratton, M. R. (2013). Signatures of mutational processes in human cancer. *Nature, 500*, 415–421. https://doi.org/10.1038/nature12477

Autrup, H. (2000). Genetic polymorphisms in human xenobiotica metabolizing enzymes as susceptibility factors in toxic response. *Mutation Research: Genetic Toxicology and Environmental Mutagenesis, 464*, 65–76. https://doi.org/10.1016/S1383-5718(99)00167-9

Balbo, S., Turesky, R. J., & Villalta, P. W. (2014). DNA adductomics. *Chemical Research in Toxicology, 27*, 356–366. https://doi.org/10.1021/tx4004352

Barnes, J. L., Zubair, M., John, K., Poirier, M. C., & Martin, F. L. (2018). Carcinogens and DNA damage. *Biochemical Society Transactions, 46*, 1213–1224. https://doi.org/10.1042/BST20180519

Barrett, C. W., Short, S. P., & Williams, C. S. (2017). Selenoproteins and oxidative stress-induced inflammatory tumorigenesis in the gut. *Cellular and Molecular Life Sciences CMLS, 74*, 607–616. https://doi.org/10.1007/s00018-016-2339-2

Bolufer, P., Barragan, E., Collado, M., Cervera, J., López, J.-A., & Sanz, M. A. (2006). Influence of genetic polymorphisms on the risk of developing leukemia and on disease progression. *Leukemia Research, 30*, 1471–1491. https://doi.org/10.1016/j.leukres.2006.01.016

Boysen, G., & Hecht, S. S. (2003). Analysis of DNA and protein adducts of benzo[a]pyrene in human tissues using structure-specific methods. *Mutation Research/Reviews in Mutation Research, 543*, 17–30. https://doi.org/10.1016/S1383-5742(02)00068-6

Boysen, G., Pachkowski, B. F., Nakamura, J., & Swenberg, J. A. (2009). The formation and biological significance of N7-guanine adducts. *Mutation Research: Genetic Toxicology and Environmental Mutagenesis, 678*, 76–94. https://doi.org/10.1016/j.mrgentox.2009.05.006

Brown, K. (2012). Methods for the detection of DNA adducts. In J. M. Parry, & E. M. Parry (Eds.), *Genetic toxicology, methods in molecular biology* (pp. 207–230). Springer New York. https://doi.org/10.1007/978-1-61779-421-6_11

Cai, Y., Geacintov, N. E., & Broyde, S. (2012). Nucleotide excision repair efficiencies of bulky carcinogen-DNA adducts are governed by a balance between stabilizing and destabilizing interactions. *Biochemistry, 51*, 1486–1499. https://doi.org/10.1021/bi201794x

Cao, M., & Zhang, X. (2024). DNA adductomics: A narrative review of its development, applications, and future. *Biomolecules, 14*, 1173. https://doi.org/10.3390/biom14091173

Carlsson, H., & Törnqvist, M. (2017). An adductomic approach to identify electrophiles in vivo. *Basic & Clinical Pharmacology & Toxicology, 121*(Suppl 3), 44–54. https://doi.org/10.1111/bcpt.12715

Cavalieri, E., Saeed, M., Zahid, M., Cassada, D., Snow, D., Miljkovic, M., & Rogan, E. (2012). Mechanism of DNA depurination by carcinogens in relation to cancer initiation. *IUBMB Life, 64*, 169–179. https://doi.org/10.1002/iub.586

Chandimali, N., Bak, S. G., Park, E. H., Lim, H.-J., Won, Y.-S., Kim, E.-K., Park, S.-I., & Lee, S. J. (2025). Free radicals and their impact on health and antioxidant defenses: A review. *Cell Death Discovery, 11*, 19. https://doi.org/10.1038/s41420-024-02278-8

Chang, Y.-J., Cooke, M. S., Hu, C.-W., & Chao, M.-R. (2018). Novel approach to integrated DNA adductomics for the assessment of in vitro and in vivo environmental exposures. *Archives of Toxicology, 92*, 2665–2680. https://doi.org/10.1007/s00204-018-2252-6

Chao, M.-R., Chang, Y.-J., Cooke, M. S., & Hu, C.-W. (2024). Multi-adductomics: Advancing mass spectrometry techniques for comprehensive exposome characterization. *TrAC, Trends in Analytical Chemistry, 180*, 117900. https://doi.org/10.1016/j.trac.2024.117900

Chappell, G., Pogribny, I. P., Guyton, K. Z., & Rusyn, I. (2016). Epigenetic alterations induced by genotoxic occupational and environmental human chemical carcinogens: A systematic literature review. *Mutation Research/Reviews in Mutation Research, 768*, 27–45. https://doi.org/10.1016/j.mrrev.2016.03.004

Charlton, T. S., Ingelse, B. A., Black, D. S., Craig, D. C., Mason, K. E., & Duncan, M. W. (1999). A covalent thymine-tyrosine adduct involved in DNA-protein crosslinks: Synthesis, characterization, and quantification. *Free Radical Biology & Medicine, 27*, 254–261. https://doi.org/10.1016/S0891-5849(99)00030-1

Chatterjee, N., & Walker, G. C. (2017). Mechanisms of DNA damage, repair, and mutagenesis. *Environmental and Molecular Mutagenesis, 58*, 235–263. https://doi.org/10.1002/em.22087

Cheng, G., Guo, J., Carmella, S. G., Lindgren, B., Ikuemonisan, J., Niesen, B., Jensen, J., Hatsukami, D. K., Balbo, S., & Hecht, S. S. (2022). Increased acrolein-DNA adducts in buccal brushings of e-cigarette users. *Carcinogenesis, 43*, 437–444. https://doi.org/10.1093/carcin/bgac026

Chovanec, M., Abu Zaid, M., Hanna, N., El-Kouri, N., Einhorn, L. H., & Albany, C. (2017). Long-term toxicity of cisplatin in germ-cell tumor survivors. *Annals of Oncology, 28*, 2670–2679. https://doi.org/10.1093/annonc/mdx360

Coldwell, K. E., Cutts, S. M., Ognibene, T. J., Henderson, P. T., & Phillips, D. R. (2008). Detection of Adriamycin–DNA adducts by accelerator mass spectrometry at clinically relevant Adriamycin concentrations. *Nucleic Acids Research, 36*, e100. https://doi.org/10.1093/nar/gkn439

Conney, A. H., Chang, R. L., Cui, X. X., Schiltz, M., Yagi, H., Jerina, D. M., & Wei, S. J. (2001). Dose-dependent differences in the profile of mutations induced by carcinogenic (R,S,S,R) bay- and fjord-region diol epoxides of polycyclic aromatic hydrocarbons. *Advances in Experimental Medicine and Biology, 500*, 697–707. https://doi.org/10.1007/978-1-4615-0667-6_102

Cooke, M. S., Evans, M. D., Dizdaroglu, M., & Lunec, J. (2003). Oxidative DNA damage: Mechanisms, mutation, and disease. *The FASEB Journal, 17*, 1195–1214. https://doi.org/10.1096/fj.02-0752rev

Cooke, M. S., Hu, C.-W., Chang, Y.-J., & Chao, M.-R. (2018). Urinary DNA adductomics—A novel approach for exposomics. *Environment International, 121*, 1033–1038. https://doi.org/10.1016/j.envint.2018.10.041

Coric, V. M., Simic, T. P., Pekmezovic, T. D., Basta-Jovanovic, G. M., Savic Radojevic, A. R., Radojevic-Skodric, S. M., Matic, M. G., Dragicevic, D. P., Radic, T. M., Bogdanovic, L. M., Dzamic, Z. M., & Pljesa-Ercegovac, M. S. (2016). Combined GSTM1-null, GSTT1-active,

GSTA1 low-activity and GSTP1-variant genotype is associated with increased risk of clear cell renal cell carcinoma. *PLoS ONE, 11*, e0160570. https://doi.org/10.1371/journal.pone.0160570

Crocco, P., Montesanto, A., Dato, S., Geracitano, S., Iannone, F., Passarino, G., & Rose, G. (2019). Inter-individual variability in xenobiotic-metabolizing enzymes: Implications for human aging and longevity. *Genes, 10*, 403. https://doi.org/10.3390/genes10050403

Dorey, A., & Howorka, S. (2024). Nanopore DNA sequencing technologies and their applications towards single-molecule proteomics. *Nature Chemistry, 16*, 314–334. https://doi.org/10.1038/s41557-023-01322-x

Drinking Water and Health, Volume 9: Selected Issues in Risk Assessment. (1989). National Academies Press. https://doi.org/10.17226/773

Eid, J., Fehr, A., Gray, J., Luong, K., Lyle, J., Otto, G., Peluso, P., Rank, D., Baybayan, P., Bettman, B., Bibillo, A., Bjornson, K., Chaudhuri, B., Christians, F., Cicero, R., Clark, S., Dalal, R., Dewinter, A., Dixon, J., … Turner, S. (2009). Real-time DNA sequencing from single polymerase molecules. *Science, 323*, 133–138. https://doi.org/10.1126/science.1162986

Ewa, B., & Danuta, M. -Š. (2017). Polycyclic aromatic hydrocarbons and PAH-related DNA adducts. *Journal of Applied Genetics, 58*, 321–330. https://doi.org/10.1007/s13353-016-0380-3

Fasullo, M. (2019). Cellular responses to aflatoxin-associated DNA adducts. In M. Mognato (Ed.), *DNA repair—An update*. IntechOpen. https://doi.org/10.5772/intechopen.81763

Feng, Z., Hu, W., Hu, Y., & Tang, M. (2006). Acrolein is a major cigarette-related lung cancer agent: Preferential binding at *p53* mutational hotspots and inhibition of DNA repair. *Proceedings of the National Academy of Sciences, 103*, 15404–15409. https://doi.org/10.1073/pnas.0607031103

Fournier, P. E., & Thomas, G. (1984). Mechanisms of chemical carcinogenesis: Recent advances. *Food Additives & Contaminants, 1*, 73–80. https://doi.org/10.1080/02652038409385828

Gammon, M. D., Sagiv, S. K., Eng, S. M., Shantakumar, S., Gaudet, M. M., Teitelbaum, S. L., Britton, J. A., Terry, M. B., Wang, L. W., Wang, Q., Stellman, S. D., Beyea, J., Hatch, M., Kabat, G. C., Wolff, M. S., Levin, B., Neugut, A. I., & Santella, R. M. (2004). Polycyclic aromatic hydrocarbon–DNA adducts and breast cancer: A pooled analysis. *Archives of Environmental Health: An International Journal, 59*, 640–649. https://doi.org/10.1080/00039890409602948

Gaskell, M., Kaur, B., Farmer, P. B., & Singh, R. (2007). Detection of phosphodiester adducts formed by the reaction of benzo[a]pyrene diol epoxide with 2′-deoxynucleotides using collision-induced dissociation electrospray ionization tandem mass spectrometry. *Nucleic Acids Research, 35*, 5014–5027. https://doi.org/10.1093/nar/gkm526

Gerdemann, A., Behrens, M., Günther, G., Ghallab, A., Hengstler, J. G., Humpf, H.-U., & Esselen, M. (2025). A simplified LC–MS-based method for sensitive analysis of DNA adducts utilizing accessible in vitro metabolism models. *Archives of Toxicology, 99*, 4021–4034. https://doi.org/10.1007/s00204-025-04125-w

Gilpatrick, T., Lee, I., Graham, J. E., Raimondeau, E., Bowen, R., Heron, A., Downs, B., Sukumar, S., Sedlazeck, F. J., & Timp, W. (2020). Targeted nanopore sequencing with Cas9-guided adapter ligation. *Nature Biotechnology, 38*, 433–438. https://doi.org/10.1038/s41587-020-0407-5

Greenman, C., Stephens, P., Smith, R., Dalgliesh, G. L., Hunter, C., Bignell, G., Davies, H., Teague, J., Butler, A., Stevens, C., Edkins, S., O'Meara, S., Vastrik, I., Schmidt, E. E., Avis, T., Barthorpe, S., Bhamra, G., Buck, G., Choudhury, B., … Stratton, M. R. (2007). Patterns of somatic mutation in human cancer genomes. *Nature, 446*, 153–158. https://doi.org/10.1038/nature05610

Groopman, J. D., Kensler, T. W., & Wild, C. P. (2008). Protective interventions to prevent aflatoxin-induced carcinogenesis in developing countries. *Annual Review of Public Health, 29*, 187–203. https://doi.org/10.1146/annurev.publhealth.29.020907.090859

Guengerich, F. P. (2008). Cytochrome P450 and chemical toxicology. *Chemical Research in Toxicology, 21*, 70–83. https://doi.org/10.1021/tx700079z

Guidolin, V., Carlson, E. S., Carrà, A., Villalta, P. W., Maertens, L. A., Hecht, S. S., & Balbo, S. (2021). Identification of new markers of alcohol-derived DNA damage in humans. *Biomolecules, 11*, 366. https://doi.org/10.3390/biom11030366

Guo, J., & Turesky, R. J. (2019). Emerging technologies in mass spectrometry-based DNA adductomics. *High-Throughput, 8*, 13. https://doi.org/10.3390/ht8020013

Hainaut, P. (2001). Patterns of p53 G→T transversions in lung cancers reflect the primary mutagenic signature of DNA-damage by tobacco smoke. *Carcinogenesis, 22*, 367–374. https://doi.org/10.1093/carcin/22.3.367

Hall, H. I., Lee, C. V., & Kaye, W. E. (1996). Cluster: A software system for epidemiologic cluster analysis. *Statistics in Medicine, 15*, 943–950. https://doi.org/10.1002/(sici)1097-0258(19960415)15:7/9%3c253C943::aid-sim264%3e253E3.0.co;2-z

Hemeryck, L. Y., Decloedt, A. I., Vanden Bussche, J., Geboes, K. P., & Vanhaecke, L. (2015). High resolution mass spectrometry based profiling of diet-related deoxyribonucleic acid adducts. *Analytica Chimica Acta, 892*, 123–131. https://doi.org/10.1016/j.aca.2015.08.019

Hernandez-Castillo, C., Termini, J., & Shuck, S. (2020). DNA adducts as biomarkers to predict, prevent, and diagnose disease—Application of analytical chemistry to clinical investigations. *Chemical Research in Toxicology, 33*, 286–307. https://doi.org/10.1021/acs.chemrestox.9b00295

Hinson, J. A., Roberts, D. W., & James, L. P. (2010). Mechanisms of acetaminophen-induced liver necrosis. In J. Uetrecht (Ed.), *Adverse drug reactions, handbook of experimental pharmacology* (pp. 369–405). Springer Berlin Heidelberg. https://doi.org/10.1007/978-3-642-00663-0_12

How, S. S., Nathan, S., Lam, S. D., & Chieng, S. (2025). ATP-binding cassette (ABC) transporters: Structures and roles in bacterial pathogenesis. *Journal of Zhejiang University Science B, 26*, 58–75. https://doi.org/10.1631/jzus.B2300641

Hu, J., Li, W., Adebali, O., Yang, Y., Oztas, O., Selby, C. P., & Sancar, A. (2019). Genome-wide mapping of nucleotide excision repair with XR-seq. *Nature Protocols, 14*, 248–282. https://doi.org/10.1038/s41596-018-0093-7

Hughes, T. B., Flynn, N., Dang, N. L., & Swamidass, S. J. (2021). Modeling the bioactivation and subsequent reactivity of drugs. *Chemical Research in Toxicology, 34*, 584–600. https://doi.org/10.1021/acs.chemrestox.0c00417

Hwa Yun, B., Guo, J., Bellamri, M., & Turesky, R. J. (2020). DNA adducts: Formation, biological effects, and new biospecimens for mass spectrometric measurements in humans. *Mass Spectrometry Reviews, 39*, 55–82. https://doi.org/10.1002/mas.21570

Ingelman-Sundberg, M., Mkrtchian, S., Zhou, Y., & Lauschke, V. M. (2018). Integrating rare genetic variants into pharmacogenetic drug response predictions. *Human Genomics, 12*, 26. https://doi.org/10.1186/s40246-018-0157-3

Jones, G. D. D., Le Pla, R. C., & Farmer, P. B. (2010). Phosphotriester adducts (PTEs): DNA's overlooked lesion. *Mutagenesis, 25*, 3–16. https://doi.org/10.1093/mutage/gep038

Kanaly, R. A., Hanaoka, T., Sugimura, H., Toda, H., Matsui, S., & Matsuda, T. (2006). Development of the adductome approach to detect DNA damage in humans. *Antioxidants & Redox Signaling, 8*, 993–1001. https://doi.org/10.1089/ars.2006.8.993

Kensler, T. W., Roebuck, B. D., Wogan, G. N., & Groopman, J. D. (2011). Aflatoxin: A 50-year odyssey of mechanistic and translational toxicology. *Toxicological Sciences, 120*, S28–S48. https://doi.org/10.1093/toxsci/kfq283

Kiyohara, C., Shirakawa, T., & Hopkin, J. M. (2002). Genetic polymorphism of enzymes involved in xenobiotic metabolism and the risk of lung cancer. *Environmental Health and Preventive Medicine, 7*, 47–59. https://doi.org/10.1007/BF02897330

Kropachev, K., Kolbanovskiy, M., Liu, Z., Cai, Y., Zhang, L., Schwaid, A. G., Kolbanovskiy, A., Ding, S., Amin, S., Broyde, S., & Geacintov, N. E. (2013). Adenine–DNA adducts derived from the highly tumorigenic dibenzo[a, l]pyrene are resistant to nucleotide excision repair while guanine adducts are not. *Chemical Research in Toxicology, 26*, 783–793. https://doi.org/10.1021/tx400080k

La Barbera, G., Nommesen, K. D., Cuparencu, C., Stanstrup, J., & Dragsted, L. O. (2022). A Comprehensive database for DNA adductomics. *Frontiers in Chemistry, 10*, 908572. https://doi.org/10.3389/fchem.2022.908572

Lai, S. W. T., Lopez Gonzalez, E. D. J., Zoukari, T., Ki, P., & Shuck, S. C. (2022). Methylglyoxal and its adducts: Induction, repair, and association with disease. *Chemical Research in Toxicology, 35*, 1720–1746. https://doi.org/10.1021/acs.chemrestox.2c00160

Lee, M.-S., Su, L., & Christiani, D. C. (2010). Synergistic effects of *NAT2* slow and *GSTM1* null genotypes on carcinogen DNA damage in the lung. *Cancer Epidemiology, Biomarkers & Prevention, 19*, 1492–1497. https://doi.org/10.1158/1055-9965.EPI-09-1195

Lewis, D. F. (2004). 57 Varieties: The human cytochromes P450. *Pharmacogenomics, 5*, 305–318. https://doi.org/10.1517/phgs.5.3.305.29827

Lindahl, T. (1993). Instability and decay of the primary structure of DNA. *Nature, 362*, 709–715. https://doi.org/10.1038/362709a0

Loeb, L. A., & Harris, C. C. (2008). Advances in chemical carcinogenesis: A historical review and prospective. *Cancer Research, 68*, 6863–6872. https://doi.org/10.1158/0008-5472.CAN-08-2852

Mangerich, A., Dedon, P. C., Fox, J. G., Tannenbaum, S. R., & Wogan, G. N. (2013). Chemistry meets biology in colitis-associated carcinogenesis. *Free Radical Research, 47*, 958–986. https://doi.org/10.3109/10715762.2013.832239

Mao, Q., Lai, Y., & Wang, J. (2018). Drug transporters in xenobiotic disposition and pharmacokinetic prediction. *Drug Metabolism and Disposition, 46*, 561–566. https://doi.org/10.1124/dmd.118.081356

Marteijn, J. A., Lans, H., Vermeulen, W., & Hoeijmakers, J. H. J. (2014). Understanding nucleotide excision repair and its roles in cancer and ageing. *Nature Reviews Molecular Cell Biology, 15*, 465–481. https://doi.org/10.1038/nrm3822

Martella, G., Motwani, N. H., Khan, Z., Sousa, P. F. M., Gorokhova, E., & Motwani, H. V. (2023). Simultaneous RNA and DNA adductomics using single data-independent acquisition mass spectrometry analysis. *Chemical Research in Toxicology, 36*, 1471–1482. https://doi.org/10.1021/acs.chemrestox.3c00041

Mikeš, P., Kořínek, M., Linhart, I., Krouželka, J., Dabrowská, L., Stránský, V., & Mráz, J. (2010). Urinary N3 adenine DNA adducts in humans occupationally exposed to styrene. *Toxicology Letters, 197*, 183–187. https://doi.org/10.1016/j.toxlet.2010.05.015

Miller, J. A., Surh, Y. J., Liem, A., & Miller, E. C. (1991). Electrophilic sulfuric acid ester metabolites of hydroxy-methyl aromatic hydrocarbons as precursors of hepatic benzylic DNA adducts in vivo. *Advances in Experimental Medicine and Biology, 283*, 555–567. https://doi.org/10.1007/978-1-4684-5877-0_71

Minchin, R. F., Reeves, P. T., Teitel, C. H., McManus, M. E., Mojarrabi, B., Ilett, K. F., & Kadlubar, F. F. (1992). N-and O-acetylation of aromatic and heterocyclic amine carcinogens by human monomorphic and polymorphic acetyltransferases expressed in COS-1 cells. *Biochemical and Biophysical Research Communications, 185*, 839–844. https://doi.org/10.1016/0006-291x(92)91703-s

Mishina, Y., Yang, C.-G., & He, C. (2005). Direct repair of the exocyclic DNA adduct $1,^{N}6$-ethenoadenine by the DNA repair AlkB proteins. *Journal of the American Chemical Society, 127*, 14594–14595. https://doi.org/10.1021/ja055957m

Mohideen, K., Sudhakar, U., Balakrishnan, T., Almasri, M. A., Al-Ahmari, M. M., Al Dira, H. S., Suhluli, M., Dubey, A., Mujoo, S., Khurshid, Z., Raj, A. T., & Patil, S. (2021). Malondialdehyde, an oxidative stress marker in oral squamous cell carcinoma—A systematic review and meta-analysis. *Current Issues in Molecular Biology, 43*, 1019–1035. https://doi.org/10.3390/cimb43020072

Moorthy, B., Chu, C., & Carlin, D. J. (2015). Polycyclic aromatic hydrocarbons: From metabolism to lung cancer. *Toxicological Sciences, 145*, 5–15. https://doi.org/10.1093/toxsci/kfv040

Muniandy, P. A., Liu, J., Majumdar, A., Liu, S., & Seidman, M. M. (2010). DNA interstrand crosslink repair in mammalian cells: Step by step. *Critical Reviews in Biochemistry and Molecular Biology, 45*, 23–49. https://doi.org/10.3109/10409230903501819

Ning, S., & Knox, S. J. (1999). G2/M-Phase arrest and death by apoptosis of HL60 cells irradiated with exponentially decreasing low-dose-rate gamma radiation. *Radiation Research, 151*, 659. https://doi.org/10.2307/3580204

Omiecinski, C. J., Vanden Heuvel, J. P., Perdew, G. H., & Peters, J. M. (2011). Xenobiotic metabolism, disposition, and regulation by receptors: From biochemical phenomenon to predictors of major toxicities. *Toxicological Sciences, 120*, S49–S75. https://doi.org/10.1093/toxsci/kfq338

Pegg, A. E. (2011). Multifaceted roles of alkyltransferase and related proteins in DNA repair, DNA damage, resistance to chemotherapy, and research tools. *Chemical Research in Toxicology, 24*, 618–639. https://doi.org/10.1021/tx200031q

Peluso, M., Munnia, A., Hoek, G., Krzyzanowski, M., Veglia, F., Airoldi, L., Autrup, H., Dunning, A., Garte, S., Hainaut, P., Malaveille, C., Gormally, E., Matullo, G., Overvad, K., Raaschou-Nielsen, O., Clavel-Chapelon, F., Linseisen, J., Boeing, H., Trichopoulou, A., … Vineis, P. (2005). DNA adducts and lung cancer risk: A prospective study. *Cancer Research, 65*, 8042–8048. https://doi.org/10.1158/0008-5472.CAN-04-3488

Perera, F. P. (1996). Molecular epidemiology: Insights into cancer susceptibility, risk assessment, and prevention. *JNCI Journal of the National Cancer Institute, 88*, 496–509. https://doi.org/10.1093/jnci/88.8.496

Phillips, D. H. (2005). DNA adducts as markers of exposure and risk. *Mutation Research - Fundamental and Molecular Mechanisms of Mutagenesis, 577*, 284–292. https://doi.org/10.1016/j.mrfmmm.2005.03.008

Phillips, D. H. (2018). Mutational spectra and mutational signatures: Insights into cancer aetiology and mechanisms of DNA damage and repair. *DNA Repair, 71*, 6–11. https://doi.org/10.1016/j.dnarep.2018.08.003

Phillips, D. H., & Arlt, V. M. (2020). 32P-Postlabeling analysis of DNA adducts. *Methods in Molecular Biology (Clifton, N.J.), 2102*, 291–302. https://doi.org/10.1007/978-1-0716-0223-2_16

Phillips, D. H., & Venitt, S. (2012). DNA and protein adducts in human tissues resulting from exposure to tobacco smoke. *International Journal of Cancer, 131*, 2733–2753. https://doi.org/10.1002/ijc.27827

Poetsch, A. R. (2020). The genomics of oxidative DNA damage, repair, and resulting mutagenesis. *Computational and Structural Biotechnology Journal, 18*, 207–219. https://doi.org/10.1016/j.csbj.2019.12.013

Poirier, M. C. (1997). DNA adducts as exposure biomarkers and indicators of cancer risk. *Environmental Health Perspectives, 105*, 907–912. https://doi.org/10.1289/ehp.97105s4907

Poirier, M. C. (2004). Chemical-induced DNA damage and human cancer risk. *Nature Reviews Cancer, 4*, 630–637. https://doi.org/10.1038/nrc1410

Poirier, M. C. (2012). Chemical-induced DNA damage and human cancer risk. *Discovery Medicine, 14*, 283–288.

Poirier, M. C. (2016). Linking DNA adduct formation and human cancer risk in chemical carcinogenesis. *Environmental and Molecular Mutagenesis, 57*, 499–507. https://doi.org/10.1002/em.22030

Poku, V. O., & Iram, S. H. (2022). A critical review on modulators of multidrug resistance protein 1 in cancer cells. *PeerJ, 10*, e12594. https://doi.org/10.7717/peerj.12594

Rappaport, S. M., Li, H., Grigoryan, H., Funk, W. E., & Williams, E. R. (2012). Adductomics: Characterizing exposures to reactive electrophiles. *Toxicology Letters, 213*, 83–90. https://doi.org/10.1016/j.toxlet.2011.04.002

Rebhan, K., Ertl, I. E., Shariat, S. F., Grollman, A. P., & Rosenquist, T. (2020). Aristolochic acid and its effect on different cancers in uro-oncology. *Current Opinion in Urology, 30*, 689–695. https://doi.org/10.1097/MOU.0000000000000806

Reddy, M. V., & Randerath, K. (1986). Nuclease P1-mediated enhancement of sensitivity of ^{32}P-postlabeling test for structurally diverse DNA adducts. *Carcinogenesis, 7*, 1543–1551. https://doi.org/10.1093/carcin/7.9.1543

Rendic, S., & Guengerich, F. P. (2012). Contributions of human enzymes in carcinogen metabolism. *Chemical Research in Toxicology, 25*, 1316–1383. https://doi.org/10.1021/tx300132k

Ring, C. L., Pearce, R. G., Setzer, R. W., Wetmore, B. A., & Wambaugh, J. F. (2017). Identifying populations sensitive to environmental chemicals by simulating toxicokinetic variability. *Environment International, 106*, 105–118. https://doi.org/10.1016/j.envint.2017.06.004

Rivero, D. P., Fox, J., Skipor, A. K., Urban, R. M., & Galante, J. O. (1988). Calcium phosphate-coated porous titanium implants for enhanced skeletal fixation. *Journal of Biomedical Materials Research, 22*, 191–201. https://doi.org/10.1002/jbm.820220304

Rundle, A. (2006). Carcinogen-DNA adducts as a biomarker for cancer risk. *Mutation Research - Fundamental and Molecular Mechanisms of Mutagenesis, 600*, 23–36. https://doi.org/10.1016/j.mrfmmm.2006.05.031

Sancar, A., Lindsey-Boltz, L. A., Unsal-Kaçmaz, K., & Linn, S. (2004). Molecular mechanisms of mammalian DNA repair and the DNA damage checkpoints. *Annual Review of Biochemistry, 73*, 39–85. https://doi.org/10.1146/annurev.biochem.73.011303.073723

Santella, R. M. (1999). Immunological methods for detection of carcinogen-DNA damage in humans. *Cancer Epidemiology, Biomarkers & Prevention, 8*, 733–739.

Schärer, O. D. (2003). Chemistry and biology of DNA repair. *Angewandte Chemie (International Edition in English), 42*, 2946–2974. https://doi.org/10.1002/anie.200200523

Schmeiser, H. H., Nortier, J. L., Singh, R., Gamboa da Costa, G., Sennesael, J., Cassuto-Viguier, E., Ambrosetti, D., Rorive, S., Pozdzik, A., Phillips, D. H., Stiborova, M., & Arlt, V. M. (2014). Exceptionally long-term persistence of DNA adducts formed by carcinogenic aristolochic acid I in renal tissue from patients with aristolochic acid nephropathy. *International Journal of Cancer, 135*, 502–507. https://doi.org/10.1002/ijc.28681

Schreier, W. J., Schrader, T. E., Koller, F. O., Gilch, P., Crespo-Hernández, C. E., Swaminathan, V. N., Carell, T., Zinth, W., & Kohler, B. (2007). Thymine dimerization in DNA is an ultrafast photoreaction. *Science, 315*, 625–629. https://doi.org/10.1126/science.1135428

Seo, K.-Y., Nagalingam, A., Tiffany, M., & Loechler, E. L. (2005). Mutagenesis studies with four stereoisomeric N2-dG benzo[a]pyrene adducts in the identical 5′-CGC sequence used in NMR studies: G→T mutations dominate in each case. *Mutagenesis, 20*, 441–448. https://doi.org/10.1093/mutage/gei061

Sharma, R., Lewis, S., & Wlodarski, M. W. (2020). DNA repair syndromes and cancer: Insights into genetics and phenotype patterns. *Frontiers in Pediatrics, 8*, 570084. https://doi.org/10.3389/fped.2020.570084

Shiizaki, K., Kawanishi, M., & Yagi, T. (2017). Modulation of benzo[a]pyrene–DNA adduct formation by CYP1 inducer and inhibitor. *Genes and Environment, 39*, 14. https://doi.org/10.1186/s41021-017-0076-x

Shimada, T. (2006). Xenobiotic-metabolizing enzymes involved in activation and detoxification of carcinogenic polycyclic aromatic hydrocarbons. *Drug Metabolism and Pharmacokinetics, 21*, 257–276. https://doi.org/10.2133/dmpk.21.257

Shirabe, K., Toshima, T., Taketomi, A., Taguchi, K., Yoshizumi, T., Uchiyama, H., Harimoto, N., Kajiyama, K., Egashira, A., & Maehara, Y. (2011). Hepatic aflatoxin B1-DNA adducts and TP53 mutations in patients with hepatocellular carcinoma despite low exposure to aflatoxin B1 in southern Japan. *Liver International: Official Journal of the International Association for the Study of the Liver, 31*, 1366–1372. https://doi.org/10.1111/j.1478-3231.2011.02572.x

Shrivastav, N., Li, D., & Essigmann, J. M. (2010). Chemical biology of mutagenesis and DNA repair: Cellular responses to DNA alkylation. *Carcinogenesis, 31*, 59–70. https://doi.org/10.1093/carcin/bgp262

Siantar, D. P., Trucksess, M. W., Scott, P. M., & Herman, E. M. (Eds.). (2008). *Food contaminants: Mycotoxins and food allergens, ACS symposium series*. American Chemical Society. https://doi.org/10.1021/bk-2008-1001

Smith, L. E., Denissenko, M. F., Bennett, W. P., Li, H., Amin, S., Tang, M., & Pfeifer, G. P. (2000). Targeting of lung cancer mutational hotspots by polycyclic aromatic hydrocarbons. *Journal of the National Cancer Institute, 92*, 803–811. https://doi.org/10.1093/jnci/92.10.803

Souliotis, V. L., Dimopoulos, M. A., & Sfikakis, P. P. (2003). Gene-specific formation and repair of DNA monoadducts and interstrand cross-links after therapeutic exposure to nitrogen mustards. *Clinical Cancer Research, 9*, 4465–4474.

Stratton, M. R., Campbell, P. J., & Futreal, P. A. (2009). The cancer genome. *Nature, 458*, 719–724. https://doi.org/10.1038/nature07943

Swenberg, J. A., Lu, K., Moeller, B. C., Gao, L., Upton, P. B., Nakamura, J., & Starr, T. B. (2011). Endogenous versus exogenous DNA adducts: Their role in carcinogenesis, epidemiology, and risk assessment. *Toxicological Sciences, 120*, S130–S145. https://doi.org/10.1093/toxsci/kfq371

Takeshita, T., Tao, F., Kojima, N., & Kanaly, R. A. (2019). Triple quadrupole mass spectrometry comparative DNA adductomics of Hep G2 cells following exposure to safrole. *Toxicology Letters, 300*, 92–104. https://doi.org/10.1016/j.toxlet.2018.10.023

Thier, R., Pemble, S. E., Kramer, H., Taylor, J. B., Guengerich, F. P., & Ketterer, B. (1996). Human glutathione S-transferase T1–1 enhances mutagenicity of 1,2-dibromoethane, dibromomethane and 1,2,3,4-diepoxybutane in Salmonella typhimurium. *Carcinogenesis, 17*, 163–166. https://doi.org/10.1093/carcin/17.1.163

Torrecilla, I., Ruggiano, A., Kiianitsa, K., Aljarbou, F., Lascaux, P., Hoslett, G., Song, W., Maizels, N., & Ramadan, K. (2024). Isolation and detection of DNA–protein crosslinks in mammalian cells. *Nucleic Acids Research, 52*, 525–547. https://doi.org/10.1093/nar/gkad1178

Tretyakova, N., Villalta, P. W., & Kotapati, S. (2013). Mass spectrometry of structurally modified DNA. *Chemical Reviews, 113*, 2395–2436. https://doi.org/10.1021/cr300391r

Tretyakova, N. Y., Groehler, A., & Ji, S. (2015). DNA–protein cross-links: Formation, structural identities, and biological outcomes. *Accounts of Chemical Research, 48*, 1631–1644. https://doi.org/10.1021/acs.accounts.5b00056

Turesky, R. J., & Le Marchand, L. (2011). Metabolism and biomarkers of heterocyclic aromatic amines in molecular epidemiology studies: Lessons learned from aromatic amines. *Chemical Research in Toxicology, 24*, 1169–1214. https://doi.org/10.1021/tx200135s

Usman, M., & Volpi, E. V. (2018). DNA damage in obesity: Initiator, promoter and predictor of cancer. *Mutation Research/Reviews in Mutation Research, 778*, 23–37. https://doi.org/10.1016/j.mrrev.2018.08.002

van Vugt-Lussenburg, B. M. A., Capinha, L., Reinen, J., Rooseboom, M., Kranendonk, M., Onderwater, R. C. A., & Jennings, P. (2022). "Commandeuring" xenobiotic metabolism: Advances in understanding xenobiotic metabolism. *Chemical Research in Toxicology, 35*, 1184–1201. https://doi.org/10.1021/acs.chemrestox.2c00067

Vodicka, P., Vodenkova, S., Horak, J., Opattova, A., Tomasova, K., Vymetalkova, V., Stetina, R., Hemminki, K., & Vodickova, L. (2023). An investigation of DNA damage and DNA repair in chemical carcinogenesis triggered by small-molecule xenobiotics and in cancer: Thirty years with the comet assay. *Mutation Research, Genetic Toxicology and Environmental Mutagenesis, 885*, 503564. https://doi.org/10.1016/j.mrgentox.2022.503564

Vongsutilers, V., & Gannett, P. M. (2018). C8-Guanine modifications: Effect on Z-DNA formation and its role in cancer. *Organic & Biomolecular Chemistry, 16*, 2198–2209. https://doi.org/10.1039/c8ob00030a

Wang, D., & Lippard, S. J. (2005). Cellular processing of platinum anticancer drugs. *Nature Reviews. Drug Discovery, 4*, 307–320. https://doi.org/10.1038/nrd1691

Wang, Y.-K., Li, W. Q., Xia, S., Guo, L., Miao, Y., & Zhang, B.-K. (2021). Metabolic activation of the toxic natural products from herbal and dietary supplements leading to toxicities. *Frontiers in Pharmacology, 12*, 758468. https://doi.org/10.3389/fphar.2021.758468

Wei, S. J., Chang, R. L., Hennig, E., Cui, X. X., Merkler, K. A., Wong, C. Q., Yagi, H., Jerina, D. M., & Conney, A. H. (1994). Mutagenic selectivity at the HPRT locus in V-79 cells: Comparison of mutations caused by bay-region benzo[a]pyrene 7,8-diol-9,10-epoxide enantiomers with high and low carcinogenic activity. *Carcinogenesis, 15*, 1729–1735. https://doi.org/10.1093/carcin/15.8.1729

Wilkens, S. (2015). Structure and mechanism of ABC transporters. *F1000Prime Reports, 7*. https://doi.org/10.12703/P7-14

Wogan, G. N., Hecht, S. S., Felton, J. S., Conney, A. H., & Loeb, L. A. (2004). Environmental and chemical carcinogenesis. *Seminars in Cancer Biology, 14*, 473–486. https://doi.org/10.1016/j.semcancer.2004.06.010

Xu, H., Li, J., Peng, S., Bai, Y., & Tang, W. (2025). From exposure to innovation: Decoding aromatic amines' role in bladder cancer mechanisms. *Discover Oncology, 16*, 888. https://doi.org/10.1007/s12672-025-02730-w

Yoshikura, H., & Taguchi, F. (1979). Induction of lytic plaques by murine leukemia virus in murine sarcoma virus-transformed nonproducer mouse cells persistently infected with mouse hepatitis virus MHV-S. *Intervirology, 11*, 69–73. https://doi.org/10.1159/000149015

Chapter 6
Biomarker and Xenobiotics—Prospects for the Assessment of Exposure and Effects in Humans

Teresa Perez-Piñar

Abstract Xenobiotics are chemical substances that don't belong to the human body. These include pollutants, pesticides, pharmaceuticals, among others. These compounds may enter the body through different pathways, such as inhalation or ingestion, and once they are inside, they may undergo diverse processes, such as biotransformation. Therefore, exposure to xenobiotics has raised as a critical concern in the field of health. This chapter aims to provide a comprehensive overview of current knowledge and emerging perspectives on the use of biomarkers in xenobiotic exposure in humans, since the metabolic evaluation requires sensitive and specific tools for detection and assessment. There are three types of biomarkers for xenobiotics. These are biomarkers of exposure, biomarkers of effect and biomarkers of susceptibility. Their detection and quantification rely on classical approaches such as gas and liquid chromatography, as well as cutting-edge methods based on omics technologies (genomics, proteomics, and metabolomics). In addition, the role of biosensors and wearable monitoring devices is discussed. Furthermore, particular attention must be given to emerging biomarkers, including oxidative stress markers, indicators of hepatic enzyme induction, and markers of immune activation. In practical terms, the chapter analyzes the integration of such biomarkers into occupational exposure monitoring, environmental risk assessment, and epidemiological studies. Nevertheless, current evidence shows that several challenges and limitations still remain, including sensitivity, specificity, or interindividual variability. Ethical concerns related to biomarker data collection in vulnerable populations are also examined. Finally, the chapter brings future perspectives that explore how artificial intelligence, machine learning, and predictive modeling can enhance biomarker interpretation and risk prediction.

Keywords Xenobiotics · Biomarkers of exposure · Biomarkers of effect · Biomarkers of susceptibility · Biosensors · Wearable monitoring devices

T. Perez-Piñar (✉)
Facultad de Ciencias Biomédicas y de la Salud, Universidad Alfonso X el Sabio (UAX), Avenida de la Universidad, 1, 28691 Villanueva de la Cañada, Madrid, Spain
e-mail: tperezpi@uax.es

K. Priya and S. K. Giri (eds.), *Environmental Toxicology: Xenobiotics and Effects on Human Genetic Material*, https://doi.org/10.1007/978-3-032-28619-2_6

1 Introduction

1.1 Definition and Relevance of Xenobiotics

Xenobiotics are defined as chemicals not naturally produced by organisms, including pharmaceuticals, pollutants, and pesticides, which may enter the body both intentionally or through environmental exposure (Štefanac et al., 2021). Usually, their metabolism differs between different species and in vitro models. This brings challenges for regulatory evaluation (Oesch & Hengstler, 2021). Even trace or chronic exposures can disrupt cellular and physiological processes, leading to toxicity (Singh et al., 2023). Regarding their metabolization, many xenobiotics are poorly metabolized and may form reactive intermediates. Therefore, understanding their mechanisms is crucial for developing safer, more sustainable compounds (Baird et al., 2005).

On the other hand, persistent substances have been reported to affect signaling, endocrine function, and gene expression, such as organochlorine pesticides (Diamanti-Kandarakis et al., 2009). Consequently, xenobiotics represent a major concern for toxicology, environmental health, and public policy. Different regulatory bodies, including the World Health Organization (WHO), Environmental Protection Agency (EPA), and European Chemicals Agency (ECHA), monitor their classification and control, emphasizing the need for sensitive biomarkers to detect exposure and support effective risk assessment (Angerer et al., 2007).

2 Human Exposure Pathways

Humans are usually exposed to xenobiotics through different pathways. These chemicals can enter the body through the skin, the gastrointestinal system, and the lungs, among others (Fig. 1). The nature of their effects is influenced by the concentration of the substance in specific target organs (Lu & Kacew, 2013).

In the process of absorption of a substance, it must trespass the cell membrane. This can be achieved through passive diffusion, membrane pores, carriers, or endocytosis (Lu & Kacew, 2013). The route of entry depends on the substance's solubility, size, charge, and other biochemical characteristics. Substances with high lipid solubility, for example, tend to cross cell membranes more easily via passive diffusion, making dermal and gastrointestinal absorption more likely (Klaassen, 2019). On the other hand, hydrophilic or charged compounds may require specific transport mechanisms or may follow paracellular routes, especially across epithelial tissues like the intestines or lungs (Hayes, 2014). Molecular weight can also influence absorption. Small molecules can penetrate biological barriers more readily than larger or bulky ones, which may face reduced membrane permeability (Lu & Kacew, 2013).

The route of entry affects the distribution of biomarkers within the body and determines which biological matrices, such as blood, urine, saliva, or exhaled breath,

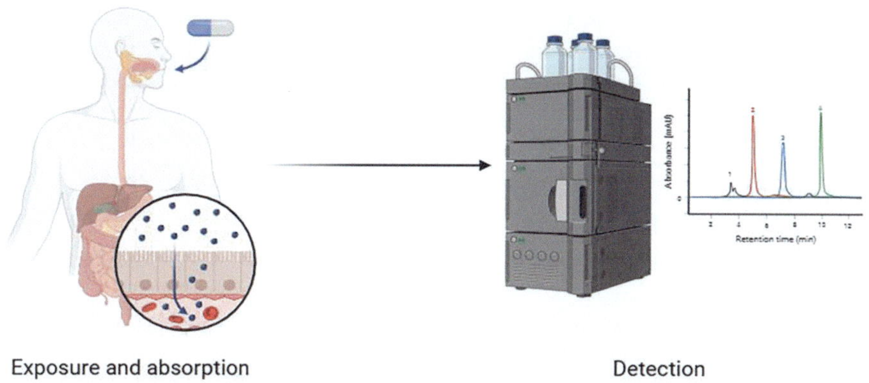

Fig. 1 Exposure, absorption, and detection of xenobiotics. Xenobiotics can enter the body through the skin, the gastrointestinal tract, or the respiratory system. Their presence can be detected using analytical techniques such as HPLC, which enable the identification and quantification of these compounds. *Created with BioRender.com*

are most suitable for biomonitoring purposes (WHO, 2023). The primary pathways of xenobiotics' absorption are the lungs, skin and gastrointestinal tract (Lu & Kacew, 2013).

Therefore, recognizing the preferred route of absorption based on a compound's chemical profile helps in selecting relevant biomarkers and improves the accuracy of exposure evaluation.

3 Health Risks and the Need for Biomarkers

The widespread presence of xenobiotics in the environment has brought significant concerns about their impact on human health. One of the main concerns lies in the biological effects of these substances. Numerous xenobiotics have been associated with endocrine disruption, immunotoxicity, carcinogenicity, and neurodevelopmental disorders (Gore et al., 2015). For instance, persistent organic pollutants (POPs) and heavy metals have been linked to hormonal imbalances and increased risks of metabolic diseases, such as diabetes and obesity (Lee et al., 2014). Moreover, certain xenobiotics may interfere with genetic and epigenetic regulation, potentially contributing to transgenerational effects (Skinner, 2014).

In addition, there is a critical need for early detection tools. These can provide insight into internal exposure and biological responses. Therefore, biomarkers emerge as essential instruments. They offer a means to detect early biological changes before clinical symptoms appear. Additionally, the necessity of such tools becomes even more apparent when considering vulnerable populations, such as children, pregnant women, and individuals with pre-existing health conditions.

The Centers for Disease Control and Prevention (CDC) have defined human biomonitoring as "the method for assessing human exposure to chemicals or their effects by measuring these chemicals, their metabolites or reaction products in human specimens" (CDC, 2005). We can benefit from human biomonitoring because (Angerer et al., 2007; CDC, 2005):

- It provides direct evidence of chemical exposure.
- We can detect health risks early.
- We can assess individual variability that can influence the metabolization of the chemical.
- We can identify vulnerable populations that are either more exposed or more susceptible to the effects of the xenobiotics.
- It can give support for public health policy in standards, laws and decisions.
- It allows for an assessment of environmental interventions.
- It improves risk assessment, as it offers more precise data.

Despite extensive research on xenobiotic exposure, there remains a lack of standardized and integrative frameworks to link biomarker data with health outcomes. Current approaches often assess exposure, effect, or susceptibility in isolation, limiting their predictive power for chronic, low-dose exposures. This chapter addresses these gaps by synthesizing existing biomarker evidence and highlighting novel strategies for exposure assessment.

Finally, the following sections detail the classification of biomarkers, analytical methodologies for their detection, and recent advances that enhance their applicability in environmental and occupational health contexts.

4 Classification of Biomarkers

Safe exposure levels to chemicals are usually expressed in intake units, such as acceptable or tolerable daily intake or reference dose (CDC, 2005). Because chemicals can accumulate or be eliminated through different tissues, various biological matrices are used in human biomonitoring, including blood, urine, hair, breast milk, saliva, and sweat (CDC, 2005; Esteban & Castaño, 2009). The selected matrix determines the type and interpretation of biomarkers detected.

According to the BEST classification (FDA-NIH, 2016), biomarkers can serve different purposes:

- Diagnostic: Detect the presence of a disease (e.g., sweat chloride for cystic fibrosis).
- Monitoring: Assess disease status or progression (e.g., PSA in prostate cancer).
- Predictive/Prognostic: Indicate susceptibility or likely outcomes (e.g., BRCA1/2 mutations).
- Pharmacodynamic/Response and Safety: Confirm biological or toxic responses (e.g., serum creatinine in nephrotoxicity).

- Susceptibility/Risk: Reflect an individual's predisposition (e.g., ApoE in Alzheimer's disease).

Alternatively, Perera and Weinstein (2000) classified biomarkers according to stages from exposure to disease, later simplified into three main categories commonly used in public health and toxicology (CDC, 2024; WHO, 2015):

- Biomarkers of exposure.
- Biomarkers of effect.
- Biomarkers of susceptibility.

4.1 *Biomarkers of Exposure*

It is defined as a metabolite or an exogenous substance that is measured in a biological matrix and reflects that an exposure to the chemical has occurred (NRC, 1989; US EPA, 2012).

In order to translate this internal data into useful information about the exposure, it is possible to perform an exposure reconstruction. Exposure reconstruction refers to the process of estimating a person's prior external exposure to a chemical based on biomonitoring data collected from the body. Often called reverse dosimetry, this approach uses pharmacokinetic (PK) models that incorporate physiological and metabolic information along with measured biomarker concentrations to calculate the likely dose or exposure (US EPA, 2012). This method is applied after exposure has occurred and is useful for estimating internal doses. However, it does not allow for the identification of specific exposure sources or pathways. Exposure reconstruction can offer one of the most accurate ways to assess total chemical exposure. A basic example of a PK model is the creatinine correction approach, which can be used for substances that are eliminated in the urine just a few hours after being ingested. More complex PK models, known as physiologically based pharmacokinetic (PBPK) models, simulate how chemicals are absorbed, distributed, metabolized, and stored in different organs and tissues. To function accurately, PBPK models require specific parameters related to physiology, chemistry, and biochemistry (US EPA, 2012). This reconstruction is not always done routinely, since that depends on the goals of the study or the data available.

A strong link between exposure levels and the biomarker is not always guaranteed, and individual responses to the same exposure can differ. For this reason, it is essential to recognize and control for variables that may affect how exposure relates to biological response. Other methods to validate biomarkers involve evaluating their correlation with levels in key target organs (such as cadmium in the kidney compared to blood or urine) or with significant health outcomes (like neurotoxic effects related to blood lead levels) (WHO, 2001).

Biomarkers of exposure to different xenobiotics are well reported. For example, mercury is found in a wide range of contexts, such as environmental or dietary. This xenobiotic can be detected as a biomarker either in blood or in urine (Fig. 2), with

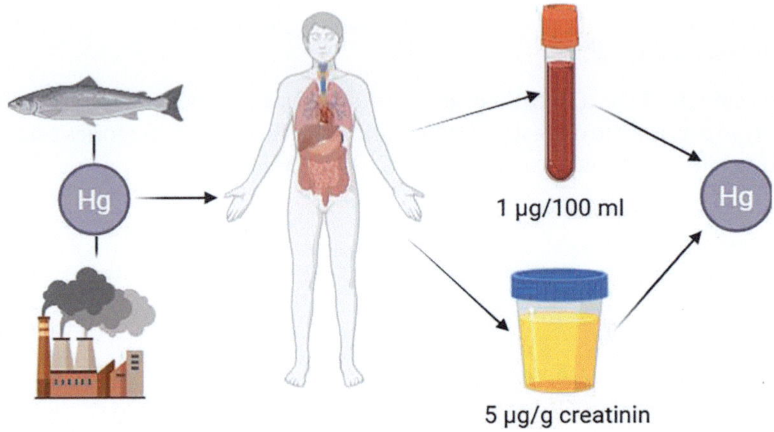

Fig. 2 Mercury as a xenobiotic and its biomarkers detected in blood and urine. Human exposure to mercury mainly occurs through the consumption of contaminated fish and emissions from industrial sources. Reference values are approximately 1 μg/100 mL in blood and 5 μg/g creatinine in urine (Gil & Pla, 2001). *Created with BioRender.com*

reference values of <5 μg/g of creatinine in urine and <1 μg/100 mL in blood (Gil & Pla, 2001). Another example is found in styrene, an organic compound found mainly in industrial environments. There are at least three biomarkers of exposure to this xenobiotic that can be detected in different matrices depending on the compound. Thus, mandelic acid is detected in urine, with a reference value of 1 g/g of creatinine. Phenylglyoxylic acid is also found in urine, with a reference value of 350 mg/g of creatinine. Finally, styrene can be detected as a biomarker either in blood or exhaled air, with a value of 18 ppm (Gil & Pla, 2001).

4.2 *Biomarkers of Effect*

Biomarkers of effect indicate measurable changes in an organism due to chemical exposure (WHO, 2015), linking environmental chemicals to health alterations (Rodríguez-Carrillo et al., 2023). Unlike exposure biomarkers, which reflect absorption, effect biomarkers capture molecular or cellular changes, aiding risk assessment and prevention. To be useful, they must be selective, sensitive, reliable, reproducible, and validated (National Academies of Sciences, Engineering, and Medicine, 2020). Validation is particularly challenging for intermediate effect biomarkers in risk assessment but improves study power and allows earlier results, especially in prospective epidemiological studies. The predictive value of a biomarker is assessed by how well it forecasts disease and the attributable proportion of cases, depending on assay sensitivity and relative risk (WHO, 2001).

The early and late effect biomarkers are distinguished depending on the time when the key events after the exposure occur (Rodríguez-Carrillo et al., 2023). At this point, it must also be differentiated between the effect of a single xenobiotic or of a mixture of them (Vinggaard et al., 2021). In addition, some of them are evaluated at an individual level, but others are evaluated at the population level (Jeddi et al., 2021).

The H2020 Human Biomonitoring for Europe (HBM4EU) project (2021) aimed to coordinate and harmonize human biomonitoring in Europe in order to assess the population's exposure to chemical substances, to support the development of evidence-based policies, and to provide data on the relationship between chemical exposure and potential health effects (Jeddi et al., 2021). Therefore, they identified some effect biomarkers, such as PFAS (perfluoroalkyl substances) related to immunologic and endocrine effects. In addition, they validated and standardized existing biomarkers, such as bisphenol S as a substitute for bisphenol A, or 4-hydroxyphenylpyruvic acid as an alternative to β_2-microglobulin for renal effects (Jeddi et al., 2021).

Rodríguez-Carrillo et al. (2023) classified effect biomarkers depending on their previous use in epidemiological and clinical settings as classical effect biomarkers, classical effect biomarkers and novel effect biomarkers.

Jeddi et al. (2021) classified effect biomarkers according to the effect they evaluate:

- Anthropometric biomarkers, such as body mass index for bisphenols.
- Cardiovascular biomarkers, such as blood pressure for inorganic arsenic.
- Serum lipids, such as low-density lipoprotein (LDL) for mycotoxins.
- Glucose homeostasis, such as glycated hemoglobin (HbA1c) for acrylamide.
- Hepatic biomarkers, such as aspartate transaminase (AST) for bisphenols.
- Renal function, such as albumin for hexavalent chromium.
- Glucocorticoid hormones, such as corticosterone, for inorganic arsenic.
- Reproductive hormones, such as estradiol, for phthalates.
- Sperm quality, such as sperm concentration for hexavalent chromium.
- Thyroid homeostasis, such as thyroid-stimulating hormone (TSH) for bisphenols.
- Cancer biomarkers, such as plasma total homocysteine for hexavalent chromium.
- Genotoxicity biomarkers, such as chromosomal aberrations for acrylamide.
- Oxidative stress, such as 8-isoprostane for phthalates.
- Epigenetics and gene expression biomarkers, such as H3K9me2 for lead.
- Immunology and allergy biomarkers, such as IgE, for inorganic arsenic.
- Neuropsychological biomarkers, such as cholinesterase activity for pesticides.

These biomarkers could also provide information about the severity or the potential reversibility.

4.3 Biomarkers of Susceptibility

A susceptibility biomarker informs about the probability of developing a disease in a healthy individual (FDA-NIH Biomarker Working Group, 2020).

In clinical settings, such biomarkers are primarily applied to inform and guide preventive measures. An elevated concentration of low-density lipoprotein (LDL) cholesterol is a well-known example of a biomarker indicating susceptibility or risk, as it signals a higher likelihood of developing coronary artery disease (Field & Vasan, 2018).

Although they have similar features with prognostic biomarkers, the key difference lies in their context of use: prognostic markers apply to patients already diagnosed with a specific illness, whereas susceptibility or risk markers can be employed in individuals who otherwise show no signs of disease (FDA-NIH Biomarker Working Group, 2020).

Susceptibility biomarkers are mainly genetic, epigenetic, or related to a person's metabolism or immune system (Dehghani & Yunesian, 2024). One limitation of using genetic mutations as biomarkers is their low occurrence rate in genome-wide association studies (GWASs), often showing, at most, about a 1% correlation between affected individuals and the disease. By contrast, epigenetic modifications demonstrate a much stronger association, typically exceeding 90–95%, with disease presence in epigenome-wide association studies (EWASs). Numerous human disorders have been linked to epigenetic biomarkers that are specific to particular diseases and can reveal an individual's susceptibility to develop the condition (Skinner, 2024). Blood has been widely used to develop genetic biomarkers, since every cell type in the human body shares the same genetic material. Epigenetics, however, presents a different scenario: each cell type possesses its own distinct epigenomic profile. When multiple cell types are present, the resulting epigenetic readout becomes a complex composite, influenced both by the proportion of each cell type and by their individual epigenetic signatures. Therefore, blood is not efficient for epigenetic assessment. An alternative is to analyze plasma, which contains cell-free DNA and can serve as a valuable source for detecting epigenetic alterations (Skinner, 2024).

Cytochrome P450 enzymes often play a key role in the effects observed after exposure to xenobiotics. They generate a primary metabolite, which is subsequently inactivated in later steps. These enzymes are encoded by different genes, and genetic polymorphisms in those genes have therefore been investigated as potential biomarkers (Cauteren et al., 1996). These are called polymorphisms of activated systems (Gil & Pla, 2001) (Fig. 3).

In addition, there is another group of biomarkers of susceptibility; these polymorphism markers in detoxification systems are assessed by measuring the activity of conjugating enzymes, including glutathione-S-transferases, acetyltransferases, sulfotransferases, glucuronyltransferases, and paraoxonase (Gil & Pla, 2001).

Thus, susceptibility biomarkers serve to detect individuals who have a heightened sensitivity to chemical agents. Gaining deeper insight into the information these biomarkers provide supports more accurate risk evaluations and aids in

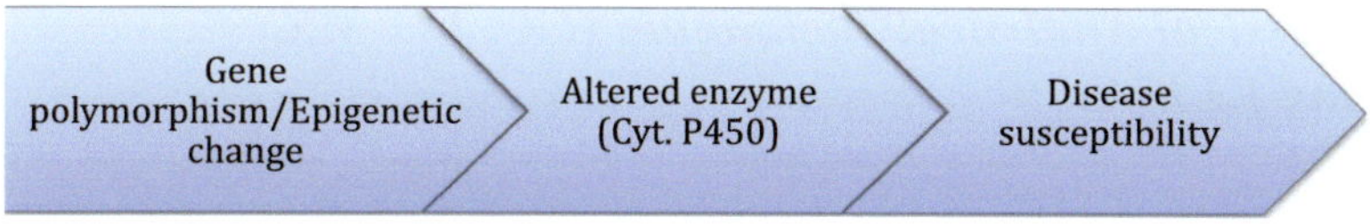

Fig. 3 Schematic representation of susceptibility biomarkers. Genetic polymorphisms or epigenetic modifications can lead to altered enzyme activity, such as changes in cytochrome P450, ultimately increasing disease susceptibility

defining susceptibility patterns related to human developmental toxicity (Dehghani & Yunesian, 2024).

5 Analytical Methods for Biomarker Detection

Selection of an appropriate biomarker requires understanding the xenobiotic's metabolism and excretion (Li, 2012). Its concentration can be affected by the sample matrix, collection time, and ADME processes, absorption, distribution, metabolism, and excretion (Aylward et al., 2014). Toxico-kinetics must be considered to evaluate how quickly a xenobiotic enters and moves through the body (DeBord et al., 2022). At this point, the choice of matrix is critical. For instance, urine biomarker levels often reflect average plasma concentrations over the bladder retention time (Rosenberg et al., 1989). Some chemicals are excreted unchanged, while others undergo metabolic transformations and conjugation (DeBord et al., 2022).

When all the above-mentioned factors are considered, the selection of the appropriate method comes next (Fig. 4).

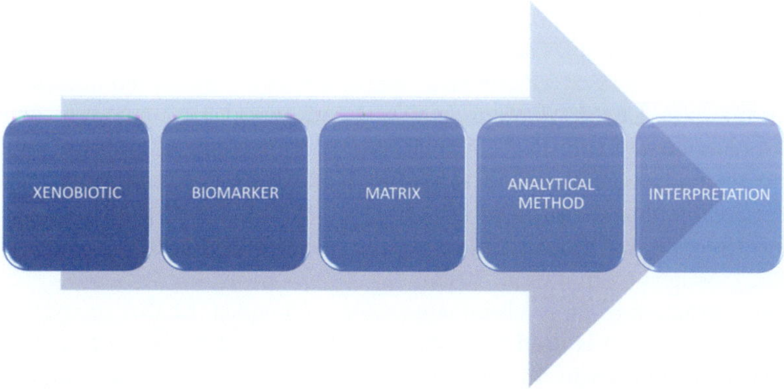

Fig. 4 Conceptual workflow of xenobiotic biomonitoring. The process starts with the identification of the xenobiotic and its corresponding biomarker, followed by the selection of the appropriate biological matrix and analytical method, leading to data interpretation for risk assessment

5.1 Classical Analytical Techniques (GC, LC, MS, and ELISA)

In all the techniques, sample preparation is a crucial first step for the success of the detection. Such preparation usually includes sample pretreatment, isolation from the matrix, a purification process, and concentration of the compound. Additional procedures depend on the sample's composition and the study's objectives. For example, serum, which contains both aqueous metabolites and abundant lipids, demands thorough protein removal and the use of phase separation techniques (Zhou et al., 2012). Regarding xenobiotics, for instance, pesticides, mycotoxins, hormones, among others, have been reported to be better extracted using acetonitrile in QuECHERS (Quick, Easy, Cheap, Effective, Rugged, and Safe) method, where sample isolation is achieved in one step (Anastassiades et al., 2003). Other extraction methods include solid-phase extraction, solid-phase microextraction, hollow-fiber liquid-phase microextraction, liquid–liquid extraction, solid–liquid extraction, and more (Tuzimski & Sherma, 2019). For example, solid-phase microextraction is used for chlorinated-bisphenols (Lashgari et al., 2019), or focused ultrasonic solid–liquid extraction is used for phthalate esters (Errekatxo et al., 2008).

Chromatography techniques are used to separate chemicals with similar structures. Gas chromatography (GC) has been used to detect volatile and thermostable substances, such as volatile pesticides. High-performance liquid chromatography (HPLC) has been used to detect polar or thermolabile non-volatile substances, such as amino acids. A variant of this method is found in the UPLC (ultra-performance liquid chromatography) (Fig. 5).

For decades, mass spectrometry (MS) has been widely used for xenobiotic metabolite analysis. When it is coupled with UPLC and quadrupole time-of-flight MS, it provides high peak resolution, low ion detection limits, and precise mass measurements (Johnson et al., 2012) and has been applied to detect metabolites of compounds such as fenofibrate (Liu et al., 2009) and aminoflavone (Chen et al., 2006). In these techniques, injected samples are first separated, then ionized—by electron impact in GC/MS or by electrospray or chemical ionization in LC/MS—and detected based on their m/z ratio. LC/MS is often combined with multivariate analyses, including principal component analysis (PCA) and orthogonal projections to latent structures discriminant analysis (OPLS-DA), to identify biomarkers, discriminate samples, and characterize complex chemical profiles in fields such as metabolomics, counterfeit cosmetic detection, and bioactive plant metabolite studies (Plumb et al., 2003; Turova et al., 2021; Waters Corporation, 2015).

As an alternative to these chromatographic techniques, ELISA (enzyme linked immunosorbent assay) is used to identify substances that can be detected by antibodies, such as some toxins. In this method, an antibody recognizes an antigen which is then detected by a marked-secondary antibody which will generate a signal (color, fluorescence, chemiluminescence). For instance, it has been used to study pharmaceuticals in aquatic environments or xenobiotics with endocrine effects (Estévez-Alberola & Marco, 2004; Jaria et al., 2020). In these examples, xenobiotics have

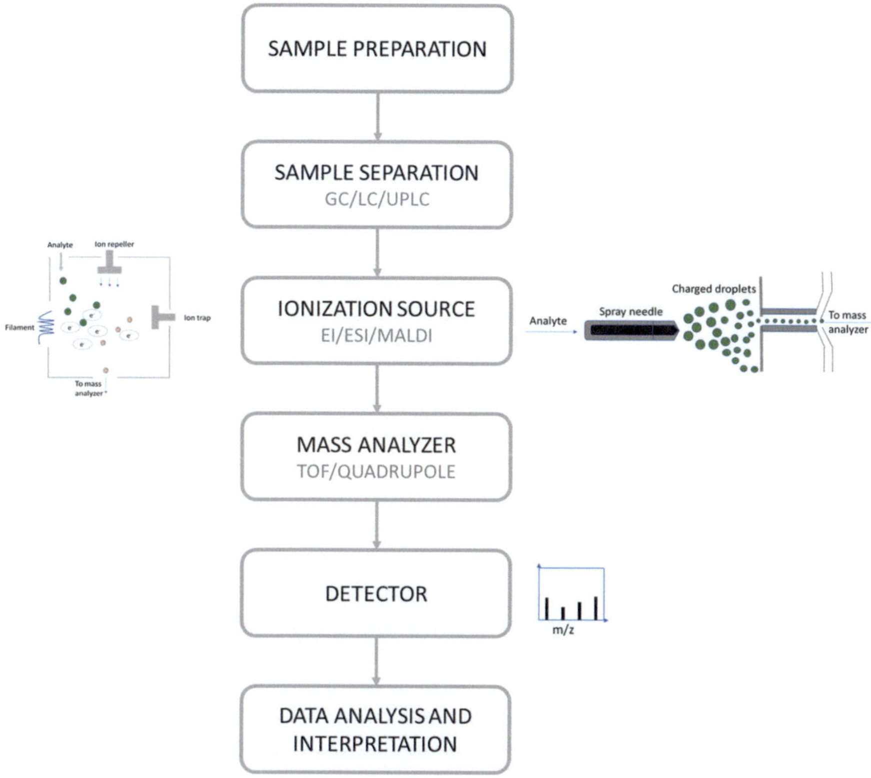

Fig. 5 General workflow of mass spectrometry (MS) analysis. The process includes sample preparation, chromatographic separation (GC, LC, or UPLC), ionization (EI, ESI, or MALDI), mass analysis (TOF or quadrupole), detection, and data interpretation

been detected, such as bisphenol A, alkylphenol or pesticides, among others. The high sensitivity and specificity together with the low cost of the method make it a rapid screening technique (Jaria et al., 2020). However, the use of antibodies can be a limiting factor.

5.2 *Advances in Omics Technologies*

Omics approaches enable the systematic study of complex molecular groups within an organism, including metabolomics, proteomics, genomics, and transcriptomics. Metabolomics has been widely applied to investigate the contribution of reactive metabolites and the disruption of endogenous metabolism following xenobiotic exposure, providing insights into xenobiotic-induced toxicity (Chen & Kim, 2013; Johnson et al., 2012; Kaddurah-Daouk et al., 2008; Lindon & Nicholson, 2008). For instance, 3,3′-biacetaminophen, a novel acetaminophen dimer formed via oxidative coupling, was identified in urine of wild-type mice treated with 400 mg/kg APAP using high-resolution mass spectrometry (Qi et al., 2002).

Mass spectrometry (MS) and nuclear magnetic resonance (NMR) are the primary detection methods, with MS offering high sensitivity and NMR providing a non-destructive alternative (Chen & Kim, 2013). Standard GC/MS and LC/MS remain widely used, though they require extensive sample preparation. Emerging methods, such as ion mobility spectrometry coupled with MS (IMS/MS), allow rapid separation, low sample consumption, and isomer distinction, having been applied to over 4000 xenobiotics (Teri et al., 2025). Elemental analysis is achievable via inductively coupled plasma MS (ICP/MS). Combinations of techniques, such as HR-MAS-NMR with GC/MS or GC $\times$ GC/TOF/MS, further enhance resolution and quantification, while stable isotope labeling (^{2}H, ^{13}C, ^{15}N, ^{18}O) facilitates discrimination from endogenous metabolites (Chan et al., 2009; Chen & Kim, 2013; Johnson et al., 2012; Shi et al., 2012).

Pharmacometabolomics integrates metabolic pathway analysis to predict or assess drug response and can be combined with pharmacogenomics. For example, Ji et al. (2011) identified basal glycine levels and an SNP in glycine dehydrogenase as biomarkers predicting response to citalopram/escitalopram in major depression. Proteomics and transcriptomics complement these analyses; enzyme profiling in human skin via LC/MS/MS (Eijl et al., 2012) and RNA sequencing in rat liver (Annalora et al., 2024) illustrate their utility in xenobiotic biomarker discovery.

No single analytical platform can cover the entire metabolome due to the diverse physicochemical properties of metabolites. Integrative omics, combining approaches such as metabolomics and transcriptomics, has been successfully applied in plants (Pérez-Piñar et al., 2024) and in xenobiotic biomarker research, providing a more comprehensive overview of exposure and biological response (Karkossa et al., 2020).

6 Emerging Biosensor Technologies

Biosensors, composed of nanomaterials, recognition elements, and a signal transduction system (Fig. 6), are effective tools for detecting xenobiotics, including pesticides, heavy metals, PAHs, toxins, pharmaceuticals, hormones, and emerging contaminants (Kimmel et al., 2012; Mohamed, 2020). Their advantages include portability, low cost, sustainability, and ease of use. Commercial sensors employ electrochemistry, surface plasmon resonance (SPR), nanowires, and other microstructures; SPR offers high sensitivity and low detection limits, while electrochemical sensors provide fast, high-throughput screening (Liu et al., 2020).

Biosensors consist of three elements: a bioreceptor that specifically binds the target molecule (often antibodies on 2D or 3D platforms), a transducer converting the binding into a measurable signal, and a processing system to amplify and analyze the output (Liu et al., 2020). These operate on nanoscale adaptations of ELISA principles. Materials used include graphene, ITO, carbon, and titanium dioxide electrodes,

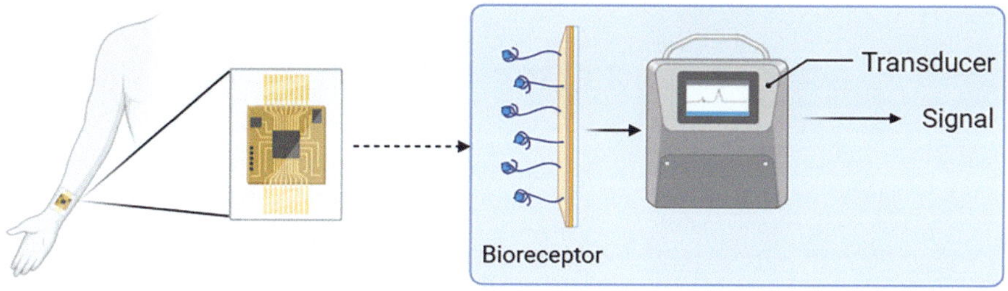

Fig. 6 Biosensor components include bioreceptor, transducer and processing system. *Created with BioRender.com*

enabling detection of diverse biomarkers such as AFP, PSA, and viral antigens (Arya et al., 2017; Feng et al., 2018; Hu et al., 2018; Klukova et al., 2016; Li et al., 2017; Seenivasan et al., 2017; Yang et al., 2018; Ye et al., 2020).

Nanowire-based sensors, characterized by high surface-to-volume ratios, immobilize bioreceptors on nanowires to rapidly transduce signals, with applications in cancer biomarker and environmental pollutant detection (Daoudi et al., 2021; Gao et al., 2017; Zhang et al., 2015).

Wearable biosensors integrate flexible devices such as skin patches, eyeglasses, or contact lenses for real-time monitoring of metabolites including lactate, cortisol, uric acid, glucose, and xenobiotics like caffeine (Kim et al., 2018, 2019; Tai et al., 2018). Challenges remain in biocompatibility, signal interference, power supply, and long-term stability, requiring further development for routine xenobiotic monitoring (Zheng et al., 2024).

7 Selected Case Studies and Applications

7.1 Biomarkers of Oxidative Stress

Commonly, biomarkers of oxidative stress are 8-isoprostane, 4-hydroxy-2-nonenal, 8-oxo-7,8-dihydro-2′-deoxyguanosine (8-OHdg), bilirubin, and others (Fig. 7) detected in urine, plasma or serum, mainly by chromatography coupled to MS and ELISA (Steffensen et al., 2020).

One noteworthy case is found in the study of Kim et al. (2004), levels of 8-OHdG in urine were measured in boilermaker workers exposed during industrial maintenance operations. The goal was to assess the exposition to fine particulate matter derived from residual oil ash and metal fumes, studying the association with DNA damage. For 5 days measures were done in 27 workers, finding that levels of 8-OHdG were increased at the end of the day. They used ICP/MS to detect metals, and ELISA to detect 8-OHdG. Analyses revealed that there is a relation between the exposure and the response, increasing 1.67 μg/g of 8-OHdG for every increase of 1 mg/m^3

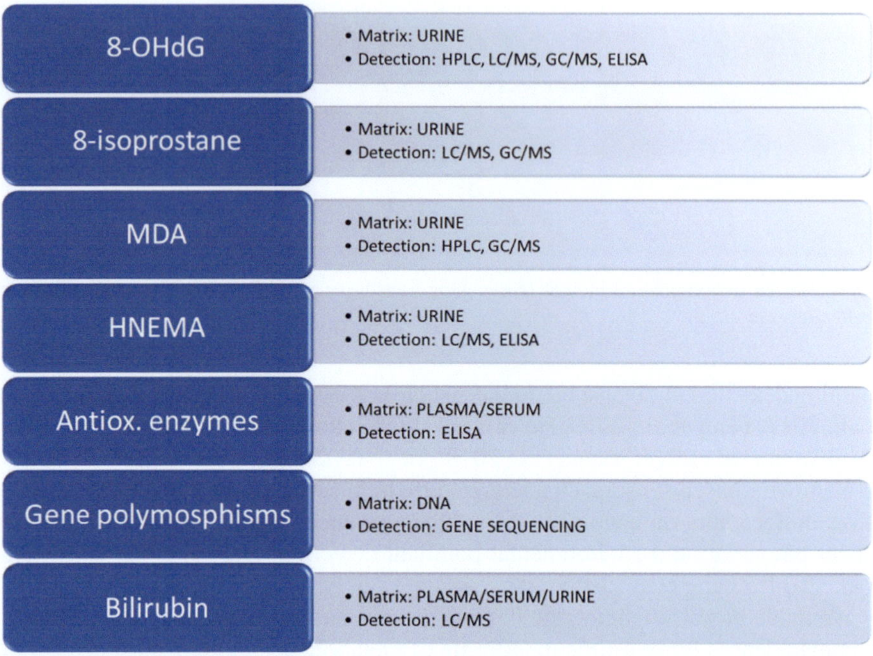

Fig. 7 Oxidative stress biomarkers, with their matrix and detection technique. These are detected in urine, plasma or serum, with HPLC, LC/MS, GC/MS or ELISA. Data compiled from studies on oxidative stress biomarkers and analytical methods

of exposure $PM_{2.5}$, supporting that 8-OHdG might work as a biomarker of oxidative damage induced by residual oil ash and metal fumes.

This compound has been reported and validated as a biomarker of oxidative damage to DNA in humans, animals and in vitro models for the European Standards Committee on Urinary (DNA) Lesion Analysis (ESCULA) (Cooke et al., 2008).

8 Hepatic and Metabolic Biomarkers

Gonzalez et al. (2015) reviewed metabolic markers of xenobiotic-induced hepatotoxicity, using genetically modified mice with human genes in the mouse genome. Most metabolites were detected by using LC/MS. For instance, acetaminophen, a widely used analgesic drug, has several biomarkers, such as serum acylcarnitines. Ethanol xenobiotic can be detected through indole-3-lactic acid in urine. Rifampin with isoniazid, typically used against tuberculosis, has hepatotoxicity that can be detected through protoporphyrin IX in bile ducts or alkaline phosphatase in serum.

Biomarkers of drug-induced injury are found in serum and blood, such as miR-122, GLDH (glutamate dehydrogenase), HMGB1 (high mobility group protein B1), cleaved/full-length cytokeratin 18 ratio, MCSFR (macrophage colony-stimulating factor receptor), osteopontin, total keratin 18, caspase-cleaved keratin 18, apoptotic index, cytokines, albumin, total bilirubin, alanine aminotransferase, aspartate aminotransferase, alkaline phosphatase, among others (Andrade et al., 2019; Church & Watkins, 2017).

A noteworthy case is found in the study of Sadler et al. (2016). They took advantage of a multiomics analysis to characterize the ontogeny of the cytochrome P450 enzymes in the human liver through all stages of human life. Thus, transcriptomics, proteomics and enzymatic activity assays were integrated to study human liver at prenatal, postnatal, infantile, juvenile and adult stages. Different isoforms of the enzymes P450 were analyzed with RT-PCR and LC/MS. They found some biomarkers to be specific to a stage. For example, P450 class 3A7 was identified as a specific biomarker of the fetal stage during liver development. After birth, this isoenzyme was replaced by others of the same family, which have a role in metabolizing xenobiotics.

9 Inflammatory and Immune Biomarkers

Biomarkers of inflammation are well known, including cytokines, acute phase proteins, soluble receptors, coagulation factors and others (Julián-Jiménez et al., 2013). Among them, interleukins indicate inflammation, C-reactive protein reflects systemic inflammation or infection, CD receptors show immune activation, and lactate indicates tissue damage.

A relevant study was performed by Liu et al. (2022), where they assessed how the combined exposure to environmental chemicals affects immunoinflammatory biomarkers of peripheral blood. With more than 1700 participants, 21 compounds were evaluated, including perfluoroalkyl substances, aromatic hydrocarbons and metallic elements. Biomarkers studied were IL-6, TNF-α, CRP, antioxidant enzymes (superoxide dismutase, glutathione peroxidase and catalase). They found that xenobiotic exposure, especially to cadmium, perfluorooctane sulfonate and perfluorodecanoic acid, is associated with the mentioned biomarkers.

10 Applications in Occupational and Environmental Health

The relationship between xenobiotic exposure and health is primarily assessed through occupational epidemiology (de Vocht et al., 2012). Exposure can be categorized as available, intake, absorbed, or biologically effective dose. Public health repositories such as the CDC (2022) and HBM4EU (n.d.) provide validated biomarker data, with the CDC offering online tables including demographics, survey years, concentrations, and sample sizes. Other resources include the Comparative Toxicogenomics Database (CTD, 2004).

Biomarker validation follows guidelines from the US Environmental Protection Agency (2018) and NRC, requiring that biomarkers reflect exposure, effect, or susceptibility, be measured with validated methods such as HPLC/MS, and correlate with known outcomes. The WHO (1993) proposed a framework emphasizing analytic and toxicologic validation and public health utility. For example, de Vocht et al. (2012) reported the use of biomarkers as thiocyanate, carbon monoxide, and cotinine to assess tobacco smoke exposure across different matrices, typically detected via chromatography coupled to mass spectrometry.

Another example of an occupational study is found in the evaluation of exposure to glyphosate in urinary biomarkers of oxidative stress in agricultural workers (Chang et al., 2023). Glyphosate is a commonly used herbicide in agriculture. Such exposure could represent a risk for the health of the workers due to the oxidative stress that it produces, which can evolve to cancer or chronic diseases. The analyzed biomarkers were 8-OHdg, malondialdehyde, isoprostanes, and 8-isoprostanic acid, which were detected using HPLC/MS. The results showed a significant association between glyphosate exposure and the studied biomarkers, indicating the necessity of protection in the agricultural environment (Fig. 8).

In a Parisian subway study, it was assessed the relation between the exposure to metallic particles in the air and the oxidative stress biomarkers in exhaled air and in urine. Such particles are generated by the train brakes, remaining in suspension in the air. These can lead to respiratory and cardiovascular diseases. Malondialdehyde, 8-OHdg and 8-isoprostane were studied as biomarkers during 10 days. Associations between the biomarkers malondialdehyde and 8-OHdg and specific metals were shown, such as copper or zinc, recommending these biomarkers to be used in large field studies.

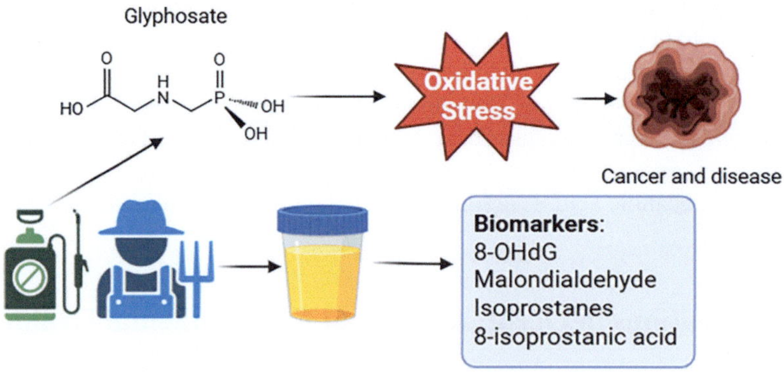

Fig. 8 Biomarkers of glyphosate exposure in an agricultural context. Glyphosate, present in herbicides used by farmers, can cause severe diseases through oxidative stress. Different biomarkers can be detected in urine, such as 8-OhdG, malondialdehyde, isoprostanes or 8-isoprostanic acid. *Created with BioRender.com*

11 Challenges, Limitations, and Ethical Considerations

Biomonitoring provides valuable data on human exposure to chemicals but faces several limitations. Rapidly eliminated substances can only be detected shortly after exposure (Esteban & Castaño, 2009). Analytical selectivity, distinguishing target analytes from similar compounds or background noise, remains challenging, as does differentiating between occupational and environmental exposure. Carryover effects can also lead to false results (CDC, 2022).

No biological matrix is universally suitable; each has advantages and drawbacks depending on the analyte and context. Ideal matrices should be easily obtainable, non-invasive, and accurately reflect internal chemical load (Esteban & Castaño, 2009). Whole blood is suitable for metals and metabolites but is invasive, while serum is preferred for water-soluble compounds with short half-lives. Non-invasive matrices such as urine, placenta, or breast milk are practical but may introduce matrix effects that may alter analytical responses (Zhu et al., 2025). These effects arise from endogenous or exogenous components such as salts, lipids, or anticoagulants (Garcia, 2005; Little et al., 2006; Mei et al., 2003; Tang & Kebarle, 1993). Ion suppression is common in LC/MS, while signal enhancement occurs in GC/MS (Panuwet et al., 2016). Strategies to minimize these issues include chromatographic optimization, sample dilution, solid-phase extraction, and use of isotopically labeled standards (Panuwet et al., 2016; CDC, 2022). Thus, determining detection and quantitation limits (LOD, LLOQ, and ULOQ) is essential to ensure accuracy, particularly for low-level exposures (CDC, 2022).

Additionally, interindividual variability affects biomonitoring outcomes. Genetic polymorphisms in xenobiotic-metabolizing enzymes (XMEs) influence detoxification efficiency and susceptibility to disease (Crocco et al., 2019; Dash et al., 2006; Spriggs et al., 2018). Thus, variants in genes such as *CYP2B6*, *CYP3A5*, and COMT have been associated with longevity and differential enzyme activity by age, sex, and metabolic rate. Stability testing further ensures sample integrity under varying storage and handling conditions (CDC, 2022).

Consequently, ethical, legal, and social considerations are fundamental in human biomonitoring. These include privacy, data ownership, and preventing misuse of exposure information (Schulte, 1992; Zhu et al., 2025). Transparent governance, informed consent, and confidentiality are critical to maintaining public trust. Moreover, translating biomonitoring findings into health policy and risk communication is essential for effective public health action. In low- and middle-income countries (LMICs), limited infrastructure and resources constrain biomarker application, but these tools hold promise for early disease detection and cost-effective healthcare improvement (Basholli et al., 2018; Jayathissa & Rupasinghe, 2024).

12 Future Directions

Predictive toxicology is evolving toward more personalized and data-driven approaches. In this regard, Gaurav (2024) identifies three key areas for progress: genetic-based predictive modeling to assess susceptibility to drugs and environmental toxicants, enhanced biomonitoring that combines exposure data with individual biology, and individualized intervention strategies for prevention and treatment. The FDA Predictive Toxicology Roadmap outlines a strategy to better anticipate toxicological risks through advanced predictive models, personalized approaches, collaborative innovation, and transparent evaluation (FDA, 2017).

The integration of artificial intelligence (AI) and machine learning (ML) transforms toxicology from traditional empirical testing into a data-intensive discipline (Hartung, 2023). AI and deep learning enable efficient screening of chemical structures, biological targets, and toxicity outcomes (Zhang et al., 2025). Therefore, future directions include AI-driven risk modeling, probabilistic risk assessment, development of implementation guidelines, and ensuring ethical and regulatory compliance (Hartung, 2023; Tonoyan & Siraki, 2024; Zhang et al., 2025).

13 Conclusion

Biomarkers are essential for studying human exposure to xenobiotics, providing insight into both the presence of toxic compounds and their biological effects, with classifications into exposure, effect, and susceptibility biomarkers linking environmental agents to physiological responses and health outcomes. Advances in analytical techniques, including omics technologies, biosensors, and wearable devices, have enhanced detection sensitivity and enabled real-time monitoring, supporting more accurate and individualized risk assessments, though challenges remain in sensitivity, specificity, interindividual variability, and standardization. In addition, the integration of artificial intelligence and predictive models promises improved interpretation and personalized risk profiling. However, future research must not forget to address ethical considerations, robust regulatory frameworks, and the translation of findings into public health policy.

Ultimately, biomarkers represent a vital link between environmental exposure and clinical outcomes, contributing to prevention strategies, regulatory decisions, and the promotion of public health (Fig. 9).

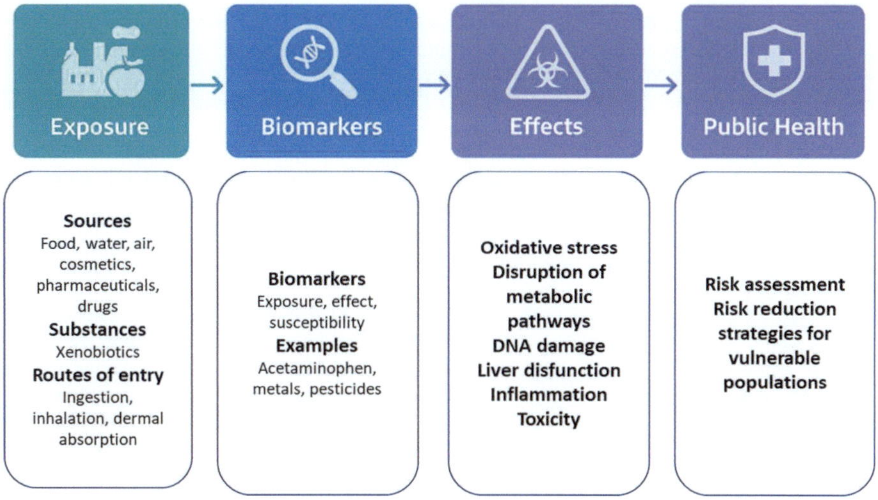

Fig. 9 Exposure to xenobiotics may occur through various sources, including food, water, and air, entering the body via ingestion, inhalation, or dermal absorption. Three main types of biomarkers can be distinguished (exposure, effect, and susceptibility), each providing insight into biological responses such as oxidative stress, metabolic disruption, DNA damage, liver dysfunction, inflammation, or general toxicity. Public health implications are addressed through comprehensive risk assessment and the development of mitigation strategies, particularly for vulnerable populations

References

Anastassiades, M., Lehotay, S. J., Tajnbaher, D., & Schenck, F. J. (2003). Fast and easy multiresidue method employing acetonitrile extraction/partitioning and "dispersive solid-phase extraction" for the determination of pesticide residues in produce. *Journal of AOAC International, 86*(2), 412–431.

Andrade, R. J., Chalasani, N., Björnsson, E. S., Suzuki, A., Kullak-Ublick, G. A., Watkins, P. B., Devarbhavi, H., Merz, M., Lucena, M. I., Kaplowitz, N., & Aithal, G. P. (2019). Drug-induced liver injury. *Nature Reviews Disease Primers, 5*(1), 58. https://doi.org/10.1038/s41572-019-0105-0

Angerer, J., Ewers, U., & Wilhelm, M. (2007). Human biomonitoring: State of the art. *International Journal of Hygiene and Environmental Health, 210*(3–4), 201–228. https://doi.org/10.1016/j.ijheh.2007.01.024

Annalora, A. J., Coburn, J. L., Jozic, A., Iversen, P. L., & Marcus, C. B. (2024). Global transcriptome modulation by xenobiotics: The role of alternative splicing in adaptive responses to chemical exposures. *Human Genomics, 18*, 127. https://doi.org/10.1186/s40246-024-00694-6

Arya, S. K., Kongsuphol, P., & Park, M. K. (2017). Off surface matrix based on-chip electrochemical biosensor platform for protein biomarker detection in undiluted serum. *Biosensors and Bioelectronics, 92*, 542–548. https://doi.org/10.1016/j.bios.2016.10.063

Aylward, L. L., Hays, S. M., Smolders, R., Koch, H. M., Cocker, J., Jones, K., Warren, N., Levy, L., & Bevan, R. (2014). Sources of variability in biomarker concentrations. *Journal of Toxicology and Environmental Health, Part B, 17*(1), 45–61. https://doi.org/10.1080/10937404.2013.864250

Baird, W. M., Hooven, L. A., & Mahadevan, B. (2005). Carcinogenic polycyclic aromatic hydrocarbon–DNA adducts and mechanism of action. *Environmental and Molecular Mutagenesis, 45*(2–3), 106–114. https://doi.org/10.1002/em.20095

Basholli, A., Lagkas, T., Bath, P. A., & Eleftherakis, G. (2018). Feasibility of sensor-based technology for monitoring health in developing countries—Cost analysis and user perception aspects. *arXiv*. https://doi.org/10.48550/arXiv.1804.08661

Centers for Disease Control and Prevention (CDC). (2005). *Third national report on human exposure to environmental chemicals*. Centers for Disease Control and Prevention.

Centers for Disease Control and Prevention. (2022). *Biomonitoring data tables for environmental chemicals*. National Center for Environmental Health, U.S. Department of Health and Human Services. https://www.cdc.gov/exposurereport/data_tables.html

Centers for Disease Control and Prevention, National Center for Environmental Health. (2024). *About the national biomonitoring program*. CDC. https://www.cdc.gov/biomonitoring/about/index.html

Chan, E. C., Koh, P. K., Mal, M., Cheah, P. Y., Eu, K. W., Backshall, A., Cavill, R., Nicholson, J. K., & Keun, H. C. (2009). Metabolic profiling of human colorectal cancer using high-resolution magic angle spinning nuclear magnetic resonance (HR-MAS NMR) spectroscopy and gas chromatography mass spectrometry (GC/MS). *Journal of Proteome Research, 8*, 352–361.

Chang, V. C., Andreotti, G., Ospina, M., Parks, C. G., Liu, D., Shearer, J. J., Rothman, N., Silverman, D. T., Sandler, D. P., Calafat, A. M., & Beane Freeman, L. E. (2023). Glyphosate exposure and urinary oxidative stress biomarkers in the agricultural health study. *Journal of the National Cancer Institute, 115*(4), 394–404. https://doi.org/10.1093/jnci/djac242

Chen, C., & Kim, S. (2013). LC-MS-based metabolomics of xenobiotic-induced toxicities. *Computational and Structural Biotechnology Journal, 4*, e201301008. https://doi.org/10.5936/csbj.201301008

Chen, C., Meng, L., Ma, X., Krausz, K. W., Pommier, Y., Idle, J. R., & Gonzalez, F. J. (2006). Urinary metabolite profiling reveals CYP1A2-mediated metabolism of NSC686288 (aminoflavone). *Journal of Pharmacology and Experimental Therapeutics, 318*, 1330–1342.

Church, R. J., & Watkins, P. B. (2017). The transformation in biomarker detection and management of drug-induced liver injury. *Liver International, 37*, 1582–1590.

Comparative Toxicogenomics Database (CTD). (2004). *A public resource of curated chemical–gene–disease networks for toxicogenomic research*. North Carolina State University & MDI Biological Laboratory. https://ctdbase.org/

Cooke, M. S., Olinski, R., Loft, S., European Standards Committee on Urinary (DNA) Lesion Analysis (ESCULA). (2008). Measurement and meaning of oxidatively modified DNA lesions in urine. *Cancer Epidemiology, Biomarkers & Prevention, 17*(1), 3–14. https://doi.org/10.1158/1055-9965.EPI-07-0751

Crocco, P., Montesanto, A., Dato, S., Geracitano, S., Iannone, F., Passarino, G., & Rose, G. (2019). Inter-individual variability in xenobiotic-metabolizing enzymes: Implications for human aging and longevity. *Genes (Basel), 10*(5), 403. https://doi.org/10.3390/genes10050403

Daoudi, K., Gaidi, M., Columbus, S., Shameer, M., & Alawadhi, H. (2021). Highly sensitive silver decorated graphene oxide silicon nanowires hybrid SERS sensors for trace level detection of environmental pollutants. *arXiv*. https://arxiv.org/abs/2101.07955

Dash, B., Afriyie-Gyawu, E., Huebner, H. J., Porter, W., Wang, J. S., Jolly, P. E., & Phillips, T. D. (2006). Noninvasive identification of interindividual variation in xenobiotic-metabolizing enzymes: Implications for cancer epidemiology and biomarker studies. *Journal of Toxicology and Environmental Health, Part A, 69*(13), 1203–1216. https://doi.org/10.1080/15287390500354938

DeBord, D. G., Shoemaker, D., B'Hymer, C., & Snawder, J. (2022). Application of biological monitoring methods for chemical exposures in occupational health. In *NIOSH manual of analytical methods* (5th ed.). Centers for Disease Control and Prevention, National Institute for Occupational Safety and Health.

de Vocht, F., Vlaanderen, J. J., Povey, A. C., Balbo, S., & Vermeulen, R. (2012). Environmental and occupational toxicants. In N. Rothman, P. Hainaut, P. Schulte, M. Smith, P. Boffetta, & F. Perera (Eds.), *Molecular epidemiology: Principles and practices* (pp. 163–174). International Agency for Research on Cancer Scientific Publications No. 163. World Health Organization.

Dehghani, S., & Yunesian, M. (2024). Combining biomarkers of susceptibility with toxicodynamic models: A novel approach for assessing modifying factors. *Journal of Environmental Health Science and Engineering, 22*, 361–364. https://doi.org/10.1007/s40201-024-00896-z

Diamanti-Kandarakis, E., Bourguignon, J.-P., Giudice, L. C., Hauser, R., Prins, G. S., Soto, A. M., Zoeller, R. T., & Gore, A. C. (2009). Endocrine-disrupting chemicals: An Endocrine Society scientific statement. *Endocrine Reviews, 30*(4), 293–342. https://doi.org/10.1210/er.2009-0002

Errekatxo, A., Prieto, A., Zuloaga, O., Usobiaga, A., Etxebarria, N., & Fernández, L. A. (2008). Simultaneous extraction of several persistent organic pollutants in sediment using focused ultrasonic solid-liquid extraction. *Analytical and Bioanalytical Chemistry, 392*, 1471–1478. https://doi.org/10.1007/s00216-008-2441-x

Esteban, M., & Castaño, A. (2009). Non-invasive matrices in human biomonitoring: A review. *Environment International, 35*(2), 438–449. https://doi.org/10.1016/j.envint.2008.09.003

Estévez-Alberola, M.-C., & Marco, M.-P. (2004). Immunochemical determination of xenobiotics with endocrine disrupting effects. *Analytical and Bioanalytical Chemistry, 378*, 563–575. https://doi.org/10.1007/s00216-003-2228-z

FDA-NIH Biomarker Working Group. (2016–2020). *BEST (Biomarkers, EndpointS, and other Tools) resource*. U.S. Food and Drug Administration and National Institutes of Health.

Feng, J., Li, Y., Gao, Z., Lv, H., Zhang, X., Fan, D., & Wei, Q. (2018). Visible-light driven label-free photoelectrochemical immunosensor based on TiO_2/S-$BiVO_4$@Ag_2S nanocomposites for sensitive detection OTA. *Biosensors and Bioelectronics, 99*, 14–20. https://doi.org/10.1016/j.bios.2017.07.029

Field, P. A., & Vasan, R. S. (2018). LDL-cholesterol is not the only clinically relevant biomarker for coronary artery disease or acute coronary syndrome. *Clinical Pharmacology & Therapeutics, 104*(2), 232–234. https://doi.org/10.1002/cpt.1126

Gao, A., Yang, X., Tong, J., Zhou, L., Wang, Y., Zhao, J., Mao, H., & Li, T. (2017). Multiplexed detection of lung cancer biomarkers in patients serum with CMOS-compatible silicon nanowire arrays. *Biosensors and Bioelectronics, 91*, 482–488. https://doi.org/10.1016/j.bios.2016.12.072

Garcia, M. C. (2005). The effect of the mobile phase additives on sensitivity in the analysis of peptides and proteins by high-performance liquid chromatography-electrospray mass spectrometry. *Journal of Chromatography B: Analytical Technologies in the Biomedical and Life Sciences, 825*(2), 111–123. https://doi.org/10.1016/j.jchromb.2005.03.041

Gaurav, K. (2024). Personalized toxicology advancements applications and future directions. *Toxicology Open Access, 10*, 299. https://doi.org/10.4172/2476-2067.1000299

Gil, F., & Pla, A. (2001). Biomarkers as biological indicators of xenobiotic exposure. *Journal of Applied Toxicology, 21*, 245–255. https://doi.org/10.1002/jat.769

Gonzalez, F. J., Fang, Z. Z., & Ma, X. (2015). Transgenic mice and metabolomics for study of hepatic xenobiotic metabolism and toxicity. *Expert Opinion on Drug Metabolism & Toxicology, 11*(6), 869–881. https://doi.org/10.1517/17425255.2015.1032245

Gore, A. C., Chappell, V. A., Fenton, S. E., Flaws, J. A., Nadal, A., Prins, G. S., Toppari, J., & Zoeller, R. T. (2015). EDC-2: The Endocrine Society's second scientific statement on endocrine-disrupting chemicals. *Endocrine Reviews, 36*(6), E1–E150. https://doi.org/10.1210/er.2015-1010

Hartung, T. (2023). Artificial intelligence as the new frontier in chemical risk assessment [Perspective]. *Frontiers in Artificial Intelligence, 6*. https://doi.org/10.3389/frai.2023.1269932

Hayes, A. W. (2014). *Principles and methods of toxicology* (6th ed.). CRC Press.

Hu, Y., Huang, Y., Wang, Y., Li, C., Wong, W., Ye, X., & Sun, D. (2018). A photoelectrochemical immunosensor based on gold nanoparticles/ZnAgInS quaternary quantum dots for the high-performance determination of hepatitis B virus surface antigen. *Analytica Chimica Acta, 1035*, 136–145. https://doi.org/10.1016/j.aca.2018.06.019

Human Biomonitoring for Europe (HBM4EU). (n.d.). *Science and policy for a healthy future.* European Environment Agency; HBM4EU. Retrieved August 17, 2025, from https://www.hbm4eu.eu/result/publications-2/

Jaria, G., Calisto, V., Otero, M., & Esteves, V. I. (2020). Monitoring pharmaceuticals in the aquatic environment using enzyme-linked immunosorbent assay (ELISA)—A practical overview. *Analytical and Bioanalytical Chemistry, 412*, 3983–4008. https://doi.org/10.1007/s00216-020-02509-8

Jayathissa, P., & Rupasinghe, A. (2024). Review on biomarkers for disease diagnosis and disease prevention: Use case in low and middle-income countries (LMICs). *International Journal of Research Publication and Reviews, 5*(8), 3224–3245. https://ijrpr.com/uploads/V5ISSUE8/IJRPR32477.pdf

Jeddi, M. Z., Hopf, N. B., Viegas, S., Price, A. B., Paini, A., van Thriel, C., Benfenati, E., Ndaw, S., Bessems, J., Behnisch, P. A., Leng, G., Duca, R. C., Verhagen, H., Cubadda, F., Brennan, L., Ali, I., David, A., Mustieles, V., Fernandez, M. F., Louro, H., & Pasanen-Kase, R. (2021). Towards a systematic use of effect biomarkers in population and occupational biomonitoring. *Environment International, 146*, 106257. https://doi.org/10.1016/j.envint.2020.106257

Ji, Y., Hebbring, S., Zhu, H., Jenkins, G. D., Biernacka, J., Snyder, K., Drews, M., Fiehn, O., Zeng, Z., Schaid, D., & Mrazek, D. A. (2011). Glycine and a glycine dehydrogenase (GLDC) SNP as citalopram/escitalopram response biomarkers in depression: Pharmacometabolomics-informed pharmacogenomics. *Clinical Pharmacology & Therapeutics, 89*, 97–104.

Johnson, C. H., Patterson, A. D., Idle, J. R., & Gonzalez, F. J. (2012). Xenobiotic metabolomics: Major impact on the metabolome. *Annual Review of Pharmacology and Toxicology, 52*, 37–56. https://doi.org/10.1146/annurev-pharmtox-010611-134748

Julián-Jiménez, A., Candel-González, F. J., & González del Castillo, J. (2013). Utilidad de los biomarcadores de inflamación e infección en los servicios de urgencias [Usefulness of inflammation and infection biomarkers in the emergency department]. *Enfermedades Infecciosas y Microbiología Clínica, 31*(6), 389–397. https://doi.org/10.1016/j.eimc.2013.01.005

Kaddurah-Daouk, R., Kristal, B. S., & Weinshilboum, R. M. (2008). Metabolomics: A global biochemical approach to drug response and disease. *Annual Review of Pharmacology and Toxicology, 48*, 653–683.

Karkossa, I., Raps, S., von Bergen, M., & Schubert, K. (2020). Systematic review of multi-omics approaches to investigate toxicological effects in macrophages. *International Journal of Molecular Sciences, 21*(24), 9371. https://doi.org/10.3390/ijms21249371

Kim, J., Campbell, A. S., & Wang, J. (2018). Wearable non-invasive epidermal glucose sensors: A review. *Talanta, 177*, 163–170.

Kim, J., Campbell, A. S., de Ávila, B. E. F., & Wang, J. (2019). Wearable biosensors for healthcare monitoring. *Nature Biotechnology, 37*, 389–406. https://doi.org/10.1038/s41587-019-0045-y

Kim, J. Y., Mukherjee, S., Ngo, L. C., & Christiani, D. C. (2004). Urinary 8-hydroxy-2′-deoxyguanosine as a biomarker of oxidative DNA damage in workers exposed to fine particulates. *Environmental Health Perspectives, 112*(6), 666–671. https://doi.org/10.1289/ehp.6827

Kimmel, D. W., LeBlanc, G., Meschievitz, M. E., & Cliffel, D. E. (2012). Electrochemical sensors and biosensors. *Analytical Chemistry, 84*, 685–707. https://doi.org/10.1021/ac202878q

Klaassen, C. D. (Ed.). (2019). *Casarett & Doull's toxicology: The basic science of poisons* (9th ed.). McGraw-Hill Education.

Klukova, L., Filip, J., & Tkac, J. (2016). Graphene-based lectin biosensor for ultrasensitive detection of glycan structures applicable in early diagnostics. In *2015 9th international conference on sensing technology (ICST)* (pp. 40–45).

Lashgari, M., Singh, V., & Pawliszyn, J. (2019). A critical review on regulatory sample preparation methods: Validating solid-phase microextraction techniques. *TrAC Trends in Analytical Chemistry, 119*, 115618. https://doi.org/10.1016/j.trac.2019.07.029

Lee, D. H., Porta, M., Jacobs, D. R., Jr., & Vandenberg, L. N. (2014). Chlorinated persistent organic pollutants, obesity, and type 2 diabetes. *Endocrine Reviews, 35*(4), 557–601. https://doi.org/10.1210/er.2013-1084

Li, F. (2012). Xenobiotic metabolomics: An ideal tool for drug metabolism. *Journal of Chromatography & Separation Techniques, 3*(7).

Li, F., Li, Y., Feng, J., Dong, Y., Wang, P., Chen, L., Chen, Z., Liu, H., & Wei, Q. (2017). Ultrasensitive amperometric immunosensor for PSA detection based on $Cu_2O@CeO_2$-Au nanocomposites as integrated triple signal amplification strategy. *Biosensors and Bioelectronics, 87*, 630–637. https://doi.org/10.1016/j.bios.2016.09.018

Lindon, J. C., & Nicholson, J. K. (2008). Spectroscopic and statistical techniques for information recovery in metabonomics and metabolomics. *Annual Review of Analytical Chemistry, 1*, 45–69.

Little, J. L., Wempe, M. F., & Buchanan, C. M. (2006). Liquid chromatography-mass spectrometry/mass spectrometry method development for drug metabolism studies: Examining lipid matrix ionization effects in plasma. *Journal of Chromatography B: Analytical Technologies in the Biomedical and Life Sciences, 833*(2), 219–230. https://doi.org/10.1016/j.jchromb.2006.02.011

Liu, R., Ye, X., & Cui, T. (2020). Recent progress of biomarker detection sensors. *Research* (Wash DC), *2020*, 7949037. https://doi.org/10.34133/2020/7949037

Liu, Y., Zhang, Z., Han, D., Zhao, Y., Yan, X., & Cui, S. (2022). Association between environmental chemicals co-exposure and peripheral blood immune-inflammatory indicators. *Frontiers in Public Health, 10*, 980987. https://doi.org/10.3389/fpubh.2022.980987

Liu, A., Chen, Y., Yang, Z., Feng, Y., Rui, W., Luo, W., Liu, Y., Gonzalez, F. J., & Dai, R. (2009). New metabolites of fenofibrate in Sprague-Dawley rats by UPLC-ESI-QTOF-MS-based metabolomics coupled with LC-MS/MS. *Xenobiotica, 39*, 345–354.

Lu, F. C., & Kacew, S. (2013). *Lu's basic toxicology: Fundamentals, target organs and risk assessment* (6th ed.). CRC Press. https://doi.org/10.1201/9781003026976

Mei, H., Hsieh, Y. S., Nardo, C., Xu, X., Wang, S., Ng, K., & Korfmacher, W. A. (2003). Investigation of matrix effects in bioanalytical high-performance liquid chromatography/tandem mass spectrometric assays: Application to drug discovery. *Rapid Communications in Mass Spectrometry, 17*(1), 97–103. https://doi.org/10.1002/rcm.876

Mohamed, H. M. (2020). Sensors and biosensors for environment contaminants. In Inamuddin, & A. M. Asiri (Eds.), *Nanosensor technology for environmental monitoring* (pp. 109–135). Springer Nature.

National Academies of Sciences, Engineering, and Medicine. (2020). *Predicting human health effects from environmental exposures: Applying translatable and accessible biomarkers of effect: Proceedings of a workshop—In brief*. National Academies Press.

National Research Council (NRC). (1989). *Biological markers in reproductive toxicology*. National Academy Press.

Oesch, F., & Hengstler, J. G. (2021). Importance of xenobiotic metabolism: Mechanistic considerations relevant for regulation. In F. X. Reichl, & M. Schwenk (Eds.), *Regulatory toxicology*. Springer. https://doi.org/10.1007/978-3-030-57499-4_71

Panuwet, P., Hunter, R. E., Jr., D'Souza, P. E., Chen, X., Radford, S. A., Cohen, J. R., Marder, M. E., Kartavenka, K., Ryan, P. B., & Barr, D. B. (2016). Biological matrix effects in quantitative tandem mass spectrometry-based analytical methods: Advancing biomonitoring. *Critical Reviews in Analytical Chemistry, 46*(2), 93–105. https://doi.org/10.1080/10408347.2014.980775

Perera, F. P., & Weinstein, I. B. (2000). Molecular epidemiology: Recent advances and future directions. *Carcinogenesis, 21*, 517–524.

Pérez-Piñar, T., Hartmann, A., Bössow, S., Gnad, H., & Mock, H.-P. (2024). Metabolic changes during wheat microspore embryogenesis induction using the highly responsive cultivar Svilena. *Journal of Plant Physiology, 294*, 154193. https://doi.org/10.1016/j.jplph.2024.154193

Plumb, R. S., Stumpf, C. L., Granger, J. H., Castro-Perez, J., Haselden, J. N., & Dear, G. J. (2003). Use of liquid chromatography/time-of-flight mass spectrometry and multivariate statistical analysis shows promise for the detection of drug metabolites in biological fluids. *Rapid Communications in Mass Spectrometry, 17*, 2632–2638.

Qi, J. F., Honda, Y., & Mineshita, S. (2002). Simultaneous HPLC determination of acetaminophen and four of its major metabolites in small-volume biologic fluids. *Rinsho Yakuri/Japanese Journal of Clinical Pharmacology and Therapeutics, 33*(6), 263–275. https://doi.org/10.3999/jscpt.33.6_263

Rodríguez-Carrillo, A., Mustieles, V., Salamanca-Fernández, E., Olivas-Martínez, A., Suárez, B., Bajard, L., Baken, K., Blaha, L., Bonefeld-Jørgensen, E. C., Couderq, S., & d'Cruz, S. C. (2023). Implementation of effect biomarkers in human biomonitoring studies: A systematic approach synergizing toxicological and epidemiological knowledge. *International Journal of Hygiene and Environmental Health, 249*, 114140. https://doi.org/10.1016/j.ijheh.2023.114140

Rosenberg, J., Fiserova-Bergerova, V., & Lowry, L. K. (1989). Biological monitoring IV: Measurements in urine. *Applied Industrial Hygiene, 4*(4), F16–F21. https://doi.org/10.1080/08828032.1989.10390349

Sadler, N. C., Nandhikonda, P., Webb-Robertson, B. J., Ansong, C., Anderson, L. N., Smith, J. N., Corley, R. A., & Wright, A. T. (2016). Hepatic cytochrome P450 activity, abundance, and expression throughout human development. *Drug Metabolism and Disposition, 44*(7), 984–991. https://doi.org/10.1124/dmd.115.068593

Schulte, P. A. (1992). Biomarkers in epidemiology: Scientific issues and ethical implications. *Environmental Health Perspectives, 98*, 143–147. https://doi.org/10.1289/ehp.9298143

Seenivasan, R., Singh, C. K., Warrick, J. W., Ahmad, N., & Gunasekaran, S. (2017). Microfluidic-integrated patterned ITO immunosensor for rapid detection of prostate-specific membrane antigen biomarker in prostate cancer. *Biosensors and Bioelectronics, 95*, 160–167. https://doi.org/10.1016/j.bios.2017.04.004

Shi, X., Yao, D., & Chen, C. (2012). Identification of N-acetyltaurine as a novel metabolite of ethanol through metabolomics-guided biochemical analysis. *Journal of Biological Chemistry, 287*, 6336–6349.

Singh, M., Mishra, R. C., Shah, I., Wadhwa, V., & Mor, V. (2023). Xenobiotics: Sources, pathways, degradation, and risk associated with major emphasis on pharmaceutical compounds. In R. Singh, P. Singh, S. Tripathi, K. K. Chandra, & R. Bhadouria (Eds.), *Xenobiotics in urban ecosystems*. Springer. https://doi.org/10.1007/978-3-031-35775-6_5

Skinner, M. K. (2014). Environmental stress and epigenetic transgenerational inheritance. *BMC Medicine, 12*, 153. https://doi.org/10.1186/s12916-014-0153-y

Skinner, M. K. (2024). Epigenetic biomarkers for disease susceptibility and preventative medicine. *Cell Metabolism, 36*(2), 263–277. https://doi.org/10.1016/j.cmet.2023.11.015

Spriggs, S., Cubberley, R., Loadman, P., Sheffield, D., & Wierzbicki, A. (2018). A study of inter-individual variability in the phase II metabolism of xenobiotics in human skin. *Toxicology Letters, 292*, 63–72. https://doi.org/10.1016/j.toxlet.2018.03.016

Štefanac, T., Grgas, D., & Landeka Dragičević, T. (2021). Xenobiotics-division and methods of detection: A review. *Journal of Xenobiotics, 11*(4), 130–141. https://doi.org/10.3390/jox11040009

Steffensen, I.-L., Dirven, H., Couderq, S., David, A., D'Cruz, S. C., Fernández, M. F., Mustieles, V., Rodríguez-Carrillo, A., & Hofer, T. (2020). Bisphenols and oxidative stress biomarkers—Associations found in human studies, evaluation of methods used, and strengths and weaknesses of the biomarkers. *International Journal of Environmental Research and Public Health, 17*(10), 3609. https://doi.org/10.3390/ijerph17103609

Tai, L. C., Gao, W., Chao, M., Bariya, M., Ngo, Q. P., Shahpar, Z., Nyein, H. Y. Y, Park, H., Sun, J., Jung, Y., Wu, E., Fahad, H. M., Lien, D.-H., Ota, H., Cho, G., & Javey, A. (2018). Methylxanthine drug monitoring with wearable sweat sensors. *Advanced Materials, 30*, e1707442.

Tang, L., & Kebarle, P. (1993). Dependence of ion intensity in electrospray mass spectrometry on the concentration of the analytes in the electrosprayed solution. *Analytical Chemistry, 65*(24), 3654–3668.

Teri, D., Aly, N. A., Dodds, J. N., Zhang, J., Thiessen, P. A., Bolton, E. E., Joseph, K. M., Williams, A. J., Schymanski, E. L., Rusyn, I., & Baker, E. S. (2025). Reference library for suspect screening of environmental toxicants using ion mobility spectrometry-mass spectrometry. *Communications Chemistry, 8*(1), 224. https://doi.org/10.1038/s42004-025-01619-7

Tonoyan, L., & Siraki, A. G. (2024). Machine learning in toxicological sciences: Opportunities for assessing drug toxicity. *Frontiers in Drug Discovery, 4*. https://doi.org/10.3389/fddsv.2024.1336025

Turova, P., Styles, I., Timashev, V., Kravets, K., Grechnikov, A., Lyskov, D., Samigullin, T., Podolskiy, I., Shpigun, O., & Stavrianidi, A. (2021). Unsupervised methods in LC-MS data treatment: Application for potential chemotaxonomic markers search. *Journal of Pharmaceutical and Biomedical Analysis, 206*, 114382. https://doi.org/10.1016/j.jpba.2021.114382

Tuzimski, T., & Sherma, J. (2019). *Determination of target xenobiotics and unknown compound residues in food, environmental, and biological samples.* Taylor & Francis Group.

U.S. Environmental Protection Agency (US EPA). (2012). Exposure assessment tools by approaches: Exposure reconstruction (biomonitoring and reverse dosimetry). In *EPA ExpoBox*. U.S. EPA.

U.S. Environmental Protection Agency (US EPA). (2018). *Development and distribution of ToxCast and Tox21 HTS assay method descriptions.* https://www.epa.gov/sites/production/files/2018-04/documents/cop_sflood25jan2018-final.pdf

U.S. Food and Drug Administration (FDA). (2017, December). *FDA's predictive toxicology roadmap.* U.S. Department of Health and Human Services.

Van Cauteren, H., de Kok, M. C. M., & Van Schooten, F. J. (1996). Cancer risk evaluation. In R. J. M. Niesink, J. de Vries, & M. A. Hollinger (Eds.), *Toxicology: Principles and applications.* CRC Press.

van Eijl, S., Zhu, Z., Cupitt, J., Gierula, M., Götz, C., Fritsche, E., & Edwards, R. J. (2012). Elucidation of xenobiotic metabolism pathways in human skin and human skin models by proteomic profiling. *PLoS ONE, 7*(7), e41721. https://doi.org/10.1371/journal.pone.0041721

Vinggaard, A. M., Bonefeld-Jørgensen, E. C., Jensen, T. K., Fernandez, M. F., Rosenmai, A. K., Taxvig, C., Rodriguez-Carrillo, A., Wielsøe, M., Long, M., Olea, N., Antignac, J.-P., Hamers, T., & Lamoree, M. (2021). Receptor-based in vitro activities to assess human exposure to chemical mixtures and related health impacts. *Environment International, 146*, 106191. https://doi.org/10.1016/j.envint.2020.106191

Waters Corporation. (2015). *Application of multivariate analysis with LC-MS for detection of counterfeit cosmetics.* Waters Application Note. https://www.waters.com/nextgen/pl/en/library/application-notes/2015/application-multivariate-analysis-lc-ms-detection-counterfeit-cosmetics.html

World Health Organization (WHO). (1993). *Biomarkers and risk assessment: Concepts and principles.* https://apps.who.int/iris/bitstream/handle/10665/39037/9241571551-eng.pdf

World Health Organization (WHO). (2001). *Biomarkers in risk assessment: Validity and validation* (Environmental Health Criteria No. 222). International Programme on Chemical Safety. https://www.inchem.org/documents/ehc/ehc/ehc222.htm

World Health Organization (WHO). (2015). *Human biomonitoring.* WHO Regional Office for Europe. https://iris.who.int/handle/10665/164588

World Health Organization (WHO). (2023). *Human biomonitoring: Assessment of exposure to chemicals and their health risks: Summary for decision makers.* https://www.who.int/europe/publications/i/item/WHO-EURO-2023-7574-47341-69480

Yang, S., Zhang, F., Wang, Z., & Liang, Q. (2018). A graphene oxide-based label-free electrochemical aptasensor for the detection of alpha-fetoprotein. *Biosensors and Bioelectronics, 112*, 186–192. https://doi.org/10.1016/j.bios.2018.04.026

Ye, M., Jiang, M., Cheng, J., Li, X., Liu, Z., Zhang, W., Mugo, S. M., Jaffrezic-Renault, N., & Guo, Z. (2020). Single-layer exfoliated reduced graphene oxide-antibody tau sensor for detection in human serum. *Sensors and Actuators B: Chemical, 308*, 127692. https://doi.org/10.1016/j.snb.2020.127692

Zhang, R., Wen, H., Lin, Z., Li, B., & Zhou, X. (2025). Artificial intelligence-driven drug toxicity prediction: Advances, challenges, and future directions. *Toxics, 13*(7), 525. https://doi.org/10.3390/toxics13070525

Zhang, Y., Chen, R., Xu, L., Ning, Y., Xie, S., & Zhang, G. J. (2015). Silicon nanowire biosensor for highly sensitive and multiplexed detection of oral squamous cell carcinoma biomarkers in saliva. *Analytical Sciences, 31*(2), 73–78. https://doi.org/10.2116/analsci.31.73

Zheng, Z., Zhu, R., Peng, I., Xu, Z., & Jiang, Y. (2024). Wearable and implantable biosensors: Mechanisms and applications in closed-loop therapeutic systems. *Journal of Materials Chemistry B*. https://doi.org/10.1039/D4TB00782D

Zhou, B., Xiao, J. F., Tuli, L., & Ressom, H. W. (2012). LC-MS-based metabolomics. *Molecular Biosystems, 8*, 470–481.

Zhu, X., Tang, X., Ng, C., Li, L., Lai, Y., Miller, G. W., Jiang, C., Barchowsky, A., Buchanich, J., & Gao, P. (2025). Biomonitoring xenobiotics in human biospecimens: Challenges, advances, and the future of exposome characterization. *Reviews of Environmental Contamination (formerly: Residue Reviews), 263*, 15. https://doi.org/10.1007/s44169-025-00088-2

Chapter 7
Xenobiotics' Epigenetic Effects: Consequences for Human Health and the Environment

M. G. Hariprasad, Rahul Kumar, and Apurwa Rahul Salunkhe

Abstract Xenobiotics are foreign chemical agents that are common in today's environment. They create significant challenges for biological systems. More research shows that many harmful effects of these substances are linked to epigenetic mechanisms. These are inherited changes to the genome that affect gene expression without changing the DNA sequence itself. This chapter summarizes what we know about the relationship between xenobiotics and the epigenome. We start by reviewing key principles of epigenetics, such as DNA methylation, changes in histones, and non-coding RNAs. Then, we look at the molecular pathways through which xenobiotics, like persistent organic pollutants, heavy metals, and endocrine disruptors, interfere with these epigenetic changes. Next, we discuss the consequences for human health, including cancer, neurodevelopmental disorders, and metabolic diseases, as well as the effects on ecosystems. The chapter also covers the methods needed to design and interpret studies in environmental epigenetics and the implications for risk assessment and public policy. Finally, we point out the main challenges and future directions for this quickly evolving field.

Keywords Epigenetic inheritance · Chemical risk assessment · Environmental exposures · Neurodevelopmental disorders · Risk assessment

Learning Objectives

Upon completing this chapter, the reader should be able to:

1. Define xenobiotics and key epigenetic mechanisms, explaining why the epigenome is a critical interface between environmental exposures and biological responses.

M. G. Hariprasad (✉) · R. Kumar
Department of Pharmacology, KLE College of Pharmacy, Rajajinagar, Bangalore, India
e-mail: hariprasadmg2004@yahoo.com

A. R. Salunkhe
Department of Quality Assurance, Krupanidhi College of Pharmacy, Bangalore, India

K. Priya and S. K. Giri (eds.), *Environmental Toxicology: Xenobiotics and Effects on Human Genetic Material*, https://doi.org/10.1007/978-3-032-28619-2_7

2. Describe the primary molecular pathways through which xenobiotics can alter DNA methylation, histone modifications, and non-coding RNA expression.
3. Critically evaluate the evidence linking xenobiotic-induced epigenetic changes to major human diseases, distinguishing between different types of evidence (epidemiological, animal, in vitro).
4. Analyze the challenges in designing and interpreting epigenome-wide association studies (EWAS), including confounding by cell-type heterogeneity, batch effects, and issues of causality.
5. Explain the difference between intergenerational and transgenerational epigenetic inheritance and assess the current state of evidence for each in toxicology.
6. Discuss how epigenetic data can be integrated into chemical risk assessment and public health policy, and identify key research gaps in the field.

1 Introduction

1.1 Scope and Rationale

The main argument of modern environmental health science is that the genome does not work alone. It constantly interacts with the environment, primarily through the epigenome. The epigenome consists of various chemical changes on DNA and its surrounding proteins. It serves as an important link between a changing environment and the relatively stable genome we inherit from our parents (Bressan et al., 2014; Xavier et al., 2019). The epigenome transmits environmental information into changes in gene expression, which affects how cells behave, grow, and react to diseases (Ogunjobi et al., 2024). Xenobiotics, or chemical agents that are foreign to biological systems, are common in the environment (Rieger et al., 2002). Therefore, understanding how they affect the epigenome is essential for toxicology, risk assessment, and public health in the twenty-first century.

This chapter provides a detailed look at this connection. It goes beyond the traditional toxicological approach, which usually focuses on immediate toxicity and simple dose–response relationships. Instead, it takes a more complex, process-oriented view on toxicity. In this view, the timing, duration, and developmental context of exposure are just as important as the dose itself. Epigenetic processes support this shift in thinking, showing how temporary environmental exposures, especially during key developmental stages, can lead to lasting changes in gene regulation that cause health issues to appear years or even decades later.

1.2 Definition of Xenobiotics

A xenobiotic, derived from the Greek words "xenos" (strange) and "bios" (life), is a chemical substance within an organism that the organism does not produce naturally or is not expected to have. This term also includes substances present at levels significantly higher than normal physiological amounts. While often associated with synthetic pollutants, the definition is broad and covers a wide variety of compounds from many sources (Stefanac et al., 2021). Key categories of xenobiotics relevant to epigenetic toxicology include:

A xenobiotic, which originates from the Greek words "xenos" (stranger) and "bios" (life), refers to a chemical compound found in an organism that is not naturally produced by that organism and is not typically expected to be present. This term also includes substances that exist at levels that significantly surpass their usual physiological concentrations. Although commonly linked to artificial pollutants, the definition is expansive and incorporates a wide variety of compounds from different origins. Important categories of xenobiotics pertinent to epigenetic toxicology consist of:

- **Persistent Organic Pollutants (POPs)**: These are extremely stable, man-made compounds that resist breakdown, accumulate in food chains, and can be transported over long distances. Examples include dioxins, polychlorinated biphenyls (PCBs), and organochlorine pesticides such as dichlorodiphenyltrichloroethane (DDT) (Vasseur & Cossu-Leguille, 2006).
- **Heavy Metals**: Metals and metalloids that can be harmful even in minimal amounts. Arsenic, lead, cadmium, and mercury are particularly concerning due to their prevalent contamination in the environment and their strong biological impacts (Schröder, et al., 2009).
- **Endocrine Disrupting Chemicals (EDCs)**: These substances interfere with the endocrine (hormonal) system of the body. This group includes bisphenol A (BPA), which is found in polycarbonate plastics and epoxy resins; phthalates, utilized as plasticizers in numerous consumer goods; and certain pesticides like vinclozolin (Zhang et al., 2011).
- **Pharmaceuticals**: Medicinal drugs and their breakdown products can enter ecosystems and function as xenobiotics in non-target organisms. Some medications, including the anticonvulsant valproic acid, have also been shown to produce direct epigenetic effects in patients (Singh et al., 2023a).
- **Air Pollutants**: This category comprises a complicated mixture of gases and particulates, with significant epigenetic implications stemming from components such as polycyclic aromatic hydrocarbons (PAHs) produced by the incomplete combustion of organic materials, as well as fine particulate matter (Lang et al., 2025).
- **Pesticides**: This is a wide-ranging category of chemicals intended to manage pests, which includes insecticides (e.g., organophosphates), herbicides (e.g., glyphosate), and fungicides (Wang et al., 2022).

1.3 Why Epigenetics Matters for Toxicology and Environment

The exploration of epigenetics is changing the field of toxicology for several important reasons. First, it provides a clear molecular mechanism to explain ongoing toxicological mysteries. One example is the developmental origins of health and disease (DOHaD) hypothesis. This hypothesis suggests that early-life exposures influence the risk of disease later in life (Silveira et al., 2007). Epigenetic changes made during key developmental stages can remain surprisingly stable. They act like a type of cellular memory that maintains altered gene expression throughout a person's life.

Second, the wide variety of xenobiotic compounds shows that they do not all work through a specific receptor (Ma, 2008). However, their toxic effects often overlap on a small number of basic cellular functions, including those that regulate the epigenome (Sobral et al., 2025). This indicates that the epigenetic regulatory system may be a common and sensitive target for many environmental threats. This idea has significant implications for understanding the cumulative risk of exposure to chemical mixtures.

Third, unlike DNA changes, epigenetic changes can be reversible in theory (Holliday, 1989). This flexibility presents an interesting possibility for interventions, such as nutrition, medications, or simply reducing exposure, that could prevent or even reverse the harmful effects of environmental exposure.

Finally, epigenetic changes can serve as strong biomarkers (Bock, 2009). They can indicate past exposures or signal an early biological response to a toxic substance well before clinical symptoms appear. In its most significant context, the concept of "generational toxicology" suggests that epigenetic changes caused by xenobiotics in the germline can be passed down to future, unexposed generations. This idea fundamentally expands our understanding of environmental damage.

2 Foundations of Epigenetics

Epigenetics includes mechanisms that cause lasting changes in gene function without altering the DNA sequence itself. These mechanisms control the complex patterns of gene expression needed for development, cell differentiation, and responses to the environment (Al Aboud, 2023). The primary epigenetic mechanisms work within an interconnected regulatory system.

Key Mechanisms

- **DNA Methylation**: This is the most studied epigenetic modification. It involves adding a methyl group to the 5-carbon position of a cytosine base, usually found in a cytosine-guanine dinucleotide (CpG site). A group of enzymes called DNA methyltransferases (DNMTs) facilitates this process. DNMT1 serves as a "maintenance" methyltransferase, copying existing methylation patterns onto new DNA strands during cell division. This ensures that methylation states are accurately passed down within a cell lineage. DNMT3A and DNMT3B act as "de novo" methyltransferases, creating new methylation patterns during development or when responding to external signals. Although methylation is generally present throughout the genome, regions rich in CpG sites, known as CpG islands, are often located in gene promoter areas and tend to remain unmethylated in active genes. Methylation of these promoter CpG islands acts as a stable marker closely associated with gene repression (Bender, 2004).

- **Histone Modifications**: Eukaryotic DNA is organized in the nucleus by wrapping around groups of histone proteins to form nucleosomes, which are the basic repeating units of chromatin. The N-terminal tails of these histone proteins extend from the nucleosome core and can undergo several post-translational modifications (PTMs), such as acetylation, methylation, phosphorylation, and ubiquitination. These modifications change the physical structure of chromatin. For example, the acetylation of lysine residues neutralizes their positive charge, which loosens chromatin and makes it more accessible for transcription. They can also serve as docking sites for "reader" proteins that recognize specific modifications and attract other regulatory elements. The "histone code" hypothesis suggests that specific combinations of these modifications can be understood as a language that influences gene expression (Biel et al., 2005).
- **Chromatin Remodeling**: This mechanism involves large protein complexes that depend on ATP to shift, remove, or change the composition of nucleosomes. By sliding nucleosomes along the DNA, these remodeling complexes can uncover or hide regulatory elements like promoters and enhancers. They play a crucial role in controlling gene accessibility and expression (Geiman et al., 2002).
- **Non-coding RNAs (ncRNAs)**: A large part of the genome is transcribed into RNA that does not code for proteins but has important regulatory functions. MicroRNAs (miRNAs) are short ncRNAs, around 22 nucleotides long, that act as vital post-transcriptional regulators. They typically bind to the 3′ untranslated region (3′ UTR) of target messenger RNAs (mRNAs), causing their degradation or blocking their translation, thereby finely tuning gene expression. Long non-coding RNAs (lncRNAs) include a diverse group of transcripts longer than 200 nucleotides. They can serve as scaffolds to help assemble proteins, guide epigenetic modifying complexes to specific genomic sites, or act as decoys to capture transcription factors or miRNAs (Costa, 2008).
- **Transgenerational Epigenetic Inheritance**: This refers specifically to the transfer of epigenetic states through the germline (sperm or egg) to future generations that were never directly exposed to the original environmental trigger. For this to happen, the induced epigenetic mark must be established in the germ cells of the affected individual and must survive the two main waves of epigenetic reprogramming that typically remove most marks after fertilization and during the formation of new germ cells (Skinner, 2016).

2.1 Molecular Readouts and Commonly Used Assays

The choice of technology for assessing epigenetic markers is a key part of study design. It greatly affects the questions researchers can explore and brings specific analytical challenges (Table 1).

2.1.1 DNA Methylation Analysis

- **Bisulfite Sequencing**: This method is known as the gold standard for methylation assessment. It treats DNA with sodium bisulfite, which turns unmethylated cytosines into uracil while leaving methylated cytosines intact. The following PCR amplification and sequencing process reads uracil as thymine. This enables precise identification of methylated locations at the single-base level (Reed et al., 2010).
- **Whole-Genome Bisulfite Sequencing (WGBS)**: This method extends bisulfite conversion and sequencing across the entire genome. It provides the most thorough and objective representation of the methylome. However, its high cost can limit its use in large epidemiological studies (Suzuki et al., 2018).
- **Reduced Representation Bisulfite Sequencing (RRBS)**: This is a more affordable method that uses a methylation-insensitive restriction enzyme, such as MspI, to digest the genome. It focuses on CpG-rich regions like promoters and CpG islands before bisulfite sequencing. While it offers detailed coverage of these important regulatory areas, it overlooks about 85–90% of CpG sites found elsewhere in the genome (Meissner et al., 2005).
- **Epigenome Arrays**: Technologies like the Illumina Infinium MethylationEPIC BeadChip use microarray technology to examine the methylation status of over 850,000 pre-selected CpG locations. They present a reliable and cost-effective option for large-scale Epigenome-Wide Association Studies (EWAS). However, they are limited to the analyzed sites and cannot detect new methylation sites (Hurd & Nelson, 2009).

2.2 Chromatin Analysis

- Chromatin Immunoprecipitation sequencing (ChIP-seq): This is the main technique used to find the genome-wide locations of histone modifications or DNA-binding proteins. It employs an antibody specific to the target protein or histone mark to enrich related DNA fragments, which are then identified using high-throughput sequencing.
- Assay for Transposase-Accessible Chromatin using sequencing (ATAC-seq): This effective method maps open chromatin regions throughout the genome. It uses a hyperactive Tn5 transposase enzyme that selectively cuts and integrates sequencing adapters into accessible areas of the genome. This gives a quick

Table 1 The investigation of xenobiotic-induced epigenetic effects requires a strategic selection of molecular assays tailored to the specific biological question, available sample types, and project scope. The following table provides a comprehensive, comparative overview of the principal assays discussed in this chapter. The panel is organized by the major classes of epigenetic modifications: DNA methylation, histone modifications, chromatin accessibility, and non-coding RNA expression

Assay	Biological question answered	Sample type	Strengths	Limitations
DNA methylation analysis				
Whole-Genome Bisulfite Sequencing (WGBS) (Suzuki et al., 2018)	What is the complete, single-base resolution methylation status of the entire genome (CpG, CHG, CHH)? The gold standard for de novo discovery of all differentially methylated regions (DMRs)	High-quality genomic DNA. Can be adapted for various sources including cells, tissues, and blood	– Comprehensive: unbiased, whole-genome coverage – High resolution: single-nucleotide resolution – Discovery power: the ultimate tool for identifying novel methylation patterns	– High cost: expensive due to deep sequencing requirements – High input: requires substantial amounts of high-quality DNA – DNA degradation: harsh bisulfite treatment fragments DNA – Complex bioinformatics: reduced sequence complexity complicates alignment

(continued)

Table 1 (continued)

Assay	Biological question answered	Sample type	Strengths	Limitations
Reduced Representation Bisulfite Sequencing (RRBS) (Meissner et al., 2005)	What is the methylation status of CpG-rich regions (e.g., promoters, CpG islands) across the genome? A cost-effective method for discovery in functionally relevant genomic areas	Genomic DNA (low input, 10–300 ng). Compatible with fresh, frozen, and FFPE samples	– Cost-effective: reduces sequencing costs by focusing on a fraction of the genome – High resolution: single-nucleotide resolution within targeted regions – Enriches for regulatory regions: captures most promoters and CpG islands	– Coverage bias: interrogates only regions containing specific restriction sites (e.g., MspI), missing CpG-poor areas like some enhancers – Incomplete Representation: provides only a partial view of the methylome – PCR Bias: can be prone to amplification bias after bisulfite treatment
Infinium methylationepic array (McEwen et al., 2018)	What is the methylation level at ~900,000 pre-selected CpG sites across the genome? Ideal for large-scale epigenome-wide association studies (EWAS) and biomarker validation	Genomic DNA (250–500 ng). Validated for a wide range, including whole blood, saliva, and FFPE tissue	– High throughput and scalable: cost-effective for large epidemiological cohorts – High reproducibility: >98% reproducibility – Standardized workflow: user-friendly protocol and established analysis pipelines – Comprehensive content: covers CpG islands, enhancers, and other regulatory elements	– No novel discovery: limited to the fixed probes on the array – SNP confounding: probes can be affected by genetic variants, leading to false signals – Bias in probe design: only interrogates a fraction of all CpGs – Human-specific: standard arrays are primarily for human studies

(continued)

Table 1 (continued)

Assay	Biological question answered	Sample type	Strengths	Limitations
Histone modification and protein binding analysis				
Chromatin Immunoprecipitation-Sequencing (ChIP-seq) (Hino et al., 2022)	Where in the genome is a specific histone modification (e.g., H3K27ac, H3K9me3) located, or where does a specific transcription factor bind?	Cross-linked chromatin from cells or tissues. Typically requires large cell numbers ($10^6–10^7$), though low-input protocols exist	– High specificity: provides a map for a single protein or modification of interest – Genome-wide: unbiased mapping across the entire genome – Mechanistic insight: directly links specific regulatory factors to genomic locations	– Antibody dependent: success is critically dependent on a high-quality, specific antibody – High input requirement: standard protocols require millions of cells – Labor-intensive: complex and lengthy protocol – Potential for artifacts: cross-linking and sonication can introduce biases
Chromatin accessibility analysis				

(continued)

Table 1 (continued)

Assay	Biological question answered	Sample type	Strengths	Limitations
Assay for Transposase-Accessible Chromatin (ATAC-seq) (Raurell-Vila et al., 2018)	Which regions of the genome are in an "open" or accessible chromatin state? Provides a global map of all potential active regulatory elements (promoters, enhancers, etc.)	Fresh or cryopreserved cells or nuclei (low input, 500–50,000). Adaptable for tissues	– Fast and simple: protocol can be completed in hours – Very low sample input: ideal for precious or limited samples – High signal-to-noise ratio: generates high-quality data – Single-cell compatible: well-established protocols for single-cell analysis (scATAC-seq)	– Non-specific: identifies accessible regions but not the specific factors bound there – Tn5 insertion bias: the transposase has a known sequence preference that must be corrected for bioinformatically – Sensitive to sample quality: requires high cell viability for optimal results
Non-coding RNA analysis				

(continued)

Table 1 (continued)

Assay	Biological question answered	Sample type	Strengths	Limitations
Small RNA sequencing (small RNA-seq) (Benesova et al., 2021; Giraldez et al., 2018)	What is the complete profile and abundance of all small non-coding RNAs (e.g., miRNAs, piRNAs) in a sample? Used for discovery of novel small RNAs and differential expression analysis	Total RNA or enriched small RNA fraction from cells, tissues, or biofluids (e.g., plasma, exosomes)	– Comprehensive discovery: identifies known and novel small RNAs without prior sequence knowledge – High resolution: can detect isoforms (isomiRs) and single-nucleotide variations – High sensitivity and dynamic range: accurately quantifies both low and high abundance transcripts	– Ligation bias: enzymatic steps in library preparation can introduce sequence-specific biases – Complex bioinformatics: requires specialized pipelines for analysis of short reads – Higher cost: more expensive per sample than targeted approaches
Targeted qPCR (e.g., TaqMan Assays) (Schleinitz et al., 2010; Lappan et al., 2021)	What is the expression level of a specific, known non-coding RNA (e.g., a particular miRNA or lncRNA)? Primarily used for validation of sequencing data or for targeted analysis of a few candidates	Total RNA or enriched RNA fractions from any source	– High sensitivity and specificity: the gold standard for accurate quantification of specific transcripts – Cost-effective for validation: Inexpensive for analyzing a few targets across many samples – Fast turnaround: results can be obtained quickly	– No discovery: can only measure pre-selected, known ncRNAs – Low throughput: not suitable for profiling hundreds or thousands of targets simultaneously – Primer/Probe design: requires careful design and optimization for each target

overview of the regulatory landscape. While ChIP-seq looks for "Where is this specific mark?", ATAC-seq investigates "Which regions are broadly open and active?" (Minard et al., 2009).

2.3 ncRNA Analysis

- Small RNA-seq: This specialized RNA-sequencing protocol is designed to capture and measure the levels of small non-coding RNAs, especially miRNAs. It includes size selection of small RNA fragments, followed by library preparation and sequencing (Costa, 2008; Kalvari et al., 2018).

The interconnectedness of these epigenetic layers is a critical part of gene regulation. For instance, certain histone modifications can attract DNMTs to induce DNA methylation. Similarly, methyl-binding domain (MBD) proteins can recognize methylated DNA and draw in histone deacetylase (HDAC) complexes, creating a stable, silenced chromatin state. A xenobiotic that disrupts one component, such as a specific miRNA, can trigger widespread changes in both histone modifications and DNA methylation at the target genes of that miRNA. This emphasizes the limitations of single-omic studies and the importance of multi-omic strategies for a complete understanding of a xenobiotic's effects.

3 How Xenobiotics Modify the Epigenome

Xenobiotics can affect the epigenome through various molecular mechanisms. These pathways are connected and often overlap, creating a complex network where a single chemical can trigger multiple processes that influence the epigenetic machinery.

Mechanistic Pathways: Oxidative Stress and Disruption of One-Carbon Metabolism: Many xenobiotics, like heavy metals, persistent organic pollutants (POPs), and air contaminants, strongly induce reactive oxygen species (ROS). This leads to oxidative stress, which has both direct and indirect effects on the epigenome (Fig. 1) (Saves et al., 1998).

i. **Direct DNA Damage**: ROS can directly oxidize DNA bases. For example, the oxidation of guanine to 8-oxoguanine in a CpG dinucleotide can physically block the binding of DNMT1. This prevents maintenance methylation after DNA replication and causes passive, locus-specific hypomethylation (Finnberg et al., 2000; Saves & Masson, 1998).
ii. **Cofactor Depletion**: Oxidative stress significantly increases the demand on the cell's antioxidant systems, especially for glutathione production. This process diverts precursors away from the methionine cycle, which is vital for producing S-adenosylmethionine (SAM), the universal methyl donor for all

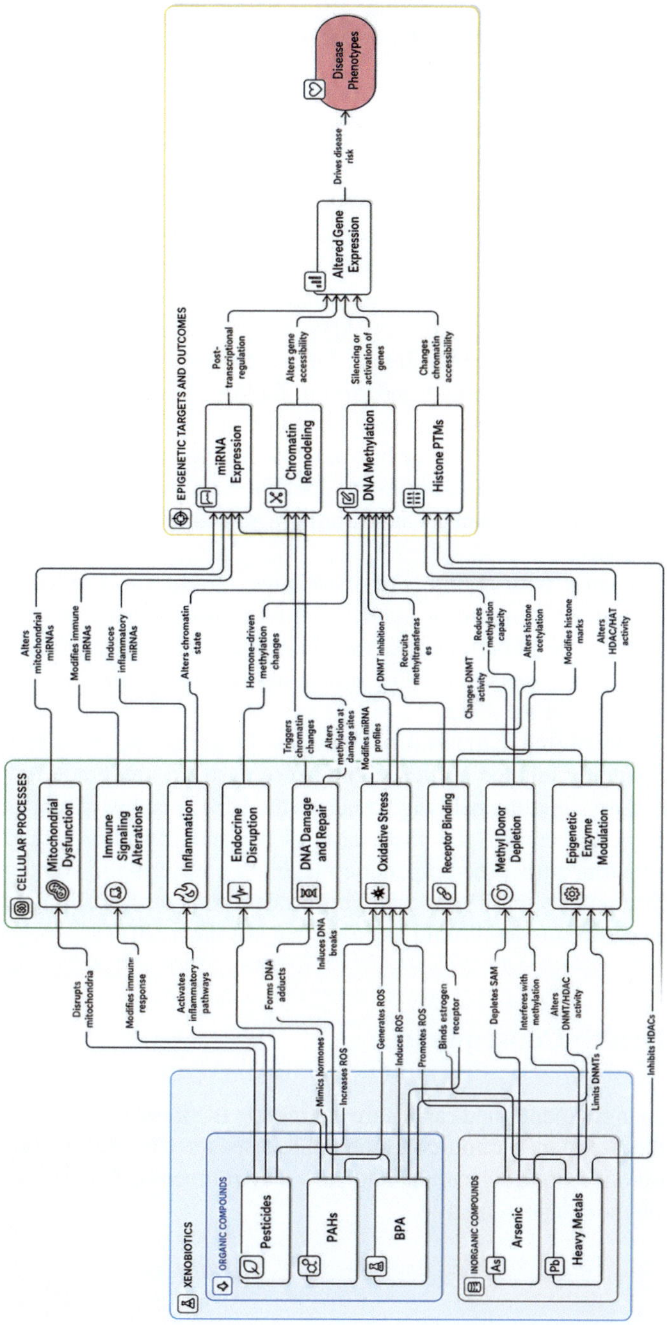

Fig. 1 Mechanistic pathways of xenobiotic-induced epigenetic modification

methylation reactions, including those involving DNA and histones. At the same time, oxidative stress can inhibit SAM synthetase activity, the enzyme that makes SAM. As a result, the cellular SAM pool is depleted, and its byproduct, S-adenosylhomocysteine (SAH), increases. SAH strongly inhibits most methyltransferases. This change in the SAM/SAH ratio can lead to global hypomethylation of DNA and histones (Darnell & Weidolf, 2013).

- **Receptor-Mediated Signaling**: Many xenobiotics work by mimicking natural substances and binding to cellular receptors. Endocrine-disrupting chemicals (EDCs), for example, can attach to nuclear hormone receptors like estrogen receptors (ERα, ERβ) or the androgen receptor (AR). Other pollutants, such as dioxins and polycyclic aromatic hydrocarbons (PAHs), act as ligands for the Aryl Hydrocarbon Receptor (AHR). Once the ligand binds, these receptors typically move into the nucleus and function as transcription factors. They recruit various co-activator or co-repressor proteins, many of which are epigenetic enzymes like histone acetyltransferases (HATs) or histone deacetylases (HDACs). This selective recruitment of epigenetic modifiers to specific gene promoter and enhancer regions leads to localized changes in chromatin structure and gene expression (Nakata et al., 2006).
- **Direct Interference with Epigenetic Enzymes**: Some xenobiotics or their metabolites can directly inhibit or alter the activity of epigenetic enzymes. For instance, certain heavy metals can take the place of essential metal cofactors like zinc in the catalytic regions of DNMTs or histone-modifying enzymes, changing their function (Sobra et al., 2025).
- **Inflammation-Mediated Chromatin Changes**: Ongoing exposure to specific pollutants, especially inhaled particles, can cause chronic, low-grade inflammation. Inflammatory signaling pathways, such as the NF-κB pathway, activate transcription factors that control the expression of pro-inflammatory genes. These transcription factors attract the epigenetic machinery, resulting in stable remodeling of the chromatin at these gene loci. This process cements a pro-inflammatory state that may contribute to chronic diseases (Bayarsaihan, 2011).

4 Human Health Consequences

The molecular changes mentioned earlier are not merely theoretical; they are increasingly linked to the rise of major non-communicable diseases. Changes in epigenetics caused by xenobiotics provide insight into how environmental factors can affect disease risk throughout a person's life.

4.1 Connections Between Epigenetic Changes Induced by Xenobiotics and Disease

i. **Cancer**: Disruption of epigenetics is a key feature of cancer. Xenobiotics can contribute to cancer development through various epigenetic processes (Saves & Masson, 1998; Williams & Phillips, 2000). A common pathway involves hypermethylation, which silences tumor suppressor genes that typically prevent cell growth (for example, arsenic-induced methylation of p53) (Williams & Phillips, 2000). On the other hand, global or specific gene hypomethylation can result in the inappropriate activation of oncogenes or the reactivation of transposable elements, potentially leading to genomic instability. Additionally, the expression of enzymes that process xenobiotics, like those in the cytochrome P450 family, is regulated by epigenetics and can be altered by exposure. This, in turn, affects how pro-carcinogens are activated (Singh et al., 2023b).

ii. **Neurodevelopmental and Neurodegenerative Disorders**: The developing brain is particularly susceptible to environmental harm (Aschner et al., 2015). Early exposure to neurotoxic substances such as lead and certain persistent organic pollutants is associated with a higher risk of attention-deficit/hyperactivity disorder, autism spectrum disorder, and lower cognitive abilities (Tran et al., 2017). These effects connect to epigenetic changes in genes that are crucial for neuron development, synaptic flexibility, and neurotransmitter signaling, such as BDNF and the glucocorticoid receptor gene. Furthermore, early epigenetic programming may create hidden vulnerabilities, linking early lead exposure to a higher risk of neurodegenerative diseases like Alzheimer's later in life (Aschner et al., 2015; Santos et al., 2022).

iii. **Metabolic Disease**: The growing epidemic of metabolic disorders, including obesity and type 2 diabetes, is linked to exposure to endocrine-disrupting chemicals, often called metabolic disruptors (Petriello et al., 2017). Substances like BPA, phthalates, and certain persistent organic pollutants can disrupt hormonal systems that regulate appetite, fat storage, and blood sugar levels. These effects likely occur partly through epigenetic changes in critical metabolic tissues, such as the liver, fat tissue, and pancreas (Banerjee & Vats, 2014).

iv. **Immune Dysregulation**: The rising rates of allergic and autoimmune disorders in developed countries point to an environmental factor. Exposure to pollutants like polycyclic aromatic hydrocarbons and diesel exhaust can epigenetically change immune cells, especially T-lymphocytes. This change can skew the immune response toward a pro-allergic phenotype, leading to asthma, or impair the function of regulatory T-cells, which are vital for preventing autoimmunity (Mokarizadeh et al., 2015).

v. **Reproductive Effects**: Endocrine-disrupting chemicals are well-known for their harmful effects on reproductive health, such as infertility, reproductive organ malformations, and hormone-sensitive cancers (Mattison, 1983). These negative impacts often start during fetal development when reproductive organs are forming. Epigenetic changes in genes related to the hypothalamic-pituitary–gonadal axis are a primary way these substances disrupt normal reproductive development (Lafuente et al., 2013).

4.2 Evidence Types: Strengths and Limitations

Establishing a causal relationship between a xenobiotic, an epigenetic change, and a disease requires a solid body of evidence from various study types, each with its own strengths and weaknesses.

4.2.1 Human Epidemiology (e.g., Cohort and Case–Control Studies) (Rothman et al., 2008)

- **Advantages**: These studies are directly related to human health and realistic exposure situations. They can reveal associations at the population level.
- **Disadvantages**: They can be affected by factors like genetics, lifestyle, and other environmental influences, which are hard to fully account for. Measuring exposure may often be inaccurate and subject to recall bias. Determining causality is often difficult since the observational nature of the data means an observed correlation might come from reverse causation (where the disease process influences the epigenome) or an unmeasured shared cause. Moreover, most human studies rely on easily accessible tissues like peripheral blood, which may not accurately reflect the epigenetic state of the target organ (like the brain). This raises significant concerns about using surrogate tissues.

4.2.2 Animal Models (e.g., Rodent Studies) (Boverhof et al., 2011; Hawksworth, 1994)

- **Advantages**: These models allow for precise control over dose, timing, and duration of exposure. They enable causality to be established through experimental manipulation. They also provide direct analysis of target tissues that are not accessible in humans.
- **Disadvantages**: There can be large species differences in metabolism, physiology, and epigenetic regulation, making it tough to directly apply findings to humans. Laboratory settings might not adequately replicate the complexities of human exposures and lifestyles.

4.2.3 Cell-Based (in Vitro) Mechanistic Studies (Hawksworth, 1994; Janošek et al., 2006):

- **Advantages**: These studies provide a controlled environment for exploring specific molecular pathways and mechanisms of action. They are also suitable for screening a large number of chemicals.
- **Disadvantages**: They lack the complex systemic context of an entire organism, including interactions between multiple organs, metabolic processes, and endocrine feedback loops. The relevance of findings from a single-cell line to a complex disease in a whole organism can be unclear.

4.3 Transgenerational and Intergenerational Effects

The idea that environmental exposures can affect the health of future generations is a key concept in environmental epigenetics. However, the terminology is specific and important.

4.3.1 Intergenerational Effects

These refer to effects on generations that were directly exposed to the initial insult (Schmidt, 2013). For example, when a pregnant female (the F0 generation) is exposed to a xenobiotic, her developing fetus (the F1 generation) is directly impacted. Furthermore, the germ cells within that fetus, which will generate the F2 generation, also experience direct exposure. Thus, negative effects in the F1 and F2 generations are classified as intergenerational (Viluksela & Pohjanvirta, 2019).

4.3.2 Transgenerational Effects

True transgenerational epigenetic inheritance (TEI) refers to effects seen in generations that were never directly exposed, whether as an organism or as a germ cell (Jiménez-Chillarón et al., 2015). Using the previous example of maternal exposure, this relates to the F3 generation and beyond. TEI requires stable transmission of an epimutation through the germline, surviving epigenetic reprogramming (Schmidt, 2013).

Evidence for TEI due to xenobiotic exposure in animal models is strong for certain substances, especially the fungicide vinclozolin, which has been shown to cause reproductive issues in male rats that can last for at least four generations. However, demonstrating TEI in humans is quite difficult because of long generation times, challenges in accounting for confounding exposures over generations, and ethical concerns (Mennigen et al., 2018).

While suggestive evidence exists from historical cohorts like the Dutch Famine Study, the current understanding of widespread xenobiotic-induced TEI in humans is still uncertain and remains an area of active research and discussion. This distinction is crucial; if a chemical's effects are truly transgenerational, existing risk assessments, which focus only on the exposed individual, greatly underestimate societal costs. This possibility necessitates a more cautious approach to chemical regulation (Painter et al., 2008).

5 Environmental and Ecological Consequences

The principles of epigenetic regulation are highly conserved across taxa, and the impacts of xenobiotics are not limited to human health. Environmental epigenetics is a rapidly growing field within ecotoxicology, providing new tools to understand how pollution affects wildlife and ecosystem stability.

5.1 Epigenetic Impacts Across Species

Xenobiotic pollutants are present in both aquatic and terrestrial ecosystems. Endocrine-disrupting compounds (EDCs) and polycyclic aromatic hydrocarbons (PAHs) affect the reproductive biology and development of various aquatic organisms, including fish, amphibians, reptiles, and invertebrates. When male fish are exposed to estrogenic substances in wastewater, they can become feminized. This is shown by the production of vitellogenin, a precursor protein for egg yolk, and the development of intersex gonads. These physiological changes occur due to alterations in gene expression, with epigenetic mechanisms now recognized as important factors in these effects (Ansari et al., 2023; Buha et al., 2022).

Using sentinel species, which are organisms that indicate environmental health, can significantly improve by adding epigenetic measures. Sampling fish, bivalves (like mussels), or earthworms from both contaminated and clean areas can give insights into their epigenetic traits, such as overall DNA methylation levels or the methylation of specific stress-response genes. These can serve as sensitive, sublethal signs of exposure and impact. This molecular information might indicate environmental stress long before visible signs, such as population declines, become apparent. It acts as an early detection system for ecosystem decline (Anastasiadi et al., 2023; Sedeño-Díaz & López-López, 2012).

5.2 *Implications at the Ecosystem Level*

Epigenetic changes in individual organisms can lead to significant effects at the population and ecosystem levels. A widespread reproductive failure in a key fish species caused by EDC-related disruptions may reduce that population. This decline can affect both predator and prey species throughout the food chain. Similarly, epigenetic changes impacting behaviors, such as feeding patterns or predator avoidance, can alter species interactions and community structures (Embrandiri et al., 2016).

Many xenobiotics remain and build up biologically; their epigenetic effects can be entrenched in an ecosystem for long periods. This raises the possibility that environmental pollution may have evolutionary consequences. A pollutant that causes a heritable epigenetic change affecting a selectable trait introduces a new layer of non-genetic variation within the population. This could speed up adaptation to changing environments or, on the other hand, lead to a maladaptive phenotype that makes the population less able to handle additional stressors, such as climate change. The relationship between chemical pollution and other environmental factors, like rising temperatures—which can also affect the epigenome—has become a key focus in ecological research. The idea that human actions not only create toxicity but also influence the evolutionary paths of other species through epigenetic processes has important implications for conservation biology and environmental stewardship (Abdolmaleky et al., 2025; Vogt, 2021).

6 Methods for Assessing Epigenetic Effects (Study Design and Analysis)

The potential of environmental epigenetics comes with its own set of challenges. Researchers must carefully consider various biases and confounders when designing, conducting, and interpreting these studies. The Epigenome-Wide Association Study (EWAS) is a common research framework that highlights these issues (Fig. 2). It aims to identify links between specific exposures or diseases and epigenetic markers across the genome (Michels et al., 2013).

6.1 *Study Design Recommendations*

A successful study in environmental epigenetics starts with careful design.

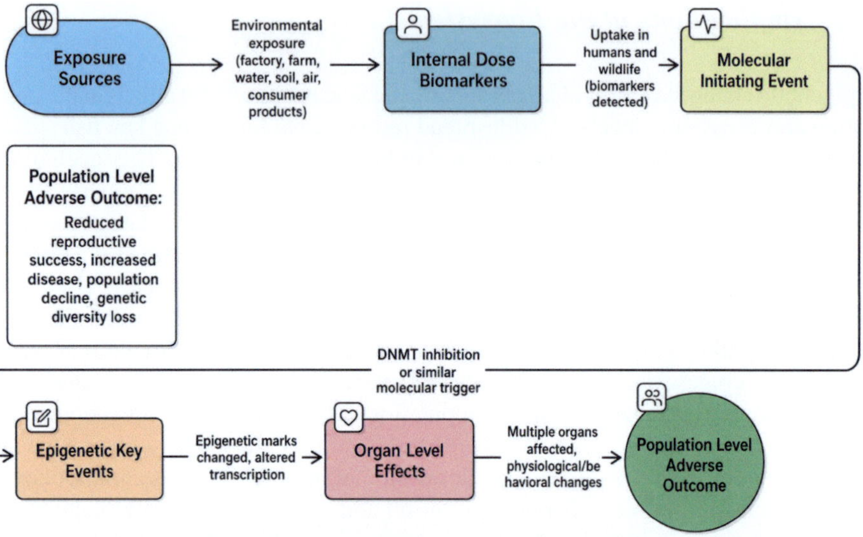

Fig. 2 Flowchart for a Robust Epigenome-Wide Association Study (EWAS) Design and Analysis. A flowchart showing the key steps in an EWAS: starting with the research question, moving to cohort selection and exposure assessment, then lab analysis, followed by a detailed data analysis pipeline including quality control, normalization, batch correction, statistical modeling with adjustment for confounders and cell type, and finally, validation and interpretation of results

6.1.1 Sample Selection and Tissue Choice

Choosing the study population and the type of biological sample is crucial. Research groups should be well defined, providing detailed information about potential confounders. Selecting a tissue involves balancing biological relevance with accessibility. The ideal target tissue, such as the brain, is rarely available in human studies. Therefore, researchers often use surrogate tissues like peripheral blood, buccal cells, or cord blood. The validity of this approach relies on how well the epigenetic marks in these surrogate tissues reflect those in the target tissue, which requires careful evaluation (Kanter et al., 2002; Teschendorff & Zheng, 2017).

6.1.2 Exposure Assessment

Accurate measurement of exposure is important but can be challenging. Biomarkers that show exposure, such as chemical levels in blood or urine, provide a snapshot of internal dose but may represent only recent exposure for chemicals that don't persist. To evaluate past exposures, particularly during critical developmental phases, researchers may use questionnaires, residential history, or environmental modeling techniques. Epigenetic marks are also being studied as indicators of cumulative, long-term exposure (Estévez et al., 2019).

6.1.3 Confounding and Co-exposure

Many factors beyond the chemical of interest can significantly affect epigenetic patterns. Important confounders include age, sex, genetic background, diet, smoking habits, and socioeconomic conditions. These must be measured and controlled statistically. Additionally, real-world environmental exposures often involve complex mixtures of chemicals, making it harder to analyze the effects of a single agent (Altenburger, 2011; Estévez et al., 2019).

6.1.4 Longitudinal Versus Cross-Sectional Design

Cross-sectional studies collect data on exposure and epigenetics at a single point in time. While they are efficient, they cannot clarify the order of events. A correlation between an exposure and an epigenetic mark may suggest the exposure caused the change, or it could indicate that individuals with a specific pre-existing epigenetic profile are more susceptible to the exposure (selection bias). Longitudinal studies track participants and collect samples over time, which provides a better foundation for establishing causality by showing that the exposure occurred before the epigenetic change (Mustieles et al., 2017; Xie et al., 2019).

6.2 *How to Evaluate Epigenetic Evidence in an Exposure Study*

This box provides a practical checklist for critically appraising a paper in environmental epigenetics.

Critical Questions to Ask:

i. **Study Design**: Is the study cross-sectional or longitudinal? A longitudinal design provides stronger evidence for the temporal relationship between exposure and epigenetic change.
ii. **Exposure Assessment**: How was exposure measured? Was it a biomarker (e.g., blood level) or an indirect estimate (e.g., questionnaire)? Was the timing of exposure assessment relevant to the hypothesized window of susceptibility?
iii. **Tissue Relevance**: What tissue was analyzed? Is it biologically relevant to the outcome of interest? If a surrogate tissue (e.g., blood) was used for a non-blood-based disease (e.g., a neurodevelopmental disorder), do the authors provide evidence or a strong rationale for its validity?
iv. **Confounding**: Did the statistical models adjust for key potential confounders, including at a minimum: age, sex, and some measure of socioeconomic status or lifestyle (e.g., smoking)?
v. **Cell-Type Heterogeneity**: For studies using mixed-cell tissues like blood, was cell-type composition accounted for? Was it measured directly or estimated

computationally? Did the authors consider whether cell composition might be a mediator rather than just a confounder?

vi. **Statistical Rigor**: Was a stringent correction for multiple testing (e.g., FDR or Bonferroni) applied? Is the reported effect size biologically meaningful, or just statistically significant?

vii. **Validation and Replication**: Was the primary finding validated using a different technology (e.g., pyrosequencing) or replicated in an independent cohort? Replication is the cornerstone of robust science.

6.3 Typical Quality Control (QC) Metrics to Expect

For microarray data: Reports on sample call rates, probe filtering (removal of poor-quality probes, cross-reactive probes, and probes containing SNPs), and normalization methods.

For sequencing data: Reports on read depth (coverage), mapping efficiency, and bisulfite conversion efficiency.

6.3.1 What Strengthens Causal Inference?

- A longitudinal design showing exposure precedes the epigenetic change, which in turn precedes the health outcome.
- Evidence of a dose–response relationship.
- Supportive evidence from experimental animal or in vitro models demonstrating a plausible biological mechanism.
- Use of advanced statistical methods like mediation analysis or Mendelian Randomization.

7 Regulatory, Public Health, and Remediation Implications

Addressing the complex and often uncertain results of environmental epigenetics to put practical measures into place for public health and environmental policy is a significant challenge (Fig. 3). It requires bridging the gap between fundamental research and the current methods of risk assessment and policy development (Bharti et al., 2023).

- **Integration into Risk Assessment**: The usual four-step process of chemical risk assessment, which includes (1) hazard identification, (2) dose–response assessment, (3) exposure assessment, and (4) risk characterization, can be improved and informed by insights from epigenetics (Ortiz et al., 2022).

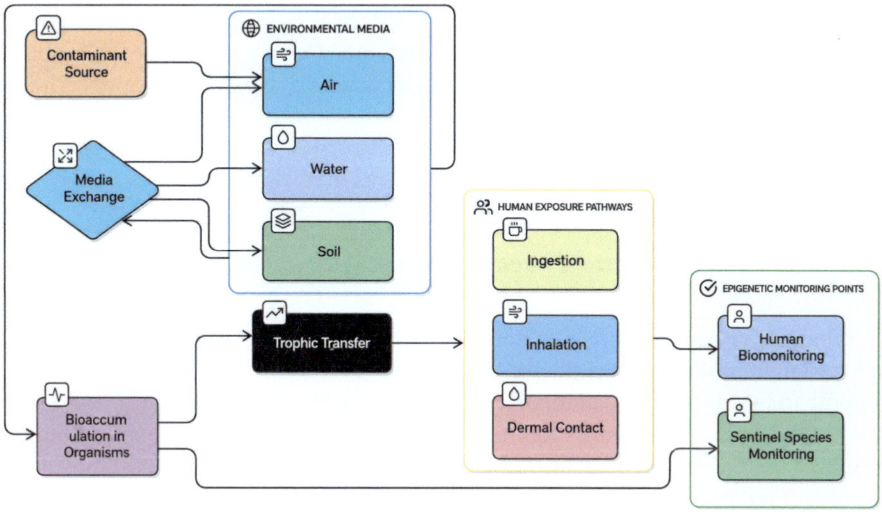

Fig. 3 Flowchart illustrating the exposure-to-effect pathway of environmental xenobiotics: contaminants enter the environment (source → media exchange) and partition into air, water, and soil, leading to bioaccumulation in organisms and trophic transfer. Resulting human exposure routes (ingestion, inhalation, dermal contact) are shown, with recommended epigenetic monitoring points (human biomonitoring and sentinel-species monitoring). Purpose: clarify where exposures occur, how they move through food webs, and where epigenetic biomarkers can be sampled to link exposure with effect

- **Hazard Identification**: Changes in epigenetics can serve as sensitive, early signs of a chemical's biological effects. An Epigenome-Wide Association Study (EWAS) can identify a chemical as a potential risk by showing its ability to alter DNA methylation at doses lower than those causing visible clinical or pathological effects, thus offering an early warning of possible harm (Baun et al., 2006).
- **Dose–Response Assessment**: Epigenetic findings can clarify a chemical's mode of action (MoA), which is crucial for understanding the shape of the dose–response curve. For endocrine-disrupting chemicals (EDCs) that show non-linear dose-responses, identifying the epigenetic changes that occur at lower doses but not at higher ones can provide the mechanistic evidence needed to move away from default linear extrapolation models.
- **Exposure Assessment**: Epigenetic changes, particularly stable modifications like DNA methylation, can act as biomarkers of cumulative exposure, reflecting long-term exposure patterns better than a single measurement of a chemical in blood or urine (Estévez et al., 2019).

However, despite this potential, significant obstacles remain. A major challenge is translating molecular epigenetic information into the quantitative measures that regulators need. For example, what level of change in methylation at a specific site qualifies as an "adverse effect"? How does a 10% increase in methylation relate to a measurable risk of disease? Addressing these questions will need a combined effort to establish and validate "epigenetic points of departure" and to integrate key epigenetic events into quantitative Adverse Outcome Pathways (AOPs). This creates a notable gap between current scientific understanding, which is often descriptive and correlational, and the demands of regulatory toxicology, which needs clear, quantitative thresholds for safety standards.

7.1 Public Health and Precautionary Approaches

The findings from environmental epigenetics, especially regarding critical developmental vulnerability and the possibility of transgenerational effects, provide strong support for using the precautionary principle in public health policy (Lang et al., 2025). This principle suggests that when an activity poses a risk of harm to human health or the environment, cautionary actions should be taken even if some cause-and-effect relationships have not been definitively proven. Given the chance of irreversible developmental programming and multi-generational damage from certain xenobiotics, waiting for solid human evidence of harm could be too late. This calls for stricter regulations on chemicals known to cause epigenetic or endocrine-disrupting effects, particularly in products meant for pregnant women and children (Ortiz et al., 2022).

7.2 Mitigation and Remediation Strategies

While reducing exposure is the main goal, research is also exploring ways to lessen epigenetic harm.

7.2.1 Nutritional Intervention

The epigenome can respond strongly to dietary factors. The one-carbon metabolism pathway, essential for producing the universal methyl donor, SAM, depends on dietary components like folate, vitamin B12, choline, and methionine. Animal studies suggest that supplementing with methyl donors can counteract the hypomethylating effects of certain xenobiotics, such as BPA. This opens the door to targeted nutritional strategies that could improve epigenetic resilience in affected populations. However, this approach also highlights an important environmental justice issue: if good nutrition can reduce the impacts of pollution, communities facing both high exposure and food insecurity may be in a "double jeopardy" of increased vulnerability. This means that effective public health strategies must tackle both chemical exposures and the social factors that influence health (Cosola et al., 2018).

7.2.2 Pharmacological Intervention

The ability to reverse epigenetic marks has led to the creation of "epi-drugs," like HDAC inhibitors and DNMT inhibitors. These drugs are now used in cancer treatment to reactivate silenced tumor suppressor genes. While using them to reverse environmental damage isn't currently practical due to their lack of specificity and possible side effects, they do offer a long-term treatment option (Carmody & Turnbaugh, 2014; Li et al., 2007).

8 Current Challenges, Gaps, and Future Directions

The field of environmental epigenetics is moving quickly, but it faces significant challenges that need to be addressed to fully unlock its potential for improving human and environmental health.

8.1 Technical Issues and Reproducibility

Even with advances in technology, there is still a need for better standardization of experimental and bioinformatics protocols. This will help ensure that results can be replicated across different laboratories. The high costs associated with thorough methods like whole-genome bisulfite sequencing (WGBS) limit their use in large human studies, making it necessary to develop affordable, targeted sequencing panels or improved array designs. Furthermore, the lack of complete reference epigenomes for the wide range of human cell types and important wildlife species makes it harder to interpret new data (Anzenbacher & Zanger, 2012; Zhu et al., 2025).

8.2 Requirement for Longitudinal, Multi-omics Investigations

So far, most human studies have been cross-sectional, focusing mainly on a single epigenetic marker, usually DNA methylation. The field needs to shift toward longitudinal cohort studies that collect samples multiple times over time. This approach will provide deeper insights into the changes in epigenetics in response to various exposures. Additionally, combining epigenomics with other "omics" areas—such as genomics, transcriptomics, proteomics, metabolomics, and the microbiome—is crucial for building a complete understanding of how xenobiotics affect health (Gao et al., 2022; Papaioannou et al., 2021).

8.3 Understanding Exposure Mixtures

In reality, living organisms are rarely exposed to just one chemical. They encounter a complex and changing mix of pollutants, dietary components, and medicines. A major challenge for toxicology is to figure out the combined and possibly interactive effects of these mixtures on epigenetics. This will require new experimental designs and advanced computational models that can handle such complexity (Silins & Högberg, 2011).

8.4 Low-Dose and Non-monotonic Responses

We need to look more closely at the mechanisms and prevalence of low-dose effects and non-monotonic dose–response (NMDR) relationships. Incorporating these non-linear concepts into official risk assessments is a tough but vital job (Kohn & Melnick, 2002).

8.5 Equity and Environmental Justice Considerations

It is essential to conduct research in environmental epigenetics that benefits communities bearing the brunt of environmental pollution. These often include low-income populations and communities of color. This means studying diverse groups and considering how chemical and non-chemical stressors interact. The idea of the "exposome" should be expanded to include the epigenetic impacts of psychosocial stress, poverty, discrimination, and poor nutrition, as well as understanding how these social factors affect biological responses to chemical exposures. As the field moves toward developing effective new exposure and risk biomarkers, like "epigenetic clocks" that measure accelerated biological aging, it must also deal with the ethical, legal, and social implications (ELSI) of this knowledge to avoid misuse and discrimination (Hayes et al., 1983; Haga, 2019).

9 Conclusion

The relationship between xenobiotics and the epigenome represents a significant area in toxicology and environmental health. It is clear that the epigenome is a main target for many environmental chemicals and plays an important role in their effects on human health and ecosystems. Strong evidence shows that exposures, especially during critical developmental periods, can lead to lasting epigenetic changes linked to cancer, neurodevelopmental disorders, metabolic issues, and immune dysfunction.

These mechanisms provide a biological basis for how environmental factors can affect long-term health outcomes.

However, turning scientific findings into public health initiatives is challenging. The complex methods needed to design and interpret epigenetic studies, particularly in human populations, require a high level of rigor. Researchers must deal with challenges related to confounding variables, establishing causality, and ensuring tissue specificity. Translating these complex molecular findings into measurable parameters for risk assessment is a major challenge for regulatory bodies.

Looking to the future, the field must recognize and address this complexity. Researchers should focus on longitudinal, multi-omics study designs, develop robust methods for establishing causal links, and tackle the practical issues posed by chemical mixtures and their interaction with social contexts. Clinicians need to consider a patient's environmental background as a key part of their overall health assessment, aiming for a future where epigenetic markers can guide preventive and treatment strategies. Policymakers should take into account findings from environmental epigenetics, especially concerning developmental vulnerability and possible transgenerational effects. This highlights the need for a more cautious approach to chemical regulation that protects both current and future generations. By promoting collaboration among molecular biologists, toxicologists, epidemiologists, clinicians, and policymakers, we can turn the insights gained from the connection between xenobiotics and the epigenome into effective actions that improve global public health.

References

Abdolmaleky, H. M., Nohesara, S., Zhou, J. R., & Thiagalingam, S. (2025). Epigenetics in evolution and adaptation to environmental challenges: Pathways for disease prevention and treatment. *Epigenomics, 17*(5), 317–333.

Al Aboud, N. M., Tupper, C., Genetics, J. I., & Mechanism, E. (2023). *In: StatPearls [Internet].* StatPearls Publishing.

Altenburger, R. (2011). Understanding combined effects for metal co-exposure in ecotoxicology. *Metal Ions in Life Sciences, 8*, 1–26. PMID: 21473374.

Anastasiadi, D., & Beemelmanns, A. (2023). Development of epigenetic biomarkers in aquatic organisms. *Epigenetics in Aquaculture, 2*, 413–438.

Ansari, R. A., Alfuraih, S., Shekh, K., Omidi, Y., Javed, S., & Shakil, S. A. (2023). Endocrine disruptors: Genetic, epigenetic, and related pathways. *Impact of Engineered Nanomaterials in Genomics and Epigenomics, 15*, 41–82.

Anzenbacher, P., & Zanger, U. M. (Eds.). (2012). *Metabolism of drugs and other xenobiotics.* Wiley-VCH.

Aschner, M., & Costa, L. G. (2015). *Environmental factors in neurodevelopmental and neurodegenerative disorders.* Academic Press

Banerjee, M., & Vats, P. (2014). Reactive metabolites and antioxidant gene polymorphisms in type 2 diabetes mellitus. *Redox Biology, 1*(2), 170–177.

Baun, A., Eriksson, E., Ledin, A., & Mikkelsen, P. S. (2006). A methodology for ranking and hazard identification of xenobiotic organic compounds in urban stormwater. *Science of the Total Environment, 370*(1), 29–38.

Bayarsaihan, D. (2011). Epigenetic mechanisms in inflammation. *Journal of Dental Research, 90*(1), 9–17.
Bender, J. (2004). DNA methylation and epigenetics. *Annual Review of Plant Biology, 55*(1), 41–68.
Benesova, S., Kubista, M., & Valihrach, L. (2021). Small RNA-sequencing: Approaches and considerations for miRNA analysis. *Diagnostics, 11*(6), 964.
Bharti, S., & Kumar, A. (2023). Xenobiotics in urban soils and water: Remediation strategies, socioeconomic impacts, and regulatory provisions. In *Xenobiotics in urban ecosystems: Sources, distribution and health impacts* (pp. 399–430). Cham: Springer International Publishing.
Biel, M., Wascholowski, V., & Giannis, A. (2005). Epigenetics—An epicenter of gene regulation: Histones and histone-modifying enzymes. *Angewandte Chemie International Edition, 44*(21), 3186–3216.
Bock, C. (2009). Epigenetic biomarker development. *Epigenomics, 1*(1), 99–110.
Boverhof, D. R., Chamberlain, M. P., Elcombe, C. R., Gonzalez, F. J., Heflich, R. H., Hernández, L. G., Jacobs, A. C., Jacobson-Kram, D., Luijten, M., Maggi, A., & Manjanatha, M. G. (2011). Transgenic animal models in toxicology: Historical perspectives and future outlook. *Toxicological Sciences, 121*(2), 207–233.
Bressan, R. A., Zhu, J. K., Van Oosten, M. J., Maggio, A., Bohnert, H. J., & Chinnusamy, V. (2014). Epigenetics connects the genome to its environment. *Plant Breeding Reviews, 38*(3), 69–142.
Buha, A., Manic, L., Maric, D., Tinkov, A., Skolny, A., Antonijevic, B., & Hayes, A. W. (2022). The effects of endocrine-disrupting chemicals (EDCs) on the epigenome—A short overview. *Toxicology Research and Application, 11*(6), 23978473221115816.
Carmody, R. N., & Turnbaugh, P. J. (2014). Host-microbial interactions in the metabolism of therapeutic and diet-derived xenobiotics. *The Journal of Clinical Investigation, 124*(10), 4173–4181.
Cosola, C., Sabatino, A., Di Bari, I., Fiaccadori, E., & Gesualdo, L. (2018). Nutrients, nutraceuticals, and xenobiotics affecting renal health. *Nutrients, 10*(7), 808.
Costa, F. F. (2008). Non-coding RNAs, epigenetics and complexity. *Gene, 410*(1), 9–17.
Darnell, M., & Weidolf, L. (2013). Metabolism of xenobiotic carboxylic acids: Focus on coenzyme a conjugation, reactivity, and interference with lipid metabolism. *Chemical Research in Toxicology, 26*(8), 1139–1155.
Embrandiri, A, Kiyasudeen, S. K., Rupani, P. F., & Ibrahim, M. H. (2016). Environmental xenobiotics and its effects on natural ecosystem. In *Plant responses to xenobiotics* (pp. 1–18). Singapore: Springer Singapore.
Estévez, J., Vilanova, E., & Sogorb, M. A.(2019). Biomarkers for testing toxicity and monitoring exposure to xenobiotics. In *Biomarkers in toxicology* (pp. 1165–1174). Academic Press.
Finnberg, N., Stenius, U., & Högberg, J. (2000). Xenobiotics modulate the p53 response to DNA damage in preneoplastic enzyme-altered foci in rat liver; effects of diethylnitrosamine and phenobarbital. *Toxicological Sciences, 54*(1), 95–103.
Gao, P., Shen, X., Zhang, X., Jiang, C., Zhang, S., Zhou, X., Rose, S. M., & Snyder, M. (2022). Precision environmental health monitoring by longitudinal exposome and multi-omics profiling. *Genome Research, 32*(6), 1199–1214.
Geiman, T. M., & Robertson, K. D. (2002). Chromatin remodeling, histone modifications, and DNA methylation—How does it all fit together? *Journal of Cellular Biochemistry, 87*(2), 117–125.
Giraldez, M. D., Spengler, R. M., Etheridge, A., Godoy, P. M., Barczak, A. J., Srinivasan, S., De Hoff, P. L., Tanriverdi, K., Courtright, A., Lu, S., & Khoory, J. (2018). Comprehensive multi-center assessment of small RNA-seq methods for quantitative miRNA profiling. *Nature Biotechnology, 36*(8), 746–757.
Haga, S. B. (2019). Pharmacogenomic testing in pediatrics: Navigating the ethical, social, and legal challenges. *Pharmacogenomics and Personalized Medicine, 14*, 273–285.
Hawksworth, G. M. (1994). Advantages and disadvantages of using human cells for pharmacological and toxicological studies. *Human & Experimental Toxicology, 13*(8), 568–573.
Hayes Jr., W. J. (1983). Ethical considerations involving studies of pesticides and other xenobiotics in man. In *Mode of action, metabolism and toxicology* (pp. 387–394). Pergamon.

Hino, S., Sato, T., & Nakao, M. (2022). Chromatin immunoprecipitation sequencing (ChIP-seq) for detecting histone modifications and modifiers. In *Epigenomics: Methods and protocols* (pp. 55–64, 202). New York, NY: Springer US.

Holliday, R. (1989). DNA methylation and epigenetic mechanisms. *Cell Biophysics, 15*(1), 15–20.

Hurd, P. J., & Nelson, C. J. (2009). Advantages of next-generation sequencing versus the microarray in epigenetic research. *Briefings in Functional Genomics and Proteomics, 8*(3), 174–183.

Janošek, J., Hilscherová, K., Bláha, L., & Holoubek, I. (2006). Environmental xenobiotics and nuclear receptors—Interactions, effects and in vitro assessment. *Toxicology in Vitro, 20*(1), 18–37.

Jiménez-Chillarón, J. C., Nijland, M. J., Ascensão, A. A., Sardao, V. A., Magalhães, J., Hitchler, M. J., Domann, F. E., & Oliveira, P. J. (2015). Back to the future: Transgenerational transmission of xenobiotic-induced epigenetic remodeling. *Epigenetics, 10*(4), 259–273.

Kalvari, I., Nawrocki, E. P., Argasinska, J., Quinones-Olvera, N., Finn, R. D., Bateman, A., & Petrov, A. I. (2018). Non-coding RNA analysis using the Rfam database. *Current Protocols in Bioinformatics, 62*(1), Article e51.

Kanter, R. D., Monshouwer, M., Meijer, D. K., & Groothuis, G. M. (2002). Precision-cut organ slices as a tool to study toxicity and metabolism of xenobiotics with special reference to non-hepatic tissues. *Current Drug Metabolism, 3*(1), 39–59.

Kohn, M. C., & Melnick, R. L. (2002). Biochemical origins of the non-monotonic receptor-mediated dose-response. *Journal of Molecular Endocrinology, 29*(1), 113–124.

Lafuente, A. (2013). The hypothalamic–pituitary–gonadal axis is target of cadmium toxicity. An update of recent studies and potential therapeutic approaches. *Food and Chemical Toxicology, 59*, 395–404.

Lang, T., Lipp, A. M., & Wechselberger, C. (2025). Xenobiotic toxicants and particulate matter: Effects, mechanisms, impacts on human health, and mitigation strategies. *Journal ofXenobiotics, 15*(4), 131.

Lappan, R., Henry, R., Chown, S. L., Luby, S. P., Higginson, E. E., Bata, L., Jirapanjawat, T., Schang, C., Openshaw, J. J., O'Toole, J., & Lin, A. (2021). Monitoring of diverse enteric pathogens across environmental and host reservoirs with TaqMan array cards and standard qPCR: A methodological comparison study. *The Lancet Planetary Health, 5*(5), e297-308.

Li, H., Ni, Y., Su, M., Qiu, Y., Zhou, M., Qiu, M., Zhao, A., Zhao, L., & Jia, W. (2007). Pharmacometabonomic phenotyping reveals different responses to xenobiotic intervention in rats. *Journal of Proteome Research, 6*(4), 1364–1370.

Ma, Q. (2008). Xenobiotic-activated receptors: From transcription to drug metabolism to disease. *Chemical Research in Toxicology, 21*(9), 1651–1671.

Mattison, D. R. (1983). The mechanisms of action of reproductive toxins. *American Journal of Industrial Medicine, 4*(1–2), 65–79.

McEwen, L. M., Jones, M. J., Lin, D. T., Edgar, R. D., Husquin, L. T., MacIsaac, J. L., Ramadori, K. E., Morin, A. M., Rider, C. F., Carlsten, C., & Quintana-Murci, L. (2018). Systematic evaluation of DNA methylation age estimation with common preprocessing methods and the Infinium MethylationEPIC BeadChip array. *Clinical Epigenetics, 10*(1), 123.

Meissner, A., Gnirke, A., Bell, G. W., Ramsahoye, B., Lander, E. S., & Jaenisch, R. (2005). Reduced representation bisulfite sequencing for comparative high-resolution DNA methylation analysis. *Nucleic Acids Research, 33*(18), 5868–5877.

Mennigen, J. A., Thompson, L. M., Bell, M., Tellez Santos, M., & Gore, A. C. (2018). Transgenerational effects of polychlorinated biphenyls: 1. development and physiology across 3 generations of rats. *Environmental Health, 17* (1), 18.

Michels, K. B., Binder, A. M., Dedeurwaerder, S., Epstein, C. B., Greally, J. M., Gut, I., Houseman, E. A., Izzi, B., Kelsey, K. T., Meissner, A., & Milosavljevic, A. (2013). Recommendations for the design and analysis of epigenome-wide association studies. *Nature Methods, 10*(10), 949–955.

Minard, M. E., Jain, A. K., & Barton, M. C. (2009). Analysis of epigenetic alterations to chromatin during development. *Genesis, 47*(8), 559–572.

Mokarizadeh, A., Faryabi, M. R., Rezvanfar, M. A., & Abdollahi, M. (2015). A comprehensive review of pesticides and the immune dysregulation: Mechanisms, evidence and consequences. *Toxicology Mechanisms Methods, 25*(4), 258–278.

Mustieles, V., Fernández, M. F., Martin-Olmedo, P., González-Alzaga, B., Fontalba-Navas, A., Hauser, R., Olea, N., & Arrebola, J. P. (2017). Human adipose tissue levels of persistent organic pollutants and metabolic syndrome components: Combining a cross-sectional with a 10-year longitudinal study using a multi-pollutant approach. *Environment International, 1*(104), 48–57.

Nakata, K., Tanaka, Y., Nakano, T., Adachi, T., Tanaka, H., Kaminuma, T., & Ishikawa, T. (2006). Nuclear receptor-mediated transcriptional regulation in Phase I, II, and III xenobiotic metabolizing systems. *Drug Metabolism and Pharmacokinetics, 21*(6), 437–457.

Ogunjobi, T. T., Gbayisomore, T.J., Nneji, P.O., Olofin, O. O., Olowe, E. N., Gigam-Ozuzu, C. D., Afolabi, J. I., Okwuokei, N. B., Boluwaji, V. A., Ojeniran, T. P., & Ogini, I. O. (Eds.). (2024) Environmental epigenetics and its impacts on disease susceptibility: A comprehensive review. *Medinformatics.*

Ortiz, P., Torres-Sánchez, A., López-Moreno, A., Cerk, K., Ruiz-Moreno, Á., Monteoliva-Sánchez, M., Ampatzoglou, A., Aguilera, M., & Gruszecka-Kosowska, A. (2022). Impact of cumulative environmental and dietary xenobiotics on human microbiota: Risk assessment for one health. *Journal of Xenobiotics., 12*(1), 56–63.

Painter, R. C., Osmond, C., Gluckman, P., Hanson, M., Phillips, D. I, & Roseboom, T. J. (2008). Transgenerational effects of prenatal exposure to the Dutch famine on neonatal adiposity and health in later life. *BJOG: An International Journal of Obstetrics & Gynaecology, 115*(10), 1243–1249.

Papaioannou, N., Distel, E., de Oliveira, E., Gabriel, C., Frydas, I. S., Anesti, O., Attignon, E. A., Odena, A., Díaz, R., Aggerbeck, M, & Horvat, M. (2021). Multi-omics analysis reveals that co-exposure to phthalates and metals disturbs urea cycle and choline metabolism. *Environmental Research, 1*(192), Article 110041.

Petriello, M. C., Hoffman, J. B., Morris, A. J., & Hennig, B. (2017). Emerging roles of xenobiotic detoxification enzymes in metabolic diseases. *Reviews on Environmental Health, 32*(1–2), 105–110.

Raurell-Vila, H., Ramos-Rodríguez, M., & Pasquali, L. (2018). Assay for transposase accessible chromatin (ATAC-Seq) to chart the open chromatin landscape of human pancreatic islets. In *CpG Islands* (pp. 197–208). New York, NY: Springer New York.

Reed, K., Poulin, M. L., Yan, L., & Parissenti, A. M. (2010). Comparison of bisulfite sequencing PCR with pyrosequencing for measuring differences in DNA methylation. *Analytical Biochemistry, 397*(1), 96–106.

Rieger, P. G., Meier, H. M., Gerle, M., Vogt, U., Groth, T., & Knackmuss, H. J. (2002). Xenobiotics in the environment: Present and future strategies to obviate the problem of biological persistence. *Journal of Biotechnology, 94*(1), 101–123.

Rothman, K. J., Greenland, S., & Lash, T. L. (2008). *Case-control studies.*

Santos, J. X., Rasga, C., Marques, A. R., Martiniano, H., Asif, M., Vilela, J., Oliveira, G., Sousa, L., Nunes, A., & Vicente, A. M. (2022). A role for gene-environment interactions in autism spectrum disorder is supported by variants in genes regulating the effects of exposure to xenobiotics. *Frontiers in Neuroscience, 19*(16), Article 862315.

Saves, I., & Masson, J. M. (1998). Mechanisms of resistance to xenobiotics in human therapy. *Cellular and Molecular Life Sciences CMLS, 54*(5), 405–426.

Schleinitz, D., DiStefano, J. K., & Kovacs, P. (2010). Targeted SNP genotyping using the TaqMan® assay. In *Disease gene identification: Methods and protocols* (pp. 77–87). Totowa, NJ: Humana Press.

Schmidt, C. W. (2013). Uncertain inheritance transgenerational effects of environmental exposures. *Environmental Health Perspectives, 121*(10), A298-303. https://doi.org/10.1289/ehp.121-A298. PMID:24218659;PMCID:PMC3801462

Schröder, P., Lyubenova, L., & Huber, C. (2009). Do heavy metals and metalloids influence the detoxification of organic xenobiotics in plants? *Environmental Science and Pollution Research, 16*(7), 795–804.

Sedeño-Díaz, J. E., & López-López, E. (2012). Freshwater fish as sentinel organisms: From the molecular to the population level. *New Advances and Contributions to Fish Biology, 21*, 151.

Silins, I., & Högberg, J. (2011). Combined toxic exposures and human health: Biomarkers of exposure and effect. *International Journal of Environmental Research and Public Health, 8*(3), 629–647.

Silveira, P. P., Portella, A. K., Goldani, M. Z., & Barbieri, M. A. (2007). Developmental origins of health and disease (DOHaD). *Jornal De Pediatria, 83*, 494–504.

Singh, R. D., Avadhesh, A., Sharma, G., Dholariya, S., Shah, R. B., Goyal, B., & Gupta, S. C. (2023a). Potential of cytochrome P450, a family of xenobiotic metabolizing enzymes, in cancer therapy. *Antioxidants & Redox Signaling, 38*(10–12), 853–876.

Singh, M., Mishra, R. C., Shah, I., Wadhwa, V., & Mor, V. (2023b). Xenobiotics: Sources, pathways, degradation, and risk associated with major emphasis on pharmaceutical compounds. In *Xenobiotics in urban ecosystems: Sources, distribution and health impacts* (pp. 87–106). Cham: Springer International Publishing.

Skinner, M. K. (2016). Epigenetic transgenerational inheritance. *Nature Reviews Endocrinology, 12*(2), 68–70.

Sobral, A. F., Cunha, A., Costa, I., Silva-Carvalho, M., Silva, R., & Barbosa, D. J. (2025). Environmental xenobiotics and epigenetic modifications: Implications for human health and disease. *Journal of Xenobiotics, 15*(4), 118.

Štefanac, T., Grgas, D., & Landeka, D. T. (2021). Xenobiotics—division and methods of detection: A review. *Journal of Xenobiotics, 11*(4), 130–141.

Suzuki, M., Liao, W., Wos, F., Johnston, A. D., DeGrazia, J., Ishii, J., Bloom, T., Zody, M. C., Germer, S., & Greally, J. M. (2018). Whole-genome bisulfite sequencing with improved accuracy and cost. *Genome Research, 28*(9), 1364–1371.

Teschendorff, A. E., & Zheng, S. C. (2017). Cell-type deconvolution in epigenome-wide association studies: A review and recommendations. *Epigenomics, 9*(5), 757–768.

Tran, N. Q., & Miyake, K. (2017). Neurodevelopmental disorders and environmental toxicants: Epigenetics as an underlying mechanism. *International Journal of Genomics, 2017*(1), 7526592.

Vasseur, P., & Cossu-Leguille, C. (2006). Linking molecular interactions to consequent effects of persistent organic pollutants (POPs) upon populations. *Chemosphere, 62*(7), 1033–1042.

Viluksela, M., & Pohjanvirta, R. (2019). Multigenerational and transgenerational effects of dioxins. *International Journal of Molecular Sciences, 20*(12), 2947.

Vogt, G. (2021). Epigenetic variation in animal populations: Sources, extent, phenotypic implications, and ecological and evolutionary relevance. *Journal of Biosciences, 46*(1), 24.

Wang, X., Sial, M. U., Bashir, M. A., Bilal, M., Raza, Q. U., Ali Raza, H. M., Rehim, A., & Geng, Y. (2022). Pesticides xenobiotics in soil ecosystem and their remediation approaches. *Sustainability, 14*(6), 3353.

Williams, J. A., & Phillips, D. H. (2000). Mammary expression of xenobiotic metabolizing enzymes and their potential role in breast cancer. *Cancer Research, 60*(17), 4667–4677.

Xavier, M. J., Roman, S. D., Aitken, R. J., & Nixon, B. (2019). Transgenerational inheritance: How impacts to the epigenetic and genetic information of parents affect offspring health. *Human Reproduction Update, 25*(5), 519–541.

Xie, Y., Lin, T., Yang, M., Zhang, Z., Deng, N., Tang, M., Xiao, Y., Guo, H., & Deng, Q. (2019). Co-exposure to polycyclic aromatic hydrocarbons and metals, four common polymorphisms in microRNA genes, and their gene-environment interactions: Influences on oxidative damage levels in Chinese coke oven workers. *Environment International, 1*(132), Article 105055.

Zhang, Y. Z., Song, X. F., Kondoh, A., Xia, J., & Tang, C. Y. (2011). Behavior, mass inventories and modeling evaluation of xenobiotic endocrine-disrupting chemicals along an urban receiving wastewater river in Henan Province. *China. Water Research, 45*(1), 292–302.

Zhu, X., Tang, X., Ng, C., Li, L., Lai, Y., Miller, G. W., Jiang, C., Barchowsky, A., Buchanich, J., & Gao, P. (2025). Biomonitoring xenobiotics in human biospecimens: Challenges, advances, and the future of exposome characterization. *Reviews of Environmental Contamination and Toxicology, 263*(1), 1–25.

Chapter 8
Genetic Polymorphism and Xenobiotic Metabolism: In Relation to Human Health

Ayan Mukherjee, Sachinandan De, Sukhen Samanta, Indrajit Kar, and Shyam Sundar Kesh

Abstract Xenobiotics are chemical compounds that infiltrate the body from the external environment but are neither naturally synthesized nor anticipated to exist in humans. A diverse type of chemical compounds, including pharmaceuticals, recreational substances, environmental pollutants, food additives and preservatives, industrial chemicals, and toxins from microorganisms, are the common xenobiotics affecting humans worldwide. These chemicals frequently require biotransformation (mostly in the liver) for detoxification and excretion. Their metabolites, however, have the ability to interact with genetic material and cause mutagenic, genotoxic, and cancerous consequences. Xenobiotic compounds leave an impression at the level of the genome, transcriptome, and epigenome. The damages caused by these compounds include distortion of the DNA helix, DNA adduct formation, chromosomal aberrations, oxidative DNA damage, base modifications, and mispairing of bases. A handful of xenobiotic compounds can alter the epigenetic signature of the cells, including DNA methylation patterns or histone modifications. All these structural changes in genome and epigenome lead to alteration in global transcriptome, resulting in various human diseases like cancer, neurodegenerative disorders, reproductive and developmental issues, and accelerated aging by telomere shortening from DNA damage. This chapter highlights the different xenobiotic compounds inducing instability in human genome, the molecular mechanism of damage, and the different diseases caused due to damages.

Keywords Transcriptome · Epigenome · DNA damage · Human genome · Chromosomal aberrations

---

A. Mukherjee · I. Kar · S. S. Kesh
Faculty of Veterinary and Animal Sciences, West Bengal University of Animal and Fishery Sciences, Mohanpur, Nadia, West Bengal, India

S. De (✉)
Animal Biotechnology Center, National Dairy Research Institute, Karnal, Haryana, India
e-mail: sachinandan@gmail.com

S. Samanta
Department of Microbiology, University of Kalyani, Kalyani, West Bengal, India

K. Priya and S. K. Giri (eds.), *Environmental Toxicology: Xenobiotics and Effects on Human Genetic Material*, https://doi.org/10.1007/978-3-032-28619-2_8

1 Introduction

The environment refers to the physical world we live in, and all the natural elements: air we breathe, water that we drink, soil used for growing our food, and living things (biotic factors) as well as things that are non-living (abiotic factors), which come into contact within ecosystems. Just like people, every organism on this planet needs a safe place to live and grow. The technological progress has been a significant factor in enhancing human life and driving the growth of the world over the last century. Nevertheless, such rapid development has generated serious environmental problems, pollution, and resource depletion first among them. Increased emissions of hazardous gases and chemicals into the atmosphere as a result of industrialization, a major outcome of technological growth contribute to environmental deterioration. Furthermore, the increasing number of electronic products available on the market has brought a major issue: e-waste. Discarding batteries and electronic equipment as garbage pollutes soil and water with dangerous chemicals like lead and mercury. Better access to medicine (including in the case of livestock) and everyday personal care products/insecticides are also adding new chemicals into the environment. Like all other living organisms, humans are also exposed to these chemicals some of which are detrimental to their health. Xenobiotics are chemicals that are alien to living things (xenos means foreign, and bios means life in Greek). The body does not naturally produce or anticipate these chemicals. The toxicity, persistence, and effects of these substances on human physiological processes are determined by the capacity of the human metabolic machinery to metabolize them.

The liver is the primary site of xenobiotic metabolism (XM), which is aided by enzymes such as glutathione S-transferases (GSTs), UDP-glucuronosyltransferases (UGTs), and cytochrome P450s (CYPs). Over the time, interest in exploring XM has deepened among the researchers, particularly in understanding how these xenobiotic metabolism enzymes (XME) of human metabolic pathways interact with the xenobiotic compounds to detoxify them. Genetic polymorphisms, i.e., structural variations (SVs) in the genes encoding these XME result in diversity in the expression pattern of the enzymes among individuals and populations. This ultimately leads to altered metabolism resulting in unfavorable drug responses, carcinogenesis, or organ-specific toxicity.

Numerous SVs have been discovered after the Human Genome Project was finished, thanks to advancements in genome sequencing technologies and bioinformatics techniques. Close association has also been found with these SV and the onset or progress of various diseases. Comprehending these molecular variations is essential for enhancing toxicological risk assessment, pharmacological safety, and personalized treatment.

This chapter summarizes recent studies designed to elucidate the genetic polymorphisms of XME genes and their association with diseases in humans.

2 What is Xenobiotic?

Xenobiotics are characterized as "substances that are not generated by organisms or the environment yet are found within them. Certain chemicals that occur naturally (endobiotics) can be classified as xenobiotics when they exist in the environment at high concentrations". These substances can be found in the water, soil, air, plants, animals, and people. Xenobiotic chemicals include a wide range of molecules, such as medications, pesticides, industrial goods, e-wastes, drugs, cosmetics, food additives, personal care items, nuclear wastes, and endobiotics (Fig. 1). A variety of human activities introduce xenobiotics into the environment. These include human food intake and waste, wastewater management plants, livestock handling, manufacturing, and farming practices. Xenobiotics are challenging to degrade due to their intricate structure, which leads to their accumulation in living organisms. Partially degraded products of xenobiotics are even more detrimental to the various environmental components. To protect the body from the detrimental effects of xenobiotics, these chemicals undergo different biochemical transformations and are eliminated by the metabolic and excretory machinery of living organisms. Although XM primarily takes place in the liver, other organs like the intestine, kidneys, lungs, and skin participate in the process.

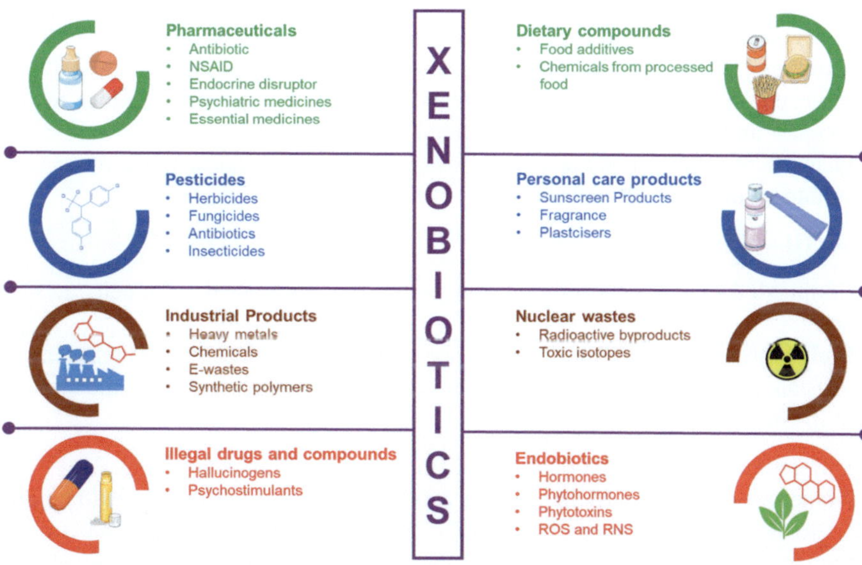

Fig. 1 Categorization of different xenobiotics. Xenobiotics comprise pharmaceuticals, pesticides, industrial products, illegal drugs, food additives, cosmetics, nuclear wastes, and endobiotics. Image created by Servier Medical Art (https://smart.servier.com/) and Pubchem Sketcher

3 Xenobiotic Metabolism and Associated Enzymes—A Brief Overview

XM refers to the biochemical transformation of foreign compounds by living organisms, mostly to enhance their water solubility and facilitate excretion. This involves three major phases:

Phase I—Functionalization Reactions

- In xenobiotics, phase I reactions either reveal an already-existing hydrophilic group or add a polar hydrophilic group.
- Oxidation, reduction, hydroxylation, and hydrolysis processes are all part of phase I metabolism.
- Cytochrome P450 monooxygenases (CYPs) are the major enzyme system intricate in phase I.
- CYPs introduce oxygen atoms into hydrocarbons that are not activated. This process can lead to the addition of hydroxyl groups or the removal of alkyl groups from nitrogen, oxygen, or sulfur atoms of substrates.

Phase II—Conjugation Reactions

- The product of phase I reaction undergoes phase II metabolic reaction.
- To make the xenobiotic molecule more hydrophilic, an endogenous hydrophilic molecule is coupled to a pharmaceutical molecule during phase II metabolic activities.
- Molecules that undergo phase II metabolic processes lose their affinity for the receptor and are subsequently eliminated from the body.
- Phase II metabolic reactions include conjugation of adducts (glucuronic acid, sulfate, amino acid, glutathione, etc.), acetylation, and methylation.

Phase III—Elimination Reaction

- Conjugated metabolites become transported and eliminated from cells and the body.
- P-glycoprotein, the multidrug resistance protein (MRP) family, the breast cancer resistance protein (BCRP) family, and other efflux transporters aid in the removal of the metabolites.

Different steps of xenobiotic metabolism have been illustrated in Fig. 2.

Numerous enzymes participate in the above process of XM in the body to modify and detoxify foreign compounds (xenobiotics) such as drugs, pollutants, and toxins, making them easier to excrete. Although they are present in the intestines, lungs, and kidneys, these enzymes primarily function in the liver. The various XME and their roles in the XM process are explained in Table 1.

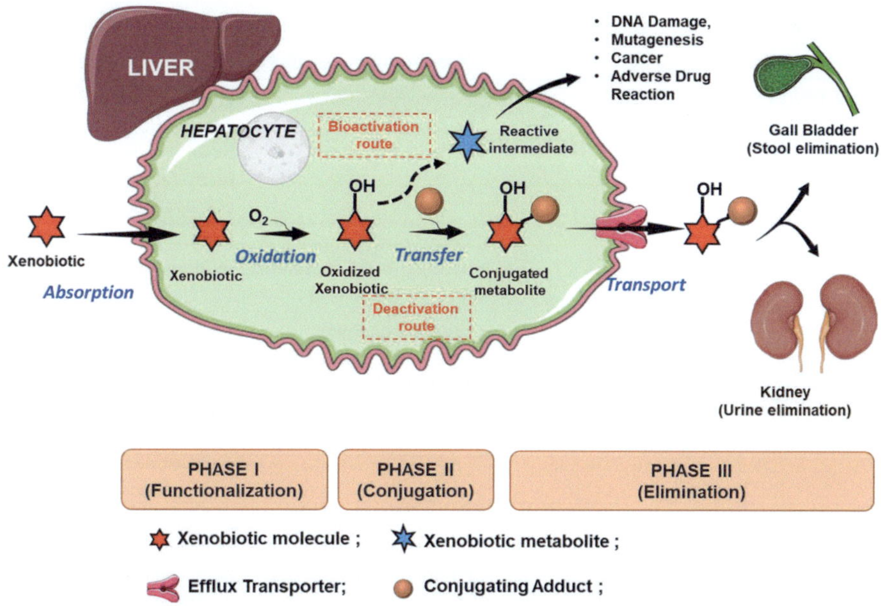

Fig. 2 Steps of xenobiotic processing in liver cells and the key role of CYP enzymes in chemical transformation. Phase I reaction (functionalization) introduces a polar hydrophilic group or exposes an existing hydrophilic group in xenobiotics. Different CYPs participate in this reaction. In phase II reaction, to make the xenobiotic chemical more hydrophilic, an endogenous hydrophilic molecule is attached to it. Finally, the xenobiotic compound is extruded from cells via different transporters (phase III reaction, elimination). OH: hydroxyl group. Image created by Servier Medical Art (https://smart.servier.com/)

4 Characteristics of Major XME Genes

4.1 Genetic Structure

The majority of XME genes have several exons and introns, and tissue-specific isoforms are produced by alternative splicing. Many are found in gene clusters or families (e.g., CYP gene superfamily, SULT gene family, GST gene family). Several response elements for transcriptional control (such as AhR, PXR, and CAR binding sites) are found in the promoter regions. The detail structural features of the XME genes have been enlisted in Table 2.

4.2 Genetic Polymorphism and Allozyme of XME

The function of XME differs among people because of the significant genetic variations present in XME genes. Xenobiotic metabolism rates can be affected by single

Table 1 Main enzymes involved in xenobiotic metabolism and their biochemical functions

Phase	Enzyme	Xenobiotic metabolized	Biochemical function
Phase I	CYP1A1	Polycyclic aromatic hydrocarbons like benzo[a]pyrene	Hydroxylation of xenobiotic molecule to make it more polar and ready for phase II conjugation
	CYP1B1	Polycyclic aromatic hydrocarbons, estradiol, retinoic acid	Hydroxylation resulting in bioactivation
	CYP1E1	Ethanol, acetone halothane, chloroform, nitrosamines, and benzene	• Catalyzes oxidation reactions (e.g., hydroxylation, dealkylation) • Uses NADPH and molecular oxygen to introduce oxygen into substrates
	CYP3A4	Drugs like paclitaxel, aflatoxin B, etc.	Oxidation, hydroxylation, and dealkylation reaction
	Microsomal epoxide hydrolase (EPHX)	Epoxide compounds, natural epoxides	Hydrolysis of epoxides to their corresponding dihydrodiols, using activated water
	Catalase	Hydrogen peroxide	Breaks down hydrogen peroxide into water (H_2O) and oxygen (O_2) and thus protects cells from oxidative damage by free radicals and reactive oxygen species (ROS)
	Myeloperoxidase (MPO)	Halides, hydrogen peroxide	Generates hypohalous acids in the presence of hydrogen peroxide and halide anions
	Aldo-keto reductase (AKR1C)	Aldehyde and ketone compounds	Reduction of aldehydes and ketones to alcohols
Phase II	N-acetyltransferase	Aromatic amines, hydrazine drugs	Transfer of acetyl group from acetyl-CoA to aromatic amines (arylamines and arylhydrazines), leading to their bioactivation or deactivation

(continued)

Table 1 (continued)

Phase	Enzyme	Xenobiotic metabolized	Biochemical function
	NAD(P)H: quinone oxidoreductase (NQO1)	Xenobiotic quinones	Two-electron reduction and detoxification of quinones
	Sulfotransferase (SULT)		Sulfation of xenobiotic compounds
	Glutathione S-transferase		Detoxification of xenobiotic compounds by conjugating with glutathione
	UDP-glucuronosyltransferase		Glucuronidation of xenobiotics
Phase III	P-glycoprotein		Extrusion of intracellular drugs by ATP-dependent efflux pump
	MRP1		Elimination of drugs and metabolites by ATP-dependent efflux pump
	BCRP		Elimination of drugs and metabolites by ATP-dependent efflux pump

nucleotide polymorphisms (SNPs) and gene deletions or duplications (e.g., *CYP2D6* poor vs. ultra-rapid metabolizers). These variations significantly influence toxicological sensitivity and pharmacogenetic diversity Allozymes are enzyme variants that are encoded by different alleles of the same gene. They catalyze the same biochemical reactions but may differ slightly in amino acid sequence, charge, or kinetic properties. In the context of xenobiotic-metabolizing enzymes (XMEs), allozymes play a significant role in explaining genetic variability in drug metabolism and detoxification among individuals and populations.

For example, *CYP2D6**1, *2, *3, *4 are common allozymes of *CYP2D6* gene. These allozymes have different drug oxidation rates, leading to poor or ultra-rapid metabolism. UGT1A1*1, *28 are variants of *UGT1A1* governing the bilirubin and drug glucuronidation. SULT1A1*1, *2 are common allozyme variants of *SULT1A1* gene (Fig. 3). The presence of these variants modifies the rate of sulfation of phenolic drugs and hormones.

Table 2 Details of various XME genes

Gene/gene family	Gene symbol	Chromosomal location	Gene length (kb)	mRNA length (kb)	Number of exons	Protein length (amino acids)
Cytochrome P450*	*CYP1A1*	15q24.1	~6	~2.5	7	512
	CYP1B1	2p22.2	~12	~5.1	3	543
	CYP2C8	10q23.33	~30	~2.2	9	490
	CYP2D6	19q13.2	~9.9	~1.6	9	494
	CYP3A4	7q22.1	~27	2.0–2.3	13	503
Microsomal epoxide hydrolase	*EPHX1*	1q42.12	~32	~2.0	9	455
Myeloperoxidase	*MPO*	17q23.1	~11–12	~3.3	12	745
Aldo–keto reductase*	*AKR1A*	1p33-p32	~8	~1.4	10	325
	AKR1B	7q33	~16	~1.4	10	316
	AKR1C3	10p15.1	~20	~1.6	9	323
	AKR1D1	7q33	~29	~1.5–1.6	9	326
	AKR1E2	10p15.1	~19	~1.2	9	319
	AKR6A3	3q26.33	~15	~2.3	8	367
Catalase	*CAT*	11p13	~34	~2.1	13	527
Glutathione S-transferase*	*GSTA1*	6p12.2	~11	~1.0	7	222
	GSTM	1p13.3	~8	~1.1	7	218
	GSTP1	11q13	~3.2	~1.0	7	210
	GSTT1	22q11.23	~8.1	~0.9	5	240
	GSTO1	10q25.1	~11.8	~1.3	6	241
Glutathione peroxidase*	*GPX1*	3p21.31	~10	~1.1–1.2	2	203
	GPX2	14q23.3	~6	~0.9	2	190

(continued)

Table 2 (continued)

Gene/gene family	Gene symbol	Chromosomal location	Gene length (kb)	mRNA length (kb)	Number of exons	Protein length (amino acids)
	GPX3	5q33.1	~12	~1.6	5	226–235
	GPX4	19p13.3	~7	~0.9–1.1	7	197
	GPX5	6p22.1	~8	1.4–1.7	5	221
	GPX6	6p22.1	~7	~1.4	5	223
NAD(P)H: quinone oxidoreductase	*NQO1*	16q22.1	~20	~1.4	6	274
N-acetyltransferase	*NAT1*	8p22	~10.9	~1.2	9	290
	NAT2	8p23	~9.9	~1.1	2	290
UDP-glucuronosyltransferase*	UGT1A1	2q37	~160	~2.36		533
	UGT1A6	2q37	~180	2.45 and 1.6		532 and 265
	UGT2B7	4q13	~17.5	~1.7	6	529
Sulfotransferase*	SULT1A1	16p12.1–p11.2	~20	~1.4	8	295
	SULT1A2	16p12.1	~23	~1.4	8	295
	SULT1A3	16p12.1	~20	~1.4	8	295
	SULT1B1	4q13.3	~26	~1.3	7	295
	SULT1E1	4q13.3	~23	~1.4	8	294
	SULT2A1	19q13.3	~18	~1.6	7	285
	SULT2B1a	19q13.3	~26	~1.6	7	365
	SULT2B1b	19q13.3	~26	~1.6	7	350
	SULT4A1	22q13.31	~26	~1.7	7	285

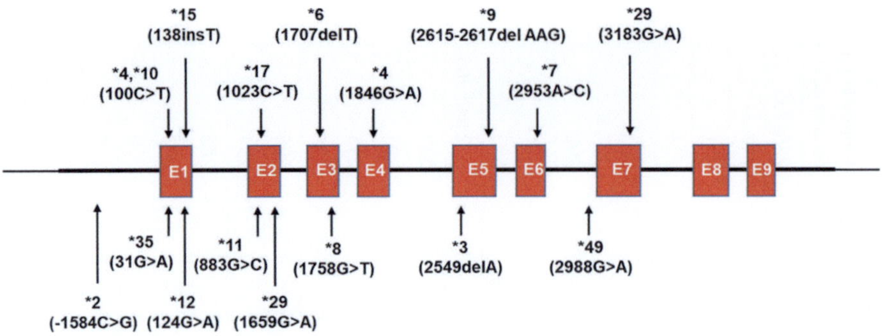

Fig. 3 Location of common polymorphisms star (*) alleles in CYP2D6 gene. Several SNPs, deletions, and insertions are present across nine exons (E1–E9) and introns. Several of these variations give rise to CYP2D6 allozymes. Figure is redrawn from Yang et al. (2017)

4.3 *Regulation of Expression*

Innumerable number of xenobiotic-sensing transcription factors regulate the expression of XME. For example, aryl hydrocarbon receptor (AhR) induces *CYP1A1, CYP1B1*; pregnane X receptor (PXR) regulates CYP3A4, UGT1A1; Constitutive androstane receptor (CAR) controls *CYP2B6*, SULTs; Nrf2 pathway regulates antioxidant and phase II enzyme genes (GSTs, NQO1).

4.4 *Tissue and Species Specificity*

XME genes show tissue-specific expression. They are high in liver, intestine, lung, and kidney (major detoxification sites). Expression patterns can differ across species, influencing comparative toxicology and pharmacokinetics.

4.5 *Inducibility*

Many XME genes are inducible upon exposure to xenobiotics, drugs, or environmental chemicals. Induction leads to adaptive upregulation of metabolic capacity, though in some cases it can enhance formation of toxic intermediates.

4.6 Evolutionary Conservation

XME gene families are evolutionarily conserved but have undergone gene duplication and diversification, allowing adaptation to diverse environmental exposures.

4.7 Physiological and Pathophysiological Roles

Beyond xenobiotic metabolism, XME genes are involved in different functions like endogenous metabolism (e.g., steroid hormones, fatty acids, bilirubin, etc.), oxidative stress response, antioxidant defense, cancer susceptibility, and drug resistance (due to altered expression or mutation).

5 Association of Different Genetic Polymorphisms of XME Genes with Human Diseases

5.1 Single Nucleotide Polymorphisms (SNPs)

Single base substitution is a typical feature of SNPs, a kind of genetic polymorphism. SNPs can occur both inside and outside of a gene's protein-coding domain. SNPs in the protein-coding region of genes may be classified into synonymous and nonsynonymous SNPs. A change in the DNA sequence in synonymous SNPs does not affect the protein's amino acid sequence. According to recent research, synonymous SNPs can also impact protein folding, which has a significant impact on protein function, as well as mRNA splicing, stability, and structure. (Hunt et al., 2009; Sharma et al., 2019). An innumerable number of SNPs have been identified in XME, which influence the metabolism of xenobiotic compounds, including various drugs, and selected findings have been tabulated in Table 3.

5.1.1 SNPs in *CYPs*

The human genome has 57 putatively functioning cytochrome P450 (CYP) genes that code for proteins. Based on sequence homology, these are divided into 18 families and 44 subfamilies. Drug and xenobiotic metabolism involve several CYP isozymes of CYP superfamilies (1A1, 1A2, 2A6, 2A13, 2B6, 2C8, 2C9, 2C18, 2C19, 2D6, 2E1, 2F1, 3A4, 3A5, 3A7). The star (*) nomenclature system is used to name CYP alleles (Fig. 4).

Table 3 Genetic variants in key enzymes involved in XM and their impact on human disease susceptibility

Gene name	Genetic variants	rsID	Amino acid change	Associated disease	Country/population	References
CYP1A1	2453C > A	rs1799814	Thr461Asn	Colorectal cancer	Polish	Kabziński et al. (2022)
CYP1A1	2455A > G	rs1048943	Ile462Val	Breast cancer	Caucasian	Sergentanis and Economopoulos (2010)
CYP1A1	c.2455A > G	rs1048943	Ile462Val	Breast cancer	Mexican	Martínez-Ramírez et al. (2013)
CYP1B1	4326C > G	rs1056836	Leu432Val	Squamous cell carcinoma of the esophagus	Iranian	Moghadam et al. (2018)
CYP1B2	1459C > T	rs3211371	Arg459Cys	Gastrointestinal cancer	Indian	Patil et al. (2024)
*CYP2D6*4*	1846G > A (Splice site variant)	rs3892097	–	Idiopathic male infertility	Turkey	Hekim et al. (2020)
CYP2E1	96-bp INDEL and 1054C > T)	rs2031920	–	Preeclampsia	Chinese	Hu et al. (2025)
CYP3A4	G > A, (Intronic variant)	rs4646437	–	Urinary bladder cancer in males and individuals exposed to pesticides	Brazil	da Silva et al. (2025)
CYP3A4	392A > G	rs2740574	–	Recurrent major depressive disorder	Finland	Świechowski et al. (2019)
EPHX1	337 T > C	rs1051740	Tyr113His	Ovarian cancer	Caucasian and Chinese	Zhong et al. (2013)
EPHX1	337 T > C	rs1051740	Tyr113His	Preeclampsia	Caucasian	Zhao et al. (2022)
	A > G	rs2234922	His139Arg		Middle Eastern	

(continued)

Table 3 (continued)

Gene name	Genetic variants	rsID	Amino acid change	Associated disease	Country/population	References
EPHX2	A > G (Intronic variant)	rs11780592		Oxidized LDL and mortality in patients with diabetic chronic kidney disease	Caucasian	Roumeliotis et al. (2021)
NAT2	857G > A	rs1799931	Arg268Lys	Lung cancer	Indian	Walia et al. (2022)
SULT1A1	638G > A	rs9282861	Arg213His	Lung cancer risk	Bangladeshi	Tasnim et al. (2017)
SULT1A1	638G > A	rs9282861	Arg213His	Breast cancer risk (especially in Asian women)	Asian (e.g., Chinese, Korean)	Forat-Yazdi et al. (2017)
SULT2B1	G > A (Promoter Variant)	rs2665580		High SULT2B1 expression, associated with coronary artery disease (unstable plaques)		Pan et al. (2024)
GSTP1	313A > G	rs1695	Ile105Val	Breast cancer	Asian women	Grussy et al. (2023)
GSTA1	69C > T (Promoter Variant)	rs3957357		Gestational hypertension	Italy	Iorio et al. (2015)
GSTA1	69C > T (Promoter Variant)	rs3957357		Schizophrenia risk	Chinese Han	Liu et al. (2020)
GSTO2	424A > G	rs156697	Asn142Asp	Diabetic nephropathy	Bosnia and Herzegovina, Serbian	Pavlović et al. (2023)
GSTZ1	94G > A, 124G > A, 245C > T	rs3177427, rs7972, rs1046428	Glu32Lys, Gly42Arg, Thr92Met	Dichloroacetate pharmacokinetics, chronic side effects		Tian et al. (2019)

(continued)

Table 3 (continued)

Gene name	Genetic variants	rsID	Amino acid change	Associated disease	Country/population	References
UGT1A4	142 T > G	rs2011425	Leu48Val	Olanzapine and lamotrigine drug metabolism and tolerability	Japanese/Chinese and Caucasian studies	Hattori et al. (2020)
ABCB1	1199G > A	rs2229109	Ser400Asn	Outcome and toxicity in childhood acute lymphoblastic leukemia	Denmark	Gregers et al. (2015)
ABCB1	1199G > A	rs2229109	Ser400Asn	Antidepressant efflux functional study		Gao and He (2018)
ABCC1 (MRP1)	2012G > T	rs45511401	Gly671Val	LDL-C response to statins in familial hypercholesterolemia	Brazilian	Dagli-Hernandez et al. (2022)
ABCG2 (BCRP)	421C > A	rs2231142	Gln141Lys	Altered drug response (ADR)	Multi-ethnic	Hoque et al. (2020)
ABCG2 (BCRP)	376C > T	rs72552713	Gln126Stop	Gout and hyperuricemia	Japanese and Chinese	Stiburkova et al. (2019)
ABCG2 (BCRP)	C > A (Intronic variant)	rs2622621	regulatory variant	Chemotherapy response, colorectal cancer outcome	Chinese CRC cohort	Hu et al. (2018)
ABCG2 (BCRP)	421C > A	rs2231142	Gln141Lys	Drug toxicity, pharmacogenomic associations	Multiple (Asian cancer cohorts)	Sun et al. (2021)

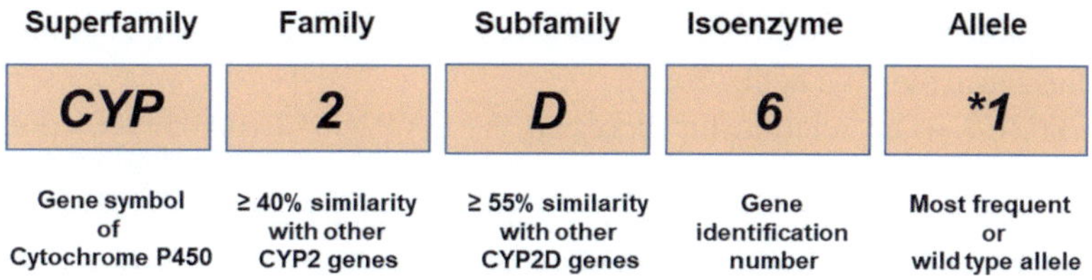

Fig. 4 Nomenclature of CYP genes. Superfamily, family, subfamily, isoenzyme type, and allele are systematically mentioned in the nomenclature

The allelic variants of CYP and other pharmacogenes (PGx) have been cataloged in a central repository named The Pharmacogene Variation (PharmVar) Consortium (https://www.pharmvar.org/htdocs/archive/index_original.htm), which focuses on haplotype structure and allelic variation (Gaedigk et al., 2018).

(i) *CYP1A1*—Cytochrome P450 1A1 (*CYP1A1*) is a crucial enzyme in the phase I metabolism of xenobiotics—especially polycyclic aromatic hydrocarbons (PAHs), heterocyclic amines, and other procarcinogens. Genetic polymorphisms in the *CYP1A1* gene can alter enzyme activity or expression, influencing individual susceptibility to various cancers. Numerous case–control studies and updated meta-analyses continue to find a link between *CYP1A1* variants (especially rs1048943/Ile462Val and the MspI/T6235C variant) and increased cancer risk; however, the strength of the association varies by cancer type, ethnicity, and exposure (e.g., smoking, HPV) (Barek et al., 2023). Functional polymorphism in *CYP1A1* modifies xenobiotic and estrogen metabolism, hence reinforcing potential biological processes for cancer risk and highlighting the necessity of considering gene-environment interactions (Kukal et al., 2023). Recent data associates *CYP1A1* polymorphisms (rs1048943) with varying treatment responses or toxicities in malignancies such as non-small cell lung cancer (NSCLC), indicating possible pharmacogenetic significance (Sito et al., 2024).

(ii) *CYP1A2—Modest* evidence suggests that *CYP1A2* polymorphisms affect cancer risk; however, the relationships are minor and context dependent, varying by cancer type, population/ethnicity, and environmental aspects such as smoking. A meta-analysis of 19 studies indicated that for the genetic polymorphism (rs762551 A > C), the CC genotype compared to AA in Caucasians had an odds ratio of around 1.29 for cancer risk. According to Tian et al. (2013), the A allele of the *CYP1A2* 1F polymorphism (rs762551) is linked to a lower risk of ovarian cancer (OR = 0.70, 95% CI: 0.54–0.89, P = 0.004) and may raise the risk of breast cancer (OR = 1.05, 95% CI: 1.01–1.10, P = 0.024).

(iii) *CYP2A6*—On chromosome 19q13.2, the CYP2A6 gene is situated in a group of CYP2 genes that most likely resulted from duplication events. It is quite similar to members of the subfamily *CYP2A7* and *CYP2A13*. The gene is made up of nine exons that span approximately 6 kb and encode a protein with 494 amino acids. A recent study found that the *CYP2A6* gene polymorphism could be a new predictive and prognostic biomarker for postmenopausal hormone receptor-positive breast cancer patients following neoadjuvant chemotherapy plus letrozole (Wu et al., 2025).

(iv) *CYP2C8*—The relationship between the CYP2C8*2 (rs1058930) polymorphism and breast cancer risk was examined in Iranian population (Golmohammadzadeh et al., 2019). The study's findings indicated that whereas the CG genotype of *CYP2C8* was associated with a protective effect, people with the GG genotype were more likely to develop breast cancer. This suggests that the susceptibility of this population to breast cancer may be influenced by *CYP2C8* polymorphisms.

(v) *CYP2D6*—Tamoxifen is a prodrug used to treat estrogen receptor-positive (ER+) breast cancer, and its metabolism depends on *CYP2D6*. *CYP2D6* enzyme activity in the liver transforms tamoxifen into its active metabolite, endoxifen. Genetic differences in *CYP2D6* can have a major impact on pharmacological efficacy and patient outcomes.

(vi) *CYP3A4*—11882 SNPs have been identified in human *CYP3A4* gene so far (dbSNP database accessed on 18th Oct, 2025). Numerous investigations investigating the pharmacogenetic impact of *CYP3A4* ∗ *22* on various pharmaceuticals, such as immunosuppressants and antidepressants, have been prompted by the SNP in intron 6 (rs35599367 C > T), which encodes the *CYP3A4* ∗ *22* allele.

5.1.2 SNPs in Microsomal Epoxide Hydrolase (mEH)

EPHX1 encodes the microsomal epoxide hydrolase (mEH) enzyme which is encoded by *EPHX1* gene. The enzyme transforms epoxide intermediates, potentially toxic and reactive, into less reactive dihydrodiols (Václavíková et al. 2015). mEH is regarded as a component of the detoxification and metabolic activation system due to the presence of epoxides as intermediates in xenobiotic metabolism (e.g., from pesticides, polycyclic aromatic hydrocarbons, and drug metabolism). Several SNPs in *EPHX1* have been reported, e.g., Y113H (rs1051740) and H139R (rs2234922) are two well-characterized amino-acid changing variants with altered enzyme activity in vitro (Hosagrahara et al., 2004). Two SNPs [rs2740168 (IVS3 + 1328G > A) and rs2292566 (Ex3-8G > A)] have been found in a nested case-control study of prostate cancer in a pesticide-exposed cohort (Koutros et al., 2011).

5.1.3 SNPs in Myeloperoxidase (MPO)

MPO is an essential enzyme that can worsen oxidative damage and promote carcinogenesis by metabolizing and activating a variety of xenobiotics and environmental pollutants into reactive intermediates (Valadez-Cosmes et al., 2022). The MPO-463G > A variant, which is located in the promoter region of the MPO gene, is the most common and thoroughly studied MPO polymorphism. By removing a transcription factor SP1 binding area in an Alu gene-responsive element, this polymorphism has a detrimental influence on the mRNA expression and transcription level of the MPO gene. Numerous studies have highlighted the impact of this polymorphism on the relationship between the incidence of lung cancer and tobacco use, as well as the relevant role that MPO expression plays in converting pro-carcinogenic substances in cigarettes into carcinogens (Valadez-Cosmes et al., 2022). The MPO-463G > A polymorphism has been connected to higher levels of helicobacter pylori infection, which is associated with the development of stomach cancer, in addition to its association with cancer risk (Hamajima et al., 2001).

5.1.4 SNPs in Aldo-Keto Reductase (AKR) Family Genes

SNPs in AKR genes may cause the enzyme's expression and activity to change, which could result in oxidative stress and estrogen-driven carcinogenesis. In women of African descent, epithelial ovarian cancer has been linked to an SNP in the intron region of the AKR1C gene (rs4525119) (Manichaikul et al., 2020). Takano et al. (2023) found that the SNP g.55101G > A in the AKR1C3 gene, which results in a Cys145Tyr amino acid change, reduces reductase function. Surprisingly, the allele was exclusive to Asians, indicating that it may have a role in interindividual differences in the therapeutic efficacy of loxoprofen, acetohexamide, and doxorubicin or toxicity in Asians when the variant reduced AKR1C3 levels.

5.1.5 SNPs in Catalase (CAT) Gene

Polymorphism of *CAT* gene alters the ability of the cell to eliminate H_2O_2 generated by xenobiotic metabolism or by xenobiotic-induced mitochondrial dysfunction. rs1001179 (C-262 T) polymorphism in *CAT* gene is the most extensively studied variant and significant association of this polymorphism with various types of cancer is well documented (Wang et al., 2016). Reduced inducible catalase activity for the T allele is supported by functional assays, providing a tenable explanation for heightened susceptibility to ROS from xenobiotics. Another common promoter SNP rs7943316 (A-21 T) has been reported in several cancer-related case–control studies, but findings are heterogeneous and need larger replication and functional data (Hoidy et al., 2023).

5.1.6 SNPs in Glutathione S-transferase (GST) Gene

As GSTs are involved in detoxification of many carcinogens, any changes that negatively impact their functions increase the risk of cancer (Singh et al., 2020). In a recent study, 126 specific GSTP1 SNPs, including fSNPs rs1695 A > G (Ile105Val) and rs1138272 C > T (Ala114Val), were evaluated in 107 colorectal cancer (CRC) patients and 107 controls of south Indian descent. While the A114V fSNP did not exhibit any relevance, the results showed a substantial correlation between the I105V fSNP and the risk of CRC. In contrast to controls, both fSNPs showed greater linkage disequilibrium in patients (Malempati et al., 2024). In a different study, Ferracini et al. (2021) found that the presence of SNPs in GSTP1 c.313A > G, ABCB1 c.1236C > T, and c.3435C > T could anticipate hematological and neurotoxicity in the clinical treatment of Brazilian EOC women receiving PTX/Carboplatin chemotherapy.

5.1.7 SNPs in Glutathione Peroxidase (GPx)

The GPx family plays a vital role in detoxifying oxygen-free radicals, such as hydrogen peroxide. A consistent interaction between the nonsynonymous SNP [rs1050450 (Pro198Leu)] in GPX1 and the connection between xenobiotic exposure smoking, alcohol, selenium status, PAHs / occupational exposure) and cancer susceptibility have been documented in various studies, so far. Reduced GPX activity caused by the GPX1 Pro198Leu SNP is linked to an increased risk of breast cancer in Danish women, according to a nested case–control study (Ravn-Haren et al., 2006). Compared to non-carriers, postmenopausal women with the variation T allele of GPX1 Pro198Leu had a 1.43-fold increased risk of breast cancer (95% CI = 1.07–1.92). Another investigation by Hansen et al. (2009) assessed the significance of the same SNP on erythrocyte GPx activity and CRC risk in a Danish population. The research noted an increased risk of CRC linked to alcohol use and smoking in homozygous GPX1 198Leu carriers.

5.1.8 SNPs in NAD(P)H: Quinone Oxidoreductase (NQO1)

The NQO1 enzyme breaks down carcinogenic substances found in tobacco smoke, such as benzo(a)pyrene, aromatic amines, and N-nitrosamines (NNA) (Stoehr et al., 2012). Variants of NQO1 with low activity, such as P187S, increase vulnerability to oxidative damage or genotoxic intermediates and decrease detoxification of quinone substrates. The most widely studied SNP in NQO1 is the C → T substitution at 609 (leading to Pro187Ser). The results differ depending on exposure, illness site, and demographics, even though several studies have shown the functional significance of the SNP in the development of cancer. A study in the Netherlands (rs1800566) found no overall association with GI cancer risk in ~1,457 cases/controls, though the CT genotype was associated with proximal colon cancer (OR ≈ 1.60) in that subgroup (Freriksen et al., 2014). A different study noted a strong correlation between the Ser/Ser variant and age category, smoking habits, tumor site, nodal status/ elevated tumor grade, as well as pesticide exposure in CRC patients (Sameer et al., 2010).

5.1.9 SNPs in N-Acetyltransferase (*NAT*) Gene

Variations in N-acetyltransferase (*NAT*) genes, particularly *NAT2*, determine distinct acetylation phenotypes in individuals, categorized as slow, intermediate, or rapid acetylators. These genetic differences have significant implications for drug metabolism and an individual's cancer risk. Specific *NAT2* alleles, such as the rapid acetylator NAT2*4 and the slow-acetylating variants like NAT2*5, NAT2*6, and NAT2*7, are highlighted for their association with altered drug processing. Furthermore, while NAT1 exhibits less variation, alleles like NAT1*4 and NAT1*10 can also influence the enzyme's overall function (Hein et al., 2000). The function of NAT enzymes has been clarified by numerous investigations, which have also demonstrated the connection between NAT variations and several cancer types, including breast, lung, and hepatocellular cancer. Significant variation exists between these studies and meta-analyses due to population/ethnic differences and variations in the definition of acetylator status (haplotype versus phenotypic inference vs. single SNPs (Zhang et al., 2015).

5.1.10 SNPs in Sulfotransferase (SULT)

Sulfotransferases (SULTs) are adaptable enzymes that control the primary pathway in phase II metabolism of several endogenous and exogenous substances, the enzymatic process of sulfonate conjugation. In the population, SNPs in SULTs are quite rare. Because of its widespread tissue distribution and abundance, SULT1A1, a member of the SULT superfamily, is thought to be the most prevalent kind of SULT. Arg to His amino acid substitution results from a nonsynonymous single nucleotide alteration at codon 213 (G > A, rs9282861) in exon 7 of the SULT1A1 gene. In order to determine the prognostic and/or predictive relevance of this SULT1A1 SNP and another SNP in the XRCC1 gene to lung cancer risk in the Bangladeshi population, Tasnim et al. (2017) undertook a study. The study found that the Arg/His and His/His genotypes, as well as the SULT1A1 rs9282861 "A" allele, are linked to an increased risk of lung cancer.

5.1.11 SNPs in UDP-Glucuronosyltransferase (UGT) Gene

The enzyme human UDP-glucuronosyltransferase was found to be genetically polymorphic. These polymorphisms lead to variations in the degree of transcriptional and functional changes and may lower the activity of the UGT enzyme. People with the polymorphisms consequently exhibit various illnesses. UGTs are essential for the detoxification of aromatic amines, which are found in tobacco smoke and a variety of industrial pollutants. An increased risk of bladder cancer is associated with the presence of certain aromatic amines. In an inclusive study, Tang et al. (2011) sequenced the first exon of *UGT1A* gene in 2000 individuals and identified 43 known coding variants (26 nonsynonymous and 17 synonymous). The study produced data for 1,170 single nucleotide polymorphisms (SNPs) in the UGT1A region using imputation from a sizable GWAS dataset and reference panels. An unusual protective genetic variation that successfully explained the initial GWAS signal was found to have the strongest connection with urinary bladder cancer, as demonstrated by a substantial odds ratio and p-value. While the protective gene (T) is unique to humans, the risk allele (G) is retained in 33 species. Additionally, the study showed that this SNP is associated with quantitative mRNA expression of a specific functional splicing variant of UGT1A6, most likely through an exonic splicing enhancer.

6 Copy Number Variations (CNVs)

The impact of copy number variants (CNVs) on the function of XME has just been acknowledged, despite SNVs being the most well-studied variants linked to illnesses. Copy number variations (CNVs), which are defined as duplications or deletions of DNA sequences ranging from 1 to 3 Kb, are thought to make up about 12% of the human genome (Redon et al., 2006). Copy number variations (CNVs) represent at least five times more variable base pairs than single nucleotide variations (SNVs) when comparing two human genomes. Several clinically relevant pharmacogenes (e.g., *CYP2A6*, *CYP2B6*, *CYP2D6*, *UGT2B17*, *UGT2B28*, *SULT1A1*) have been identified to have CNVs linked to the variability in enzyme activity and their phenotypic effect (Table 4).

Table 4 Selected studies which showed the effect of CNV on XME genes on phenotypes

Genes	CNV type	Clinical impact	References
CYP2A6	Deletion (*4), duplication	Nicotine metabolism; affects smoking cessation therapy	Jones et al. (2022)
CYP2B6	Duplication or deletion	Influences metabolism of efavirenz, cyclophosphamide	Wang et al. (2019)
CYP2D6	Deletion (*5)	Poor metabolizer phenotypes affects antidepressants, opioids, tamoxifen	Jarvis et al. (2019)
	Duplication (*1xN, *2xN)	Ultra-rapid metabolizer	
UGT2B17	Deletion(s)	Colorectal cancer (CRC)	Angstadt et al. (2013)
UGT2B28	Germline copy number loss	Esophageal squamous cell carcinoma (ESCC) risk	Hu et al. (2015)
SULT1A1	Duplication	Faster drug clearance (e.g., tamoxifen); potential decreased efficacy of some drugs; enhanced detoxification of environmental carcinogens	Hebbring et al. (2007)

6.1 CNVs in CYP2

The CYP2D6 gene has several structural variants (e.g., CNVs) that can include the deletion of an entire coding region, duplications of the coding region, and multiple copies of the coding region (Fig. 5). As a result, there is a wide variety of functions for CYP2D6. Individuals have different variations of the enzyme, which can have an effect on how the body processes drugs, and therefore experience different levels of drug response. In the case of those with low or no CYP2D6 enzyme activity, exposure to a drug that is metabolized by the CYP2D6 enzyme could result in less effective therapeutic results (i.e., drug failure) and/or adverse drug reactions (Owen et al., 2009). According to one study, there was a report that in the US, 12.6% of the general population had copy number variation in CYP2D6 (Beoris et al., 2016). A GWAS study of 11696 Japanese individuals identified a haplotype consisting of the CYP2A6 gene on chromosome 19q13, including a single nucleotide polymorphism (SNP) located approximately 30 kb downstream of the CYP2A6 gene, associated with smoking-related diseases, including lung cancer, chronic obstructive pulmonary disease, and arteriosclerosis obliterans (Kumasaka et al., 2012).

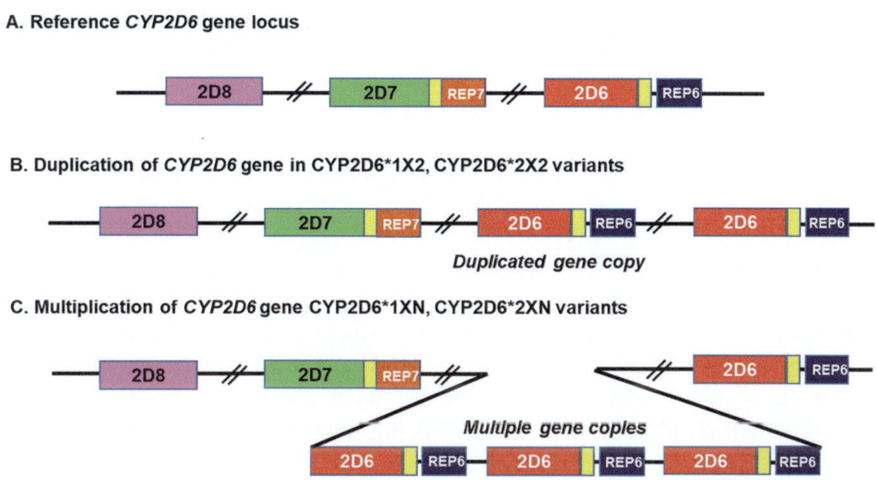

Fig. 5 Copy number variation of *CYP* genes by duplication and multiplication. **a** Reference gene locus consisting of *CYP2D6* and two pseudogenes, *CYP2D7* (green), and *CYP2D8* (pink). Yellow boxes indicate nearly identical downstream sequences and the blue and brown boxes labeled REP6 and REP7 represent regions with repetitive elements located downstream of *CYP2D6* and *CYP2D7*, respectively

6.2 CNVs in Glutathione S-Transferases

Of the eight GST enzyme classes, genetic polymorphisms in GSTM1, GSTT1, and GSTP1 eventually result in nonfunctional enzymes. The GSTM1 enzyme avoids DNA damage by catalyzing the conjugation process between electrophilic compounds and reduced glutathione, which results in less reactive metabolites (Hayes & Strange, 2000). According to a recent study, oral squamous cell carcinoma (OSCC) CNV of *GSTM1* is a promising prognostic factor (Firigato, 2022). The study showed that two copies of *GSTM1* reduced the OSCC risk by 53% (OR 0.47; 95% CI 0.24–0.92). The CNV of *GSTM1* ranged from zero to two copies. The interaction between the CNV of *GSTM1* and cigarette smoking resulted in a changed tumor risk, with the OSCC risk gradually decreasing as the number of copies of *GSTM1* increased for an annual cigarette consumption of 40 packs.

6.3 CNVs in SULT1A1

The chromosomal locus of human cytosolic sulfotransferase 1A1 (SULT1A1) is found on the short arm of chromosome 16 and is a member of the SULT1A subfamily that also includes SULT1A2, SULT1A3, and SULT1A4. The SULT1A1 gene is expressed in all human tissues and accounts for the largest percentage of SULTs in the body; additionally, it is the most widely distributed SULT in the human body. A strong relationship has been established between the number of copies of the SULT1A1 gene and levels of methyl eugenol DNA adducts in the human liver (Tremmel et al., 2017). Methyl eugenol (4-allyl-1,2-dimethoxybenzene) is a chemical produced naturally by many plants that are consumed as food, utilized to produce fragrant oils, or used in medicine. Methyl eugenol has been shown to cause cancer in the liver as well as tumors in other tissues including the bile duct, kidneys, and mammary glands when tested in laboratory animals. The amount of DNA adducts is directly linked to the carcinogenicity of the metabolically active derivative of methyl eugenol, which covalently bonds with DNA to produce a DNA adduct. According to Tremmel et al. (2017), DNA adduct production in human liver tissues is linked to SULT1A1 deletions and duplications.

7 Gene Deletion and Homozygous Null Genotype of XME Genes

A study on Caucasians (Angstad et al., 2013) found that a deletion in the UGT2B17 gene is linked to a lower chance of colorectal cancer (CRC), especially in men. This UGT2B17 deletion genotype ($0/0$) might mean people have more protective substances in their diet, such as some nonsteroidal anti-inflammatory drugs and flavonoids, which the gene processes.

8 Recent Developments to Mine the Variations in XME Genes

8.1 Next-Generation Sequencing to Mine the Variants in XME

In recent years, NGS methodologies have been widely employed on a broad scale to discover and characterize potentially functional variations. Numerous large-scale international research consortia have generated comprehensive population resources of DNA sequence data, frequently associated with clinical information. For instance, the 1000 Genomes Project mapped mutations across more than a thousand human genomes, depicting a clearer picture of genetic diversity. Researchers can mine the uncommon genetic variants by using the Exome Aggregation Consortium (ExAC) and Genome Aggregation Database (gnomAD), which compile and analyze exome and genome sequencing data from more than 60,000 individuals. In the U.S., the All of Us Research Program is working to gather genetic, lifestyle, and health information from more than a million people, all to move precision medicine forward. On a global scale, the Human Pangenome Reference Consortium (HPRC) is building a more complete and inclusive reference for human genetic variation. In order to capture the entire spectrum of human genetic variation, it incorporates genomes from several individuals with different ancestries sequenced by long-read sequencing, going beyond the single reference genome (such as GRCh38).

8.2 Targeted NGS to Identify Variations in Pharmacogenes

Genome-wide sequencing in a population is often not feasible due to significant cost, time, and data management requirements. Alternatively, exome-wide sequencing approach targets only protein-coding regions, which make up about 1% of the human genome, or approximately 20,000 genes. One disadvantage of this approach is that it overlooks functional variation in noncoding regions. Targeted exome capture technique overcomes this drawback as this technique focuses on sequencing a small number of selected genes at a high coverage, typically greater than 30-fold. This significantly reduces error rates associated with next-generation sequencing data. This method enables the precise identification of both common and uncommon genetic variations inside clinically significant genes, in contrast to more comprehensive genome-wide SNV candidate techniques. Targeted genomic region may also include particular noncoding functional areas of interest, such as those present in genes that encode phase I and phase II enzymes. As a result, this accurate and dependable approach is quite useful for applying genotype-based treatments in clinical settings.

8.3 Bioinformatic Tools to Detect Pharmacogenetic Variants

8.3.1 Aldy

- Aldy is a computational tool that analyzes high-throughput sequencing data to genotype pharmacogenes. It identifies star alleles or haplotypes of a pharmacogene harbored by a person. It doesn't just look at single nucleotide variants (SNVs) or small insertions and deletions; it also considers copy number variation and phasing.
- Aldy 4 is the newer version with improved copy number models, better phasing, and support for long-read sequencing (Hari et al., 2023).
- Aldy divides the gene region or regions into segments and calculates the number of copies per segment (in relation to reference coverage), maybe through comparison with control/reference coverage models. Next, it deduces the "structural configurations" of the gene, such as the number of copies overall, the presence of fusion breakpoints, and if parts of the gene are duplicated.
- Many pharmacogenes have alleles that are determined by a whole gene haplotype rather than a single nucleotide variant (SNV). Therefore, phasing, or haplotyping, of the entire genic area is the only way to ascertain the precise functional significance of an allele. When available, Aldy employs phasing evidence, such as linked reads, long reads, and barcoded fragments. Aldy reports one or more diplotype calls, which indicate which two star alleles are present in the genomic sample.

8.3.2 Stargazer

Stargazer is a bioinformatics tool developed for pharmacogene genotyping and phasing using sequencing data (Lee et al., 2019). It identifies star alleles by finding single nucleotide variants, insertions/deletions, and structural variants. It detects complex structural variants, such as gene deletions, duplications, and hybrids, by figuring out paralog-specific copy number from read depth.

8.3.3 Cyrius/ BCyrius

Cyrius is a powerful bioinformatics technique that uses short-read whole-genome sequencing (WGS) information to genotype the *CYP2D6* gene with high accuracy. *CYP2D6* has significant homology with pseudogenes like *CYP2D7*, or *CYP2D8* and many SVs and CNV are there across the gene. These complexities make it challenging to distinguish true variants from alignment artifacts. Cyrius addresses these issues through its ability to work with sequence data aligned to both GRCh38 and GRCh37 (Chen et al., 2021). It is also validated using reference samples from the CDC Genetic Testing Reference Material Program (GeT-RM). In GeT-RM several genotyping platforms are used to derive the consensus genotypes of key pharmacogenes.

BCyrius is an extension of the Cyrius with more accurate CYP2D6 genotyping capability. In contrast to Cyrius, BCyrius uses long-read sequencing data (PacBio/ONT) to provide a comprehensive portrait of CYP2D6 for better confirmation of complex variations and haplotypes.

8.3.4 PharmaCAT

- PharmCAT (Pharmacogenomics Clinical Annotation Tool) is an open-source computational tool that evaluates genetic variants to predict medication response and customize medical treatment according to an individual patient's genetic profile (Sangkuhl et al., 2020).
- It parses VCF files generated from WGS or whole exome sequencing (WES) and identifies PGx genotypes to infer haplotypes or star alleles.
- It employs pharmacogene diplotypes to predict pharmacogenomic phenotypes and provides the associated drug-prescribing recommendations.

8.3.5 StarPhase

- StarPhase is a phase-aware pharmacogenomic diplotyper developed by Pacific Biosciences for analyzing long-read sequencing data (specifically PacBio HiFi reads) to call pharmacogene "star (*) alleles" in diploid genomes (Holt et al., 2024).
- It diplotypes 21 pharmacogenes that the Clinical Pharmacogenetics Implementation Consortium (CPIC) ranks at Level A. (Level A—their top recommendation for using genetic info to guide drug or dose choices).
- It also sorts out tricky haplotypes for pharmacogenes like CYP2D6 or HLA-A/B, with structural complexity, duplications, deletions, or hybrid alleles.
- StarPhase is phase-aware, meaning it uses phasing information to determine which variants belong together on the same allele.
- By using long-read sequencing (PacBio HiFi), StarPhase is able to phase variants over long stretches—that is, determine which variant alleles are together on a haplotype.
- Because HiFi reads are long and accurate, StarPhase can see duplications or deletions directly (e.g., CYP2D6 duplication + deletion events). This avoids miscalls that might happen with short-read or array-based genotyping.

9 Conclusion

In summary, this chapter describes xenobiotic metabolism process, different enzymes involved in the intricate pathway, and genetic variations of the XME genes. The chapter also illustrates the different computational tools used for efficient, accurate, and large-scale pharmacogenetic variant detection from NGS reads. Development of these algorithms is essential to translate genomic data into actionable clinical insights. Numerous XME genetic variations have been linked to human health and diseases. Identifying these variants in XME genes is essential for understanding individual variability, optimizing therapy, predicting disease risk, and developing safer drugs and chemicals. Several studies have been discussed that show association between SVs of the XME genes and altered drug response and susceptibility to diseases following xenobiotic exposure. Genetic variations like SNPs and CNVs in XME may alter the expression of XME genes, which in turn affect an individual's response to exposure to drugs and other foreign compounds. Although the role of SNPs in XME genes has been well documented, further research is required to unravel the role of CNVs in XME genes.

References

Angstadt, A. Y., Berg, A., Zhu, J., Miller, P., Hartman, T. J., Lesko, S. M., et al. (2013). The effect of copy number variation in the phase II detoxification genes *UGT2B17* and *UGT2B28* on colorectal cancer risk. *Cancer, 119*(13), 2477–2485.

Barek, M. A., Basher, M. A., Aziz, M. A., Hossen, M. S., Jahan, N., Afroz, N., et al. (2023). Assessment of the association of CYP1A1 gene polymorphisms with the susceptibility of cervical cancer: A case-control study and meta-analysis. *Heliyon, 9*(7), e17712.

Beoris, M., Wilson, J. A., Garces, J. A., & Lukowiak, A. A. (2016). CYP2D6 copy number distribution in the US population. *Pharmacogenetics and Genomics, 26*(2), 96–99.

Chen, X., Shen, F., Gonzaludo, N., Malhotra, A., Rogert, C., Taft, R. J., et al. (2021). Cyrius: Accurate CYP2D6 genotyping using whole-genome sequencing data. *The Pharmacogenomics Journal, 21*(2), 251–261.

Da Silva, I. M., Maraslis, F. T., Kawasaki, J. A. I., Aida, N. K., Barcelos, G. R. M., Koike, A., et al. (2025). Allelic variants in xenobiotic metabolism genes predict susceptibility and worse prognosis of urothelial bladder cancer. *Pathology—Research and Practice, 266*, Article 155767.

Dagli-Hernandez, C., Borges, J. B., Marçal, E. D. S. R., De Freitas, R. C. C., Mori, A. A., Gonçalves, R. M., et al. (2022). Genetic variant ABCC1 rs45511401 is associated with increased response to statins in patients with familial hypercholesterolemia. *Pharmaceutics, 14*(5), 944.

Ferracini, A. C., Lopes-Aguiar, L., Lourenço, G. J., Yoshida, A., Lima, C. S. P., Sarian, L. O., et al. (2021). *GSTP1* and *ABCB1* polymorphisms predicting toxicities and clinical management on carboplatin and paclitaxel-based chemotherapy in ovarian cancer. *Clinical Translational Science, 14*(2), 720–728.

Firigato, I., López, R. V. M., Curioni, O. A., De Antonio, J., Gattás, G. J. F., & Gonçalves, F. D. T. (2022). The copy number variation of GSTM1 as a promising prognostic factor of oral squamous cell carcinoma. *Oral Surgery, Oral Medicine, Oral Pathology and Oral Radiology, 134*(5), 615–626.

Forat-Yazdi, M., Jafari, M., Kargar, S., Abolbaghaei, S. M., Nasiri, R., Farahnak, S., et al. (2017). Association between SULT1A1 Arg213His (rs9282861) polymorphism and risk of breast cancer: A systematic review and meta-analysis. *Journal of Research in Health Science, 17*(4), Article e00396.

Freriksen, J. J. M., Salomon, J., Roelofs, H. M. J., Te Morsche, R. H. M., Van Der Stappen, J. W. J., Dura, P., et al. (2014). Genetic polymorphism 609C > T in NAD(P)H: Quinone oxidoreductase 1 enhances the risk of proximal colon cancer. *Journal of Human Genetics, 59*(7), 381–386.

Gaedigk, A., Ingelman-Sundberg, M., Miller, N. A., Leeder, J. S., Whirl-Carrillo, M., Klein, T. E., et al. (2018). The pharmacogene variation (PharmVar) consortium: Incorporation of the human cytochrome P450 (*CYP*) Allele nomenclature database. *Clinical Pharmacology and Therapeutics, 103*(3), 399–401.

Gao, Y., He, B. (2018). *ABCB1* 1199G>A Polymorphism affects the intracellular accumulation of antidepressants in LLC-PK1 recombinant cell lines. *DNA and Cell Biology, 37*(12), 1055–1060.

Golmohammadzadeh, G., Mohammadpour, A., Ahangar, N., & Shokrzadeh, M. (2019). Polymorphisms in phase I (CYP450) Genes CYP1A1 (rs4646421), CYP1B1 (rs1056836), CYP19A1 (rs749292) and CYP2C8 (rs1058930) and their relation to risk of breast cancer: A case-control study in Mazandaran Province in North of Iran. *Open Access Macedonian Journal of Medical Sciences, 7*(15), 2488–2496.

Gregers, J., Gréen, H., Christensen, I. J., Dalhoff, K., Schroeder, H., Carlsen, N., et al. (2015). Polymorphisms in the ABCB1 gene and effect on outcome and toxicity in childhood acute lymphoblastic leukemia. *The Pharmacogenomics Journal, 15*(4), 372–379.

Grussy, K., Łaska, M., Moczurad, W., Król-Kulikowska, M., & Ściskalska, M. (2023). The importance of polymorphisms in the genes encoding glutathione S-transferase isoenzymes in development of selected cancers and cardiovascular diseases. *Molecular Biology Reports, 50*(11), 9649–9661.

Hamajima, N., Matsuo, K., Suzuki, T., Nakamura, T., Matsuura, A., Tajima, K., et al. (2001). Low expression myeloperoxidase genotype negatively associated with *Helicobacter pylori* infection. *Japanese Journal of Cancer Research., 92*(5), 488–493.

Hansen, R. D., Krath, B. N., Frederiksen, K., Tjønneland, A., Overvad, K., Roswall, N., et al. (2009). GPX1 Pro198Leu polymorphism, erythrocyte GPX activity, interaction with alcohol consumption and smoking, and risk of colorectal cancer. *Mutation Research—Fundamental and Molecular Mechanisms of Mutagenesis, 664*(1–2), 13–19.

Hari, A., Zhou, Q., Gonzaludo, N., Harting, J., Scott, S. A., Qin, X., et al. (2023). An efficient genotyper and star-allele caller for pharmacogenomics. *Genome Research, 33*(1), 61–70.

Hattori, S., Suda, A., Miyauchi, M., Shiraishi, Y., Saeki, T., Fukushima, T., et al. (2020). The association of genetic polymorphisms in CYP1A2, UGT1A4, and ABCB1 with autonomic nervous system dysfunction in schizophrenia patients treated with olanzapine. *BMC Psychiatry, 20*(1), 72.

Hayes, J. D., & Strange, R. C. (2000). Glutathione S-transferase polymorphisms and their biological consequences. *Pharmacology, 61*(3), 154–166.

Hebbring, S. J., Adjei, A. A., Baer, J. L., Jenkins, G. D., Zhang, J., Cunningham, J. M., et al. (2007). Human SULT1A1 gene: Copy number differences and functional implications. *Human Molecular Genetics, 16*(5), 463–470.

Hein, D. W., Doll, M. A., Fretland, A. J., Leff, M. A., Webb, S. J., Xiao, G. H., et al. (2000). Molecular genetics and epidemiology of the NAT1 and NAT2 acetylation polymorphisms. *Cancer Epidemiology, Biomarkers & Prevention, 9*(1), 29–42.

Hekim, N., Gure, M. A., Metin Mahmutoglu, A., Gunes, S., Asci, R., & Henkel, R. (2020). SNP's in xenobiotic metabolism and male infertility. *Xenobiotica, 50*(3), 363–370.

Hoidy, W., Nubgan, A., & Alsaadi, M. (2023). Development of a T-ARMS-PCR assay for detecting genetic polymorphism in the catalase (rs7943316) Gene in the Iraqi population with breast cancer. *Asian Pacific Journal of Cancer Prevention, 24*(9), 3283–3289.

Holt, J. M., Harting, J., Chen, X., Baker, D., Saunders, C. T., Kronenberg, Z., et al. (2024). StarPhase: Comprehensive phase-aware pharmacogenomic diplotyper for long-read sequencing data. Retrieved January 22, 2026, from http://biorxiv.org/lookup/doi/10.1101/2024.12.10.627527

Hoque, K. M., Dixon, E. E., Lewis, R. M., Allan, J., Gamble, G. D., Phipps-Green, A. J., et al. (2020). The ABCG2 Q141K hyperuricemia and gout associated variant illuminates the physiology of human urate excretion. *Nature Communications, 11*(1), 2767.

Hosagrahara, V. P., Rettie, A. E., Hassett, C., & Omiecinski, C. J. (2004). Functional analysis of human microsomal epoxide hydrolase genetic variants. *Chemico-Biological Interactions, 150*(2), 149–159.

Hu, K., Liu, Q., Liu, X., Bai, H., Wu, Y., & Fan, P. (2025). Association of xenobiotic-metabolizing gene cytochrome P450 2E1 variants with preeclampsia in Chinese women. *Reproductive Sciences, 32*(10), 3419–3428.

Hu, L., Wu, Y., Guan, X., Liang, Y., Yao, X., Tan, D., et al. (2015). Germline copy number loss of UGT2B28 and gain of PLEC contribute to increased human esophageal squamous cell carcinoma risk in Southwest China. *American Journal of Cancer Research, 5*(10), 3056–3071.

Hu, X., Qin, W., Li, S., He, M., Wang, Y., Guan, S., et al. (2018). Polymorphisms in DNA repair pathway genes and ABCG2 gene in advanced colorectal cancer: Correlation with tumor characteristics and clinical outcome in oxaliplatin-based chemotherapy. *CMAR, 11*, 285–297.

Hunt, R., Sauna, Z. E., Ambudkar, S. V., Gottesman, M. M., Kimchi-Sarfaty, C. (2009). Silent (Synonymous) SNPs: Should we care about them? In A.A. Komar (Ed.), *Single nucleotide polymorphisms, methods in molecular biology* (vol. 578, pp. 23–39). Humana Press. Retrieved January 22, 2026, from http://link.springer.com//10.1007/978-1-60327-411-1_2

Iorio, A., Spinelli, M., Polimanti, R., Lorenzi, F., Valensise, H., Manfellotto, D., et al. (2015). GSTA1 gene variation associated with gestational hypertension and its involvement in pregnancy-related pathogenic conditions. *European Journal of Obstetrics & Gynecology and Reproductive Biology, 194*, 34–37.

Jarvis, J. P., Peter, A. P., & Shaman, J. A. (2019). Consequences of CYP2D6 copy-number variation for pharmacogenomics in psychiatry. *Front Psychiatry, 20*(10), 432.

Jones, S. K., Wolf, B. J., Froeliger, B., Wallace, K., Carpenter, M. J., & Alberg, A. J. (2022). Nicotine metabolism predicted by *CYP2A6* genotypes in relation to smoking cessation: A systematic review. *Nicotine & Tobacco Research, 24*(5), 633–642.

Kabziński, J., Gogolewska, M., Nowakowski, M., Kucharska, E., Dziki, Ł., Mik, M., et al. (2022). Polymorphism association of NIL1, NIL2, CYP1A1 xenobiotic metabolism genes and their expression with the risk of colorectal cancer in the Polish population. *Polski Przegląd Chirurgiczny, 94*(2).

Koutros, S., Andreotti, G., Berndt, S. I., Hughes Barry, K., Lubin, J. H., Hoppin, J. A., et al. (2011). Xenobiotic-metabolizing gene variants, pesticide use, and the risk of prostate cancer. *Pharmacogenetics and Genomics, 21*(10), 615–623.

Kukal, S., Thakran, S., Kanojia, N., Yadav, S., Mishra, M. K., Guin, D., et al. (2023). Genic-intergenic polymorphisms of CYP1A genes and their clinical impact. *Gene, 857*, Article 147171.

Kumasaka, N., Aoki, M., Okada, Y., Takahashi, A., Ozaki, K., Mushiroda, T., et al. (2012). Haplotypes with copy number and single nucleotide polymorphisms in CYP2A6 locus are associated with smoking quantity in a Japanese population. R. Medeiros (Ed.) *Plos One, 7*(9), e44507.

Lee, S., Wheeler, M. M., Patterson, K., McGee, S., Dalton, R., Woodahl, E. L., et al. (2019). Stargazer: A software tool for calling star alleles from next-generation sequencing data using CYP2D6 as a model. *Genetics in Medicine, 21*(2), 361–372.

Liu, C., Song, S., Zhang, J., Li, X., & Gao, H. (2020). Effects of GSTA1 and GPX3 polymorphisms on the risk of Schizophrenia in Chinese Han population. *NDT, 16*, 113–118.

Malempati, S., Agrawal, N., Ravisankar, D., Kothapalli, V. S. R. T., Cheemanapalli, S., Tamanam, R. R., et al. (2024). Role of GSTP1 functional polymorphisms in molecular pathogenesis of colorectal cancer. *Human Gene, 42*, Article 201335.

Manichaikul, A., Peres, L. C., Wang, X., Barnard, M. E., Chyn, D., Sheng, X., et al. (2020). Identification of novel epithelial ovarian cancer loci in women of African ancestry. *International Journal of Cancer, 146*(11), 2987–2998.

Martínez-Ramírez, O. C., Pérez-Morales, R., Castro, C., Flores-Díaz, A., Soto-Cruz, K. E., Astorga-Ramos, A., et al. (2013). Polymorphisms of catechol estrogens metabolism pathway genes and breast cancer risk in Mexican women. *The Breast, 22*(3), 335–343.

Moghadam, A. R., Mehramiz, M., Entezari, M., Aboutalebi, H., Kohansal, F., Dadjoo, P., et al. (2018). A genetic polymorphism in the *CYP1B1* gene in patients with squamous cell carcinoma of the esophagus: An Iranian Mashhad Cohort study recruited over 10 years. *Pharmacogenomics, 19*(6), 539–546.

Owen, R. P., Sangkuhl, K., Klein, T. E., & Altman, R. B. (2009). Cytochrome P450 2D6. *Pharmacogenetics and Genomics, 19*(7), 559–562.

Pan, H., Wu, T., Huang, K., Guo, Z., Liang, H., Lyu, P., et al. (2024). Reducing SULT2B1 promotes the interaction of LncRNAgga3-204 with SMAD4 to inhibit the macrophage inflammatory response and delay atherosclerosis progression. *Translational Research, 268*, 13–27.

Patil, M. N., Datkhile, K. D., Gudur, A. K., Gudur, R. A., & Patil, S. R. (2024). Single-nucleotide polymorphism in CYP1A1, CYP1B1, CYP2B6, CYP2C8, and CYP2C9 genes and their association with gastrointestinal cancer: A hospital-based case-control study. *Journal of Cancer Research and Therapeutics, 20*(1), 216–223.

Pavlovic, D., Ristic, S., Djukanovic, L., Matic, M., Kovacevic, M., Pljesa-Ercegovac, M., et al. (2023). The GSTO2 (rs156697) polymorphism modifies diabetic nephropathy risk. *Medicina, 59*(1), 164.

Ravn-Haren, G., Olsen, A., Tjønneland, A., Dragsted, L. O., Nexø, B. A., Wallin, H., et al. (2006). Associations between GPX1 Pro198Leu polymorphism, erythrocyte GPX activity, alcohol consumption and breast cancer risk in a prospective cohort study. *Carcinogenesis, 27*(4), 820–825.

Redon, R., Ishikawa, S., Fitch, K. R., Feuk, L., Perry, G. H., Andrews, T. D., Fiegler, H., Shapero, M. H., Carson, A. R., Chen, W., & Cho, E. K. (2006). Global variation in copy number in the human genome. *Nature, 444*(7118), 444–454.

Roumeliotis, S., Roumeliotis, A., Stamou, A., Panagoutsos, S., Manolopoulos, V. G., Tsetsos, F., et al. (2021). Association of rs11780592 polymorphism in the human soluble epoxide hydrolase gene (EPHX2) with oxidized LDL and mortality in patients with diabetic chronic kidney disease. J. Ghose (Ed.) *Oxidative Medicine and Cellular Longevity, 2021*(1), 8817502.

Sameer, A. S., Shah, Z. A., Syeed, N., Rasool, R., Afroze, D., & Siddiqi, M. A. (2010). NAD(P)H: Quinone oxidoreductase 1 (NQO1) Pro87Ser polymorphism and colorectal cancer predisposition in the ethnic Kashmiri population. *Asian Pacific Journal of Cancer Prevention, 11*(1), 209–213.

Sangkuhl, K., Whirl-Carrillo, M., Whaley, R. M., Woon, M., Lavertu, A., Altman, R. B., et al. (2020). Pharmacogenomics clinical annotation tool (Pharm CAT). *Clinical Pharmacology and Therapeutics, 107*(1), 203–210.

Sergentanis, T. N., & Economopoulos, K. P. (2010). Four polymorphisms in cytochrome P450 1A1 (CYP1A1) gene and breast cancer risk: A meta-analysis. *Breast Cancer Research and Treatment, 122*(2), 459–469.

Stoehr, C. G., Nolte, E., Wach, S., Wieland, W. F., Hofstaedter, F., Hartmann, A., & Stoehr, R. (2012). NAD (P) H: Quinone oxidoreductase 1 (NQO1) P187S polymorphism and prostate cancer risk in Caucasians. *International Journal of Molecular Sciences, 13*(9), 10959–10969.

Sharma, Y., Miladi, M., Dukare, S., Boulay, K., Caudron-Herger, M., Groß, M., et al. (2019). A pan-cancer analysis of synonymous mutations. *Nature Communications, 10*(1), 2569.

Singh, R. R., Mohammad, J., Orr, M., & Reindl, K. M. (2020). Glutathione S-transferase pi-1 knockdown reduces pancreatic ductal adenocarcinoma growth by activating oxidative stress response pathways. *Cancers, 12*(6), 1501.

Sito, H., Sharzehan, M. A. K., Islam, M. A., & Tan, S. C. (2024). Genetic variants associated with response to platinum-based chemotherapy in non-small cell lung cancer patients: A field synopsis and meta-analysis. *British Journal of Biomedical Science, 21*(81), 11835.

Stiburkova, B., Pavelcova, K., Pavlikova, M., Ješina, P., & Pavelka, K. (2019). The impact of dysfunctional variants of ABCG2 on hyperuricemia and gout in pediatric-onset patients. *Arthritis Research & Therapy, 21*(1), 77.

Sun, F., Chen, Z., Yao, P., Weng, B., Liu, Z., & Cheng, L. (2021). Meta-analysis of ABCG2 and ABCB1 polymorphisms with sunitinib-induced toxicity and efficacy in renal cell carcinoma. *Frontiers in Pharmacology, 8*(12), Article 641075.

Świechowski, R., Jeleń, A., Mirowski, M., Talarowska, M., Gałecki, P., Pietrzak, J., et al. (2019). Estimation of *CYP3A4*1B* single nucleotide polymorphism in patients with recurrent major depressive disorder. *Molecular Genetics & Genomic Medicine, 7*(6), Article e669.

Takano, S., Fukami, T., Ichida, H., Suzuki, K., Nakano, M., & Nakajima, M. (2023). In vitro evaluation of the reductase activities of human AKR1C3 allelic variants. *Drug Metabolism and Disposition, 51*(9), 1188–1195.

Tang, W., Fu, Y. P., Figueroa, J. D., Malats, N., Garcia-Closas, M., Chatterjee, N., et al. (2011). An unusual suspect: an uncommon human-specific synonymous coding variant within the UGT1A6 gene explains a GWAS signal and protects against bladder cancer. *Genome Biology, 12*(S1), P19, gb-2011-12-s1-p19.

Tasnim, T., Al-Mamun, M. M. A., Nahid, N. A., Islam, M. R., Apu, M. N. H., Bushra, M. U., et al. (2017). Genetic variants of SULT1A1 and XRCC1 genes and risk of lung cancer in Bangladeshi population. *Tumour Biology, 39*(11), 101042831772927.

Tian, D. D., Bennett, S. K., Coupland, L. A., Forwood, K., Lwin, Y., Pooryousef, N., et al. (2019). *GSTZ1* genotypes correlate with dichloroacetate pharmacokinetics and chronic side effects in multiple myeloma patients in a pilot phase 2 clinical trial. *Pharmacology Research & Perspectives, 7*(6), Article e00526.

Tian, Z., Li, Y. L., Zhao, L., & Zhang, C. L. (2013). RETRACTED: Role of CYP1A2*1F polymorphism in cancer risk: Evidence from a meta-analysis of 46 case–control studies. *Gene, 524*(2), 168–174.

Tremmel, R., Herrmann, K., Engst, W., Meinl, W., Klein, K., Glatt, H., et al. (2017). Methyleugenol DNA adducts in human liver are associated with SULT1A1 copy number variations and expression levels. *Archives of Toxicology, 91*(10), 3329–3339.

Václavíková, R., Hughes, D. J., & Souček, P. (2015). Microsomal epoxide hydrolase 1 (EPHX1): Gene, structure, function, and role in human disease. *Gene, 571*(1), 1–8.

Valadez-Cosmes, P., Raftopoulou, S., Mihalic, Z. N., Marsche, G., & Kargl, J. (2022). Myeloperoxidase: Growing importance in cancer pathogenesis and potential drug target. *Pharmacology & Therapeutics, 1*(236), Article 108052.

Walia, H. K., Singh, N., & Sharma, S. (2022). Association of *NAT-2* gene polymorphisms toward lung cancer susceptibility and prognosis in north indian patients treated with platinum-based chemotherapy. *Pharmacogenomics, 23*(2), 97–118.

Wang, C. D., Sun, Y., Chen, N., Huang, L., Huang, J. W., Zhu, M., et al. (2016). The role of catalase C262T gene polymorphism in the susceptibility and survival of cancers. *Science and Reports, 6*(1), 26973.

Wang, P. F., Neiner, A., & Kharasch, E. D. (2019). Efavirenz metabolism: Influence of polymorphic CYP2B6 variants and stereochemistry. *Drug Metabolism and Disposition, 47*(10), 1195–1205.

Wu, Y., Wang, Y., Sheng, X., Lin, Y., Zhou, L., Lu, J., & Yin, W. (2025). Genetic variations in CYP2A6 predict efficacy of neoadjuvant letrozole concurrent with chemotherapy in locally advanced hormone receptor-positive breast cancer. *Annals of Medicine, 57*(1), 2557510.

Zhang, K., Gao, L., Wu, Y., Chen, J., Lin, C., Liang, S., et al. (2015). NAT1 polymorphisms and cancer risk: A systematic review and meta-analysis. *International Journal of Clinical and Experimental Medicine, 8*(6), 9177–9191.

Zhao, G., Liu, J., & Meng, T. (2022). Oxidative stress-related genes (*EPHX1* and *MnSOD*) polymorphism and risk of pre-eclampsia: A meta-analysis. *The Journal of Maternal-Fetal & Neonatal Medicine, 35*(25), 5526–5538.

Zhong, J. H., Zhang, Z. M., & Li, L. Q. (2013). MEH Tyr113His polymorphism and the risk of ovarian cancer development. *Journal of Ovarian Research, 6*(1), 40.

Chapter 9
Effects of Genetic Polymorphism on Xenobiotic-Induced Genetic Damage and Diseases

Anita Saini

Abstract Genetic polymorphisms are natural variations in DNA sequence. These variations are important determinants of how individuals respond to xenobiotics such as drugs, environmental pollutants, pesticides, and dietary chemicals. Such variations influence the metabolism, detoxification, and elimination of foreign compounds, which helps in regulating the body's exposure to reactive intermediates. The Polymorphisms in Phase I enzymes, particularly cytochrome P450s, can modify the rate of xenobiotic bioactivation. Phase II enzymes, including glutathione S-transferases and N-acetyltransferases, influence detoxification efficiency and clearance of xenobiotics from the body. As a result, reactive intermediates may persist longer within the body, which increases the rate of cellular damage caused by mutations and oxidative stress. Genetic variability in DNA repair enzymes and antioxidant defense systems further determines the cellular capacity to manage DNA lesions and oxidative stress. The repair system and the balance between activation and detoxification are disrupted when metabolic genes or DNA repair genes are altered due to polymorphisms. Such modifications can increase the likelihood of DNA damage persistence, thereby contributing to mutations, genomic instability, and disease progression. In response to xenobiotic exposure, a comprehensive understanding of the genetic and metabolic factors is essential for gaining insight. This chapter therefore presents a general overview of the relationship between genetic polymorphisms, xenobiotic-induced genetic damage, and disease outcomes.

Keywords Metabolic genes · Genomic instability · DNA damage · Xenobiotic bioactivation · Genetic polymorphism

A. Saini (✉)
Department of Biotechnology and Microbiology, School of Basic and Applied Sciences, Maharaja Agrasen University, Baddi, Himachal Pradesh, India
e-mail: sannie10@gmail.com

K. Priya and S. K. Giri (eds.), *Environmental Toxicology: Xenobiotics and Effects on Human Genetic Material*, https://doi.org/10.1007/978-3-032-28619-2_9

1 Introduction

Xenobiotics are chemical substances which are not produced naturally by the human body. They enter the biological systems as a result of exposure to the environment or through medicines, diet, or occupational work. The common examples of xenobiotics include agricultural pesticides, industrial and environmental pollutants, food additives, and pharmaceuticals. Many xenobiotics, at low exposure, are eliminated efficiently from the body. However, some enter metabolic pathways and are converted into reactive intermediates. These compounds interact with cellular macromolecules, particularly DNA, and cause damage to the genetic material, ultimately leading to various health problems.

Different individuals show different responses to the xenobiotic exposure, even when conditions are nearly similar. This may be attributed largely to the genetic polymorphisms, i.e., common variations in genome sequence that are inherited in populations. These alterations influence the activities of metabolic enzymes and mechanisms of cellular defense. When these variations occur in genes involved in xenobiotic metabolism, DNA repair, and antioxidant defense, they often result in an imbalance between activation, detoxification, and repair processes in exposed cells. As a consequence, due to inherited variants, individuals tend to accumulate higher levels of reactive metabolites and show a reduced capacity to repair altered DNA. Such factors increase the susceptibility of individuals to genetic mutations and disease. Research has revealed that variability in xenobiotic metabolism and/or DNA repair pathways is associated with a broad spectrum of diseases, including cancer, cardiovascular, respiratory, neurodegenerative, and reproductive disorders. This study also explains the differences in the occurrence of disease among exposed populations and control cohorts.

The present chapter discusses the role of genetic polymorphisms in xenobiotic-induced genetic damage and disease development. It will focus on key xenobiotic-metabolizing gene polymorphisms and DNA repair, along with the diseases associated with them. The content will help in understanding the biomarker of effect and susceptibility on genetic variability to the disease upon exposure to the xenobiotic. Such insights are crucial for risk assessment and have significant implications for public health, occupational safety, and disease prevention.

2 Xenobiotic-Metabolizing Enzymes

Xenobiotic metabolism involves two phases that convert foreign compounds mostly into polar forms which can be further easily eliminated from the body. Phase I involves enzymatic reactions such as oxidation, reduction, or hydrolysis that catalyze the addition or exposure of functional groups on xenobiotic compounds. Phase 1 reactions are further catalyzed the cytochrome P450 (CYP) enzyme family. Phase I reactions play a role in facilitating detoxification, which results in generating highly reactive intermediates, that can induce mutations through DNA damage (Nebert & Dalton, 2006; Guengerich, 2008). Phase II metabolism mainly involves conjugation reactions which attach endogenous molecules to xenobiotics or their metabolites produced in Phase I. This reduces the chemical reactivity of intermediates and enhances their water solubility, which are essential for their excretion from the body (Iyanagi, 2007). Table 1 provides an overview of the major Phase I and Phase II xenobiotic-metabolizing enzymes and their functions in xenobiotic metabolism. Genetic polymorphisms in these xenobiotic-metabolizing enzymes can result in altered expression of their genes and enzyme activity. This causes an imbalance between xenobiotic activation and detoxification, which ultimately influences DNA damage and susceptibility to disease. The relationship between xenobiotic exposure, genetic polymorphisms, and disease susceptibility is summarized in Fig. 1.

3 Polymorphisms in Xenobiotic-Metabolizing Enzymes and Disease Risk

3.1 Cytochrome P450 (CYP) Polymorphisms

Cytochrome P450 (CYP450) enzymes form a large family of heme-containing monooxygenases that are essential for the metabolism of a wide array of xenobiotics, including drugs, environmental pollutants, industrial chemicals, and naturally occurring compounds. These heme-thiolate monooxygenases catalyze phase I oxidation reactions such as hydroxylation, epoxidation, and dealkylation, which increase the water solubility of xenobiotics and facilitate their elimination from the body after absorption and distribution to tissues (Esteves et al., 2021). Genetic polymorphisms in CYP450 genes give rise to substantial inter-individual variability in enzyme expression, catalytic activity, substrate specificity, and inducibility, ultimately affecting the efficiency of xenobiotic clearance and the internal dose that tissues are exposed to (Chilipweli et al., 2025; Prost & Płaziński, 2025).

CYP1A1, *CYP1A2*, *CYP2D6*, *CYP2E1*, *CYP3A4*, and *CYP3A5* are among the most studied polymorphisms of CYP450 that result from different variations, including single-nucleotide substitutions, insertions, deletions, and copy number variations. The enzymes produced from the expression of these isoforms may show modifications in their structure and substrate affinity that ultimately alter the regulation of xenobiotic metabolic processes. Some variations result in reduced enzyme activity or expression that in turn reduces the rate of biotransformation of xenobiotics. As a consequence, these xenobiotics or their metabolic intermediates are retained for longer durations in the cells. Accumulation of these potentially harmful compounds often results in diseases. On the other hand, several variants enhance the catalytic activity of xenobiotic-metabolizing enzymes, which can accelerate the bioconversion of xenobiotics into reactive intermediates and reactive oxygen species (ROS). These products interact with different cellular components, including lipids, proteins, and

Table 1 Key xenobiotic-metabolizing enzymes, their functions, and representative substrates

Enzyme class and examples	Function in xenobiotic metabolism	Representative substrates/ reactions
Phase I—cytochrome P450s (CYPs)	Oxidative modification of xenobiotics (hydroxylation, epoxidation, dealkylation) to introduce polar groups for further metabolism (Esteves et al., 2021)	CYP1A1/1A2—Oxidation of aromatic hydrocarbons and heterocyclic amines; CYP2E1—metabolism of ethanol, nitrosamines, solvents; CYP3A4—oxidative metabolism of many drugs and environmental xenobiotics
Phase I—auxiliary enzymes	Complement CYP metabolism by oxidation/ reduction/hydrolysis of xenobiotics (Esteves et al., 2021)	Flavin-containing monooxygenases (FMOs), alcohol and aldehyde dehydrogenases
Phase II—glutathione S-transferases (GSTs)	Conjugation of reduced glutathione (GSH) to electrophilic xenobiotic metabolites, neutralizing reactivity and increasing water solubility (Stoian et al., 2025)	Conjugation of electrophilic intermediates from Phase I (e.g., epoxides) to GSH
Phase II—UDP-glucuronosyltransferases (UGTs)	Glucuronidation of xenobiotics and their metabolites (lipophilic) to form water-soluble (hydrophilic) glucuronides for excretion (Mano et al., 2018)	Glucuronidation of bilirubin, drugs (e.g., opioid analgesics), xenobiotic phenols
Phase II—N-acetyltransferases (NAT1, NAT2)	Acetylation of aromatic amines and hydrazines, reducing toxicity and aiding elimination (Vats et al., 2015)	Acetylation of carcinogenic amines (e.g., aromatic amines)
Phase II—sulfotransferases (SULTs)	Sulfate conjugation of phenolic and hydroxyl-containing xenobiotics, enhancing water solubility (Suiko et al., 2017)	Sulfation of hormones, phenols, and xenobiotics

nucleic acids, causing oxidative stress and alterations in the cells (Esteves et al., 2021; Chilipweli et al., 2025). Some polymorphisms also show marked impact on inducibility of metabolic enzymes by environmental chemicals, which further influences the balance between detoxification and bioactivation depending on the extent of exposure.

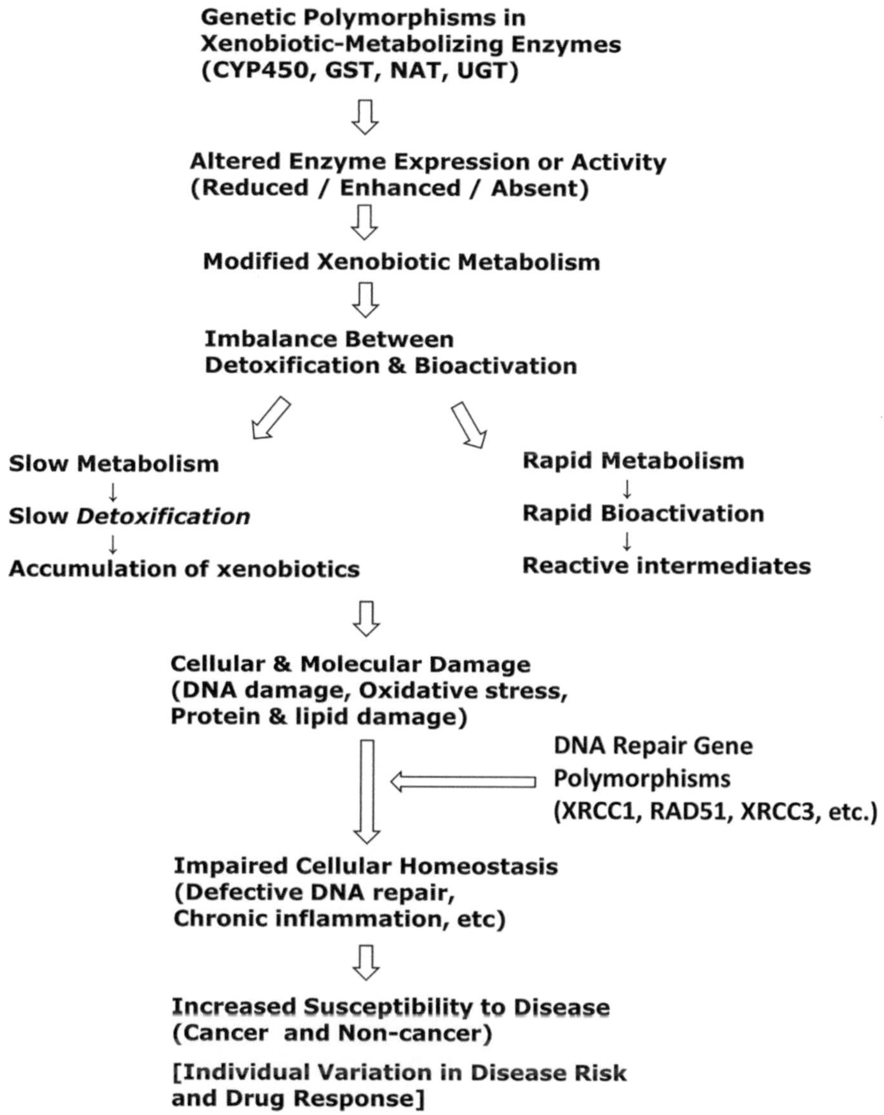

Fig. 1 Impact of genetic polymorphisms in xenobiotic metabolism and disease susceptibility

Collectively, these functional variations in CYP450 enzymes contribute to significant differences in metabolism of xenobiotics in different individuals. Their understanding explains the reason for rapid elimination of foreign compounds by some people compared to others who accumulate reactive metabolites and develop diseases. Understanding these polymorphisms is, therefore, critical to assess risk of chemical exposure, individual susceptibility, and to aid in developing strategies for personalized prevention and intervention (Esteves et al., 2021; Prost & Płaziński, 2025).

3.2 Glutathione S-Transferase (GST) Polymorphisms

Glutathione S-transferases (GSTs) are a superfamily of Phase II detoxification enzymes, which catalyze the conjugation of reduced glutathione (GSH) to electrophilic xenobiotic metabolites. This conjugation increases the solubility of these metabolites, which facilitates their elimination from the body (Alnasser, 2025). Conjugation is a key step in neutralizing reactive intermediates generated from the xenobiotic metabolism in Phase I that plays a critical role in protecting cells from electrophilic attack and oxidative damage (Alnasser, 2025; Villegas Sánchez, 2025). Genetic polymorphisms in GST genes, therefore, alter the metabolism of xenobiotics. For example, homozygous deletions of *GSTM1* and *GSTT1* result in complete absence of the corresponding enzymes. This reduces cellular capacity for glutathione conjugation, which ultimately impairs the detoxification of electrophilic xenobiotic metabolites (Sadafi et al., 2024). Another common variant, *GSTP1* Ile105Val (rs1695), alters the substrate binding in the *GSTP1* enzyme and its catalytic efficiency, which results in an altered rate of conjugation of xenobiotics to glutathione and their subsequent elimination (Kassogue et al., 2020). *GSTP1* Ile105Val is known to reduce catalytic efficiency for many substrates. This may increase the exposure of cellular constituents to the accumulated reactive metabolites that disrupt the redox homeostasis and increase oxidative stress. Additionally, some polymorphisms may affect regulatory elements that control expression of genes for enzymes, which further modify detoxification responses among individuals depending on the level of exposure to xenobiotics (Villegas Sánchez, 2025). Together, these functional variations determine how effectively different individuals can metabolize xenobiotics to limit oxidative and electrophilic damage caused by them.

3.3 N-Acetyltransferase (NAT) Polymorphisms

N-acetyltransferases (NATs) are Phase II xenobiotic-metabolizing enzymes that help the body process and detoxify foreign chemicals by transferring an acetyl group from acetyl-CoA to aromatic and heterocyclic amines (Habil & Hein, 2025). In humans, this function is mainly carried out by two isoforms, NAT1 and NAT2, which are encoded by separate genes located on chromosome 8 and act on a wide range of xenobiotic substrates. Among these, NAT1 exhibits relatively lower functional variability. NAT2, however, shows high genetic polymorphism, resulting in marked differences in acetylation capacity among individuals (Lopes et al., 2023). These differences arise from specific combinations of single-nucleotide polymorphisms in the NAT2 gene that influence enzyme structure, stability, and catalytic efficiency

(Gutierrez-Virgen et al., 2023; Lopes et al., 2023). Several NAT2 allelic variants, such as NAT25 (*T341C*), *NAT26* (G590A), and NAT27 (*G857A*), are linked to reduced acetylation activity, whereas the reference *NAT24* allele is associated with a more efficient acetylation phenotype (Dutra et al., 2024).

These polymorphisms influence xenobiotic metabolism by altering the balance between detoxification and metabolic activation. They affect the rate of acetyl group transfer. Slow acetylator variants show reduced acetylation activity. This results in slower conjugation and elimination of arylamine and hydrazine compounds. Reactive intermediates formed during Phase I metabolism therefore persist for longer periods. These intermediates can interact with cellular macromolecules. This interaction increases oxidative stress and genetic damage. In contrast, rapid acetylator genotypes metabolize certain substrates more efficiently. This supports faster conjugation and clearance. However, in specific pathways, such as O-acetylation of N-hydroxylated intermediates, increased enzyme activity can enhance the formation of reactive electrophiles. These electrophiles are capable of forming DNA adducts. The effect depends on substrate type and exposure conditions. Genotype-dependent differences in NAT1 and NAT2 expression and catalytic activity have been demonstrated in human hepatocytes. NAT2 acetylation capacity correlates with NAT2 protein levels. This relationship is observed across rapid, intermediate, and slow acetylator genotypes (Salazar-González et al., 2020). These findings show how NAT genetic polymorphisms shape individual metabolic responses to xenobiotic exposure. They also emphasize the importance of considering acetylation phenotype in studies of xenobiotic metabolism and chemical exposure.

3.4 UGT, SULT and Other Phase-II Enzyme Polymorphisms

Phase II enzymes such as UDP-glucuronosyltransferases (UGTs) and sulfotransferases (SULTs) carry out conjugation reactions. These reactions attach endogenous molecules, such as glucuronic acid or sulfate, to xenobiotics or their Phase I metabolites. This process increases their water solubility and supports their efficient excretion from the body. Within the UGT family, UGT1A1 is a key isoform. It plays a major role in the glucuronidation of many foreign compounds. It also acts on endogenous substrates, including bilirubin. A well-characterized polymorphism in this gene, UGT1A1*28, consists of an extra thymine-adenine (TA) repeat in the promoter region, which reduces gene transcription and lowers glucuronidation activity relative to the reference allele (with six TA repeats) (Alkharfy et al., 2013; Dean, 2012). This reduction in UGT1A1 activity leads to slower conjugation of xenobiotics and endogenous compounds, prolonging their presence in circulation and increasing exposure of tissues to reactive intermediates. Other UGT variants, such as those in UGT1A6 or UGT1A9, also influence glucuronidation efficiency for specific substrates, demonstrating that promoter and coding region polymorphisms can modify enzyme kinetics and overall detoxification capacity (Sugatani, 2013).

Sulfotransferases (SULTs) are another class of Phase II enzymes that transfer a sulfonate group from 3′-phosphoadenosine-5′-phosphosulfate (PAPS) to hydroxyl or amino groups of xenobiotics and endogenous compounds, a reaction that generally promotes excretion and reduces electrophilic reactivity (Kurogi et al., 2021). Among the cytosolic SULT isoforms, SULT1A1 is highly expressed and functionally polymorphic; variants such as SULT1A1*2 (Arg213His) alter protein conformation and catalytic properties, often reducing sulfation activity compared with the reference allele (Kurogi et al., 2021). Reduced SULT1A1 activity decreases the efficiency of sulfate conjugation, which can allow electrophilic metabolites of phenolic xenobiotics to persist longer and interact with cellular macromolecules. Other SULT isoforms (e.g., SULT1A2, SULT1E1) also exhibit functional genetic variation that influences the metabolism of distinct classes of xenobiotics, hormones, and endogenous metabolites (Babalola et al., 2023).

In addition to UGTs and SULTs, other Phase II enzymes such as catechol-O-methyltransferase (COMT) contribute to xenobiotic metabolism by catalyzing methylation of catechol-containing substrates, including xenobiotics (Hall et al., 2019). The common Val158Met (rs4680) polymorphism in the COMT gene causes an amino acid substitution. This change reduces enzymatic activity. It affects the ability of COMT to methylate catechol substrates. It also alters the balance between detoxification and the accumulation of reactive catechol intermediates (Hall et al., 2019; Mir et al., 2018). Variants with reduced COMT activity slow the methylation of catechol xenobiotics. This prolongs their presence inside the cell. The extended exposure increases the likelihood of redox cycling. It also increases oxidative stress.

4 DNA Repair Gene Polymorphisms and Disease Risk

Xenobiotic-induced genetic damage may not only result from direct interaction of reactive metabolites with DNA. It also occurs through secondary mechanisms such as oxidative stress. Cells use DNA repair systems to prevent DNA damage. However, genetic polymorphisms in DNA repair genes can reduce the efficiency of repair systems, which can lead to the persistence of DNA damage, depending on the extent of damage. This in turn increases the risk of diseases (Zhang et al., 2014). When DNA repair capacity is impaired, even normal levels of DNA damage can become harmful. This can result in mutations and chromosomal instability. Figure 2 presents a schematic overview of how DNA repair gene polymorphisms contribute to DNA damage.

4.1 Nucleotide Excision Repair (NER) Polymorphisms

The nucleotide excision repair (NER) pathway is one of the major DNA repair systems that work against bulky DNA damage upon xenobiotic exposure. Exposure to

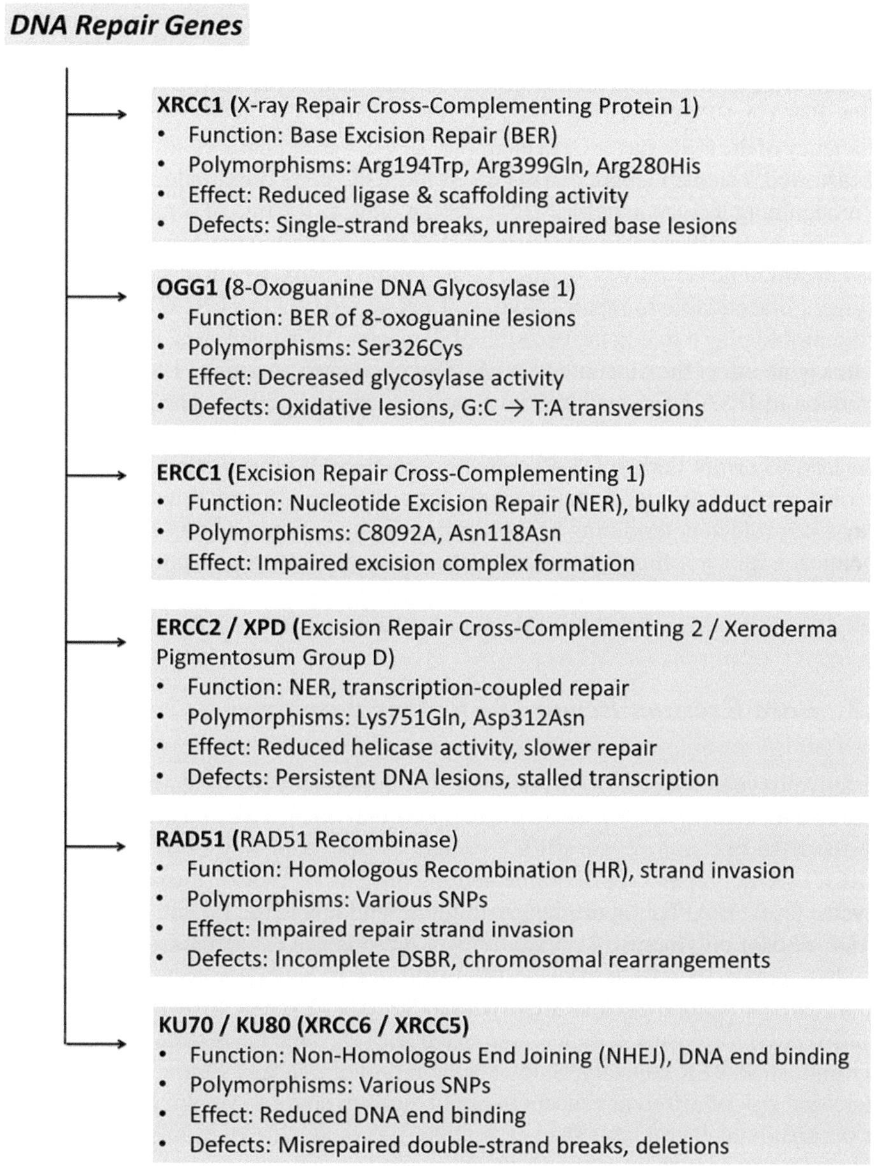

Fig. 2 Polymorphisms in DNA repair genes and their role in DNA damage

some environmental chemicals, such as polycyclic aromatic hydrocarbons, aromatic amines, and industrial pollutants, causes the formation of large DNA adducts in cells that distort the DNA helical structure and interfere with its normal replication and transcription. NER systems play a critical role in restoring genomic integrity by removing these lesions using a well-coordinated process involving recognition of damage and unwinding of DNA at the recognized site, followed by excision of the lesion and its replacement with new DNA (Marteijn et al., 2014; Spivak, 2015).

However, genetic polymorphisms in several core NER genes, including XPC, XPA, ERCC1, ERCC2 (XPD), ERCC4 (XPF), and ERCC5 (XPG), can alter the repair efficiency of the NER system. For example, XPC is essential for the initial recognition of distorted lesions. Genetic variations in the XPC gene often reduce its efficiency in recognizing lesions in helical DNA, which reduces the rate of repair initiation (Li et al., 2019). Similarly, variants in ERCC2 (XPD), including the Asp312Asn variant, may impair helicase activity required for unwinding DNA, which makes the damaged segment inaccessible to repair enzymes (Li et al., 2019). The ERCC1 gene encodes proteins that play a role in the excision of damaged DNA fragments. Polymorphisms in this gene affect the outcome of NER. Together, such variations lead to prolonged retention of DNA lesions within the genome (Lee et al., 2015). The replication and transcription of damaged DNA may initiate before completion of its repair. This can lead to errors during DNA replication, increased transcriptional stress, and a gradual accumulation of mutations over time. As a result, variations in NER genes play a key role in determining how an individual's genome integrity is affected after chemical exposure, highlighting the importance of DNA repair polymorphisms in shaping cellular responses to environmental genotoxins.

4.2 Base Excision Repair (BER) Polymorphisms

Reactive oxygen species (ROS) or other xenobiotic-induced oxidative damage can give rise to small DNA lesions in the exposed individuals. Such damage is repaired by the base excision repair (BER) system of the cells. Key BER genes include XRCC1 (X-ray rcpair cross-complementing protein 1), OGG1 (8-oxoguanine DNA glycosylase 1), APE1 (apurinic/apyrimidinic endonuclease 1), and PARP1 (poly-(ADP-ribose) polymerase 1). Variations in these genes can decrease the efficiency of DNA repair, thereby increasing the likelihood of mutation accumuation (Gohil et al., 2023). Among various BER genes, XRCC1 Arg399Gln is one of the most studied polymorphisms, which is associated with elevated DNA strand breaks owing to diminished BER capacity. Studies have reported that this variant is linked to an increased risk of different cancers in populations exposed to smoking, air pollution, or occupational chemicals (Shen et al., 2005; Yi et al., 2013). Similarly, the OGG1 Ser326Cys variant reduces the ability to repair 8-oxoguanine, a common oxidative DNA lesion. This leads to accumulation of mutations and increased risk of diseases (Wang et al., 2023). When BER is impaired, exposure to environmental toxins that generate oxidative stress can lead to a more rapid onset or progression of various diseases.

4.3 Double-Strand Break Repair Gene Polymorphisms and Disease Risk

Double-strand breaks (DSBs) are among the most severe forms of DNA damage. They can occur indirectly due to exposure to chemical mutagens that generate reactive intermediates or stalled replication forks. In human cells, DSBs are mainly repaired through two pathways: homologous recombination (HR) and non-homologous end joining (NHEJ). Both pathways are essential for maintaining genomic stability. Key HR proteins include RAD51 and its paralogs, XRCC2 and XRCC3. These proteins help repair breaks accurately by promoting strand invasion and pairing with an undamaged sister chromatid. Genetic polymorphisms in HR genes can reduce repair efficiency. They may alter protein function or disrupt interactions within repair complexes. This can lower homologous recombination capacity and allow DSBs to persist longer (Nowacka-Zawisza et al., 2015; Rajagopal et al., 2022). For example, RAD51 G135C and XRCC3 Thr241Met variants affect repair activity. Certain genotypes with these variants show higher levels of DNA damage markers in human populations (Rajagopal et al., 2022). In the NHEJ pathway, core proteins such as XRCC4, LIG4, and the Ku70/Ku80 complex recognize DNA ends and ligate them. Polymorphisms in these genes can also affect DSB repair efficiency. This contributes to differences in chromosomal stability among individuals (Sishc & Davis, 2017).

5 Antioxidant Defense Gene Polymorphisms

The oxidative stress produced during xenobiotic metabolism can be counteracted by the cell's antioxidant defense systems. Many xenobiotics, especially those processed through cytochrome P450-dependent pathways, generate reactive oxygen species (ROS) as by-products of bioactivation. Enzymatic antioxidants, including superoxide dismutases (SODs), catalase (CAT), and glutathione peroxidases (GPXs), play a key role in neutralizing these ROS and maintaining cellular redox balance.

Genetic polymorphisms in antioxidant defense genes can alter gene expression, cellular distribution and catalytic efficiency of antioxidant enzymes. One of the most studied variants is SOD2 Val16Ala (rs4880), which modulates the transport of manganese superoxide dismutase to mitochondria and alters the efficiency of superoxide radical conversion within these organelles. Experimental and population studies suggest that this variant can affect mitochondrial oxidative stress levels, especially under conditions of elevated ROS, such as during xenobiotic exposure (Bastaki et al., 2006; Bresciani et al., 2015). Polymorphisms in the CAT gene have also been shown to modify catalase activity, which impacts the breakdown of hydrogen peroxide into water and oxygen. Reduced catalase activity can lead to hydrogen peroxide accumulation, increasing oxidative pressure on cellular macromolecules during xenobiotic metabolism (Hernández et al., 2013). Similarly, the GPX1 Pro198Leu (rs1050450) variant has been associated with decreased glutathione peroxidase activity, reducing the cell's capacity to detoxify hydrogen peroxide generated under oxidative stress (Wang et al., 2017).

6 Role of Gene Polymorphisms in Xenobiotic-Associated Human Diseases

6.1 Cancer

6.1.1 Lung Cancer

Lung cancer has been widely studied in relation to xenobiotic metabolism. Cigarette smoke contains many procarcinogens, including polycyclic aromatic hydrocarbons (PAHs) and nitrosamines. These compounds are first activated by Phase I enzymes, such as *CYP1A1*. They are then detoxified by Phase II enzymes, like *GSTM1* and *GSTT1*, and finally excreted from the body. Genetic variations in these enzymes can change their activity. This can affect how well the body processes these carcinogens. As a result, these polymorphisms can influence an individual's risk of developing lung cancer (Sun et al., 2020). For example, a North Indian study reported that the *GSTT1* null genotype was associated with an approximately threefold increased risk of lung cancer, and this risk further increased (up to 4.8-fold) among smokers carrying the null genotype. Combined null genotypes of *GSTM1* and *GSTT1*, as well as interactions with *GSTP1* variants, were also found to elevate cancer susceptibility (Tiwari et al., 2019).

Similarly, a South Indian cohort study demonstrated that *CYP1A1* polymorphisms (m1 and m2) were independently associated with lung cancer risk, with reported odds ratios reaching nearly 8.8-fold. In the same population, the *GSTT1* null genotype contributed to disease susceptibility when combined with smoking and elevated oxidative stress markers (Peddireddy et al., 2016). Studies conducted in Greek populations also observed a higher prevalence of rare genotype combinations involving *CYP1A1*, *GSTM1*, and *GSTT1* among smokers with lung cancer, suggesting cumulative effects of multiple gene variants (Dialyna et al., 2003). However, several studies have reported no significant associations for these polymorphisms, indicating population-specific effects. Polymorphisms in other xenobiotic metabolism genes, such as UGT1A7, have also been proposed as additional risk factors for lung cancer (Araki et al., 2005). Minina et al. (2022) further examined polymorphisms in *GSTM1* (deletion), APEX1 (rs1130409), XPD (rs13181), and NBS1 (rs1805794) among coal mine workers and reported that certain combinations of these gene variants increased susceptibility to lung cancer.

6.1.2 Colorectal Cancer

Colorectal cancer (CRC) has been associated with genetic polymorphisms in xenobiotic-metabolizing genes. The risk of this disease is influenced by interactions between dietary procarcinogens (e.g., heterocyclic amines in well-done meat) and xenobiotic metabolism genes. In this context, an association study examined single-nucleotide polymorphisms in Phase I and Phase II xenobiotic-metabolizing genes. The results suggested that *CYP1A1* and *CYP1B1* may be involved in colorectal cancer development and support the role of aromatic amine metabolism in colorectal carcinogenesis (Landi et al., 2005). A case-control study conducted among Caucasians, Japanese, and Native Hawaiians investigated the association between heterocyclic amine exposure from well-done meat and tobacco smoke and colorectal adenoma risk. The results revealed that individuals with rapid NAT2 activity and high heterocyclic amine exposure had an increased risk of colorectal cancer (Voutsinas et al., 2013). Similarly, another case–control study conducted in Iraq, including 200 colorectal cancer patients and 200 cancer-free controls, found that polymorphisms in *CYP1A1* and *CYP1B1* were associated with colorectal cancer risk, with the *CYP1A1* rs1048943 variant emerging as a potential risk marker (Ibrahem et al., 2021). The *GSTM1* polymorphism has also been associated with an increased risk of colorectal cancer, as observed in nonsmoking Saudi individuals (Sindi et al., 2022).

6.1.3 Breast Cancer

Studies have evaluated the effect of polymorphisms in Phase I and Phase II enzymes on the development of breast cancer. Variants in *CYP1A1* and *CYP1B1* have been shown to increase exposure to activated carcinogens. At the same time, null variants of *GSTM1* and *GSTT1* reduce the body's detoxification capacity. Together, these changes can increase the risk of breast cancer (Justenhoven, 2012). A large meta-analysis of studies from several countries found that *GSTM1* and *GSTT1* null polymorphisms were statistically linked to higher breast cancer risk. This supports the idea that gene variants involved in xenobiotic metabolism can affect cancer susceptibility (Hussain et al., 2018). Another study examined the prevalence of *GSTM1*, *GSTT1*, and *CYP2E1* polymorphisms in agricultural workers exposed to pesticides. The results suggest that changes in these detoxifying genes, combined with increased oxidative stress, may contribute to cancer development in this population (Pandiyan et al., 2023). Polymorphisms in DNA repair genes have also been associated with breast cancer risk. For instance, significant correlations have been observed for XRCC1 Arg399Gln, XPD Lys751Gln, and RAD51–4719A/T with increased susceptibility to the disease (Smolarz et al., 2019).

6.1.4 Leukemia and Lymphoma

Benzene, a common environmental pollutant and industrial solvent, is a potent Group 1 human carcinogen and has been associated with blood disorders such as acute myeloid leukemia, lymphocytic leukemia, and non-Hodgkin lymphoma. Genetic polymorphisms in *CYP2E1*, *GSTT1*, and *GSTM1* have been reported to influence benzene metabolism and the generation of toxic metabolites, thereby affecting individual susceptibility to DNA damage and the development of these diseases (Ramírez-Lopera et al., 2021). Among pesticide-exposed agricultural workers, null variants of *GSTT1 and GSTM1* have been found to be associated with significantly increased risks of lymphoma and breast cancer, suggesting that detoxification pathways modify the oncogenic effects of occupational chemical exposure (Pandiyan et al., 2023). Genes encoding the arylamine N-acetyltransferases NAT1 and NAT2 show high genetic variability across populations. Specific single-nucleotide polymorphisms, particularly haplotypes resulting in rapid NAT1 and slow NAT2 activity, have been linked to an increased risk of acute lymphoblastic leukemia (ALL) in children (Hernández-González et al., 2018).

6.1.5 Other Cancers

It has been observed that mutations in genes involved in xenobiotic biotransformation and providing antioxidant defense can also increase the risk of other cancers. For example, Yu et al. (2014) have reported that CYP2E1 gene variations are associated with the risk of esophageal cancer (Yu et al., 2014). Further, NAT2 polymorphisms have been found to be linked to bladder cancer, which is due to reduced detoxification of harmful compounds (Zhang et al., 2022). In addition, SOD2 and CAT gene variations have been linked with prostate cancer, highlighting the importance of antioxidant defense in preventing cancer development (Zhang et al., 2023). GST gene polymorphisms are also associated with renal cell carcinoma, further supporting their role in detoxification and cancer protection (Zhu et al., 2019). Overall, these genetic variations may increase cancer risk by weakening detoxification and antioxidant defense systems (Zhang et al., 2014).

6.1.6 Non-cancer Diseases: Neurodegenerative Diseases

Neurodegenerative diseases, such as Parkinson's disease (PD), may result from complex causes involving both genetics and environmental factors. Studies have shown that exposure to pesticides and solvents is associated with a higher risk of Parkinson's disease. This risk may be partly due to variations in genes involved in xenobiotic detoxification. A case-control study by Elbaz et al. (2004) reported that individuals with poor metabolizer genotypes of CYP2D6 who were exposed to pesticides had a higher risk of developing PD. Another study examined the link between defects in detoxifying and DNA repair enzymes and the risk of multiple sclerosis

(MS). It found that individuals with OGG1 Ser/Cys or Ser/Cys + Cys/Cys genotypes, as well as the XRCC1 Arg/Gln + Gln/Gln genotype, had increased susceptibility to MS (Karahalil et al., 2015). Cytochrome P450 2D6 (CYP2D6) is a highly active and polymorphic enzyme that metabolizes Parkinsonian toxins and anti-Parkinson's drugs. Individuals with the poor metabolizer phenotype of CYP2D6 were found to have higher genetic susceptibility to PD (ur Rasheed et al., 2017). Studies have also shown that null genotypes or single-nucleotide polymorphisms in glutathione S-transferases (GSTs) can influence the risk of neurological disorders. For example, the GSTM1 null genotype and GSTO1-1 variant have been linked to an increased risk of Alzheimer's disease. Deletions of GSTM1 and GSTT1 were also associated with lower survival rates in glioma patients (Dasari et al., 2018).

6.2 Reproductive and Developmental Disorders

Research has shown that maternal smoking is linked to variations in xenobiotic metabolism genes, such as *GSTM1* and *GSTT1*. These genetic differences may increase the risk of low birth weight (LBW) and intrauterine growth restriction (IUGR). Several studies have reported that children of smoking mothers who carry the *GSTM1* null genotype experience more developmental problems. This is likely because their bodies are less able to detoxify harmful chemicals from smoke (Grazuleviciene et al., 2009). Genetic variations in xenobiotic-metabolizing enzymes, such as CYP1A1, can also affect male reproductive health. Certain variants have been found more frequently in infertile men (Schuppe et al., 2000). In addition, single-nucleotide polymorphisms (SNPs) in genes that process environmental endocrine disruptors (EEDs) may increase the risk of male genital malformations, including cryptorchidism and hypospadias (Qin et al., 2012).

6.3 Pulmonary Disorders

Gene polymorphisms can contribute to the development of several respiratory diseases. For example, studies have shown that variations in *GSTT1*, *GSTM1*, and *GSTP1* are linked to both atopic and non-atopic asthma (Tamer et al., 2004). This may be due to reduced GST activity, which allows reactive metabolites to build up, causing oxidative stress and cellular damage in affected individuals. Pulmonary emphysema, often caused by smoking, may also be influenced by genetic differences in enzymes that process oxidative compounds. For instance, *GSTT1* polymorphisms have been associated with variations in the risk and severity of emphysema (Kukkonen et al., 2011). A case-control study found that *GSTM1* and *GSTT1* null genotypes were more common in children with asthma. The absence of these genes was linked to higher oxidative stress, which may increase the risk or severity of bronchial asthma (Kumar et al., 2022). Genetic variations in xenobiotic-metabolizing enzymes may also affect susceptibility to chronic rhinosinusitis (CRS). Variants in CYP1A and GST genes have been shown to be significantly associated with allergic CRS, especially in patients with asthma (Jangala et al., 2023).

6.4 Cardiovascular Diseases

Mutations in genes involved in xenobiotic biotransformation may also influence the development of heart diseases. In a cross-sectional study of people exposed to polycyclic aromatic hydrocarbons near a natural gas plant, researchers examined the impact of polymorphisms in *CYP1A1*, *GSTM1*, and *GSTT1* on cardiovascular risk and lipid levels. The study found that carriers of the functional *GSTT1* allele had higher total cholesterol and triglycerides compared to null carriers. This suggests that *GSTT1* may affect lipid metabolism, which is an important factor in cardiovascular risk (Pašalić & Marinković, 2017). Another case-control study reported that specific genetic combinations involving *GSTP1*, *CYP1A1*, and *CYP1A2* were associated with disease risk in women and their children, indicating that interactions between these genes may contribute to congenital heart disease (Tsepokina et al., 2021). In addition, the *GSTP1* Val/Val genotype and *GSTT1* and *GSTM1* null variants were linked to metabolic disturbances and changes in lipid profiles, including HDL, LDL, apolipoprotein A1, triglycerides, total cholesterol, and lipoprotein(a). These lipid changes are recognized cardiovascular risk factors and may increase the likelihood of coronary heart disease, heart failure, and myocardial infarction (Grussy et al., 2023).

7 Other Diseases Influenced by Xenobiotic Metabolism Gene Polymorphisms

In addition to cancer and other well-known non-cancer conditions, variants in xenobiotic metabolism genes are increasingly recognized for their role in modifying susceptibility to a wide range of diseases. These gene variations can influence the risk of autoimmune disorders by affecting how the body processes foreign compounds. For instance, polymorphisms in CYP1A1 (such as the m1 and m4 alleles), the COMT H108L variant, and the GSTT1 null genotype have been shown to impair both Phase I and Phase II detoxification pathways, potentially altering the body's ability to manage harmful substances. These impaired pathways have been linked to a higher risk of systemic lupus erythematosus (SLE), suggesting that reduced ability to metabolize environmental chemicals may contribute to autoimmunity (Rupasree et al., 2013). A meta-analysis of *CYP2D6* polymorphisms in Caucasian populations showed that the *CYP2D64* and *CYP2D63* alleles and associated poor/intermediate metabolizer phenotypes were associated with significantly increased susceptibility to several autoimmune diseases, particularly SLE and ankylosing spondylitis (AS) (Lee and Bae, 2017).

Beyond autoimmunity, variants in xenobiotic detoxification pathways may modify susceptibility to infectious diseases as well. For example, a case-control study demonstrated that polymorphisms in *GSTM1*, *GSTT1*, *NAT2*, *CYP2E1*, and ABCB1 were associated with an altered risk for pulmonary tuberculosis. These findings showed that *GSTM1* deletion was linked to increased susceptibility and *GSTT1* deletion to reduced risk in one Russian population, indicating that host xenobiotic processing capacity may influence infection susceptibility or progression (Alymenko et al., 2024). Reproductive and developmental outcomes have also been connected to xenobiotic gene variants. Polymorphisms in *CYP2E1* have been linked to conditions like preeclampsia. In this disorder, changes in the detoxification of environmental pollutants and oxidative stress products may play a role in its development (Hu et al., 2025). Similarly, genetic variants influencing estrogen-metabolizing genes and hormone regulation have been linked to increased risk of adenomyosis, a gynecological disorder. For example, polymorphisms in estrogen-metabolizing genes, including *CYP1A1 (M1)*, *CYP1A2*1F*, and *CYP19* (aromatase), were found more frequently in women with adenomyosis than in healthy controls. This suggests that changes in estrogen metabolism may increase susceptibility to the disease (Artymuk et al., 2012). Furthermore, mutations in genes involved in xenobiotic biotransformation have also been found to be linked to the development of diabetes. This is because mutations in these genes alter the body's ability to manage oxidative stress and detoxification processes. A study by Yadav et al. (2020) has reported that polymorphisms in glutathione S-transferase genes (GSTM1 and GSTT1) is linked to an increased risk of type 2 diabetes mellitus, which is attributed to the reduced antioxidant defense and impaired detoxification capacity. In addition, genetic variations in antioxidant enzyme genes such as SOD2, GPX1, and CAT have also been found to be associated with susceptibility to type 2 diabetes mellitus (Vats et al., 2015).

In addition to these disease risks, genetic variations in xenobiotic-metabolizing enzymes, such as CYPs, can strongly influence how individuals process drugs. This effect can be further modified by environmental exposures. For example, clopidogrel, an antiplatelet prodrug, must be activated by *CYP2C19* to work effectively. People carrying poor metabolizer alleles, a type of genetic variant, may not activate the drug enough, reducing its effectiveness and increasing the risk of blood clots. Conversely, ultra-rapid metabolizer variants can activate it too quickly, raising the risk of bleeding. Similar patterns have been seen with other CYP enzymes, including *CYP1A2* and *CYP3A4/5*, where factors such as smoking, diet, and chemical exposure can further influence drug metabolism (Ali et al., 2025).

8 Gene–Environment Interaction and Population Risk

Individuals exposed to the same xenobiotics may show very different health responses due to genetic differences. Variations in xenobiotic-metabolizing genes can change how the body processes chemicals, which can alter disease risk. For example, a case-control study of communities exposed to organochlorine pesticides found that SNPs in *CYP1A1*, *CYP2B6*, *CYP2D6*, *CYP2C19*, *GSTM1*, and *GSTT1* were linked to adverse health effects. People carrying high-risk variants were more vulnerable when exposure levels were high (Altynova et al., 2023). Similarly, a systematic review reported that polymorphisms in *CYP3A4*, *CYP2C9*, *CYP2B6*, and *CYP1A2* can influence how pesticide exposure affects child neurodevelopment, indicating that genetic differences can have significant impacts even early in life (Chilipweli et al., 2025). These studies emphasize the critical role of gene–environment interactions in determining individual susceptibility to xenobiotics and highlight the importance of considering genetic profiles in personalized health risk assessment and disease prevention strategies.

At the population level, disease risk depends on both the frequency of high-risk genetic variants and the extent of environmental exposure. Populations in which high-risk alleles are common and exposure to pollutants is high tend to show greater susceptibility to diseases. For instance, a case-control study by Elbaz et al. (2004) reported that individuals with a poor metabolizer *CYP2D6* genotype who had high exposure to pesticides had nearly double the risk of developing Parkinson's disease compared to non-exposed carriers. This shows that both the genetic variant and environmental exposure contribute to higher disease risk. Similarly, a case-control study in Algeria found that *GSTT1 null* and *GSTM1/GSTT1* double-null genotypes demonstrated an association with prostate cancer risk, showing how specific gene variants can amplify the effects of environmental toxins on population health (Medjani et al., 2020). Population-based studies on air pollution also demonstrate that genetic differences in oxidative stress and detoxification genes influence disease susceptibility. Individuals carrying variants in *GSTM1*, *GSTP1*, *NQO1*, *CAT*, and *TNF* experience stronger negative effects on lung function and higher rates of asthma when exposed to pollutants such as ozone and particulate matter (Romieu et al., 2010). Understanding the interplay between genetic variants and environmental exposures allows the identification of susceptible individuals, assessing community risk, and implementing focused health interventions.

9 Conclusion

Genetic polymorphisms in genes related to xenobiotic metabolism and DNA repair play a key role in determining how individuals and populations respond to foreign chemicals. These variations can affect how efficiently the body processes and detoxifies xenobiotics, sometimes leading to harmful effects instead of effective protection. Changes in Phase I and Phase II metabolic genes, such as CYPs, GSTs, NATs, and UGTs, have been linked to a wide range of diseases, including cancer, cardiovascular, respiratory, neurodegenerative, and reproductive disorders. Understanding these genetic differences has important implications for both clinical practice and public health. It can help guide personalized medicine, such as tailoring drug choice and dosage, identifying individuals at high risk in occupational or environmental settings, and shaping public health policies to protect vulnerable populations.

Future research should investigate multiple genes together and integrate information from genomics, epigenomics, transcriptomics, proteomics, and metabolomics to better understand how genes and environmental exposures interact. Large population studies and AI-driven models may help improve predictions of disease risk. At the same time, strong ethical and regulatory safeguards are essential to protect privacy, prevent discrimination, and ensure fair access, so that genetic information can be used safely to improve health outcomes.

References

Ali, H., Hanna, M. M., Alziny, N., Mahmoud, S., Borham, A., Mustafa, A., & Abdelnaser, A. (2025, October). The impact of the exposome on cytochrome P450-mediated drug metabolism. *Frontiers in Pharmacology, 16*, 1639646.

Alkharfy, K. M., Alghamdi, A. M., Bagulb, K. M., Al-Jenoobi, F. I., Al-Mohizea, A. M., Al-Muhsen, S., Halwani, R., Parvez, M. K. & Al-Dosari, M. S. (2013). Distribution of selected gene polymorphisms of UGT1A1 in a Saudi population. *Archives of Medical Science, 9*(4), 731–738.

Alnasser, S. M. (2025). The role of glutathione S-transferases in human disease pathogenesis and their current inhibitors. *Genes and Diseases, 12*(4), 101482.

Altynova, N., Khamdiyeva, O., Garshin, A., Baratzhanova, G., Amirgaliyeva, A., Seisenbayeva, A., Abylkassymova, G., Yergali, K., Tolebaeva, A., Skvortsova, L., Zhunussova, G., Bekmanov, B, Cakir-Kiefer, C., Djansugurova, L. (2023, November 21). Case-control study of the association between single nucleotide polymorphisms of genes involved in xenobiotic detoxification and antioxidant protection with the long-term influence of organochlorine pesticides on the population of the Almaty region. *Toxics, 11*(12), 948.

Alymenko, M. A., Valiev, R. S., Valiev, N. R., Balobanova, N. P., Batishchev, A. V., Tragira, I. N., Lipatov, V. A., Polonikov, A. V., Kolomiets, V. M., Volkova, S. N., & Mal, G. S. (2024, May 18). Association of genotypes of biotransformation enzymes with susceptibility to chronic pulmonary tuberculosis. *Consilium Medicum, 26*(3), 193–198.

Araki, J., Kobayashi, Y., Iwasa, M., Urawa, N., Gabazza, E. C., Taguchi, O., Kaito, M., & Adachi, Y. (2005, October 01). Polymorphism of UDP-glucuronosyltransferase 1A7 gene: A possible new risk factor for lung cancer. *European Journal of Cancer, 41*(15), 2360–2365.

Artymuk, N. V., Gulyayeva, L. F., Zotova, O. A., & Khvostova, Y. P. (2012, December 15). The role of polymorphisms genes of detoxification of xenobiotics in the development of endometriosis. *Journal of Obstetrics and Women's Diseases, 61*(6), 18–24.

Babalola, C. P., Kotila, O., & Iwuchukwu, O. F. (2023). The expanding role of gene-based prescribing for phase II drug-metabolizing enzymes. *American Journal of Pharmacotherapy and Pharmaceutical Sciences, 2.*

Bastaki, M., Huen, K., Manzanillo, P., Chande, N., Chen, C., Balmes, J. R., Tager, I. B., & Holland, N. (2006). Genotype–activity relationship for Mn-superoxide dismutase, glutathione peroxidase 1 and catalase in humans. *Pharmacogenetics and Genomics, 16*(4), 279–286.

Bresciani, G., da Cruz, I. B. M., & González-Gallego, J. (2015). Manganese superoxide dismutase and oxidative stress modulation. *Advances in Clinical Chemistry, 68*, 87–130.

Chilipweli, P. M., Kidenya, B. R., Nyanza, E. C., Ngowi, A. V., & Manji, K. (2025, October 13). The role of cytochrome P450 genetic variants in pesticide metabolism and the risk of child neurodevelopment: A systematic review. *Frontiers in Pediatrics, 13*, 1651916.

Dasari, S., Gonuguntla, S., Ganjayi, M. S., Bukke, S., Sreenivasulu, B., & Meriga, B. (2018, December). Genetic polymorphism of glutathione S-transferases: Relevance to neurological disorders. *Pathophysiology, 25*(4), 285–292.

Dean, L. (2012). Irinotecan therapy and UGT1A1 genotype. In V. M. Pratt, S. A. Scott, M. Pirmohamed et al. (Eds.), *Medical genetics summaries*. National Center for Biotechnology Information (US). https://www.ncbi.nlm.nih.gov/books/NBK294473/

Dialyna, I. A., Miyakis, S., Georgatou, N., & Spandidos, D. A. (2003, November). Genetic polymorphisms of CYP1A1, GSTM1 and GSTT1 genes and lung cancer risk. *Oncology Reports, 10*(6), 1829–1835.

Dutra, C. A., Teixeira, R. L. D. F., Lopes, M. Q. P., Silva, V. D. M., Suffys, P. N., Carvalho, R. D. S., Moreira, A. R., Santos, A. R., & Kritski, A. L. (2024). Determination of NAT2 genotypes in a cohort of patients with suspected TB in the state of Rio de Janeiro. *Pharmaceutics, 16*(7), 917.

Elbaz, A., Levecque, C., Clavel, J., Vidal, J. S., Richard, F., Amouyel, P., Alpérovitch, A., Chartier-Harlin, M. C., & Tzourio, C. (2004, March). CYP2D6 polymorphism, pesticide exposure, and Parkinson's disease. *Annals of Neurology, 55*(3), 430–434.

Esteves, F., Rueff, J., & Kranendonk, M. (2021). The central role of cytochrome P450 in xenobiotic metabolism—A brief review on a fascinating enzyme family. *Journal of Xenobiotics, 11*(3), 94–114.

Gohil, D., Sarker, A. H., & Roy, R. (2023). Base excision repair: Mechanisms and impact in biology, disease, and medicine. *International Journal of Molecular Sciences, 24*(18), 14186.

Grazuleviciene, R., Danileviciute, A., Nadisauskiene, R., & Vencloviene, J. (2009, March). Maternal smoking, GSTM1 and GSTT1 polymorphism and susceptibility to adverse pregnancy outcomes. *International Journal of Environmental Research and Public Health, 6*(3), 1282–1297.

Grussy, K., Łaska, M., Moczurad, W., Król-Kulikowska, M., & Ściskalska, M. (2023, November). The importance of polymorphisms in the genes encoding glutathione S-transferase isoenzymes in development of selected cancers and cardiovascular diseases. *Molecular Biology Reports, 50*(11), 9649–9661.

Guengerich, F. P. (2008, January 21). Cytochrome P450 and chemical toxicology. *Chemical Research in Toxicology, 21*(1), 70–83.

Gutierrez-Virgen, J. E., Pina-Pozas, M., Hernandez-Tobias, E. A., Taja-Chayeb, L., Lopez-Gonzalez, M. D. L., Meraz-Rios, M. A., & Gomez, R. (2023). NAT2 global landscape: Genetic diversity and acetylation statuses from a systematic review. *PLoS One, 18*(4), e0283726.

Habil, M. R., & Hein, D. W. (2025). Mini-review: Human N-acetyltransferase genetic polymorphisms in susceptibility to aromatic amine and alkylaniline genotoxicity. *Journal of Pharmacology and Experimental Therapeutics*, 103670.

Hall, K. T., Loscalzo, J., & Kaptchuk, T. J. (2019). Systems pharmacogenomics–gene, disease, drug and placebo interactions: A case study in COMT. *Pharmacogenomics, 20*(7), 529–551.

Hernández, A. F., Lacasaña, M., Gil, F., Rodríguez-Barranco, M., Pla, A., & López-Guarnido, O. (2013). Evaluation of pesticide-induced oxidative stress from a gene–environment interaction perspective. *Toxicology, 307*, 95–102.

Hernández-González, O., Ortiz-Zamudio, J. J., Rodríguez-Pinal, C. J., Alvarado-Morales, I., Martínez-Jiménez, V. D. C., Salazar-González, R. A., Correa-González, L. C., Gómez, R., Portales-Pérez, D. P., & Milán-Segovia, R. D. C. (2018). Genetic polymorphisms of arylamine N-acetyltransferases 1 and 2 and the likelihood of developing pediatric acute lymphoblastic leukemia. *Leukemia & lymphoma, 59*(8), 1968–1975.

Hu, K., Liu, Q., Liu, X., Bai, H., Wu, Y., & Fan, P. (2025, June). Association of xenobiotic-metabolizing gene cytochrome P450 2E1 variants with preeclampsia in Chinese women. *Reproductive Sciences, 11*, 1.

Hussain, T., Alrokayan, S., Upasna, U., Pavithrakumari, M., Jayapriya, J., Kutala, V. K., & Naushad, S. M. (2018, June). Meta-analysis of genetic polymorphisms in xenobiotic metabolizing enzymes and their association with breast cancer risk. *Journal of Genetics, 97*(2), 523–537.

Ibrahem, S. Q., Al-Dalawi, Z. T., & Bahaaldin, A. S. (2021, April). Sequence polymorphism in xenobiotic metabolising genes in Iraqi colorectal cancer patients. *Asian Pacific Journal of Cancer Prevention, 22*(4), 1203.

Iyanagi, T. (2007, January 01). Molecular mechanism of phase I and phase II drug-metabolizing enzymes: Implications for detoxification. *International Review of Cytology, 1*(260), 35–112.

Jangala, M., Manche, S. K., Katika, M. M., Koralla, R. M., & Akka, J. (2023, March 07). Association of CYP1A2 and GST gene variants with asthma in cases presenting with allergic chronic rhinosinusitis. *Egyptian Journal of Medical Human Genetics, 24*(1), 22.

Justenhoven, C. (2012). Polymorphisms of phase I and phase II enzymes and breast cancer risk. *Frontiers in Genetics, 3*, 258.

Karahalil, B., Orhan, G., & Ak, F. (2015, December 01). The impact of detoxifying and repair gene polymorphisms and the levels of serum ROS in the susceptibility to multiple sclerosis. *Clinical Neurology and Neurosurgery, 1*(139), 288–294.

Kassogue, Y., Diakite, B., Kassogue, O., Konate, I., Tamboura, K., Diarra, Z., Dehbi, H., Nadifi, S., Traore, C. B., Dao, S., & Doumbia, S. (2020). Genetic polymorphism of drug metabolism enzymes (GSTM1, GSTT1 and GSTP1) in the healthy Malian population. *Molecular Biology Reports, 47*(1), 393–400.

Kukkonen, M. K., Hämäläinen, S., Kaleva, S., Vehmas, T., Huuskonen, M. S., Oksa, P., Vainio, H., Piirilä, P., & Hirvonen, A. (2011, December 01). Genetic polymorphisms of xenobiotic-metabolizing enzymes influence the risk of pulmonary emphysema. *Pharmacogenetics and Genomics, 21*(12), 876–883.

Kumar, A., Ahmed, T., Parvez, M. K., Banerjee, B. D., Maqusood, M., & Jagirdar, R. M. (2022, May 01). Glutathione S-transferase gene polymorphism and asthma: A case-control study in a pediatric population. *Pharmacogenomics, 23*(7), 405–413.

Kurogi, K., Rasool, M. I., Alherz, F. A., El Daibani, A. A., Bairam, A. F., Abunnaja, M. S., Yasuda, S., Wilson, L. J., Hui, Y., & Liu, M. C. (2021). SULT genetic polymorphisms: physiological, pharmacological and clinical implications. *Expert Opinion on Drug Metabolism & Toxicology, 17*(7), 767–784.

Landi, S., Gemignani, F., Moreno, V., Gioia-Patricola, L., Chabrier, A., Guino, E., Navarro, M., De Oca, J., Capellà, G., Canzian, F., & Bellvitge Colorectal Cancer Study Group. (2005, August 01). A comprehensive analysis of phase I and phase II metabolism gene polymorphisms and risk of colorectal cancer. *Pharmacogenetics and Genomics, 15*(8), 535–546.

Lee, Y. H., & Bae, S. C. (2017, February 17). Association between functional CYP2D6 polymorphisms and susceptibility to autoimmune diseases: A meta-analysis. *Immunological Investigations, 46*(2), 109–122.

Lee, M. S., Liu, C. Y., Su, L., & Christiani, D. C. (2015). Polymorphisms in ERCC1 and ERCC2/XPD genes and carcinogen DNA adducts in human lung. *Lung Cancer, 89*(1), 8–12.

Li, Q., Damish, A. W., Frazier, Z., Liu, D., Reznichenko, E., Kamburov, A., Bell, A., Zhao, H., Jordan, E. J., Gao, S. P., & Ma, J. (2019). ERCC2 helicase domain mutations confer nucleotide excision repair deficiency and drive cisplatin sensitivity in muscle-invasive bladder cancer. *Clinical Cancer Research, 25*(3), 977–988.

Lopes, M. Q. P., Teixeira, R. L. F., Cabello, P. H., Nery, J. A. C., Sales, A. M., Nahn Jr., E. P., Moreira, M. V., Stahlke, E. V. R., Possuelo, L. G., Rossetti, M. L. R., & Rabahi, M. F. (2023). Human N-acetyltransferase 2 (NAT2) gene variability in Brazilian populations from different geographical areas. *Frontiers in Pharmacology, 14*, 1278720.

Mano, E. C., Scott, A. L., & Honorio, K. M. (2018, August 01). UDP-glucuronosyltransferases: Structure, function and drug design studies. *Current Medicinal Chemistry, 25*(27), 3247–3255.

Marteijn, J. A., Lans, H., Vermeulen, W., & Hoeijmakers, J. H. (2014). Understanding nucleotide excision repair and its roles in cancer and ageing. *Nature Reviews Molecular Cell Biology, 15*(7), 465–481.

Medjani, S., Chellat-Rezgoune, D., Kezai, T., Chidekh, M., Abadi, N., & Satta, D. (2020, December). Association of CYP1A1, GSTM1 and GSTT1 gene polymorphisms with risk of prostate cancer in Algerian population. *The African Journal of Urology, 26*(1), 44.

Minina, V., Timofeeva, A., Torgunakova, A., Soboleva, O., Bakanova, M., Savchenko, Y., Voronina, E., Glushkov, A., Prosekov, A., & Fucic, A. (2022, February 09). Polymorphisms in DNA repair and xenobiotic biotransformation enzyme genes and lung cancer risk in coal mine workers. *Life, 12*(2), 255.

Mir, R., Bhat, M., Javid, J., Jha, C., Saxena, A., & Banu, S. (2018). Potential impact of COMT-rs4680 G> A gene polymorphism in coronary artery disease. *Journal of Cardiovascular Development and Disease, 5*(3), 38.

Nebert, D. W., & Dalton, T. P. (2006, December 01). The role of cytochrome P450 enzymes in endogenous signalling pathways and environmental carcinogenesis. *Nature Reviews Cancer, 6*(12), 947–960.

Nowacka-Zawisza, M., Wiśnik, E., Wasilewski, A., Skowrońska, M., Forma, E., Bryś, M., Różański, W., & Krajewska, W. M. (2015). Polymorphisms of homologous recombination RAD51, RAD51B, XRCC2, and XRCC3 genes and the risk of prostate cancer. *Analytical Cellular Pathology (Amsterdam), 2015*, 828646.

Pandiyan, A., Lari, S., Vanka, J., Kumar, B. S., Ghosh, S., Jee, B., & Jonnalagadda, P. R. (2023). Genetic polymorphism in xenobiotic metabolising genes and increased oxidative stress among pesticides exposed agricultural workers diagnosed with cancers. *Asian Pacific Journal of Cancer Prevention, 24*(11), 3795.

Pašalić, D., & Marinković, N. (2017, March 23). Genetic polymorphisms of the CYP1A1, GSTM1, and GSTT1 enzymes and their influence on cardiovascular risk and lipid profile in people who live near a natural gas plant. *Arhiv za Higijenu Rada i Toksikologiju, 68*(1), 46–52.

Peddireddy, V., Badabagni, S. P., Gundimeda, S. D., Mamidipudi, V., Penagaluru, P. R., & Mundluru, H. P. (2016, April 18). Association of CYP1A1, GSTM1 and GSTT1 gene polymorphisms with risk of non-small cell lung cancer in Andhra Pradesh region of South India. *European Journal of Medical Research, 21*(1), 17.

Prost, R., & Płaziński, W. (2025). Natural polymorphic variants in the CYP450 superfamily: A review of potential structural mechanisms and functional consequences. *International Journal of Molecular Sciences, 26*(16), 7797.

Qin, X. Y., Kojima, Y., Mizuno, K., Ueoka, K., Massart, F., Spinelli, C., Zaha, H., Okura, M., Yoshinaga, J., Yonemoto, J., & Kohri, K. (2012, July). Association of variants in genes involved in environmental chemical metabolism and risk of cryptorchidism and hypospadias. *Journal of Human Genetics, 57*(7), 434–441.

Rajagopal, T., Seshachalam, A., Rathnam, K. K., Talluri, S., Venkatabalasubramanian, S., & Dunna, N. R. (2022). Homologous recombination DNA repair gene RAD51, XRCC2 & XRCC3 polymorphisms and breast cancer risk in South Indian women. *Plos One, 17*(1), e0259761.

Ramírez-Lopera, V., Uribe-Castro, D., Bautista-Amorocho, H., Silva-Sayago, J. A., Mateus-Sánchez, E., Ardila-Barbosa, W. Y., & Pérez-Cala, T. L. (2021, September). The effects of genetic polymorphisms on benzene-exposed workers: A systematic review. *Health Science Reports, 4*(3), e327.

Romieu, I., Moreno-Macias, H., & London, S. J. (2010, May 01). Gene by environment interaction and ambient air pollution. *Proceedings of the American Thoracic Society, 7*(2), 116–122.

Rupasree, Y., Naushad, S. M., Rajasekhar, L., & Kutala, V. K. (2013, October 01). Association of genetic variants of xenobiotic metabolic pathway with systemic lupus erythematosus. *Indian Journal of Biochemistry & Biophysics, 50*(5), 447–452.

Sadafi, S., Choubsaz, P., Kazemeini, S. M. M., Imani, M. M., & Sadeghi, M. (2024). Glutathione S-transferase theta 1 (GSTT1) deletion polymorphism and susceptibility to head and neck carcinoma: A systematic review with five analyses. *BMC Cancer, 24*(1), 885.

Salazar-González, R. A., Doll, M. A., & Hein, D. W. (2020). Human arylamine N-acetyltransferase 2 genotype-dependent protein expression in cryopreserved human hepatocytes. *Science and Reports, 10*(1), 7566.

Schuppe, H. C., Wieneke, P., Donat, S., Fritsche, E., Köhn, F. M., & Abel, J. (2000, September). Xenobiotic metabolism, genetic polymorphisms and male infertility. *Andrologia, 32*(4–5), 255–262.

Shen, J., Gammon, M. D., Terry, M. B., Wang, L., Wang, Q., Zhang, F., Teitelbaum, S. L., Eng, S. M., Sagiv, S. K., Gaudet, M. M., & Neugut, A. I. (2005). Polymorphisms in XRCC1 modify the association between polycyclic aromatic hydrocarbon-DNA adducts, cigarette smoking, dietary antioxidants, and breast cancer risk. *Cancer Epidemiology Biomarkers & Prevention, 14*(2), 336–342.

Sindi, I. A., Babalghith, A. O., Tayeb, M. T., Mufti, A. H., Naffadi, H., Ekram, S. N., Elhawary, E. N., Alenezi, M., & Elhawary, N. A. (2021, May). Risk of colorectal carcinoma may predispose to the genetic variants of the GST, CYP450, and TP53 genes among nonsmokers in the Saudi community. *International Journal of General Medicine*, 1311–1323.

Sishc, B. J., & Davis, A. J. (2017). The role of the core non-homologous end joining factors in carcinogenesis and cancer. *Cancers, 9*(7), 81.

Smolarz, B., Michalska, M.M., Samulak, D., Romanowicz, H. and Wójcik, L. (2019). Polymorphism of DNA repair genes in breast cancer. *Oncotarget, 10*(4), 527.

Spivak, G. (2015). Nucleotide excision repair in humans. *DNA Repair, 36*, 13–18.

Stoian, I. A. M., Vlad, A., Gilca, M. & Dragos, D. (2025). Modulation of glutathione-s-transferase by phytochemicals: To activate or inhibit—That is the question. *International Journal of Molecular Sciences, 26*(15), 7202.

Sugatani, J. (2013). Function, genetic polymorphism, and transcriptional regulation of human UDP-glucuronosyltransferase (UGT) 1A1. *Drug Metabolism and Pharmacokinetics, 28*(2), 83–92.

Suiko, M., Kurogi, K., Hashiguchi, T., Sakakibara, Y., & Liu, M. C. (2017). Updated perspectives on the cytosolic sulfotransferases (SULTs) and SULTmediated sulfation. *Bioscience, Biotechnology, and Biochemistry, 81*(1), 63–72.

Sun, X., He, Y., Li, Y., Liu, X., Wu, Y., Zhang, J., et al. (2020, February 10). Association of CYP1A1 and GSTM1 polymorphisms with risk of lung cancer: A meta-analysis. *Gene, 10*(727), 144162.

Tamer, L., Calikoğlu, M., Ates, N. A., Yildirim, H., Ercan, B., Saritas, E., Ünlü, A. & Atik, U. (2004). Glutathione-S-transferase gene polymorphisms (GSTT1, GSTM1, GSTP1) as increased risk factors for asthma. *Respirology, 9*(4), 493–498.

Tiwari, S., Vijayaraghavalu, S. & Kumar, M. (2019). Genetic polymorphisms of xenobiotic metabolizing genes (GSTM1, GSTT1, GSTP1), gene-gene interaction with association to lung cancer risk in North India; a case control study. *Asian Pacific Journal of Cancer Prevention, 20*(9), 2707.

Tsepokina, A., Shmulevich, S., Ponasenko, A., & Shabaldin, A. (2021). Genetic predisposition to the development of congenital heart diseases: Role of xenobiotic biotransformation genes. *Birth Defects Research, 113*(7), 579–588.

ur Rasheed, M. S., Mishra, A. K. & Singh, M. P. (2017). Cytochrome P450 2D6 and Parkinson's disease: Polymorphism, metabolic role, risk and protection. *Neurochemical Research, 42*(12), 3353–3361.

Vats, P., Sagar, N., Singh, T. P., & Banerjee, M. (2015). Association of superoxide dismutases (SOD1 and SOD2) and glutathione peroxidase 1 (GPx1) gene polymorphisms with type 2 diabetes mellitus. *Free Radical Research, 49*(1), 17–24.

Villegas Sánchez, V., Chávez Pacheco, J. L., Palacios Arreola, M. I., Sierra-Vargas, M. P., Colín Godinez, L. A., Ahumada Topete, V. H., Fernández Plata, R., Higuera-Iglesias, A., Lara-Lemus, R., Aquino-Gálvez, A., & Torres-Espíndola, L. M. (2025). A systematic review of genetic variants in glutathione S-transferase genes and their dual role in SARS-CoV-2 pathogenesis: from acute respiratory complications to long COVID. *Antioxidants, 14*(8), 912.

Voutsinas, J., Wilkens, L. R., Franke, A., Vogt, T. M., Yokochi, L. A., Decker, R., & Le Marchand, L. (2013). Heterocyclic amine intake, smoking, cytochrome P450 1A2 and N-acetylation phenotypes, and risk of colorectal adenoma in a multiethnic population. *Gut, 62*(3), 416–422.

Wang, W., Xu, B., Yu, L., Zhang, L., Xu, Y., & Li, H. (2023, March). Association of SOD2 and CAT gene polymorphisms with risk of cardiovascular disease: A meta-analysis. *Cardiovascular Toxicology, 23*(3), 271–284.

Wang, X., Jia, P., Zhang, Y., Liu, Y., Wang, Y., Yu, D., et al. (2017, March 28). Phase I and II metabolic gene polymorphisms in susceptibility to prostate cancer in a Chinese population. *Oncotarget, 8*(13), 21783.

Yadav, A., Kumari, N., & Kumar, R. (2020). Association of glutathione S-transferase M1 and T1 polymorphisms with risk of type 2 diabetes mellitus: A systematic review and meta-analysis. *Meta Gene, 25*, 100779.

Yi, L., Xiao-Feng, H., Yun-Tao, L., Hao, L., Ye, S. & Song-Tao, Q. (2013). Association between the XRCC1 Arg399Gln polymorphism and risk of cancer: Evidence from 297 case–control studies. *Plos One, 8*(10), e78071.

Yu, H., Xia, X., Liu, X., & Li, D. (2014, June). Association of CYP2E1 polymorphisms with risk of esophageal cancer in the Chinese population: A meta-analysis. *Tumour Biology, 35*(6), 5539–5546.

Zhang, H., Ma, Y., Song, H., Chen, L., Liu, Y., Zhang, L., et al. (2023). Genetic polymorphisms of SOD2 and CAT and risk of prostate cancer: A meta-analysis. *Cancer Cell International, 23*, 242.

Zhang, L., Zhang, L., Xu, Y., Liu, J., Wang, X., Li, C., et al. (2022, April). Association between NAT2 polymorphisms and the risk of bladder cancer: A meta-analysis. *Environmental Science and Pollution Research International, 29*(14), 20851–20864.

Zhang, Y., Li, J., Li, D., Wang, Q., & Wang, J. (2014, January). Phase II metabolism gene polymorphisms and risk of hepatocellular carcinoma: A meta-analysis. *Tumour Biology, 35*(1), 187–195.

Zhu, F., Zhang, R., Li, C., Zhang, H., Wang, J., & He, S. (2019, May). Association between glutathione S-transferase polymorphisms and risk of renal cell carcinoma: A meta-analysis. *International Urology and Nephrology, 51*(5), 749–760.

Chapter 10
Occupational Toxicology: Exposure Pathways, Disease, Evaluation of Agents, Monitoring and Risk Assessment

Smita Jain, Akhil Gupta, Aarti, Radhika Bhadana, and Soumi Sadhu

Abstract Xenobiotics are chemical entities which are directly or indirectly released into the environment through various means such as industrial waste, excessive accumulation in agricultural soil, residue run off to fresh and sea water flora and fauna, etc. Its indiscriminate use poses serious health concerns to occupational workers and consumers world-wide, however, the latter cannot be terminated but can only be regulated to decrease their undesirable adverse effects in all living forms including humans. The major routes of exposure and intoxication includes oral (ingestion), respiratory (inhalation), and cutaneous (skin absorption). Their major toxicological effects include disruption of neuronal, endocrine, immunological, reproductive, and respiratory functions Severe xenobiotic toxicity may even lead to skin, lung, and liver disorders including organ specific cancers as well in addition to stimulation of autoimmune diseases in some cases. Identification of potent xenobiotics, regular monitoring of permissible safe limits, and adverse impacts on occupational workers and end consumers may help in abrogating the toxicity at early stages and prevent the health deterioration. Concomitantly, analytical evaluations by advanced techniques such as high-performance liquid chromatography (HPLC), atomic absorption spectroscopy (AAS) along with and biochemical evaluations of vital body functions using diagnostic methods such as complete blood count (CBC), liver function tests (LFT), kidney function tests (KFT), and pulmonary function test (PFT) help in establishing the dose-time relationship of the respective xenobiotics' exposure in conferring damage. Moreover, designing of newer and effective legislative strategies for occupational site checks to ensure the use of potential safeguards and training of workers may further contribute to exacerbate these xenobiotic-induced health hazards.

Keywords Xenobiotics · Pesticides · Heavy metals · Industrial pollutants · Occupational toxicity

---

S. Jain (✉) · A. Gupta · Aarti · R. Bhadana · S. Sadhu
Department of Life Science, Sharda School of Bioscience and Technology, Sharda University, Greater Noida, U.P., India
e-mail: smita.jain@sharda.ac.in

K. Priya and S. K. Giri (eds.), *Environmental Toxicology: Xenobiotics and Effects on Human Genetic Material*, https://doi.org/10.1007/978-3-032-28619-2_10

1 Introduction

Xenobiotics are chemical species which are unintentionally released into the environment either as agricultural run off to soil and water or via industrial waste and remain undegraded for years. They, often, remain accumulated in living beings who consume contaminated water and sea foods. Xenobiotics encompass various categories (Fig. 1) such as (i) heavy metals like lead, arsenic, cadmium, mercury, etc., (ii) industrial pollutants containing harmful chemicals such as volatile organic solvents (benzene, toluene, ethylbenzene, xylene), polycyclic aromatic hydrocarbons (PAHS), and persistent substances like dioxins and polychlorinated biphenyls (PCBS), (iii) pesticides, (iv) Natural toxins derived from plants, animals, and microbes.

1.1 Exposure Pathways

Heavy metals are dense, naturally occurring elements that cause health hazards by accumulating in the environment and living beings. Exposure occurs via inhalation,

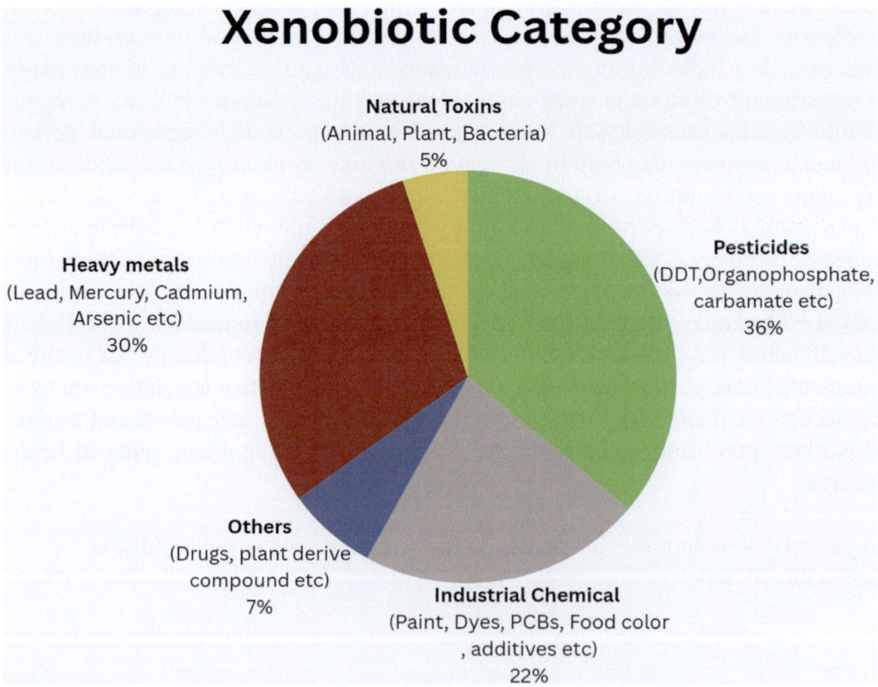

Fig. 1 Different categories of xenobiotics

ingestion and dermal contact (Fig. 2) during mining and smelting, battery manufacturing and recycling, welding and soldering, pigment and paint manufacturing, electroplating, tanning, e-waste recycling, and unauthorized metal scrap management. Industrial waste containing Hg is released into the waters. Aquatic microbes convert inorganic Hg (iHg) to organic Hg (CH_3Hg). Groundwater also shows high levels of arsenic (As), in West Bengal India, and neighboring Bangladesh, far above permissible limits exert harmful effects on the residing population (Rahman et al., 2001). The incidence and magnitude of the toxicities of individual heavy metals vary with the geographical location, natural soil content, habits, customs, location of industries, regulatory measures to contain pollution, healthcare facilities to detect heavy metal toxicity, and individual factors like nutritional status and genetics. When a heavy metal is released into the air, water, or soil, it can be absorbed by plants, crops, consumed by cattle and fish, and finally, end up in humans to complete the food chain.

Increasing urbanization and development have led to a massive growth of air-polluting industries. Exposure to heavy metals and industrial pollutants (defined as undesirable byproducts released from industries), primarily occurred at occupational sites or work places and include industrial wastes, contaminated water, hazardous gases, and chemical precipitates (Figs. 3 and 4). Secondary exposure occurs at the consumer or recipient level via the consumption of contaminated water from rivers, lakes, ponds, and seas which often contain harmful industrial effluents and ultimately affect aquatic life (Nisar et al., 2021). Additionally, sewage from industries releases a

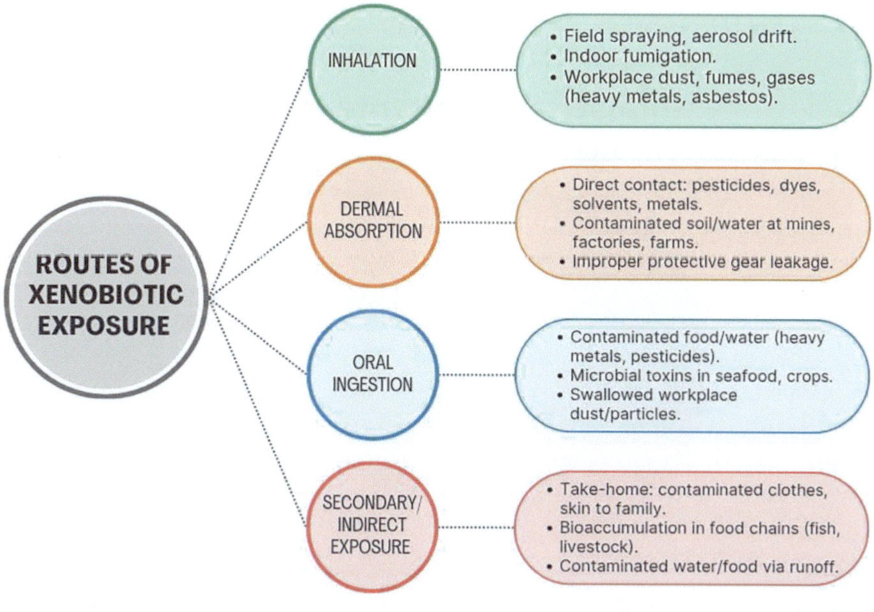

Fig. 2 Various routes of xenobiotic exposure

variety of ions into water bodies, including phosphate, sulphate, nitrate, and chlorine, which pollutes the air, water, land, and soils. Such infiltration of undesirable industrial waste leads to abnormally high bioaccumulation in different tissues or organs of aquatic organisms and eventually upset the food chains (Zhu et al., 2021).

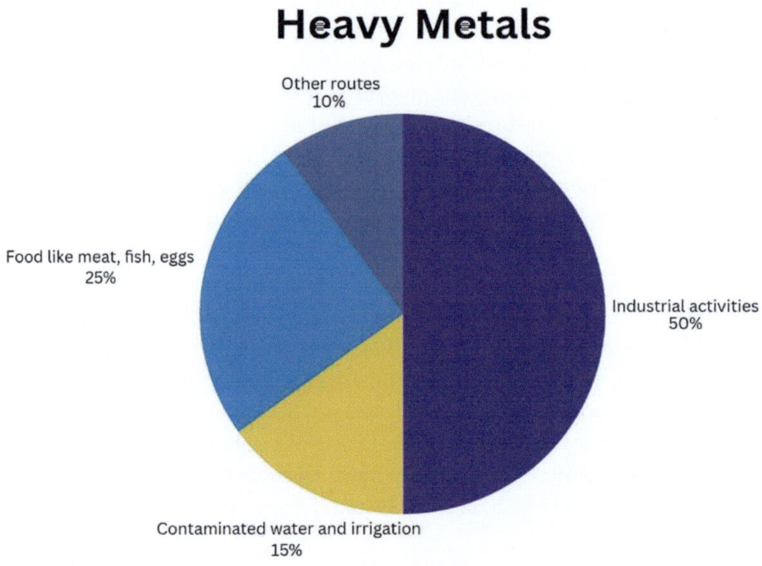

Fig. 3 Sources of heavy metals

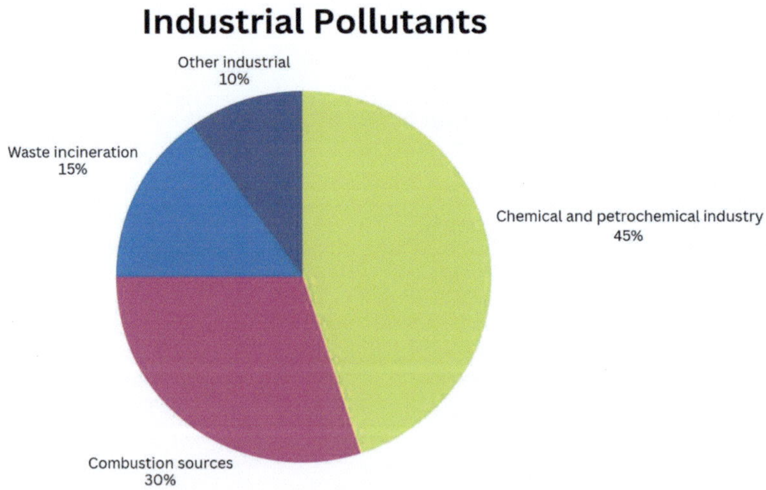

Fig. 4 Major sources of industrial pollutants

Occupational pesticide exposure is highly common in factory workers involved in synthesis, mixing of pesticides as well as to field workers and farmers. It is a large issue for agricultural workers especially in low- and middle-income countries that have inconsistent regulatory controls and minimal mechanization. This exposure happens often for long periods via skin and respiratory routes due to use of concentrated products during crop plantation and harvesting. During this period, majorly women farmers are involved who might experience hormone disruption, reduced fertility, and poor pregnancy outcomes (Sanborn et al., 2007; Singh et al., 2018). Pesticide exposure also happens. Since most of the pesticides are lipophilic, they gain access to lipid rich organs such as the brain via respiratory routes and confer extreme neurotoxic damage which may span from behavioral abnormalities to memory deficits and neurodegeneration.

Toxins derived from animals and microbes are often hazardous following the exposure via dermal contact, dietary consumption, or inhaled during slaughter or meat processing. While the severity of xenobiotic exposure depends on the proximity to sources and everyday activities like time spent inside and outdoors, airborne contaminants can spread over wide areas or stay localized close to emission sources. Many microbes reside in the gastrointestinal tract of aquatic organisms and produce different types of toxins which may pose a severe threat to primary as well as secondary target species. Sometimes these toxins are gradually accumulated without affecting the survival of the host species but later on cause adverse effects on higher consumers in the food chain.

1.2 Diseases Due to Occupational Exposure to Xenobiotics

1. Skin Disorders

Exposure to xenobiotics, particularly heavy metals, arsenic, and polycyclic aromatic hydrocarbons (PAHs) is strongly associated with skin cancer due to direct contact with the xenobiotics during handling and use (Pal & Firdous, 2024). Chronic exposure to pesticides, toxic chemicals in dyes, shampoos, bleaching agents, rubber chemicals (carbamates, thiurams), epoxy resins and nickel and chromium led to direct damage to the skin's surface and cause irritant dermatitis (immune independent) and allergic contact dermatitis (immune hypersensitivity) (Srinivas & Sethy, 2022). Skin infections such as cellulitis, caused by *Streptococcus iniae,* Fish Handler's Disease, caused by *Mycobacterium marinum,* Erysipeloid, caused by *Erysipelothrix rhusiopathiae,* are transmitted to occupational fishers and slaughter house workers by direct contact with contaminated fishes (Rahman et al., 2020). Contact Urticaria, an important skin disorder, is observed in factory workers exposed to a variety of industrial chemicals such as glutaraldehyde, quaternary ammonium compounds, latex, and nickel.

2. Liver Disorders

Aflatoxins, acetaminophen, industrial solvents, amatoxins from toxic mushrooms (like Amanita phalloides), and ciguatoxins/tetrodotoxin from contaminated fish are causes of some liver disorders, including acute hepatitis (caused by amatoxins via RNA polymerase II inhibition), hepatocellular cancer (aflatoxin- DNA adducts), veno-occlusive disease (solvent-related sinusoidal blockage), cirrhosis (ion channel disruption caused by ciguatoxin), and hepatocyte necrosis (acetaminophen overdose). Amatoxin poisoning causes fulminant liver failure with centrilobular necrosis and coagulopathy in foragers within 48 to 72 h of ingestion. Fishermen who are exposed to ciguatoxin develop cholestatic hepatitis due to oxidative lipid peroxidation, bile duct inflammation, and persistent Na + channel activation. In pufferfish handlers, tetrodotoxins induce neuromuscular inhibition, which leads to hypoperfusion and hypoxic hepatitis. Endoplasmic reticulum stress and oxidative damage are the foundation of all hepatotoxicity brought on by xenobiotics and zootoxins (Traylor et al., 2025).

3. Neurodegenerative Disorders

Neurodegenerative diseases like Parkinson's and Alzheimer's are increasingly associated with occupational and chronic exposure to heavy metals (lead, mercury), pesticides (paraquat, rotenone), industrial solvents like MPTP, and animal toxins from snake venoms (α-neurotoxins) or marine core snails (conotoxins). Shell collector conotoxins cause prolonged excitotoxicity by blocking voltage gated calcium channels. Snake α- neurotoxins block cholinergic signaling in herpetologists by attaching to nicotinic receptors. Xenobiotics also stimulate the disruption of mitochondrial function, lower ATP production, and increase oxidative stress leading to α-synuclein, aggregation, loss of dopaminergic neurons and misfolding of amyloid β and tau hyperphosphorylation (Srinivas et al., 2020, Wang et al., 2022).

4. Respiratory Disorders

Occupational exposure via inhalation of asbestos fibers, silica dust, diesel exhaust particles, cigarette smoke chemicals (nitrosamines benzene), and often animal toxins in slaughter houses can cause respiratory disorders due to occupational exposure. The long-term inflammation, fibrosis, and mutagenesis of these substances cause lung cancer (nitrosamine- DNA adducts), pulmonary fibrosis (asbestos/silica-induced fibroblast proliferation and scarring), pneumoconiosis (silica dust macrophage activation), and chronic obstructive pulmonary disease (COPD from diesel particulates via airway remodeling). Asbestos stays in the alveoli and promotes ROS-mediated fibrosis in miners. Animal toxins, such as bee venom aerosols in apiarists or marine saxitoxins from shellfish processing, can cause acute bronchospasm, eosinophilic pneumonia, and worsening of COPD histamine by release and neurogenic inflammation. Long-term exposure to zootoxins and xenobiotics causes respiratory diseases (Mazumder & Hussain, 2024; Cheng et al., 2025).

5. Endocrine Disorders

Endocrine-disrupting xenobiotics such as bisphenol A (BPA), phthalates, dioxins, pesticides (atrazine, vinclozolin), and zootoxins from amphibians (bufadienolides from bufo toads) or marine species (okadaic acid from dinoflagellates) mimic or block natural hormones by binding to estrogen, androgen, or thyroid receptors. This disruption modifies metabolic and reproductive signals, leading to thyroid dysfunction (PCB/thyroid hormone receptor blockage), infertility (Phthalate-induced Leydig cell malfunction), obesity/metabolic syndrome (BPA-adipogenesis promotion), polycystic ovarian syndrome (atrazine-aromatase upregulation), thyroid cancer (dioxin-AHR pathway activation), and hypogonadism (vinclozolin-androgen receptor antagonistic). BPA increases serum estradiol and disrupts menstruation in plastic workers by acting as a moderate estrogen agonist by blocking Na^+/K^+-ATPase, toad handlers' bufadienolides mimic digitalis toxicity by causing hypercalcemia and parathyroid hormone suppression. Okadaic acid, which is present in shellfish processors, gradually raises insulin resistance and type 2 diabetes by blocking protein phosphates-2A prolonged exposure to these chemicals has been linked to endocrine axis disruption and long-term hormonal imbalance(Sobral et al., 2025).

6. Reproductive Disorder

Xenobiotics damage gametes, interfere with hormone regulation, and impair placental function through exposure to pesticides, lead endocrine disruptors, and animal toxins at work. Pesticides (e.g., DDT), lead, bisphenol A, phthalates, and zootoxins like snake metalloproteinases can cause a number of reproductive disorders, including oligospermia (pesticide-induced Sertoli cell dysfunction), anovulation (phthalate-ovarian follicle atresia), recurrent miscarriage (lead-placental ischemia), congenital malformations (DDT-estrogen mimicry), and preterm birth (breve toxin-uterine contraction dysregulation) (Madhu et al., 2022). During pregnancy, these xenobiotics cross the placental barrier and cause teratogenic effects such neural tube defects and developmental delays. Snake venoms and zinc dependent metalloprotease cause hemorrhaging that damage the ovarian vasculature of herpetologists, leading to sterility. Brevotoxins in shellfish workers disrupt progesterone signaling, increasing fetal resorption rates. Mechanistically, these drugs alter gene expression during embryogenesis and interfere with hormone-dependent processes crucial to fetal development(Cheng et al., 2025).

7. Immunological Disorders

Dioxins, heavy metals (mercury, lead), pesticides (DDT), or toxins can cause a number of immunological disorders, including immunosuppression (dioxin T-cell apoptosis), autoimmunity (mercury induced autoantibodies), increased susceptibility to viral infections, decreased vaccine efficacy (lead B-cell impairment), and hypersensitivity (mast cell degranulation). Dioxin exposure increases the risk of cancer by reducing the activity of natural killer cells in chemical workers. In miners, mercury causes helper T cells (Th2) skewing and causes worsening of systemic lupus erythematosus. Crotoxin causes lymphopenia in herpetologists by inhibiting phospholipase A2. Melittin from bee venom hyperactivates IgE in apiarists, causing anaphylaxis. Immunotoxicity can manifest as persistent autoimmune illnesses, reduced vaccine efficacy, or increased susceptibility to infections (Yadav et al, 2025; Saha & Samuel, 2025).

8. Botulism

Occupational exposure to potent neurotoxins from the bacterium *Clostridium botulinum*, generally found in the intestines of fishes, cause severe paralytic illness, muscle weakness, difficulty swallowing and fatal respiratory paralysis in individuals exposed to toxin contaminated food.

9. Histamine (Scombroid) Poisoning

Occupational exposure through inhalation and skin contact in factories involving spoiled fish processing leads to a pseudo-allergic reaction called histamine poisoning. Bacteria inhabiting the fish such as tuna, mackerel convert histidine into histamine which gets inhaled and bioaccumulates in the handlers.

2 Agents

2.1 Heavy Metals

Heavy metals are naturally occurring elements that become toxic when they accumulate in the body, either due to chronic exposure at work or through consumption of contaminated food and water. It includes non-essential elements such as Arsenic (As; metalloid), Cadmium (Cd), Lead (Pb), Mercury (Hg), as well as essential elements such as Cobalt (Co), Copper (Cu), chromium (Cr), Iron (Fe), Manganese (Mn), Molybdenum (Mo), Nickel (Ni), Selenium (Se) and Zinc (Zn). Different heavy metals pose a differential type of adverse effects, some of them are briefly discussed in this chapter. Mercury (Hg), especially methylmercury, which binds to sulfhydryl groups in proteins to harm the kidneys and neurological system is associated with neurodevelopmental issues. The industrial release of methyl mercury into rivers and seawater resulted in poisoning and deaths in Japan, infamously known as Minamata disease. Cd accumulation in bones with resultant pains and fractures lead to the so-called "Itai-Itai (it hurts-it hurts) disease" (Rajkumar et al., 2025). Lead causes cognitive impairments, anemia, and reproductive damage by interfering with calcium metabolism and neural signaling among battery workers, smelter workers, and painters. Arsenic (As) interferes with DNA methylation and repair, causing diabetes, liver damage, skin cancer, and cardiovascular disease (Jaishankar et al., 2014). Hexavalent chromium, which is used in welding, electroplating and the production of stainless steel, has been linked to lung cancer, dermatitis, asthma, and nasal septal perforation. Cadmium (Cd) accumulation in the kidney and lung causes proximal tubular dysfunction, reduced glomerular filtration, osteopenia, and an increased risk of lung and prostate cancer in smelter workers and welders. Chromium (Cr VI), which is used in industrial processes, is a potent carcinogen associated with lung cancer. Because heavy metals impede DNA repair, produce oxidative stress, and block enzymes, they play a significant role in neurological, hepatic, renal, reproductive, and carcinogenic diseases (Dinka, 2018). Long-term exposure to arsenic in mining, metallurgy, and pesticide production can result in peripheral neuropathy, liver disease, skin hyperpigmentation and keratosis, and an increased risk of skin, lung, and bladder cancers.

2.2 *Industrial Pollutants*

Industrial pollutants are another important type of xenobiotic agents that originate from manufacturing, combustion, and waste disposal, (Fig. 4). Polycyclic aromatic hydrocarbons (PAHs), which are potent mutagens that cause DNA adduct formation and cancers of the skin, lungs, and bladder, are produced when coal, oil, and tobacco are not completely burned. Polychlorinated biphenyls (PCBs), which were once used in electrical equipment, persist in the environment and bioaccumulate in fatty tissues, affecting the immune and endocrine systems and increasing the risk of thyroid problems and breast cancer. Dioxin is one of the most hazardous byproducts of burning and producing insecticides. By attaching to the aryl hydrocarbon receptor (AhR) and changing gene transcription, they cause immunotoxicity, poor reproduction, and cancer. Volatile organic compounds (VOCs), including xylene, toluene, and benzene, are inhaled pollutants that damage the liver and bone marrow; leukemia is strongly linked to benzene. Asbestos fibers from construction lodge in the lung tissue cause chronic inflammation, which can lead to mesothelioma and pulmonary fibrosis. As a result, there is a strong link between industrial xenobiotics and respiratory conditions, cancer, endocrine issues, and immunological dysfunction (Miglani et al., 2022; Yadav et al., 2025).

Occupational exposure to crystalline silica has been linked to several autoimmune diseases including rheumatoid arthritis (RA), systemic sclerosis (SSc), systemic lupus erythematosus (SLE) and anti-neutrophil cytoplasmic antibody (ANCA)-related diseases. Solvents, or chemicals with similar structures, including vinyl chloride, epoxy resin, trichloroethylene (TCE), perchloroethylene, or mixed solvents were confidently associated with SSc. Other studies supported a likely association between solvent exposure and multiple sclerosis (MS). There is some evidence for an association of cosmetics containing several industrial chemicals with SLE, RA and primary biliary cholangitis (PBC) (Miller et al., 2012; Reeves et al., 2009).

2.3 *Pesticides*

Pesticides, a significant class of xenobiotic agents, are widely used in agriculture to control insects, weeds, fungi, and rodents. They can be classified into various categories such as insecticides, herbicides, rodenticides, etc. depending on the target species. Hazards posed by these pesticides are briefly listed in Fig. 5. Insecticides interfere with nervous system signals, usually by blocking acetylcholinesterase, and include subtypes such as pyrethroids, carbamates, organophosphates, and organochlorines (Costa et al., 2022). Herbicides such as glyphosate,

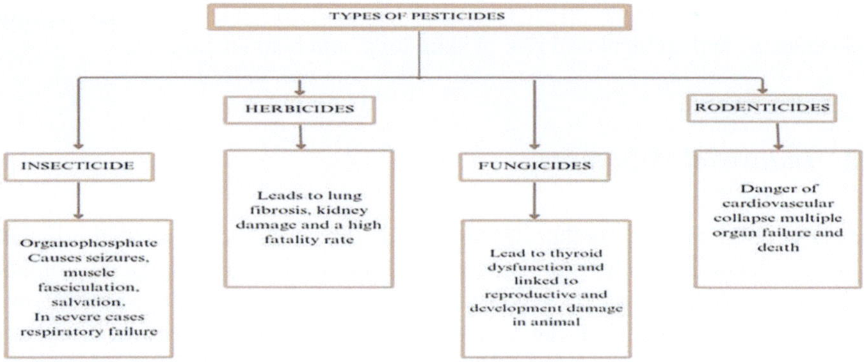

Fig. 5 Different types of pesticides and their effects

paraquat, and atrazine not only disrupt plant growth processes but also harm the mitochondria and cause oxidative stress, which can cause lung, breast, and skin cancer as well as respiratory disorders (Mostafalou et al., 2023; Ye et al., 2017). Fungicides like mancozeb and captan are linked to endocrine disruption, developmental problems, liver toxicity, thyroid problems, and reproductive toxicity (Hernández et al., 2013). On the other hand, rodenticides like warfarin and brodifacoum interfere with blood coagulation and increase the risk of bleeding in unintentional human exposure (Mishra et al., 2021).

2.4 Natural Toxins (Animal, Plant, Microbial)

Some toxins which are produced naturally by animals, plants, or microorganisms pose threat to the host and non-target species including humans. Some of the sources of these naturally occurring xenobiotics (as toxins) are shown in Fig. 6. Animal poisons like snake venom and marine toxins like tetrodotoxin damage ion channels and neurotransmission, resulting in paralysis and cardiovascular collapse. Plant toxins that interfere with protein synthesis or emit cyanide, such as ricin, atropine, and cyanogenic glycosides, can cause multi-organ failure. Aflatoxins and other microbial toxins are strong hepatocarcinogens, while botulinum toxin paralyses by blocking the release of acetylcholine. These pollutants are linked to neurological, hepatic, cardiovascular, and cancer-causing conditions (Ali, 2024). Potent endotoxins from bacterium *Clostridium botulinum*, generally found in the intestines of fishes, cause botulism in the primary and secondary host species.

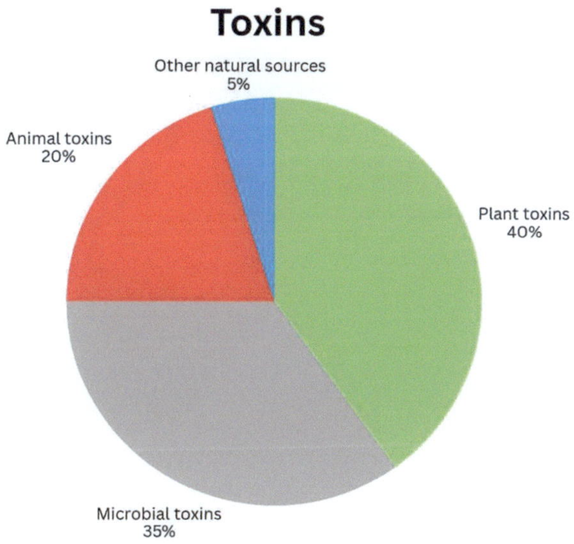

Fig. 6 Sources of natural toxins

3 Other Xenobiotics (Drugs and Plant-Derived Compounds)

Pharmaceuticals and chemicals derived from plants can act as xenobiotics when handled incorrectly or transformed into dangerous intermediates (Fig. 7). While drugs like acetaminophen cause hepatotoxicity by depleting glutathione, chemotherapeutics like cyclophosphamide generate DNA cross-links that increase the risk of cancer. NSAIDs are associated with gastric and renal toxicity, whereas antibiotics such as isoniazid can damage the liver. Plant-derived compounds like pyrrolizidine alkaloids cause hepatic veno-occlusive disease, while aristolochic acid is associated with urothelial cancers and renal failure. These compounds show how both synthetic and natural drugs can induce hepatic, kidney, reproductive, and carcinogenic issues (Kuku, 2023).

4 Monitoring and Assessment of Xenobiotics

Monitoring intends to identify the potent xenobiotic, which may be a naturally occurring toxin (origin: plant, animal or microbe) or industrially synthesized, characterize its source and its potential harmful effects. It is highly desirable to minimize the severity of imposed hazards among occupational workers and secondary consumers which otherwise shows an abnormally increasing trend due to progressive urbanization and undesirable climate change including deforestation. The monitoring process generally involves sample collection from air, water, soil, and biota (blood, urine,

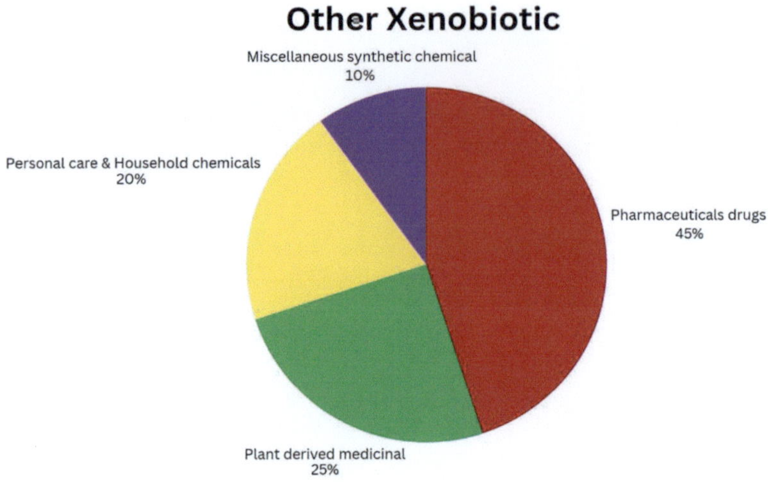

Fig. 7 Other miscellaneous xenobiotic agents

and tissue) followed by analytical evaluation of xenobiotics concentration (parent compound or metabolites) and its impact on vital metabolic processes in humans. Analytical evaluation is carried out by using advanced scientific methods like High-performance liquid chromatography (HPLC), Gas chromatography (GC) especially when coupled with mass spectrometry for estimation of pesticides, mycotoxins, etc. Heavy metals are usually quantified by using atomic absorption spectrometers (AAS) equipped with hollow cathode lamps.

Biochemical analysis to identify the xenobiotic-induced toxicity with respect to cellular morphology, count, and vital metabolic functions can be detected in occupational workers by using a battery of tests such as complete hemograms (or complete blood count CBC), liver function test (LFT), kidney function test (KFT), and pulmonary function test (PFT) respectively. Sometimes, hair or nail testing may be included for checking the intensity of prolonged bioaccumulation. Hematological changes such as anemia, cytopenia, changes in platelet values, as well as leukocytosis can be frequently diagnosed by CBC. Such symptoms indicate generation of oxidative stress and altered bone marrow function indicative of acute or chronic xenobiotic exposure (Hernández et al., 2013; Mishra et al., 2021). Abnormal LFT report representing altered levels of serum glutamic oxaloacetic transaminase (SGOT/AST), serum glutamic pyruvic transaminase (SGPT/ALT), bilirubin, and alkaline phosphatase (ALP) and inflammatory biomarkers indicate reduced functioning of the liver and impaired detoxification of the encountered xenobiotic(s). KFT results showing higher serum creatinine and blood urea nitrogen levels indicate renal tube damage and glomerular dysfunction often pointing towards xenobiotic-induced kidney impairment. PFT indicates forced vital capacity (FVC), forced expiratory volume in one second (FEV1), and a reduced peak expiratory flow rate (PEFR) which is often found altered in occupational factory workers.

Presence of an individual toxin(s) such as botulinum, cyanotoxin, etc. can be rapidly screened by using specific immunoassays such as ELISA and radioimmunoassay which represents a specific, sensitive, and cost-effective method. Evaluation of blood hormone levels, fertility biomarkers, cardiac markers, behavioral disturbances, and neuronal damage at the cellular and genetic level can also be determined by specific ELISA assays thereby demonstrating the dversity of hazards attributed to occupational xenobiotic exposure. In addition to detect specific protein antigens, the genome presence can also be identified more accurately by using advanced molecular methods such as polymerase chain reaction (PCR) and DNA sequencing. This enables the detection of complex mixtures of toxins as well as genes responsible for toxin production (e.g., *mcyE* gene for microcystins), and identification of the relative abundance and distribution of various aquatic species infected with toxin producing microbes. Additionally, remote sensing, biosensors, and machine learning models are emerging tools that can provide early warnings and cover large geographical areas, complementing traditional methods.

Urinary testing of metabolites especially 1-hydroxypyrene and other benzo[a]pyrene metabolites, together with genotoxicity markers such as DNA adducts or micronucleus frequency, are frequently used to measure PAH exposure in coke-oven and asphalt workers. Quantitative analysis of pesticides (parent and/or metabolites) and heavy metals (parent and/or metal adduct) in occupational workers at regular intervals provide helpful insights to understand the hazard severity.

Risk assessment of xenobiotic toxicity integrates the data obtained from monitoring strategies and processes to establish the hazardous nature of any xenobiotic. It involves various steps starting from identification of toxin itself or specific toxic biota based on the existing reports on its hazardous effects. Afterwards, it determines the extent, duration, and dose response curve with respect to a specific xenobiotic for a given population. The latter characterizes the dose, time, and severity relationship which would help in setting occupational exposure limits, designing efficient prevention and remedial measures for occupational xenobiotic toxicity. It also plays a role in designing regular auditing of infrastructural safety gears, engineering controls such as exhaust or ventilation units, and stringent occupational practices which ensure the minimum xenobiotic-induced toxicity among workers (Shekhar et al., 2024).

5 Conclusion

Based on the monitoring data and existing risk assessment approaches followed in industrial settings, policies and strategies can be developed to increase the awareness about hazards posed by these xenobiotics and ensure proper training modules to minimize the xenobiotic-induced risk (FAO, 2023; García-García et al., 2022). Digital surveillance can further be added as an extra cautionary step in minimizing the xenobiotic-induced health hazards in occupational workers.

References

Ali, H. B. B. M. (2024). Nutraceuticals in metabolism and xenobiotics. In S. M. Jafari, S. Kumar Jha, & V. Kumar Goswami (Eds.), *Nutraceuticals: A new millennium approach for preventive medicine*. Food Bioactive Ingredients. Springer.

Cheng, Y. F., Zhao, Y. J., Chen, C., & Zhang, F. (2025). Heavy metals toxicity: Mechanism, health effects, and therapeutic interventions. *MedComm, 6*(9), Article e70241. https://doi.org/10.1002/mco2.70241

Costa, L. G., Giordano, G., Guizzetti, M., & Vitalone, A. (2022). Exposure routes and health risks associated with pesticide use. *Toxics, 10*(6), 335.

Dinka, D. D. (2018). Environmental xenobiotics and their adverse health impacts: A general review. *Journal of Environmental Pollution and Human Health, 6*, 77–88.

FAO. (2023). *Pesticide registration toolkit—Occupational risks*. Food and Agriculture Organization of the United Nations. https://www.fao.org/pesticide-registration-toolkit/registration-tools/registration-criteria/human-health-risks/occupational-risks/en/

García-García, C. R., Parrón, T., Requena, M., Alarcón, R., Tsatsakis, A. M., & Hernández, A. F. (2022). Human biomonitoring of pesticide exposure: Current approaches and future directions. *Toxicology Letters, 368*, 42–54.

Hernández, A. F., Parrón, T., Tsatsakis, A. M., Requena, M., Alarcón, R., & López-Guarnido, O. (2013). Toxic effects of pesticide mixtures at a molecular level: Their relevance to human health. *Toxicology, 307*, 136–145.

Jaishankar, M., Tseten, T., Anbalagan, N., Mathew, B. B., & Beeregowda, K. N. (2014). Toxicity, mechanism and health effects of some heavy metals. *Interdisciplinary Toxicology, 7*(2), 60–72.

Kuku, A. (2023). Unveiling the menace: a comprehensive study of xenobiotics toxicity and their health ramifications. *Toxicology: Open Access, 9*(6), 243.

Madhu, N. R., Sarkar, B., Slama, P., Jha, N. K., Ghorai, S. K., Jana, S. K., & Roychoudhury, S. (2022). Effect of environmental stressors, xenobiotics, and oxidative stress on male reproductive and sexual health. In: *Oxidative stress and toxicity in reproductive biology and medicine: a comprehensive update on male infertility* (vol. 2, pp. 33–58).

Mazumder, M. H. H., & Hussain, S. (2024). Air-pollution-mediated microbial dysbiosis in health and disease: lung–gut axis and beyond. *Journal of Xenobiotics, 14*(4).

Miglani, R., Parveen, N., Kumar, A., Ansari, M. A., Khanna, S., Rawat, G., & Ansari, M. N. (2022). Degradation of xenobiotic pollutants: An environmentally sustainable approach. *Metabolites, 12*(9), 818.

Miller, F. W., Alfredsson, L., Costenbader, K. H., Kamen, D. L., Nelson, L. M., Norris, J. M., & De Roos, A. J. (2012). Epidemiology of environmental exposures and human autoimmune diseases: Findings from a National Institute of Environmental Health Sciences Expert Panel Workshop. *Journal of Autoimmunity, 39*(4), 259–271.

Mishra, S., Shukla, G., Yadav, R. S., & Singh, S. (2021). Pesticide exposure and health risks in India: Current scenario and future perspectives. *Environmental Science and Pollution Research, 28*(12), 14115–14130.

Mostafalou, S., Abdollahi, M., & Baeeri, M. (2023). A systematic review of pesticide exposure and associated risks in occupational settings. *Frontiers in Public Health, 11*, 11664077.

Nisar, M., Mohammad, N., Liya, G., Muhammad, H. K., Muhammad, B., Wei, P. C., & Grzegorz, L. (2021). Assessment of industrial wastewater for potentially toxic elements, human health (dermal) risks, and pollution sources: a case study of Gadoon Amazai industrial estate, Swabi, Pakistan. *Journal of Hazardous Materials, 419*.

Pal, S., & Firdous, S. M. (2024). Unraveling the role of heavy metals xenobiotics in cancer: A critical review. *Discover Oncology, 15*, 615.

Rahman, M. M., Chowdhury, U. K., Mukherjee, S. C., Mondal, B. K., Paul, K., Lodh, D., Biswas, B. K., Chanda, C. R., Basu, G. K., Saha, K. C., Roy, S., Das, R., Palit, S. K., Quamruzzaman, Q., & Chakraborti, D. (2001). Chronic arsenic toxicity in Bangladesh and West Bengal, India: A review and commentary. *Journal of Toxicology and Clinical Toxicology, 39*(7), 683–700.

Rahman, M. T., Sobur, M. A., Islam, M. S., Ievy, S., Hossain, M. J., El Zowalaty, M. E., Rahman, A. T., & Ashour, H. M. (2020). Zoonotic diseases: Etiology, impact, and control. *Microorganisms, 8*(9), 1405.

Rajkumar, N., Chitturi, C. M. K., Lavanya, K., Subhashini, V., Shamshad, S., & Seethamma, G. (2025). Consequences of toxic heavy metals on environment and human health: A review. *Uttar Pradesh Journal of Zoology, 46*(9), 198–207.

Reeves, W. H., Lee, P. Y., Weinstein, J. S., Satoh, M., & Lu, L. (2009). Induction of autoimmunity by pristane and other naturally occurring hydrocarbons. *Trends in Immunology, 30*(9), 455–464.

Saha, N., & Samuel, M. (2025). Dietary xenobiotics and their role in immunomodulation. *Food Science and Biotechnology, 34*, 1805–1817.

Sanborn, M., Kerr, K. J., Sanin, L. H., Cole, D. C., Bassil, K. L., & Vakil, C. (2007). Non-cancer health effects of pesticides: Systematic review and implications for family doctors. *Canadian Family Physician, 53*(10), 1712–1720.

Shekhar, C., Khosya, R., Thakur, K., Mahajan, D., Kumar, R., Kumar, S., & Sharma, A. K. (2024). A systematic review of pesticide exposure, associated risks, and long-term human health impacts. *Toxicology Reports, 13*, Article 101840.

Singh, R., et al. (2018). Patterns of pesticide poisoning in South India: A hospital-based study. *Indian Journal of Occupational and Environmental Medicine, 22*(2), 90–95.

Sobral, A. F., Cunha, A., Costa, I., Silva-Carvalho, M., Silva, R., & Barbosa, D. J. (2025). Environmental xenobiotics and epigenetic modifications: Implications for human health and disease. *Journal of Xenobiotics, 15*(4), 118.

Srinivas, C. R., & Sethy, M. (2022). Occupational dermatoses. *Indian Dermatology Online Journal, 14*(1), 21–31.

Traylor, J., Murray, B. P., & Singhal, M. (2025). Ciguatera toxicity. In *StatPearls* [Internet]. StatPearls Publishing. https://www.ncbi.nlm.nih.gov/books/NBK482511/

Wang, X., Sial, M. U., Bashir, M. A., Bilal, M., Raza, Q. U. A., Raza, H. M. A., & Geng, Y. (2022). Pesticides xenobiotics in soil ecosystem and their remediation approaches. *Sustainability, 14*(6), 3353.

Yadav, K., Sharma, A. K., Jena, M. K., Parashar, N. C., Singh, T., Choudhary, R., Ramniwas, S., Tuli, H. S., & Yadav, V. (2025). Xenobiotic exposure and autoimmune diseases: An update on epidemiological evidences and potential mechanisms. *Journal of Biochemical and Molecular Toxicology, 39*(12), Article e70615.

Ye, M., Beach, J., Martin, J. W., & Senthilselvan, A. (2017). Pesticide exposures and respiratory health in general populations. *Journal of Environmental Sciences (China), 51*, 361–370.

Zhu, T., Li, F., Niu, W., Gao, Z., Han, Y., & Zhang, X. (2021). Health risk assessment of toxic and harmful air pollutants discharged by a petrochemical company in the Beijing–Tianjin–Hebei region of China. *Atmosphere, 12*(12), 1604.

Chapter 11
The Prospects, Difficulties, and Research Paths for Environmental Toxicology

Minza Ilahi, Yogita, Sayan Chatterjee, and Navjyoti Chakraborty

Abstract Environmental toxicology stands at a pivotal moment, confronting a widening gap between the complexity of modern environmental challenges and the limitations of traditional testing paradigms. The field has relied in the past on whole-animal testing and single-chemical tests like the LC_{50} test, which are time-consuming, expensive, and frequently lack in translational relevance to human health, particularly considering their dependence on unrealistically high doses. Meanwhile, the increasing quantity of chemicals in use, the discovery of new contaminants, and the heterogeneity of actual exposure conditions, such as mixtures, low-dose risks, and climate-mediated effects, present challenges that can't be addressed using traditional methods. To address these needs, the discipline is transforming toward predictive, mechanistic, and human-relevant science. New Approach Methodologies (NAMs) such as high-throughput screening, organ-on-a-chip systems, in silico modeling, and omics-based technologies are key to this transformation. These technologies give molecular-level understanding, improve our capacity for modeling toxicokinetic and toxicodynamic, enabling extrapolation to complex ecological and population-level endpoints. Integration with artificial intelligence (AI) and machine learning (ML) has further strengthened predictive toxicology by enabling pattern recognition in large datasets. In this chapter, we discuss that future research needs to adopt the idea of the exposome, the comprehensiveness of environmental exposures over a lifetime, and transition to exposure-driven rather than hazard-driven paradigms. We also discuss the method to elucidate source-to-outcome pathways of contaminants, relating sub-individual responses to ecosystem function and biodiversity, and use of the "One Health" approach to link human, animal, and environmental health. Finally, the most significant challenge is not technological innovation alone but institutional adaptation and implementation of regulatory reforms, interdisciplinarity, and persistent funding. Pursuing these lines of research, environmental toxicology can realize its enormous potential to inform sound public health policy and protect ecosystems during a time of unparalleled environmental change.

M. Ilahi · Yogita · S. Chatterjee · N. Chakraborty (✉)
University School of Biotechnology, Guru Gobind Singh Indraprastha University, New Delhi, India
e-mail: nchakraborty@ipu.ac.in

K. Priya and S. K. Giri (eds.), *Environmental Toxicology: Xenobiotics and Effects on Human Genetic Material*, https://doi.org/10.1007/978-3-032-28619-2_11

Keywords Risk assessment · Chemical regulation · Predictive toxicology · Ecotoxicity · Bioinformatics

1 Introduction

Environmental toxicology, or the study of chemical, biological and physical pollutants on living organisms or ecosystems, in today's day and age has developed into a highly complex situation, harboring the proliferation of emerging contaminants (ECs) like mico-plastics and PFAS, evolving industrial practices, and global environmental change. Although the basic transport pathways of these pollutants are understood, contemporary challenges centered on the poorly characterized compounds, long-term ecological and human health impacts remain uncertain. ECs including pharmaceuticals, endocrine-disrupting compounds, nanomaterials, micro- and nanoplastics, industrial chemicals, and per- and polyfluoroalkyl substances remain in the ecosystems at low concentrations, however, they can produce subtle, cumulative, and synergistic effects (Lei et al., 2015; Manke et al., 2013; Michelangeli & East, 2015). These factors complicate traditional risk assessment approaches, which mostly rely on single-compound exposures and controlled laboratory conditions thereby underscoring the need for predictive, integrative methodologies capable of addressing mixture toxicity and low-dose chronic exposures (Han et al., 2024; Vinken, 2024). The complexity of these interactions can be more clearly depicted and reviewed through a systematic view of pollutant movement and biological impact. Figure 1 illustrates this exact conceptual framework, highlighting and visualizing the journey of contaminants from anthropogenic and natural sources to the environment and then to the living organisms, while emphasizing the continuous feedback loop among research, risk assessment, and policy interventions.

A significant obstacle for environmental toxicology is bridging laboratory observations with real and live ecosystems. Laboratory studies can provide insights into the mechanisms of toxicity but those measures rarely capture the combined effects of multiple stressors which can include chemical mixtures, microplastics, radiation, and climate-driven changes. Additionally, climate change further complicates the matter of pollutant distribution, degradation, and susceptibility of different organisms by adding different layers of uncertainty to the exposure–response relationships (Geng et al., 2023; Wang et al., 2021). At the same time, urbanization and the ever-shifting industrial practices introduce novel contaminants and exposure pathways continually, which leads to the requirement of adaptive monitoring strategies. Field-based studies, in situ biological monitoring of sentinel species, and the integration of environmental data with epidemiological observations are therefore critical to produce ecologically and clinically relevant insights (Lee et al., 2023).

Technological advancements present promising opportunities for overcoming all these challenges. High-throughput "omics" approaches like genomics, transcriptomics, proteomics, and metabolomics studies allow detection of early molecular

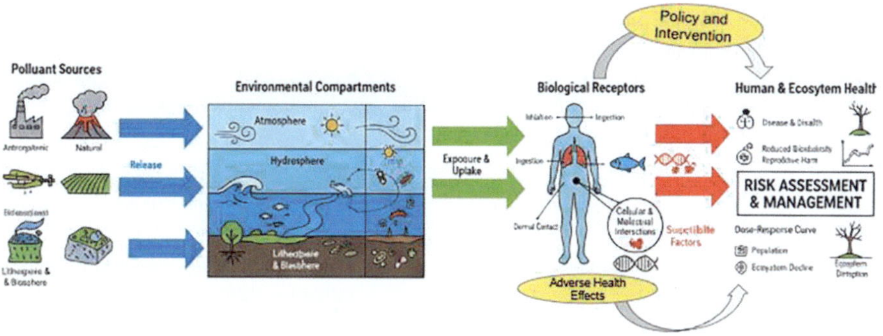

Fig. 1 Conceptual framework of environmental toxicology: schematic representation of the environmental toxicology framework, depicting the flow of pollutants from anthropogenic and natural sources through environmental compartments (air, water, soil) to biological receptors, resulting in toxic responses at molecular, organismal, and ecosystem levels. The figure also emphasizes the feedback loop between scientific research, risk assessment, and policy intervention

markers for exposure before extreme toxicity can do any damage. These tools identify biomarkers of susceptibility and cumulative effects, providing a mechanized way to predict population-level risks (Han et al., 2024; Lynch et al., 2024). Parallelly, cutting-edge technologies like computational toxicology, machine learning, and artificial intelligence have facilitated the modeling of complex toxicity pathways, simulating conditions for untested compounds, and cross-species extrapolation. Reliance on animal testing can be decreased while predictive accuracy is increased through integration with physiologically based pharmacokinetic models. Exposomics is a paradigm shift toward holistic assessment that aims to quantify all environmental exposures across an individual's lifetime. It captures the interaction between biology and environment and links chemical exposure profiles to metabolic and genetic responses (Li et al., 2024). The field's transition from descriptive observation to computationally predicted by data-driven science has been gradual and molded by key historical breakthroughs. Table 1 summarizes major milestones in environmental toxicology, highlighting how successive decades have expanded its scope—from early discoveries of bioaccumulation to present-day integration of AI and the One Health paradigm.

The adoption of standardized methodologies and harmonized global databases is another vital need. Analytical and regulatory issues arise from the variety of ECs, their presence in complex combinations, and their frequently low environmental concentrations (Li et al., 2024). Predicting dose–response can be complicated by the potential of novel drugs that may have antagonistic or synergistic interactions. It is therefore crucial to standardize analytical methods, reference materials, and reporting frameworks with the goal to improve reproducibility and comparability across research and geographic regions. The transboundary nature of many ECs is likewise important, requiring international collaboration and adaptable regulatory frameworks that can promptly incorporate new scientific findings (Wang et al.,

Table 1 Historical milestones and scope of environmental toxicology: major developments that have shaped the field of environmental toxicology, from its emergence in response to chemical pollution crises to its present integration with molecular biology, data science, and global health frameworks. These milestones illustrate the discipline's evolution from descriptive observation to predictive, systems-level science

Year/period	Key milestone	Scientific/policy impact
1960s	Discovery of bioaccumulation (DDT, PCBs)	Led to global bans and ecotoxicology birth
1980s	Rise of molecular toxicology	Mechanistic understanding of oxidative stress
2000s	Introduction of omics and PBPK models	Mechanistic and predictive toxicology
2020s	AI and one health integration	Global, systems-level environmental risk approach

2024). Together, the mechanisms depicted in Fig. 1 and the developmental trajectory summarized in Table 1 highlight how environmental toxicology now function as a cyclical system—linking discovery, regulation, and policy through iterative scientific progress.

Future directions for environmental toxicology must look to include a multidisciplinary collaboration and policy relevance between global nations. Example of this is the One Health paradigm that recognizes the interconnectedness of ecological, animal, and human health and serves as a framework for cooperative study in the fields of environmental science, medicine, veterinary science, and public health (Wang et al., 2021). To reduce the dangers that come with the production of chemicals and to get pollution under control, green chemistry and sustainable design have become increasingly preferred approaches as these are better suited than post-release mitigation. By combining cutting-edge technologies like citizen science, remote sensing, and continuous AI-driven environmental monitoring with the current reforms, the results would be early-warning capabilities, empowering data collection, and acceleration in conversion of scientific findings into practical policies (Han et al., 2024).

Long-term observational studies demonstrating a connection between environmental exposures and health outcomes are necessary to promote evidence-based regulation and verify predictive algorithms (Geng et al., 2023). Similarly, approaches for risk assessment which consider the complexity of the actual world are ought to be built on mechanistic studies into mixture toxicity, bioaccumulation, and climate-mediated exposure dynamics in order to protect human and ecological well-being (Lee et al., 2023; Vinken, 2024). This field must ultimately move from diagnostic evaluation toward predictive and preventive environmental health research, combining high-resolution data, mechanistic understanding, and policy-oriented applications (Li et al., 2024).

In conclusion, the future of environmental toxicology relies on our ability to tackle the multifaceted issues presented by emerging contaminants and environmental change by integrating cutting-edge technology, interdisciplinary approaches, and worldwide collaboration. This field is ready to transition from recording contamination to actively prevent and mitigate its impact by adopting predictive technologies, standardized procedures, and holistic frameworks like exposure genomics and the One Health model (Lynch et al., 2024; Wang et al., 2021). Research approaches that are flexible, cooperative, and predictive will be needed in the years to follow as a means to influence sustainable chemical design, inform policy, and safeguard ecosystems and public health in a world which is changing swiftly.

2 Sources and Types of Environmental Pollutants

2.1 Chemical Pollutants

Chemical pollutants represent a vast variety of synthetic substances and naturally occurring inorganic compounds that cause harm to environmental and biological systems. This category is comprised of pesticides, industrial solvents, medicines, heavy metals, and Persistent Organic Pollutants (POPs). Many of these xenobiotics are defined by their ability to lead to chronic toxicity even at minimal environmentally relevant doses, persistence (resistance to degradation), and capacity for bioaccumulation in food webs (Li et al., 2024; Mohan et al., 2020; Wang et al., 2021). For instance, the inhibition of enzymatic and hormonal systems by substances like organophosphate insecticides illustrate their broad-spectrum biological influence, while the neurotoxicity of heavy metals like lead and mercury is a globally recognized public health concern (Mohan et al., 2020). Research attention these days has shifted from traditional, well-characterized contaminants to the dynamic problem of Emerging Contaminants (ECs) and Persistent Toxic Substances (PTSs), as a result of the constant manufacturing of novel compounds brought to existence, thanks to the rapid worldwide development and innovation (Geng et al., 2023; Wang et al., 2021).

The complex presence and various types of these pollutants can be more clearly understood when viewed as a whole for all contaminant classes of chemical, physical, and biological pollutants, each of which act through distinct yet overlapping mechanisms. Table 2 provides an overview of these categories, summarizing representative examples, their primary sources, mechanisms of action, and corresponding environmental or health concerns.

Table 2 Major categories of environmental pollutants and their characteristics: overview of the primary classes of environmental pollutants—chemical, physical, and biological—along with representative examples, sources, mechanisms of action, and major ecological or health concerns. This table highlights the multidimensional nature of pollution and the overlapping impacts of different contaminant classes on environmental systems

Pollutant type	Key examples	Primary source	Mode of action/ mechanism	Major environmental concern
Chemical	Pesticides, PFAS, heavy metals	Agriculture, industry	Bioaccumulation, endocrine disruption	Persistent toxicity
Physical	Radiation, noise, microplastics	Energy sector, transport, waste	Energetic/ mechanical stress	Habitat alteration
Biological	Pathogens, ARGs, cyanotoxins	Wastewater, livestock, algal blooms	Infection, toxin release	Disease spread, ecosystem imbalance

2.1.1 Emerging Chemical Pollutants

Emerging Contaminants (ECs) are a variety of unregulated or recently discovered pollutants increasingly found in the environment. These are the molecules whose occurrence and possible ecological or human health impacts have only been recently completely understood with the aid of advancements in analytical chemistry (Li et al., 2024).

2.1.1.1 Pharmaceuticals and Personal Care Products (PPCPs)

PPCPs represent a broad and continuous ecological threat, originating from both human and animal exploitation of the environment. Along with industrial chemicals such as UV filters, parabens, and artificial perfumes; active pharmaceutical ingredients (APIs) like lipid regulators, antibiotics, antidepressants, and endocrine medicines are also included in this. Antibiotics are especially important since the global problem of antibiotic resistance genes (ARGs) is associated with antibiotic residues in aquatic environments that are an outcome of ineffective removal by the wastewater treatment plants (WWTPs) (Mohan et al., 2020) (discussed later in Sect. 2.3). The quantity of PPCPs detected in WWTP effluents is thought to be an essential indicator of human impact on the urban water chain. Non-therapeutic levels, frequently in the nanogram-per-liter range in surface and wastewater, are enough to have long-term and subtle effects on organisms that are not the target (Lee et al., 2023).

2.1.1.2 Endocrine-Disrupting Chemicals (EDCs)

EDCs are toxic compounds that cause harm by disrupting the production, secretion, circulation, binding, action, or breakdown of natural hormones like estrogen and testosterone. This alters the homeostasis functions and vital bodily mechanisms like reproduction, growth, and neurodevelopment which can be detrimental for survival. However, despite these hazards, EDCs exposure like bisphenol A (BPA), dioxins, alkylphenols and phthalates, are almost inevitable due to their widespread use in consumer goods, food packaging, and agricultural runoff (Lei et al., 2015; Wang et al., 2021). One of the ways these can cause harm is by binding to the hormone receptors, like estrogen or androgen, which would result in either blocking (antagonism) or mimicking (agonism) of the natural signal of these hormones. Pharmaceutical hormones, like ethinylestradiol which is found in oral contraceptives, are strong EDCs and they can cause major issues in the long run for the ones exposed. This then becomes a huge health risk, especially when substances like sewage sludge are used for agricultural soils (Li et al., 2024).

2.1.1.3 Industrial Chemicals and By-Products (ICBs)

ICBs are chemicals that are produced by industrial activities like large-scale production of goods, chemical manufacture of detergents or other chemicals and mineral extraction. Some of the most significant members of this group are Polycyclic Aromatic Hydrocarbons (PAHs), Polychlorinated Biphenyls (PCBs) and by-products like dioxins and furans because their unregulated disposal creates complex mixes of both organic and inorganic contaminants, for example, solvents, heavy metals, and novel flame retardants (Michelangeli & East, 2015). The rise of e-waste (electronic trash) poses another challange and is responsible for greatly contributing to isolated contamination hotspots (Geng et al., 2023).

2.1.1.4 Nanomaterials and Micro/Nanoplastics (MNPs)

MNPs are a class of pollutants with a distinctive dual identity because they act as both physical and chemical contaminants. Nanomaterials (compounds of 1–100 nm size), such as carbon nanotubes (CNTs) and synthetically engineered metal nanoparticles such as silver nanoparticles (AgNPs) and titanium dioxide (TiO2), have a high surface-area-to-volume ratios which contributes to their ability to adsorb other toxins like heavy metals or POPs. This makes them unique vectors for the transport of contaminants which can be very dangerous when speaking ecologically (Manke et al., 2013). Nowadays, microplastics (MPs) (less than 5 mm size) and their smaller equivalents like nanoplastics (NPs) are found everywhere owing to ever-increasing use of plastics. In addition to causing physical damage like obstructing the GI tract or causing abrasions to the gut in organisms of all trophic levels, they also gather other pollutants like heavy metals and PAHs from the environment which makes it much more convenient for them to enter biological tissues (Lee et al., 2023). The rapid rise of nanotechnological developments and plastic production has intensified worries regarding environmental dissemination and their biological impacts (Vinken, 2024).

2.1.1.5 Per- and Polyfluoroalkyl Substances (PFAS)

PFAS, such as Perfluoro-octane Sulfonate (PFOS) and Perfluorooctanoic Acid (PFOA), also known as "forever chemicals," can be identified by their extraordinary durability which they possess thanks to the presence of the carbon–fluorine bond, making them highly stable and extremely resistant to degradation. Due to their incredibly long environmental half-lives, they are a long-term ecological threat and may pose serious harm to public health (Wang et al., 2024). However, they are also very hard to avoid as they have a widespread use in firefighting foams, stain-resistant coatings, and non-stick cookware that has led to extensive contamination in soil and water (Li et al., 2024).

2.1.1.6 Agrochemicals (Pesticides and Herbicides)

While conventional pesticides (like DDT) are older pollutants and policies and strategies to deal with them were developed and implemented, their residues persists till today and newer and systemic varieties of pesticides available today like neonicotinoids continue to pose emerging risks. Their primary route to enter the environment is through agricultural wastewater and spray drift. Notably, some agrochemicals, such atrazine, are classified as strong EDCs validates their status as constantly emerging risks (Lei et al., 2015; Mohan et al., 2020).

2.1.2 Conventional Chemical Pollutants

These are well-known and conventional pollutants, usually inorganic, whose impact on the environment has been acknowledged for many years.

2.1.2.1 Heavy Metals (HMs)

Heavy metals like lead (Pb), mercury (Hg), cadmium (Cd), and arsenic (As) and metalloids like arsenic emerge from both natural sources like by weathering rocks and significant industrial activities like mining, industrial combustion, and electronic waste (Mohan et al., 2020). HMs, unlike organic contaminants, are non-degradable and persistent, so they accumulate in soil and sediments through which they are continuously mobilized. Food safety is directly endangered by Cd, As, and Pb, which are commonly listed to be the most prevalent contaminants in agricultural soils worldwide (Li et al., 2024).

2.1.2.2 Disinfection By-Products (DBPs)

DBPs are those toxic compounds that are formed when chemical disinfectants like chlorine, used in drinking water and pool purification react with the organic matter. There are more than 600 DBPs that are known to exist, but the most prominent ones are trihalomethanes (THMs) and haloacetic acids (HAAs), that make up a significant percentage of the entire halogenated DBP impact (Lei et al., 2015). And even though chlorination is necessary for microbiological safety in our day-to-day lives, the resulting DBPs can be mutagenic or even carcinogenic, which makes these by-products a major health concern (Michelangeli & East, 2015).

2.2 Physical Pollutants

Physical pollutants are non-chemical compounds that can alter environmental quality or disrupt biological function predominantly through mechanical or energetic mechanisms. Thermal fluctuations, ionizing radiation, and noise pollution are some examples of physical pollutants. Environmental toxicology is dominated by chemical pollutants, but physical stressors are becoming widely acknowledged for their significant ecological effects (Geng et al., 2023). For instance, chronic exposure to high-intensity noise from industrial or transportation sources can interfere with how animals communicate, reproduce, and even cardiovascular health of humans. Another significant example of this issue is presence of radionuclides from nuclear or medical waste because ionizing radiation directly damages DNA, causing mutagenesis and cancer (Vinken, 2024). Microplastics are also a noteworthy mention when it comes to physical pollutants in terms of their mechanistic effects as they interfere with feeding and reproductive processes in animals, especially in aquatic biota, and causes physical harm (such as impaction of the gut) (Lee et al., 2023).

2.3 Biological Pollutants

Biological pollutants are living organisms, their products (toxins), or their genetic material that negatively affect ecosystem functioning and human well-being. In the past, the primary emphasis was on naturally occurring infections which were largely caused by microbial pathogens, but recently, novel biological dangers brought on by human activity have come into focus (Wang et al., 2021).

2.3.1 Pathogens and Antibiotic Resistance Genes (ARGs)

When we talk about biological threats, the most significant medical issue of this century is that of Antibiotic resistance which comes about because of the development of Antibiotic Resistance Genes (ARGs), Antibiotic-Resistant Bacteria (ARB) and Mobile Genetic Elements (MGEs) in microbial pathogens. Antibiotic resistance is the selective resistance toward drugs that pathogens develop because of irresponsible use of antibiotics by humans and in animal husbandry. This results in environmental pools of ARGs in agricultural soils, sewage sludge, and wastewater from where they may gravely impact public health, as they not only make the prognosis complicated and longer but they also develop the ability to spread horizontally to other bacteria, including human commensals and pathogens (Mohan et al., 2020). The return of ARB/ARGs to the human microbiome through food and water is also accelerated by the frequent co-selection for resistance mechanisms by non-antibiotic pollutants such biocides and heavy metals (Li et al., 2024).

2.3.2 Toxins

Biological toxins, for instance cyanotoxins (produced by harmful algal and cyanobacterial blooms) and prions are produced as a result of global warming and eutrophication (Wang et al., 2021). These phenomenon lead to the blooming of cyanobacterial colonies that yield strong hepatotoxic (like microcystin) and neurotoxic (like saxitoxin) substances. These toxins not only pollute the water sources that are fit for use but also have the ability to aerosolize and deteriorate the quality of atmosphere which can be detrimental for our ecological stability. Aside from enzymatic toxins, misfolded proteins called prions, can also cause deadly neurological disorders like Chronic Wasting Disease in cervids, are extremely resistant to breaking down. They have the ability to latch on to the soil particles which gives them the ability to become stable, long-lasting reservoirs of contamination that can propagate throughout the food chain (Vinken, 2024). Together, the examples of chemical, physical, and biological contaminants discussed above reflect the interconnected structure summarized in Table 2, which captures how diverse pollutants share overlapping pathways of persistence, bioaccumulation, and ecological disruption.

3 Environmental Fate and Transport of Pollutants

3.1 Fate and Transport Mechanisms

Emerging Contaminants (ECs) and other xenobiotics' environmental fate and mobility are determined by a constant combination of physical, chemical, and biological changes that affect a pollutant's persistence, dispersion, and final bioavailability (Lei et al., 2015; Manke et al., 2013). Destructive or modifying activities like degradation and transformation and phase transfer processes (sorption, volatilization, and leaching) are important mechanisms. The physicochemical characteristics of

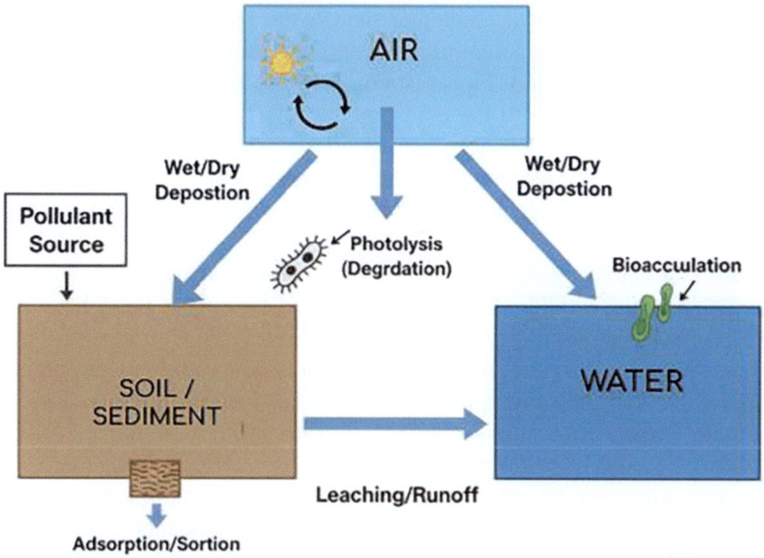

Fig. 2 Environmental fate and transport pathways: Illustration of key mechanisms governing the environmental fate and transport of emerging contaminants (ECs) across air, water, and soil compartments. Processes such as sorption, volatilization, leaching, degradation, and bioaccumulation determine pollutant persistence, mobility, and eventual bioavailability within ecosystems

a substance, like its vapor pressure, solubility, and octanol–water partition coefficient (log K_{OW}), and the environmental factors, like its pH, temperature, and redox potential, are inextricably tied to the rates of these mechanisms (Geng et al., 2023; Michelangeli & East, 2015). The overall dynamics of these processes can be best understood through a visual of how pollutants migrate and transform through the environmental compartments. Figure 2 illustrates the environmental fate and transport pathways of emerging contaminants, showing how air, water, and soil interact through processes such as volatilization, leaching, degradation, and bioaccumulation. This schematic framework underscores how physicochemical properties and environmental variables together dictate contaminant persistence and movement through ecosystems.

3.1.1 Degradation and Transformation

For many organic xenobiotics, including some pharmaceuticals and personal care products (PPCPs), biodegradation—which is mostly mediated by microbial communities—represents an essential removal method (Mohan et al., 2020; Wang et al., 2021). ECs are classified as Persistent Toxic Substances (PTSs) because of their limited biodegradability, which is frequently caused by their molecular stability and chemical complexity (Lei et al., 2015).

A significant toxicological challenge lies in the formation of Transformation Products. Prior to excretion, parent pharmaceuticals go through metabolic processes (Phase I: oxidation, reduction; Phase II: conjugation) within the organisms, which release active metabolites into the environment, often at amounts higher than the parent substance (Han et al., 2024). A substantial research gap is evident by the fact that ECs can undergo photolysis in aquatic systems which is a process in which chemical interactions triggered by sunlight result in photo-induced transformation (Li et al., 2024).

Transformation for Engineered Nanoparticles (ENPs) includes dissolution, aggregation, and chemical modification. A key factor of toxicity is the release of metal ions, or dissolution, which is regulated by fundamental characteristics (for example, surface coating and size) plus environmental factors (for example, low pH) (Manke et al., 2013). A common process that changes the fate and ecotoxicity of metal nanoparticles by reducing their surface reactivity is sulfidation which is also sometimes referred to as passivation. Since these extremely dynamic and reactive species have a major impact on the chemical pathways down the road, further research is required to be done to understand the kinetics of Environmentally Persistent Free Radicals (EPFRs) that are released during incomplete combustion or degradation processes (Vinken, 2024).

3.1.2 Sorption (Adsorption and Desorption)

The primary variable governing the mobility and capacity for leaching of contaminants is sorption processes (Lei et al., 2015; Wang et al., 2021). The concentration of the pollutant while being in water phase is effectively controlled and its bioavailability can be successfully restricted by adsorption onto solid phases, such as soil particles, suspended solids, and sediments (Michelangeli & East, 2015). Because of their large surface area, microplastics (MPs) and nanoplastics (NPs) attract large focus as they can serve as successful vectors for the adsorption and subsequent movement of other hydrophobic pollutants, including heavy metals (HMs) and polycyclic aromatic hydrocarbons (PAHs), where they can make it easier for these pollutants to enter aquatic food webs (Lee et al., 2023; Manke et al., 2013). Engineered sorption, which employs highly porous materials like activated carbon and biochar to immobilize the contaminants, is an essential subject of research into the restoration of HM-contaminated soils (Geng et al., 2023). A type of biological sorption is known to bind strongly to soil particles, resulting in restricted mobility and long-lasting localized contamination.

Taken together, the degradation, sorption, bioaccumulation, and inter-compartmental transfer of contaminants form the foundation of their fate analysis. Table 3 summarizes the major environmental fate and transport mechanisms, covering their governing factors, typical contaminant examples, and key ecological implications. Such a tabular overview helps in comparing how physical, chemical, and biological processes interact to determine pollutant persistence and mobility.

Table 3 Major mechanisms governing environmental fate and transport of pollutants: summary of key mechanisms—degradation, transformation, sorption, volatilization, and bioaccumulation—along with their controlling environmental factors, pollutant examples, and ecological outcomes. This table provides an at-a-glance understanding of how physicochemical and biological processes shape contaminant dynamics

Mechanism	Controlling factors	Representative pollutants	Key ecological/ health impacts
Degradation/ transformation	Microbial activity, light exposure, redox potential	PPCPs, PAHs, pesticides	Metabolite formation, altered toxicity
Sorption/desorption	pH, organic carbon content, particle size	Heavy metals, PFAS, microplastics	Reduced mobility, sediment accumulation
Volatilization	Temperature, vapor pressure, wind speed	VOCs, PCBs	Airborne transport, long-range exposure
Bioaccumulation	Lipophilicity (log Kow), food web structure	PFAS, mercury, dioxins	Biomagnification, trophic transfer
Leaching/runoff	Soil permeability, rainfall, land use	Pesticides, nitrates	Groundwater contamination, eutrophication

3.1.3 Bioaccumulation and Trophic Transfer

Bioaccumulation describes the uptake and retention of xenobiotics by an organism from all exposure sources (water, sediment, diet). When accumulation is magnified through successive trophic levels it is termed biomagnification or trophic transfer (Michelangeli & East, 2015). This process poses acute risks to apex predators and humans. In marine food webs, the traditional log K_{OW} is often insufficient for predicting the bioaccumulation potential of ionizable chemicals; the log DOW (distribution coefficient at a specific pH) particularly for ionizable compounds is a more accurate indicator (Wang et al., 2021). Both terrestrial soil-dwelling species and aquatic organisms bioaccumulate stubborn pollutants like PFAS and HMs, which can cause interference with their physiological processes and harm these organisms while also becoming carriers for the movement of ECs through the soil food web (Li et al., 2024).

3.2 Environmental Compartments

The environment is conceived as an interconnected multimedia system of three primary compartments, that is, air, water, and soil, through which the pollutants continuously pass through (Geng et al., 2023). Xenobiotics spread across these compartments by inter-media exchange (like by volatilization from water to air), runoff, leaching, and atmospheric deposition (Lei et al., 2015; Vinken, 2024).

3.2.1 Water Compartment (Surface Water, Groundwater, Wastewater)

Aquatic systems are often the ultimate final dumping place for pollutants (Mohan et al., 2020). ECs are continuously released through industrial discharge, agricultural runoff, and WWTP effluents (Lee et al., 2023; Wang et al., 2021). The environmental behavior of a pollutant is determined by its solubility. Hydrophilic PPCPs, like sulphonamide antibiotics, tend to remain dissolved in the water phase, whereas more hydrophobic substances like estrogens easily adhere to sediments and form long-term environmental reservoirs (Li et al., 2024). Advanced removal techniques are needed because the constant discharge puts traditional water treatment technology capabilities to the test (Han et al., 2024).

3.2.2 Soil Compartment

Soil is a complex matrix and a major sink for complex xenobiotics that are hard to degrade or remediate (Lei et al., 2015; Wang et al., 2021). Contamination in soil may come from various different sources which can include agricultural runoffs like pesticides and veterinary drugs, and biosolids which are treated sewage sludge to be used as soil amendments, and introduce a vast array of PPCPs and ARGs into the ground (Li et al., 2024. Soil contamination is a serious matter because it can influence pollutant bioavailability and trophic transfer and it is here that persistent pollutants like PFAS, ENPs, and HMs would eventually build up (Manke et al., 2013). Because contaminants may proceed to a different compartment, like groundwater or crops, causing a variety of organisms and endangering the safety of food, soil contamination is crucial (Geng et al., 2023; Michelangeli & East, 2015).

3.2.3 Air Compartment

In order to facilitate the Long-Range Transport (LRT) of semi-volatile organic compounds, like PCBs and PAHs, on a global and regional level, air is a vital transport medium (Mohan et al., 2020). Particulate matter (PM2.5) and the gas phase are used to transport pollutants that are then deposited to distant regions (atmospheric deposition) (Lee et al., 2023). The migration and deposition capabilities of airborne pollutants, which contribute to complex chemical exposomes, are heavily influenced by climatic conditions such as temperature and wind speed (Geng et al., 2023). These sophisticated PM2.5 exposomes may disrupt physiological transport and metabolism when inhaled, resulting in potential risks that are unexpected (Vinken, 2024). The interactions among air, water, and soil compartments visualized in Fig. 2 and summarized in Table 3 demonstrate that pollutant behavior is inherently systemic—any shift in one medium's dynamics (e.g., pH or temperature) can propagate across the others, influencing contaminant transport and exposure potential.

3.3 Modeling Environmental Fate and Transport

Mathematical and computational models are fundamental instruments in modern environmental toxicology, enabling the quantitative prediction of pollutant movement and impact which is a key research area (Han et al., 2024). These mathematical models provide predictive capability for proactive risk assessment and environmental management, going beyond descriptive monitoring (Wang et al., 2024).

3.3.1 Role and Goals of Modeling

Modeling presents crucial insights into EC sources, distribution kinetics, life cycles, and influencing mechanisms (Vinken, 2024). In order to predict potential outcomes under different climatic or regulatory situations, generate early warnings of possible pollution events, monitor data, and evaluate the effectiveness of remediation procedures (Geng et al., 2023). Because of this convenience, models are essential supplements (Han et al., 2024).

3.3.2 Types of Models

The complete picture about environmental behavior can be simulated using Multimedia Fate Models such as USEtox and Mackay level III models (Li et al., 2024). These combine mass balance equations and physicochemical parameters like partition coefficients and degradation rates across interconnected compartments of air, water, soil, and biota on regional to global scales. Even though Pharmacokinetic (PK) models often focus upon uptake and distribution of pollutants within an organism, they also take data from the field observations of environmental outcomes of a situation based on real-life scenarios to better predict the retention and exchange of xenobiotics across the organism–environment interface (Han et al., 2024). Quantitative Structure–Activity Relationships (QSARs) are predictive models that connect molecular structure of a chemical with its environmental behavior (for example, in volatilization and bioaccumulation potential) using sophisticated statistical and machine learning approaches (Noyes et al., 2009).

3.3.3 Challenges in Modeling

Despite their benefits, the application of EC models presents significant difficulties, leading to a major research direction (Vinken, 2024).

1. **Complexity and Universality**: Given the extreme diversity of chemical structures and complex, non-linear toxic processes, there are very limited universal, process-based models that can accurately simulate the migration of all ECs simultaneously (Han et al., 2024).

2. **Data Scarcity and Uncertainty**: For a great deal of ECs, the lack of experimental training data drastically limits the ability to build correct models (Richardson & Kimura, 2020; Wang et al., 2024). By creating issues like over-fitting in QSAR models and necessitating the use of significant variability or application factors, this limits the accuracy of modeling outputs for policy decisions (Geng et al., 2023).
3. **Multi-Pollutant Interactions**: One of the fundamental gaps in evaluating environmental risk in the real world is the lack of ability of current models to fully account for the complex, antagonistic, or synergistic interactions of multi-pollutant mixtures (like the combined effect of HMs and ARGs) (Li et al., 2024).

4 Toxicity and Mechanisms of Action

To understand the potential of environmental contaminants to cause adverse outcomes is basically rooted in two core toxicology principals: the evaluation of hazard magnitude and the explanation of fundamental molecular mechanisms. Some modern analytical frameworks are developed and used for characterizing toxicity and the key cellular and molecular pathways that mediate the adverse effects of the xenobiotics, emphasizing the indispensable role of dose–response dynamics in risk assessment.

4.1 Toxicity Testing and Assessment

The system of environmental toxicity assessment is facing a paradigm shift, rather than relying on high dose, laborious in vivo study of animals, regarding a comprehensive framework rooted in human and ecological biological understanding. This shift is often referred to as Toxicity Testing in the twenty-first century (Tox21) which highly leverages the high-throughput screening (HTS) and computational toxicology (Han et al., 2024; Lynch et al., 2024).

The modern process prioritizes the identification of toxicity pathway perturbations, focusing more on the molecular initiating events (MIEs) rather than solely relying on the apical endpoints. The major changes are as follows:

4.1.1 High-Throughput Screening (HTS)

The digital assays use cell lines from humans or from the environment to enable the rapid examination of chemicals across a broad range of toxicity pathways. These tests give concentration response data, which needs further interpretation through modeling (Lynch et al., 2024).

4.1.2 Adverse Outcome Pathways (AOPs)

This anatomy of AOP provide a conceptual and structured representation of the chronology of the events, from an MIE to a series of measurable key events (KEs) at various levels of biological system (molecular, cellular, tissue, and organ), culminating in an adverse outcome of toxicological or regulatory importance. AOPs are substantial for linking in vitro pathways data to in vivo effects and for supporting mechanical risk assessment (OECD, 2018; Vinken, 2024).

4.1.3 In Vitro to In Vivo Extrapolation (IVIVE) and PBPK Modeling

To meet the gap between in vitro concentration data and in vivo exposure levels, Physiologically Based Pharmacokinetic (PBPK) models are inseparable. This modeling can be used to simulate ADME (adsorption, distribution, metabolism, and excretion) of a toxicant in an organism, that too, mathematically, which allows the estimation of concentrations of internal exposure at target tissues, which thus facilitates the estimation of effective doses from the in vitro data with respect to environmental context (Han et al., 2024).

4.2 Mechanisms of Toxicity (Genotoxicity, Oxidative Stress)

By interfering with the host organism's internal homeostasis, the toxicants may exert their effects. It does so via interactions with macromolecules in cells specifically or through disruptions of critical signaling cascades typically. Genotoxicity and oxidative stress are the most practical and well-studied and researched mechanisms of toxicology of the environment shown in Fig. 3.

4.2.1 Oxidative Stress

Across the whole physiochemically desperate classes of the environmental pollutants (from metal ions to particulate matter), oxidative stress is without a doubt the most common mechanism. Oxidative stress is a persistant imbalance between the generation of reactive oxygen species (ROS) and the organism's endogenous antioxidant defence capacity, which leads to excess of free radicals and other reactive intermediaries (Manke et al., 2013).

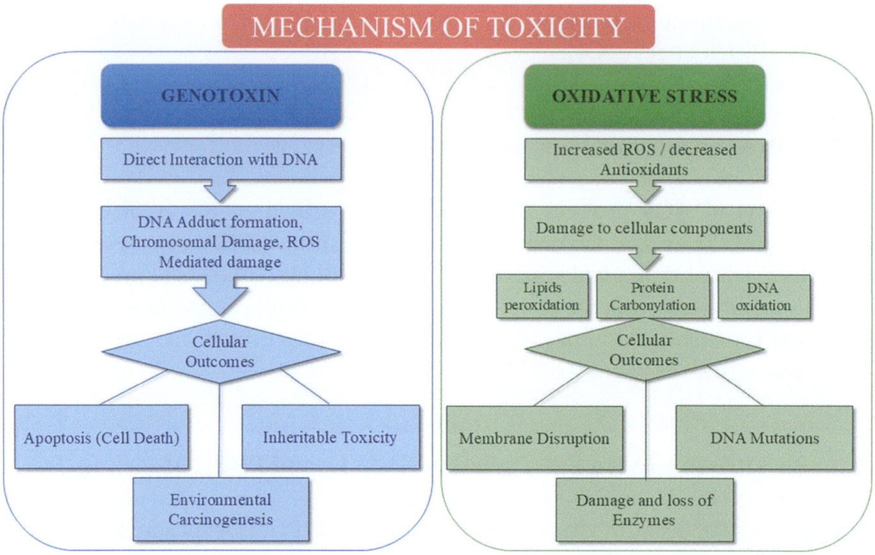

Fig. 3 Mechanisms of toxicity: diagrammatic representation of the toxicity mechanisms of toxic compounds in the host, including genotoxicity and oxidative stress as the major mechanisms. Diagram shows the actions of the toxicants upon entering the host and there effects

4.2.1.1 ROS Generation

Xenobiotics may introduce ROS directly such as redox-active metals catalyzing Fenton reactions, indirectly such as it may redox cycling intermediaries like quinones through bioactivation, or by interfering in the mitochondrial function, which may lead to electron leakage.

4.2.1.2 Macromolecular Damage

The resulting accumulation of ROS which includes superoxide anion ($O_2{}^{2-}$), hydroxyl radical (OH^-), and hydrogen peroxide (H_2O_2) leads to severe damage across critical components in a cell.

1. **Lipid Peroxidation**: ROS attack the polyunsaturated fatty acids within cellular membranes, compromising structural integrity and then releasing highly reactive secondary products.
2. **Protein Carbonylation**: It oxidizes the amino acids side chains and results in structural change and loss of enzymatic characteristics.
3. **DNA Oxidation**: It forms oxidized add-ons which are highly mutagenic, notably 8-hydroxy-2′-deoxyguanosine (8-OHdG).

4.2.1.3 Cellular Response

In response, nuclear factor erythroid 2-related factor 2 (Nrf2) signaling pathway is activated by the cell. The Nrf2 travels to the nucleus where it binds to the Antioxidant Response Element (ARE). The Antioxidant Response Elements (AREs) are present in the promoter regions of the genes encoding phase II detoxification enzymes, and antioxidant proteins such as glutathione S-transferases, heme oxygenase-1 (Deavall et al., 2012).

4.2.2 Genotoxicity

Any substance that can cause any kind of damage or effect to the genetic material (DNA or chromosomes) within the cell is called genotoxic, and may result in mutations, chromosomal aberrations, or genomic instability. While all mutagens are inherently genotoxic, meaning they can cause damage to the genetic material but all genotoxic agents don't cause mutations, i.e., genotoxic agents do not necessarily possess mutagenic characteristics (Manke et al., 2013). Their Primary Mechanism is:

4.2.2.1 Direct DNA Adduct Formation

There are certain xenobiotics or their activated metabolites that are highly electrophilic and may make covalent bonds with nucleophilic sites on the DNA bases, resulting in formation of DNA-Adducts. This kind of bulky lesions may disturb the DNA replication and its repair procedure. For instance, Polycyclic Aromatic Hydrocarbons (PAH) require metabolic activity in the reactive epoxides to become direct alkylating agents (Kirsch-Volders et al., 2003).

4.2.2.2 Chromosomal Damage

Structural changes, such as breaks, deletions, and translocations, or numerical changes, such as aneuploidy, to the chromosomes may be caused by the agents. They are often detected by assays like the micronucleus test.

4.2.2.3 ROS-Mediated Damage

The major source of genotoxicity is the oxidative stress, it produces single- and double-strand DNA breakage and also produces oxidative base modifications (Hei & Filipic, 2004).

This persisting or inaccurate repair of the genotoxic damage is the founding step in the heritable toxicity and environmental carcinogenesis (Manke et al., 2013).

4.3 *Dose–Response Relationships*

It is the quantitative description of the inter-relation between the extent of exposure or dose of a contaminant and the magnitude of the resulting response. This inter-relation is central to toxicology and it embodies the dictum, "*Sola dosis facit venenum*," which means only the dose makes the poison (Swenberg et al., 2007).

The risk assessment traditionally is based on the threshold model, which says that there is a dose level which needs to be reached in order to cause any adverse effects, below this dose level there is no adverse effects observed. This dose level is called threshold dose. This model utilizes the No Observed Adverse Effect Level (NOAEL) where the highest dose, at which no statistically significant adverse effects are observed experimentally, is considered. Modern toxicology now use more statistically advanced and comprehensive methods which are given below:

4.3.1 Benchmark Dose (BMD) Modeling

This approach has highly replaced the NOAEL approach. This method involves the application of a mathematical model to the dose–response data in order to estimate the dose with respect to a predetermined change in response, which is known as the Benchmark response (BMR) which is typically a 5–10% change from the background. The lower confidence limit of BMD is known as BMDL and it is used as the starting point for the risk assessment and characterization. This method used all of the available dose–response data to provide a parameter that is independent of experimental design which gives it an advantage (Slob & Setzer, 2013).

4.3.2 Non-monotonicity and Hormesis

The increasing data of Non-monotonic Dose–Response Curves (NMDRC) is a critical challenge to the traditional models. The problem with NMDRC is that they show effects at low concentrations but as the dose increases, the effects are reversed. A specific form of NMDRC called Hormesis is characterized by a biphasic curve in which the low doses impart a beneficial or stimulatory effect whereas the high doses lead to inhibition or toxicity. The presence of NMDRCs makes the low-dose estimation very complex and it also challenges the fundamental assumption of linear or simple sigmoidal dose–response relationships for all of the endpoints (Calabrese & Baldwin, 2003).

5 Environmental Toxicology and Human Health

5.1 Human Exposure Pathways

Human exposure to environmental toxicants is a complex and multifactorial process that involves three cardinal routes: inhalation, ingestion, and dermal absorption. Because ECs are so common because of global emissions from industries, agricultural runoff, and improper waste disposal, complex exposure pathways are created, subjecting the general public to low-dose exposures over time (Li et al., 2024). The interconnectedness of these routes can be better understood through Fig. 4, which maps how pollutants infiltrate the human body via ingestion, inhalation, and dermal contact. The schematic (Fig. 4) demonstrates how contaminants are absorbed and distributed systemically, affecting target organs such as the lungs, liver, kidneys, and brain. By visualizing exposure in this manner, the figure highlights the mechanisms behind pollutant entry and the broad spectrum of ensuing toxic outcomes.

5.1.1 Ingestion (Contaminated Water and Food)

Ingestion of contaminated substances is the most prevalent route for many persistent xenobiotics. Historical contamination incidents like the Flint Water Crisis involving Lead illustrates the broader implications of environmental pollution on human exposure pathways, as proved by the growing presence of ECs in groundwater and drinking water reservoirs (Lei et al., 2015; Wang et al., 2024). The complex network of food chains creates a powerful web that becomes the key route for the spread of these pollutants due to the possibility of bioaccumulation and biomagnification. Crops, being at the very start of almost all the terrestrial food chains, absorb toxins such as heavy metals, herbicides, and PFAS from contaminated soil and irrigation water and then from there the spread of contaminants begin as biomagnification intensifies at each step of the food web (Li et al., 2024; Michelangeli & East, 2015). In the same way, pollutants like PPCPs and microplastics from contaminated waters are bioaccumulated by aquatic species and then ingested by organisms including humans through the seafood intake, which causes us to accumulate contaminants in our bodies (Lee et al., 2023). Even recently discovered toxins, like cyanotoxins (microcystins and nodularins), can propagate and accumulate over time in the aquatic food chain, making fish and shellfish consumption potentially dangerous (Mohan et al., 2020; Lee et al., 2023). Additionally, because we frequently interact with Personal Care Products (PCPs), we are continuously exposed to EDCs including phthalates and parabens at low levels (Wang et al., 2024).

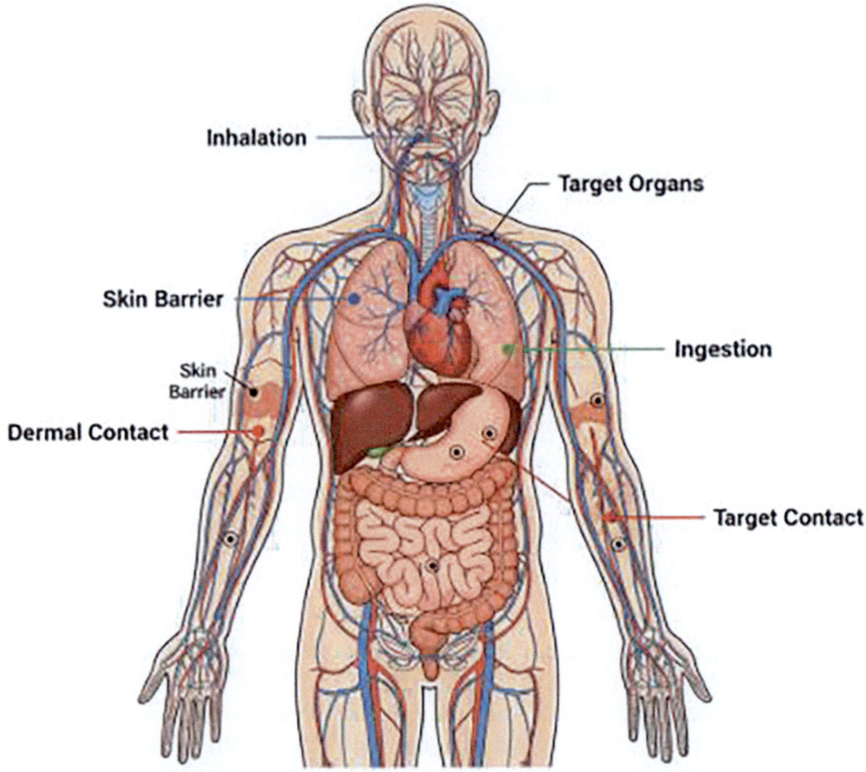

Fig. 4 Human exposure pathways and target organs: diagrammatic overview of the major exposure routes—ingestion, inhalation, and dermal absorption—showing pollutant entry points, systemic distribution, and key affected organs. The figure highlights how environmental contaminants can enter biological systems and contribute to carcinogenic, endocrine, and neurotoxic outcomes

5.1.2 Inhalation (Airborne Pollutants)

Inhalation of air pollutants introduces toxic agents swiftly into our respiratory system, where fine particles (PM2.5) can penetrate the deepest regions of the lung, maximizing systemic absorption (Wang et al., 2024). Alongside traditional pollutants, the airborne exposome also carries biological ECs like viruses (e.g., SARS-CoV-2) and nanoparticles (NPs) and POPs (Li et al., 2024). These advanced airborne ECs can disrupt physiological mobility and metabolism, leading to unpredictable toxicological outcomes. Since they may travel great distances and are produced from a variety of environmental sources and bioaerosols (particles made up of pathogenic bacteria, fungi, and viruses), they represent a particular public health concern which came to light during the COVID-19 epidemic (Mohan et al., 2020). Employees in manufacturing or waste management come into contact with large quantities of ECs by inhalation, making occupational exposure a crucial factor (Lee et al., 2023).

Table 4 Human exposure pathways and associated health risks

Exposure pathway	Typical pollutants	Primary source	Major health outcomes
Ingestion	PFAS, heavy metals, PPCPs	Contaminated water, food	Gastrointestinal, endocrine, carcinogenic effects
Inhalation	PM2.5, VOCs, NPs	Industrial emissions, aerosols	Respiratory, neurological disorders
Dermal	PAHs, EDCs	Occupational contact, cosmetics	Dermatitis, systemic absorption

5.1.3 Dermal Contact (Skin Absorption)

Dermal exposure, although less significant than inhalation or ingestion, can be drastic for certain compounds, especially during direct exposure to skin, for example, while taking a bath, swimming, or during occupational handling (Lei et al., 2015). Additionally, for some semi-volatile organic compounds, such as Polycyclic Aromatic Hydrocarbons (PAHs), direct air-to-skin dermal uptake is shown to be similar to inhalation, providing evidence of a significant and often understated load on total human exposure (Michelangeli & East, 2015). To consolidate the major exposure routes and their corresponding health outcomes, Table 4 provides a comparative summary of typical pollutants, primary sources, and key health effects associated with ingestion, inhalation, and dermal contact. This tabular overview reinforces how distinct exposure pathways collectively contribute to cumulative toxic burdens and chronic disease risks. It also emphasizes ingestion as the predominant route for long-term, low-dose exposure to ECs in humans.

5.2 Health Effects of Environmental Pollutants

Depending on the chemical type, dosage, and length of exposure, long-term exposure to environmental pollutants can have complicated health effects that are marked by delayed effects and non-specificity. A variety of non-communicable diseases have started to be linked with ECs (Mohan et al., 2020; Wang et al., 2024).

5.2.1 Carcinogenesis and Genetic Damage

Environmental carcinogens continually remain a major public health concern. Although research and epidemiological studies on the topic are still ongoing, disinfection by-products (DBPs), especially trihalomethanes (THMs) in drinking water, have been correlated to an elevated risk of colorectal and bladder malignancies (Li et al., 2024). Similarly, perfluoroalkyl compounds (PFAS) have been linked to cancers of the kidney, liver, testicles, and breast despite also being endocrine disruptors (Lei et al., 2015; Wang et al., 2024). The discovery of something called "cancer clusters" (localized increases in specific cancer types) in densely populated places, highlights the systemic risk posed by PM2.5 and toxic waste.

5.2.2 Endocrine and Metabolic Disruption

A prominent effect of exposure to ECs is the disruption of the endocrine and metabolic systems. Pharmaceutical hormones, phthalates, and BPA are prominent examples of Endocrine-Disrupting Chemicals (EDCs) that imitate or inhibit the natural hormone activity (Li et al., 2024; Wang et al., 2024). Symptoms like impaired fertility, abnormal reproductive patterns, and the emergence of diseases can be related to the Metabolic-Disrupting Chemicals (MDCs) which trigger problems with the endocrine system and lead to disorders, including obesity, Type 2 diabetes, and cardiovascular disease, all of which are correlated to this interference (Lei et al., 2015; Michelangeli & East, 2015). For instance, there is a significant relationship between exposure of PFAS and thyroid dysfunction and dyslipidemia (Wang et al., 2024). Moreover, continuous low-dose exposure to these compounds often result in such effects which may also impact epigenetics and be carried on for generations (Lei et al., 2015).

5.2.3 Neurotoxicity and Systemic Effects

Neurotoxicity, particularly from early-life exposure to ECs, is a grave outcome of EC contamination. Certain POPs, cyanotoxins, and heavy metal(loid)s such as arsenic and mercury are known neurotoxicants that increase the risk of neurodevelopmental abnormalities and neurodegenerative conditions like Parkinson's and Alzheimer's later on in life (Lee et al., 2023; Michelangeli & East, 2015). Besides the central nervous system, ECs may also seriously compromise immunity by affecting tolerance, disturbing gut health, and making people more vulnerable to autoimmune and infectious illnesses (Wang et al., 2024). They are also linked with cardiovascular diseases (CVDs) that include hypertension and atherosclerosis (Lei et al., 2015).

5.2.4 Antibiotic Resistance

The environmental introduction of Antibiotics (PPCPs) drives the evolution and dissemination of Antibiotic Resistance Genes (ARGs) and ARB. This transfer of resistance from environmental reservoirs back to the human microbiome (via water, food, and contact) represents a major global health security threat, as conventional treatments become ineffective against these emerging biological pollutants (Mohan et al., 2020; Wang et al., 2024).

5.3 *Risk Assessment and Management*

The essential, methodical framework for assessing the probability and seriousness of negative impacts is provided by Environmental Risk Assessment (ERA). The complexity of ECs mandates a continuous evolution of both assessment methodologies and management strategies—the central theme of this chapter (Li et al., 2024).

5.3.1 Advancements in Risk Assessment Methodologies

For assessing the long-term, low-dose, synergistic effects of EC combinations, traditional toxicology is often insufficient. New Approach Methodologies (NAMs), that provide predictive, high-throughput, and mechanistically informed alternatives to resource-intensive animal testing, were developed as a result of this (Lee et al., 2023; Li et al., 2024). NAMs include complex Microphysiological Systems (MPS), such as 3D cell cultures and organoids, as well as advanced in vitro systems. One example for this is the "body-on-a-chip" technology, which merges several human organ models for the simultaneous, human-relevant evaluation of pollutant impacts (Wang et al., 2024).

Predictive modeling is crucial in addition to NAMs. Artificial Intelligence (AI)/ Machine Learning techniques and Quantitative Structure–Activity Relationships (QSARs) are being used more and more to examine massive chemical and toxicological datasets. AI that specializes in identifying complex patterns, predicting contamination sources, and developing risk prediction tools to assist in setting priorities for monitoring and mitigation efforts is an excellent tool that is highly productive. However, a lack of experimental training data and the inherent issue of adequately predicting the intricate, frequently non-additive effects of multi-pollutant mixtures constrain the development of these models (Li et al., 2024).

5.3.2 Future-Oriented Management and Policy

Efficient management needs to shift away from reactive cleanup to proactive, preventative strategies.

1. **Regulatory Frameworks**: An essential strategy that encourages preventive action in the face of scientific inconsistency regarding possible danger is the precautionary principle. Chemicals are getting increasingly complex as we evolve into a modern society which makes a flexible and dynamic set of regulations necessary to keep up with emerging risks (Mohan et al., 2020).
2. **Global Cooperation and Technology**: Because ECs are a global issue and not a regional one, agencies like the WHO and UNEP which work for the global benefit advocate strongly for uniform global regulations. For the facilitation of this cause, one of the key mitigation strategies is to improve wastewater treatment

technology, for example, making oxidation processes better so as to prevent EC discharge right at its source (Li et al., 2024).

3. **The One Health Approach**: The holistic perspective associated with this model is of great importance as it emphasizes on how closely the life cycles of humans and animals along with the environmental health are intertwined. The One Health approach views environmental contamination as a shared, interconnected catastrophe, which seeks collaboration of multiple fields between environmental research, veterinary science, and medicine so to effectively evaluate and minimize the risks associated with ECs and ARGs (Lei et al., 2015; Mohan et al., 2020; Wang et al., 2024; Destoumieux-Garzón et al., 2018).

6 Ecotoxicology and Wildlife Conservation

Environmental toxicity takes the knowledge of an individual organism and scales it up to a population, community, or entire ecosystem. It is dedicated to the examination on fate and effects of the chemicals in the environment as a point of contact between the environmental science and conservation biology (Peterson et al., 2017).

6.1 Ecological Risk Assessment (EcoRA)

Ecological risk assessment (EcoRA) is a methodology that examines the likelihood of unfavorable effects on ecological receptors and, as such, to wildlife and ecosystems due to the exposure of stressors. EcoRA addresses multiple aspects of biological organization and it manages complex and interconnected endpoints (Chen & Liu, 2014). The EcoRA process has three phases which are designed by regulatory bodies:

6.1.1 Problem Formulation

Problem formulation involves defining the goals of assessment, identification of stressors, characterization of exposure pathways, and selection of specific assessment endpoints, for example, reproductive success of a sensitive bird species, or community diversity in wetland.

6.1.2 Analysis Phase

Analysis phase involves quantification of exposure as well as ecological effects like dose–response relationships for relevant species. A basic tool used in screening level assessment is the risk quotient:

$$RQ = EEC/EC$$

where RQ = risk quotient, EEC = estimated environmental concentration, and EC = effect concentration (the concentration expected to cause an adverse effect).

Here, $RQ > 1$, high risk of toxicity while $RQ \leq 1$, low risk of toxicity (Raimondo & Forbes, 2022).

6.1.3 Risk Characterization

Risk characterization integrates the exposure and effects data and uses it to estimate the magnitude and nature of the ecological risk, with the description of associated uncertainties which helps the informed decision-making by regulatory authorities.

The estimation from measurement endpoints to the desired assessment endpoints is a challenge in EcoRA because it often requires sophisticated population modeling (Chen & Liu, 2014).

6.2 Impacts on Wildlife Populations

Contaminants like Persistent Organic pollutants (POPs) such as DDT and PCBs along with certain heavy metals such as mercury and cadmium pose special danger to wildlife because of their movement and concentration throughout the food web (Barton & Ainerua, 2020) as shown in Fig. 5.

6.2.1 Bioaccumulation

The net uptake and retention of a substance within an individual, occurring when the rate of absorption is higher as compared to rate of elimination, is called bioaccumulation. Substances that have high lipid solubility and long environmental half-lives are more prone to bioaccumulation and are typically stored in an organism's fatty tissues such as adipose tissue and blubber in marine animals (Ali & Khan, 2018).

6.2.2 Biomagnification

Biomagnification is the result of bioaccumulation at the ecosystem level. Biomagnification means the progressive increase in the concentration of the contaminants with each trophic level. As an example, let's consider an algae which is a primary producer, if algae bioaccumulates a toxic substance from water, the primary consumer such as zooplankton that consumes many algae will accumulate the substance at higher concentration, similarly an apex predator like an orca or a bird of prey consumes

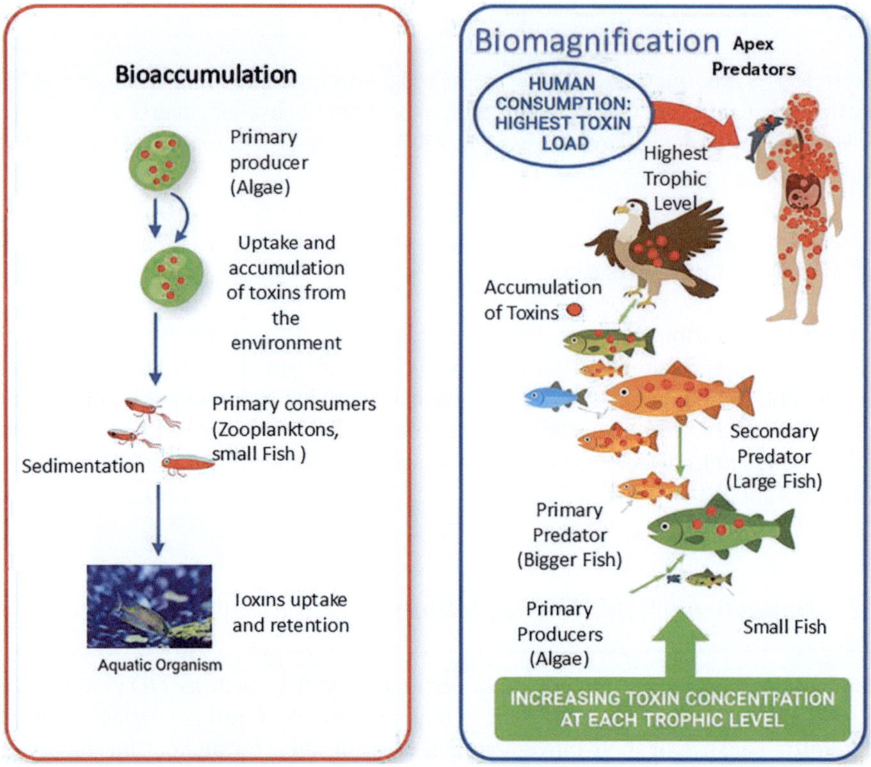

Fig. 5 Bioaccumulation and biomagnification of the contaminants. Diagram shows the bioaccumulation of the toxic substances and their biomagnification through the food chain. The concentration of toxic substances is lowest at the lowest trophic level (Algae), and as the trophic level increases, the concentration of the toxic substances also increases, reaching the maximum concentration at the highest trophic level (apex predators, humans)

many contaminated consumers over the time, which results in a very high accumulation of toxic substances in their bodies (Ali & Khan, 2018; Renuka & Patyal, 2025).

The biomagnified substances usually have very severe toxicological impacts such as chronic, sub-lethal effects like endocrine disruption, reproductive failure, immunotoxicity, and neurotoxicity, on species at the top of the food chain (Ali & Khan, 2018).

6.3 Conservation Strategies and Ecological Restoration

To conserve and restore the environment in order to mitigate and reverse the harm caused due to exposure to chemical stressors, ecological toxicology studies are conducted and conservation strategies are developed. These strategies focus upon prevention and source control:

1. Policy and Regulation: It involves application of international conventions such as Stockholm Convention and local rules and regulations, in order to stop or minimize the production of highly toxic and persistent chemicals in the environment.
2. Targeted Monitoring: It involves the establishment of monitoring programs for long term to track the contaminant levels among species in ecosystems. This provides early warning signals for the upcoming ecological risks.

Ecological Restoration includes efforts done in order to remediate the contaminated sites and recover ecosystem functionality. This include:

1. Structural restoration of ecosystem, including physical restoration of the habitat of native species after the contaminant threat is mitigated.
2. Bioremediation and phytoremediation stand for use of living organisms for remediation of environment by cleaning up contaminants. Here organisms that can metabolize the contaminants are introduced in environment in order to bring back ecological balance and function, thereby making it habitable for the native species again. In bioremediation, microbes like fungi or bacteria are used to biochemically degrade the pollutants or transform them into substances that are less toxic. On the other hand, in phytoremediation, plants are used to extract, sequester, or stabilize the metals and organic pollutants from the soil and water.
3. In bioaugmentation, specific strains of microbes are introduced to a contaminated area. This enhances the natural process of degradation of the contaminants.

The integration of toxicological-risk models such as EcoRA and ecological principles gives effective ecological conservation. This also helps in prioritized site selection, selection of appropriate remediation methods, and measurement of restoration success on the basis of recovery of assessment endpoints.

7 Regulatory Frameworks and Policies

There are various regulatory frameworks (national and international) that set-out provisions for environmental protection and also provide an action plan to govern the scientific findings of toxicology, especially those related to human health. These policies set standards for environmental quality and public safety and also govern how chemicals are developed, tested, used, and disposed of.

7.1 National and International Regulations (Toxic Substances Control Act, REACH)

Different regulatory frameworks may differ globally, but they share a common goal of chemical-risk management. Examples showing different approaches are as follows:

7.1.1 Toxic Substance Control Act (TSCA) (United States)

The TSCA was established in 1976 and was later amended by Frank R. Lautenberg Chemical Safety for the 21st Century Act in 2016, which is the primary law in the US for regulating chemical substances (Han et al., 2024). TSCA mandates the U.S. Environmental Protection Agency (EPA) to

1. **Review Existing Chemicals**: The 2016 amendments mandated the EPA to prioritize and run a systematic risk-based analysis of the chemicals that are already in the market. It is done for the assessment of the hazardous exposure and specifically focuses on the potential risk that may be caused to the vulnerable populations and the characteristics of the chemicals such as persistence, bioaccumulation, and toxicity (PBT).
2. **Evaluate New Chemicals**: The producers of these chemicals have to submit a pre-manufacture notice (PMN) for manufacturing of new chemical substances, which helps the EPA to conduct its risk assessment, and the result of this assessment should be that the substance is "not likely to cause harm" for it to enter the market. Once a substance passes this test the manufacturer is free to release the chemical into the market.
3. **Mandate Testing**: The amendment of 2016 gives powers and a broad authority to the EPA to mandate the manufacturers and importers to conduct a test on these chemicals and then submit the data on health and safety, chemical identity, production volume, potential exposure, and disposal.

After the amendment of 2016, the EPA mandated that the chemical shall cause no harm to the human health and the environment and the primary and the economic cost were considered secondary. Now the burden of proof lies upon the manufacturer to prove that his chemicals are not harmful rather than the EPA proving that it's harmful. It allows the EPA to ensure that a harmful substance doesn't reach the public (Wang et al., 2025).

7.1.2 Registration, Evaluation, Authorization and Restriction of Chemicals (REACH) (European Union)

The European Union has adopted a regulation that makes sure that any toxic substances do not reach the public and the markets. This regulation is **REACH** (Registration, Evaluation, Authorization and Restriction of Chemicals) and it is one of the most comprehensive chemical regulation globally. An independent regulatory agency ECHA is established to oversee the functions of the REACH. REACH places the responsibilities of risk management of chemicals and providing safety information about the chemical, on the company. If any company or firm does not comply with this regulation then their product may be seized or banned from the European market.

1. **Registration**: All the companies based in Europe have to mandatorily register in-depth information on its essential properties of the chemical substances if they are produced and imported more than 1 ton per year, a user help guide and safety measures guide shall be provided with it (Hansen & Blainey, 2008).
2. **Evaluation**: The national authorities and ECHA evaluate the application dossiers and test proposals and ensure that sufficient data exists to identify risks.
3. **Authorization**: This the most rigid measure to ensure safety it applies to the substances of very high concern (SVHCs). It includes those chemicals that are characterized as carcinogenic, mutagenic, or reproductive toxicants, and also those which are persistent or very persistent and bio-accumulative or very bio-accumulative. These types of chemicals can only be allowed to use if the applicant can prove that the potential risks are under control or that the use of these substances has great socio-economic advantage and also that there are no existing technically and economically suitable options.
4. **Restriction**: If the risk evaluated is unacceptably high then the manufacturing, placement in the market or usage of such hazardous substances may be restricted in all over Europe.

REACH is significant for its encompassing focus on the inherent hazardous properties of a substance and its dynamic data-driven system management through ECHAs' centralized database (Williams et al., 2009).

7.2 Policy Implications and Enforcement

7.2.1 Risk Management and Standards

The regulatory bodies use the results that they get through these assessments: Human Health Risk Assessment (HHRA) and Ecological Risk Assessment (EcoRA), and then form enforceable standards. For the protection of sensitive human subpopulation, the reference dose or concentration that is derived by the application of the safety factors to the BMDL or NOAEL is set up to get non-carcinogenic effect. For potential carcinogens, the authorities often adopt a below threshold and linear dose–response model to quantify regulatory standards, such as cleanup levels or allowable discharge, which is based on the quantified cancer slope factor. These standards set a limit for the acceptable exposure of food contaminants and occupational exposure limits, such as Maximum Contaminants Levels (MCLs) for water and Acceptable Daily Intakes (ADIs) (Lee & Choi, 2019).

7.2.2 Enforcement

Enforcement literally means compliance with the newly established standards. The enforcement of these standards requires strong environmental monitoring, regular facility inspections, and mandatory reporting programs such as the toxic release inventory, in the U.S. An effective legal framework is backed by a statute, it is to deter the no-compliance, and it does so by imposing penalties (civil and criminal) to ensure accountability of the polluter industries. The enforcement works upon the "polluter pays principle" (Lee & Choi, 2019).

7.2.3 Precautionary Principle Versus Sound Science

Regulatory policies usually involve a fundamental reasoned balancing act. The Precautionary Principle says that a preventive action shall be taken against the industry even if the scientific evidence or proof is insufficient, this places the burden of proof of the safety on the industry. The regulatory frameworks like REACH often lean toward this principle, specifically regarding SVHCs. On the other hand, Sound Science Approaches have a totally different view, it says that a regulatory restriction or penalty shall be backed by sufficient proof or evidence that clearly shows an unreasonable risk under realistic exposure scenarios (Lee & Choi, 2019; Turaga & Sugathan, 2020).

7.3 Future Directions for Regulatory Frameworks

The rapid technological and scientific advancements have led to the rapid adaptation by the regulatory toxicology, and this new change has been seen most prominently in genomics, in vitro testing, and computational modeling (Patel & Shah, 2017). The future regulations are expected to prioritize efficiency, speed, and ethical considerations:

7.3.1 Integration of New Approach Methodologies (NAMs)

Scientists are moving away from the traditional animal testing to the new approaches that are more humane, modern, and advanced, which includes high-throughput screening, in silico toxicology, and sophisticated computational toxicology. Regulators are using data from NAMs progressively rather than relying on the animal testing, new approaches such as in vitro assays for breakage of endocrine system or immune responses, to swiftly prioritize and screen chemicals. This kind of methods are rapidly validated using the Adverse Outcome Pathways (AOP) approach to make sure that the molecular changes that are measured are relevant to adverse outcomes observed in vivo (D'Amore et al., 2025).

7.3.2 Chemical Grouping and Read-Across

The regulatory sciences are working on grouping the chemicals with the same kind of structures or mechanisms of the process. To predict the toxicity of an untested target chemical within the same group, the read-across framework uses the toxicity data from a properly tested source chemical. The read-across approach is often backed by cheminformatics tools, such as quantitative structure–activity relationship, and there is a model toxicity on the bases of molecular structure, drastically improving the efficiency of the assessment process (D'Amore et al., 2025).

7.3.3 Global Harmonization

There is a strong international mandate that is advocated by the organizations like OECD, and it is to unify the testing frameworks, data standards, and chemical categorization system. The globally harmonized system of classification and labeling of chemicals (GHS) calibrates the hazard communication all across the globe. What's more is that the OECD's mutual acceptance of data (MAD) system makes sure that according to OECD guidelines, toxicity test data that is created in one member state must be accepted in all the other member states as well; it is underlying for commerce across the globe, and helps in identifying the duplicative testing and reduces further efforts as well as improves animal welfare (D'Amore et al., 2025).

8 Emerging Contaminants and Challenges

The constantly changing lifestyle of human beings keeps introducing new chemical substances and pollutants in the environment. These substances that are recently recognized as contaminants due to their persistence and their potential impact on the ecosystem and human health. These contaminants include pharmaceuticals, industrial by-products, and microplastics that accumulate and interact with the environment and pose great risks.

8.1 Pharmaceuticals and Personal Care Products (PPCPs)

PPCPs are a vast group of chemicals that are continuously being released in the environment through various ways like wastewater, air, industrial waste discharge, etc. and the wastewater treatment plants that are not designed to completely remove such substances lead to their introduction in the environment. PPCPs include the over-the-counter drugs like antibiotics, hormones, antidepressants, etc., prescription drugs, veterinary medicines, cosmetics, and many such consumer products (Rehman et al., 2024).

8.1.1 Mode of Action and Risk

PPCPs are biologically active molecules and chemicals, thus they can apply their therapeutic effects in the unintended environmental receptors even at low concentrations. This phenomenon underscores the threat of pharmacologically active environmental contaminants (PAECs) acting on non-target organisms. One of the primary concerns due to PAECs is the endocrine disruption because of the action of synthetic biomolecules like hormones. This may lead to severe effects, for example, sex reversal in male fish population even at nanogram-per-liter concentrations. Another example is of triclosan which is a common antibacterial agent that bioaccumulates in aquatic organisms and disrupts their thyroid hormone signaling pathway. On top of that, the constant presence of small amounts of antibiotics in the environment leads to the emergence of antibiotic resistance bacteria, multi-drug-resistant bacteria, and even extremely drug-resistant bacteria, which is a very serious global health crisis (Liu et al., 2025; Rehman et al., 2024).

8.1.2 Fate in Environment

Multiple PPCPs are considered to be pseudo-persistent because inspite of their short environmental half-lives and quick degradation, their persistence in the environment is ensured by their constant introduction into the sewer systems which leads to chronic low level exposure. The complex chemical structures of such compounds make their natural degradation even more difficult. Sometimes their degradation products can sometimes be even more toxic and persistent than the original compound. This leads to the requirement of toxicological assessment of a chemical and its environmental transformation product (Rehman et al., 2024; Wang et al., 2024).

8.2 Nanomaterials and Nanotoxicology

Nanomaterials are substances that are engineered by humans, with at least one of their dimensions between 1 and 100 nm. Having such exceptionally small size, these substances exhibit unique chemical and physical properties like extremely high surface-to-volume ratio, quantum effects, etc., which makes them different from their bulk forms (Akhtar et al., 2025). Some nanomaterials have also been identified as contaminants which necessitates the field of nanotoxicology.

8.2.1 Unique Toxicological Challenges

The toxicity of nanomaterials depends on their shape, size, surface coating, and aggregation state rather than on chemical composition. For example, materials like carbon nanotubes (CNTs) which are highly rigid and have high aspect ratio may exhibit toxicity pathways that are similar to asbestos fibers, which may lead to chronic inflammation in lungs. On the other hand, common metal oxide nanoparticles like TiO_2, ZnO, etc. can lead to oxidative stress because they have catalytic surface activity, thus causing cytotoxicity, membrane damage, and genotoxicity in living cells. An example is silver nanoparticles (AgNPs), which is used for its antibacterial properties (Abonyi et al., 2024). These exhibit toxicity mainly by releasing Ag^+ ions. The extent of release of Ag^+ ions decides the result, whether it will be favorable antibacterial activity or toxicity. Thus, the assessment of chemical kinetics is also very important in risk assessment of a nanoparticle (Kharrazian, 2021; Tralau et al., 2015).

8.2.2 Environmental Fate and Transport

The risk assessment of nanomaterials is very complicated because of the fact that they are very highly reactive and dynamic in the environment. The nanomaterials readily agglomerate or dissolve depending upon their environment and thus altering their bioavailability. On top of that, the small size of nanoparticles allows easy internalization by the cells and their transport across the biological membranes as compared to their bulk counterparts. This reason makes the standard regulatory tests, which are based on mass concentration, insufficient for their complete risk profile (Akhtar et al., 2025; Wang et al., 2024).

8.3 Climate Change and Environmental Toxicology

Presence of contaminants and toxicants in the environment gives rise to climate change directly or indirectly, while the climate change makes it favorable for the contaminants to stay persistent in the environment. It acts as a global modulator for the exposure to contaminants, their transport, and effects (Noyes et al., 2009). Figure 6 shows a synergy between physical and chemical stressors which becomes a challenge for the single stressor risk assessment.

8.3.1 Altered Exposure Pathways

8.3.1.1 Temperature

The temperature may have an effect on the toxicity and activity of some contaminants. Elevated temperatures can accelerate or retard an organism's metabolic rate, thereby increasing or decreasing of their toxicant uptake. On the other hand, higher temperature of water causes a decrease in the saturation of dissolved oxygen, which creates a situation of hypoxia. This condition becomes a very strong stressor, making the aquatic organisms even more susceptible to chemical harm (He et al., 2025).

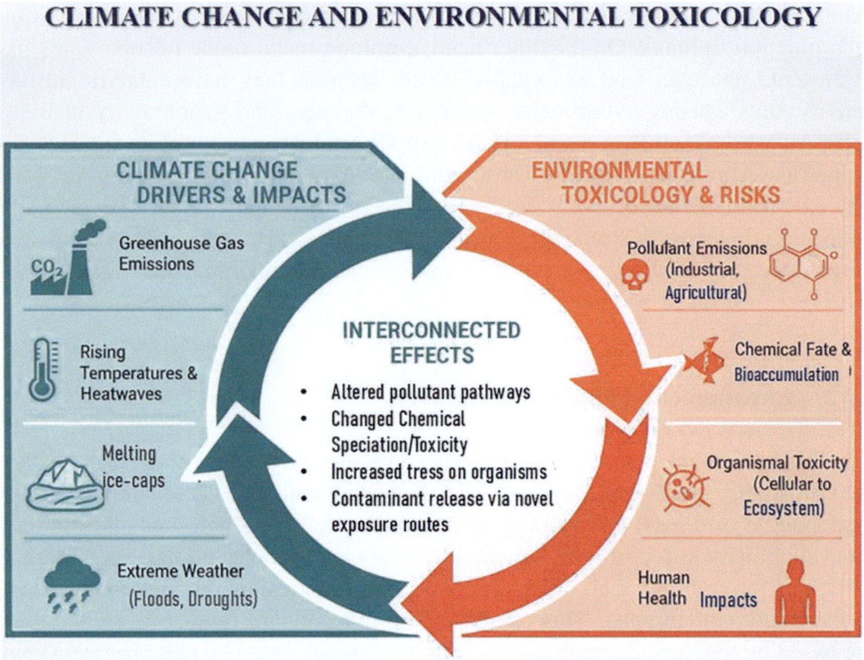

Fig. 6 Relationship between climate change and environmental toxicology. Diagram shows the inter-relatedness of the climate change and environmental toxicology. This diagram represents how environmental toxicology compliments climate change and climate change promotes the presence of toxic substances in environment

8.3.1.2 Extreme Weather and Hydrology

Disasters and calamities, such as floods, droughts, and earthquakes, affect and control the dynamics of contaminants in the environment very significantly. Floods may lead to re-dissolution of contaminants like PCBs, dioxins, heavy metals, etc. that had sedimented in the soil at contaminated sites, which may lead to sudden increase in concentration of contaminants in the water (He et al., 2025). On the other hand, droughts concentrate contaminants in shrinking water bodies, thereby increasing the local exposure concentrations.

8.3.2 Increased Bioconcentration, Bioavailability, and Speciation

Water pollution ultimately leads to the change of ocean chemistry, specifically the ocean acidification, that happens due to absorption of atmospheric CO_2. This can severely alter the chemical speciation and solubility of metal ions like copper and zinc. For example, with decreasing pH, the bioavailability and toxicity of some metals increase. Such shifts may increase their uptake by marine organisms and contribute to bioaccumulation of toxic substances. Another example is the increased production of methylmercury, which is highly toxic, by microbes because of increased global temperatures, increasing the rate of bioaccumulation and biomagnification in the food web (Kharrazian, 2021).

8.3.3 Changes in Organism Sensitivity

The change in climate may impart primary physical stresses like thermal extremes, loss of habitat, and compromised reproductive cycles. This may redirect energy within an organism from vital detoxification and repair pathways. Due to this, organisms that are already stressed by such physical changes become especially vulnerable to the chemical toxicity, which is studied under a special field known as co-stressors and multiple stressors. The incorporation of these complex interactions into development of risk prediction models is very necessary in order to accurately predict or forecast the future outcomes of a chemical in the environment (Alum, 2025; Hutton et al., 2024; Kharrazian, 2021).

9 Research Methods and Tools

9.1 Laboratory and Field Studies

Experimental toxicology offers a fundamental understanding of how pollutants interact with biological systems. While field investigations validate findings in real-world settings, controlled laboratory research isolates precise causal mechanisms of toxicity by eliminating confounding environmental variables. Enzyme inhibition, DNA damage, and oxidative stress indicators are examples of biomarkers that are essential for tracking exposure and sub-lethal impacts in a variety of wildlife species (Li et al., 2024; Wang et al., 2024).

9.1.1 Bioassays (Laboratory and In Situ)

Bioassays are quick, cost-effective methods used to evaluate the integral toxicity of environmental samples. They are particularly valuable because they capture the cumulative, mixture toxic effects (both synergistic or antagonistic) that rely on target chemical analysis alone would miss (Lei et al., 2015). For instance, growth inhibition is typically the primary aim for standard microalgal bioassays. However, because negative impacts are complicated and naturally connected, contemporary environmental assessment calls for evaluating a variety of endpoints, including photosynthetic efficiency, ROS production, and antioxidant enzymatic activity like SOD and

POD (Michelangeli & East, 2015). For sediment-overlying water, the idea of multi-species testing and in situ assays ensures that toxic responses are reviewed for species specificity and ecological relevance (Mohan et al., 2020).

9.1.2 Advanced Laboratory Techniques/New Approach Methologies (NAMs)

The expensive resource consumption, moral dilemmas, and frequently poor predictive ability of conventional in vivo animal studies, particularly for chronic, low-dose EC exposures, have caused a significant change in modern toxicology toward New Approach Methodologies (NAMs) (Wang et al., 2025).

NAMs provide human-relevant, high-throughput, and mechanistically informed data (Li et al., 2024).

1. **In Vitro Systems**: These include complex platforms like 3D cell cultures, induced pluripotent stem cell (iPSC)-derived organoids, and microphysiological systems (MPS), often referred to as "organ-on-a-chip" systems. MPS are the frontier, allowing for the concurrent assessment of pollutant effects across multiple integrated human organ models (Lee et al., 2023; Wang et al., 2024).
2. **Omics-Based Approaches**: These novel and modern-day approaches offer deep mechanistic understanding of contaminants by uncovering perturbations at a molecular level. This includes the areas of genomics, transcriptomics, proteomics, and metabolomics which evaluate how pollutants alter metabolic pathways, gene expression, and protein function. Among these, metabolomics is the study of the low-molecular-weight metabolites which is especially suited for investigating complex contaminant combos and assessing outcomes across different organisms, providing an insightful molecular-level biomarker (Michelangeli & East, 2015).

9.1.3 Molecular and Biological Studies

For determining cytotoxic effects, techniques such as flow cytometry are vital because they allow for high-throughput, multiparametric assessment of single-cell parameters like viability, DNA content, and size (Li et al., 2024). Modern detection techniques for biological ECs (pathogens, ARGs) depend upon next-generation sequencing (NGS) and high-throughput quantitative polymerase chain reaction (qPCR) techniques for metagenomic studies, which permit for a quick and thorough contamination landscape assessment (Mohan et al., 2020; Wang et al., 2024; Clippinger et al., 2018; Hollender et al., 2017).

9.2 Analytical Techniques

Advanced analytical procedures must have a high degree of sensitivity and selectivity in order to detect and precisely quantify ECs, which are frequently present at trace quantities (parts per billion or parts per trillion) (Wang et al., 2024).

9.2.1 Sample Preparation and Advanced Sampling

Meticulous sample preparation remains the bottleneck in trace analysis. Contemporary methodologies like Solid-Phase Extraction (SPE) and microextraction methods (e.g., MEPS, SPME) are designed to isolate ECs and restrict matrix interference (Lei et al., 2015). Due to its flexibility, Solid-Phase Microextraction (SPME) supports high throughput and even in vivo sample (using small-needle forms) for a direct evaluation of the chemical exposome at the sampling location (Li et al., 2024). A cumulative exposure profile that is more ecologically significant than single grab samples is also offered by passive samplers, such as POCIS, which have become the industry standard for water monitoring since they accumulate trace contaminants over long periods of time (Mohan et al., 2020).

9.2.2 Chromatography and Mass Spectrometry (MS)

Chromatography coupled with Mass Spectrometry (MS) forms the core technology for pollutant analysis (Michelangeli & East, 2015).

1. **Gas Chromatography–Mass Spectrometry (GC–MS)**: Highly efficient for volatile and semi-volatile compounds. The main limitation is that polar or large molecular weight ECs must undergo chemical derivatization, which increases complexity and error risk (Li et al., 2024).
2. **Liquid Chromatography–Tandem Mass Spectrometry (LC–MS/MS)**: For polar, non-volatile ECs like PFAS and medications, this potent instrument works better. It is a crucial necessity for Non-Targeted Screening (NTS) (Wang et al., 2024). High-Resolution MS (HRMS) equipment (e.g., Orbitrap, TOF) are crucial due to their high mass resolution, which enables the structural clarification and tentative identification of hitherto unidentified ECs and their transformation products (Lee et al., 2023). However, the vulnerability of LC–MS/MS to matrix effects, which can either boost or suppress ionization and affect accuracy, continues to be a significant concern (Lei et al., 2015).

9.2.3 Other Specialized Analytical Techniques

Some of the analytical techniques that can be categorized in earlier categories are the ones that don't follow the principle of chromatography. Inductively Coupled Plasma–Mass Spectrometry (ICP–MS) is one of the techniques which is also the gold standard method for ultra-trace detection of rare earth elements, metalloids, and heavy metals in inorganic contaminants (Wang et al., 2024). Another method for de novo structural characterization of newly discovered contaminants and their breakdown products is nuclear magnetic resonance (NMR) spectroscopy. It is a significant technique, although less sensitive than mass spectroscopy (MS) but presenting a higher scope of innovation (Michelangeli & East, 2015). Also, spectroscopic techniques like Fourier Transform-Infrared Spectroscopy (FT-IR) and Raman Spectroscopy have become essential tools when requirement is there for a quick, non-destructive characterization of elements, particularly in the expanding field of identifying and quantifying microplastics (Li et al., 2024; Mohan et al., 2020).

9.3 Computational Models and Simulations

A crucial research venture for the future lies in the application of computational toxicology and bioinformatics in risk management and hazard prediction. The convergence of experimental, analytical, and computational strategies in modern environmental toxicology is conceptually summarized in Fig. 7. This integrative model illustrates how New Approach Methodologies (NAMs), multi-omics technologies, and advanced analytical tools such as GC–MS and LC–MS/MS interface with AI/ML-based computational modeling to predict pollutant behavior and risk. By linking laboratory data with computational simulations, this framework underlines the transition toward mechanism-based, predictive toxicology.

9.3.1 Quantitative Structure–Activity Relationships (QSARs)

QSARs are advanced analytical models that establish links between the molecular makeup of a substance and its toxicological or environmental impact. They play a crucial role in swiftly forecasting the fate (volatilization or bioaccumulation) and possible hazard of novel or inadequately documented ECs (Wang et al., 2024). They are central to tiered hazard evaluation programs. However, their reliability is currently constrained by the scarcity of high-quality experimental training data, which can lead to statistical issues like over-fitting and necessitates the application of high uncertainty factors in regulatory decisions (Wang et al., 2024).

9.3.2 Fate, Transport, and Risk Models

Mathematical models that predict the flow and change of pollutants within biological and environmental compartments, like Pharmacokinetic (PK) models and multimedia fate models, enhance monitoring efforts (Li et al., 2024). To predict the Long-Range Transport (LRT) of semi-volatile ECs like PFAS and microplastics, air transport models such as 3D chemical transport models are required. The absence

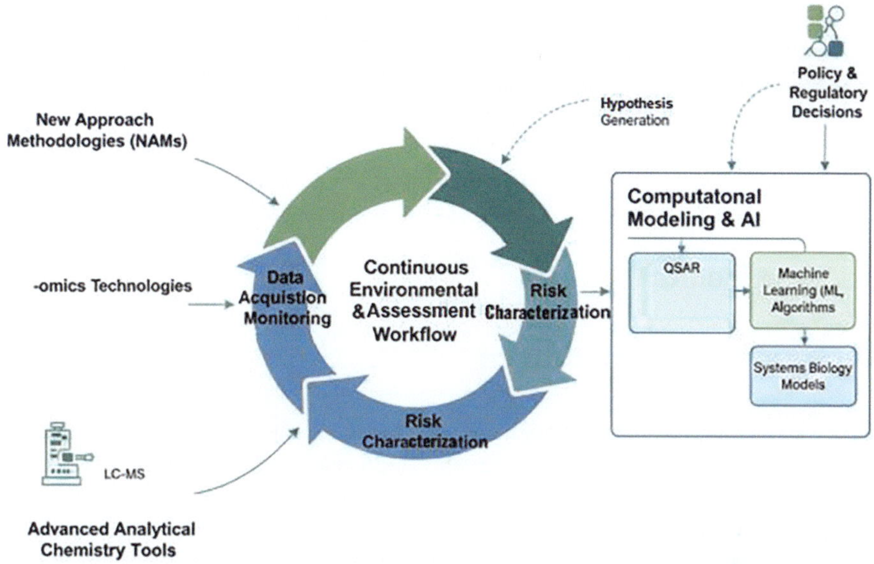

Fig. 7 Research tools and computational integration in environmental toxicology: conceptual model illustrating the convergence of experimental and computational methodologies in modern environmental toxicology. It integrates New Approach Methodologies (NAMs), multi-omics analyses, analytical techniques (GC–MS, LC–MS/MS, ICP–MS), and AI/ML-based computational modeling (QSARs, PBPK models) to enable mechanism-based risk prediction and regulatory assessment

of universal, process-based models that can faithfully capture the intricate, non-linear toxic processes and biological interactions that define EC migration in actual ecosystems is a key challenge in this study (Mohan et al., 2020).

9.3.3 Artificial Intelligence (AI) and Machine Learning (ML)

AI and ML have emerged as tools of rapidly growing significance in environmental risk research over the last decade, acting as powerful tools for pattern recognition and prediction.

1. **Predictive Modeling and Risk Assessment**: ML models (random forest, deep learning) are used extensively to forecast air contamination risk, predict the adsorption capacity of ECs onto remediation materials (like biochar), and assess the bioactivities of compounds like PFAS in human targets (Li et al., 2024). AI can extract hidden patterns from vast toxicogenomic datasets to support mechanism-based predictions and accelerate the regulatory read-across between similar compounds (Mohan et al., 2020; Wang et al., 2024).
2. **Environmental Monitoring**: AI algorithms are now analyzing sensor network data and satellite imagery (GIS) to monitor EC occurrence in real time in water

treatment plants and river systems, enabling autonomous adjustment of treatment chemicals for improved efficiency (Lee et al., 2023). The ultimate future direction of this field is the development of AI-enabled, true unsupervised Non-Targeted Screening (NTS), which will revolutionize the ability to detect and identify novel ECs instantly without prior knowledge.

10 Case Studies

10.1 Environmental Disasters

Human error, chemical mismanagement and mismanagement of resources, industrial activities, political activities, warfare, etc. often lead to large-scale disasters that become the stark remainders of the hazards caused by humans. Studying such events is very important for learning the toxicological impacts of human carelessness on wildlife, ecosystems as well as human health and lives. It is also important for improving the regulatory frameworks in order to prevent future occurrences of such disasters. Some examples are as follows:

10.1.1 Bhopal Gas Tragedy (1984)

The Bhopal Gas Tragedy occurred in December 1984 in the city of Bhopal, India and it is considered to be one of the top catastrophic industrial disasters in the history of mankind. Carelessness in handling and management caused a leak of methyl isocyanate (MIC) in a pesticide manufacturing plant. This exposed approximately 500,000 people to toxic MIC which led to severe respiratory distress, eye irritation, and thousands of deaths within days and the long-term impacts include neurological disorders, respiratory problems, and reproductive issues.

The toxic gas was released into the environment and from there it got absorbed into the local soil and water systems. Because of that, the toxic nature stayed in the environment for decades. This disaster highlighted lacking industrial safety protocols, emergency preparedness, and toxic waste management. It also led to changes and development in regulatory procedures focusing on industrial risk assessment, emergency response protocols, and worker safety standards in India as well as globally (Ellos & Bacosa, 2025; Kumar, 2025).

10.1.2 Chernobyl Nuclear Disaster (1986)

The Chernobyl nuclear disaster happened in April 1986 in Ukraine, which was a part of Soviet Union at that time. It happened during a reactor safety test when an explosion caused release of large amounts of radioactive isotopes that includes

^{131}I, ^{137}Cs, and ^{90}Sr which heavily impacted the human lives in the surrounding environment.

While, at the time of explosion, the immediate deaths were limited to the plant workers, the long-term effects like thyroid cancer, leukemia, and genetic mutations subsequently affected thousands. Another result of Chernobyl disaster was the spread of radioactive contamination across Europe, which lead to bioaccumulation in plants, animals and water systems, and biomagnification in food web. This disaster brought light to the importance of nuclear safety, environmental monitoring, and transparent communication during emergencies (Deniz, 2025; Yap & Al-Mutairi, 2025).

Although the basic natures of both the disasters were different, one was chemical and other was radiological, one common point in both the places was human intervention and mishandling toxic substances and, as a result, the ecological exposure to those substances leading to long-lasting ill-effects and consequences shown in Fig. 8.

10.2 Successful Remediation and Restoration Projects

Remediation and restoration projects point out the possibilities of mitigation and recovery of the environment after such disasters. Some examples are as follows:

1. Love Canal, USA: In New York, a chemical landfill caused highly extensive groundwater contamination. Its remediation involved removal of contaminated soil, containment system installation, long-term monitoring, etc. which, as a result, slowly restored the ground water and residential safety (Gill & Mix, 2020).
2. Minamata Bay, Japan: In Minamata Bay, industrial wastewater led to mercury contamination which caused bioaccumulation in marine life and severe neurological disorders in humans. Remediation was done by comprehensive and regular clean up, including the sediment removal and very strict regulations regarding the industrial effluent. This gradually reduced the mercury levels in the environment (Harada, 1995).

These examples represent some of the successful remediation projects that showcase the importance of the involvement of community, development of stringent regulations, and long-term monitoring of toxicants and contaminants in the environment and population.

10.3 Emerging Issues

Some of the major issues include widespread detection of emerging contaminants, such as PFAS (per- and polyfluoroalkyl substances), microplastics, and pharmaceutical wastes, which cause complex toxicological risks. They compromise

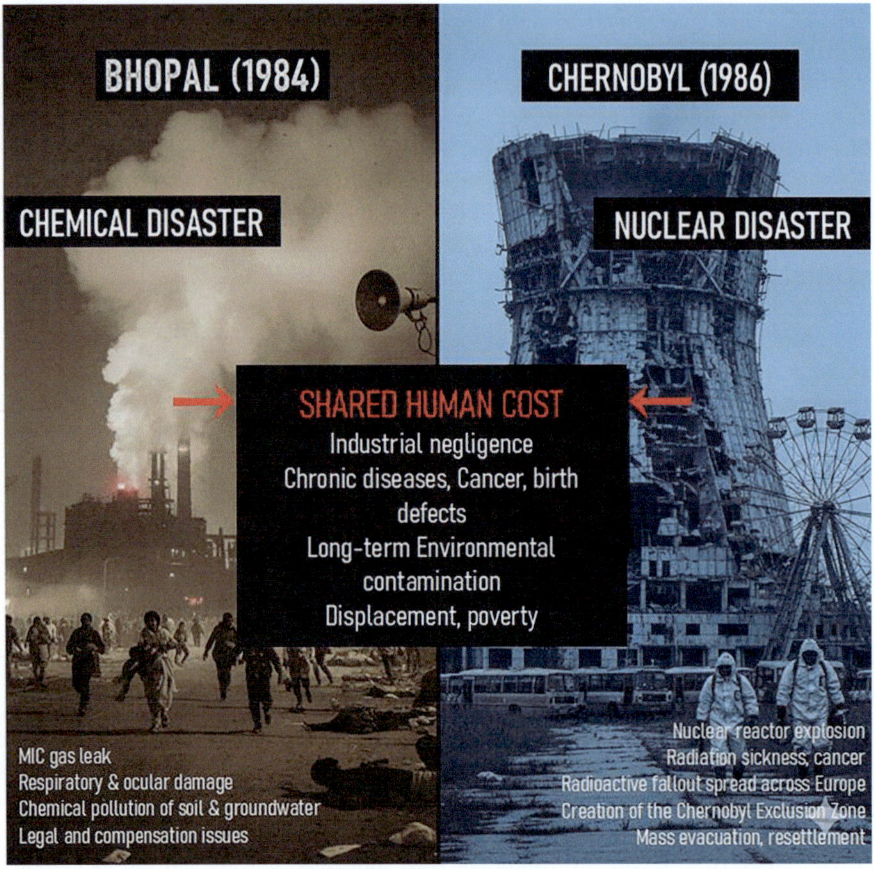

Fig. 8 Diagram showing the comparison between Bhopal Gas Tragedy (1984) and Chernobyl Nuclear Meltdown (1986). The diagram shows the results of human negligence led to two massive disasters. Diagram also shows the after effects unique to each disaster as well as common after effects

human health as well as environment by being present in the environment and bioaccumulating at every trophic level of food chain thereby disrupting ecological stability.

Rapid globalization, industrial expansion, and growth increase the risk of cross-border pollution. Therefore, managing the chemical and radioactive wastes and prevention of accidental releases of pollutants remain big challenges especially in developing areas.

Changing climate patterns also affect the behavior of contaminants, and their distribution and toxicity. Keeping up with the emerging contaminants, new distribution patterns, and new toxicity patterns of contaminants is also a challenge.

In order to tackle the emerging challenges, important steps need to be taken and reforms need to be made in regulatory practices. Some of such steps are development of early warning systems and models for prediction of contaminant behavior, promoting sustainable practices, green chemistry, and eco-friendly industrial procedures to reduce environmental burden and reinforcing global policies and frameworks, and make cooperative efforts to manage pollution and track emerging contaminants worldwide.

11 Future Directions and Research Paths

11.1 Interdisciplinary Approaches and Collaboration

Owing to the complex nature of today's issues related to the environment, primarily the pervasiveness and transboundary character of Emerging Contaminants (ECs), scholarly study needs to transition from isolated investigations to radical interdisciplinary cooperation (Richardson & Kimura, 2020). Environmental toxicology will need to integrate insights from ecology, data science, environmental genomics, analytical chemistry, and public policy in order to understand ecosystem–human health relationships at a systems level (UNEP, 2022).

11.1.1 Adopting the One Health Perspective

In order to address ECs and global environmental risk, the One Health concept is specifically acknowledged as the essential and comprehensive paradigm (Destoumieux-Garzón et al., 2018). The health of people, animals, plants, and their shared environment are all explicitly connected by this approach. To enhance planetary well-being and minimize contamination risks, One Health implementation and its success requires a combined effort from a number of disciplines including public health, environmental science, veterinary medicine, and general medicine (Destoumieux-Garzón et al., 2018). By acknowledging the inherent interconnection of all phenomena, be it animate or inanimate, within ecological webs, including the indigenous worldviews into this concept is an essential research route that provides a deeper and more complex understanding of environmental interactions (UNEP, 2022).

11.1.2 Cross-Sectoral and Global Cooperation

Given that ECs do not respect political boundaries, international collaboration is a paramount measure that the global community must inevitably adopt. For this, building strong international networks and collaborations to enable the sharing of resources that are required like cutting-edge technology and scientific information, and this will increase the world's ability to handle complicated pollution problems with multiple origins. Moreover, to manage assessments along with developing all-encompassing mitigation plans, cross-sectoral cooperation between government organizations, business, academia, and public health specialists is also necessary (Richardson & Kimura, 2020).

11.1.3 Enhancing the Research-Policy Link

A persistent difficulty in environmental management is the gap between scientific discovery and timely regulatory action. Effective environmental preservation requires strengthening the connection between research and policy. To guarantee that the most recent, mechanism-based scientific results (such as those from NAMs) directly influence policy decisions and the researchers will obtain useful knowledge of regulatory limitations and public health goals, this calls for a continuous, two-way conversation (UNEP, 2022). To make sure that the policies are evidence-based and flexible enough to reflect new scientific discoveries, this synergy implies the creation of institutional systems for merging scientific research into dynamic regulatory frameworks.

11.2 Advanced Technologies and Methodologies

Future environmental initiatives must place more emphasis on using cutting-edge technologies for thorough monitoring, proactive risk assessment, and long-term EC cleanup. This technological evolution centers on computational tools and non-animal testing methods.

11.2.1 New Approach Methodologies (NAMs) and Mechanistic Insight

The future of environmental toxicology depends on the shift to mechanistically informed, high-throughput and predictive NAMs (Schmeisser et al., 2023). Further, future studies are needed to construct human-relevant platforms that can transcend the moral and scientific constraints of conventional animal testing, such as 3D cell cultures, organoids, and Microphysiological Systems (MPS) (organ-on-a-chip) (Pamies et al., 2022). Important barriers to NAM integration must be overcome, namely, the absence of established procedures, lack of regulatory guidelines, and the limited accessibility of industry data to the general public. For high-priority chemical classes like PFAS and microplastics, proposals for centralized validation hubs should be proposed and future NAM validation should display strong reproducibility and direct relation to established Adverse Outcome Pathways (AOPs). Simultaneously, to obtain a mechanistic understanding which is also necessary to properly interpret the NAM data, it is crucial to employ multi-omics techniques like genomics, transcriptomics, and metabolomics.

11.2.2 Artificial Intelligence (AI) and Machine Learning (ML)

Since AI and ML can capture uncertainty in massive environmental big data sets and simulate the complicated and non-linear connections, they are thought to be essential tools for environmental risk research in the future (Schmeisser et al., 2023).

1. **Predictive Governance**: By combining data from satellite monitoring systems and urban sensor networks, AI models could easily estimate pollution hazards or predict environmental disturbances such as uncontrolled discharges to provide early danger warnings (Richardson & Kimura, 2020).
2. **Analytical Revolution**: True, unsupervised Non-Targeted Screening (NTS), which will eventually be made possible by future advancements in machine learning, is highly anticipated as it would enable analytical chemists to quickly and without prior knowledge detect novel ECs in complicated matrices. This breakthrough is essential for keeping up with chemical innovation. Continued research is required for the development of AI-based environmental risk assessment and control technologies.

11.2.3 Remediation and Treatment Innovations

Research when it comes to topics like innovation, sustainability, and energy-efficient remediation technologies should be prioritized so that innovations to improve remediation approaches based on nanotechnology or efforts aimed at producing nanomaterials that are more stable, reusable, and have fewer secondary hazards can be focused on (UNEP, 2022). To significantly boost the consumption of energy more efficiently and reduce hazardous by-products, research must be conducted for advanced chemical processes like Advanced Oxidation Processes (AOPs). Also, techniques involving biosorption and bioaccumulation need adequate funding along with the use of genetic engineering and biotechnological developments to improve their efficacy and selectivity for certain pollutants (Richardson & Kimura, 2020).

11.3 Policy and Regulatory Implications

To handle the chemical threat, scientific discoveries must be translated into proactive, efficient, and internationally coordinated regulatory action.

11.3.1 Establishing Proactive Regulatory Frameworks

Regulatory frameworks need to switch from reactive observation to dynamic, adaptive, and predictive governance. This involves promoting NAM-derived data from auxiliary to primary evidence, while accelerating formal regulatory acceptance pathways. The Precautionary Principle, which requires preventive measures in the context of scientific uncertainty, needs to be generally accepted (UNEP, 2022). The regulatory lag, which is a significant issue where the rate of chemical innovation continuously evolves with the rate of hazard assessment and policy creation, needs to be addressed. Before chemicals are included into consumer or commercial goods, regulators must require thorough testing.

11.3.2 Global Governance and International Bodies

Given the worldwide scale and intricacy of chemical contamination, a centralized, authorized organization is required (UNEP, 2022). Researchers have demanded the creation of a comprehensive global Science-Policy Panel (SPP) on Waste and Chemicals (UNEP, 2022). This body would guarantee that global initiatives are based on the most recent scientific data, address the wide range of "chemicals of emerging concern," and allow important bidirectional science-policy connections. A rigorous conflict-of-interest policy that keeps specialists with financial conflicts out of key decision-making processes is necessary for such an SPP to remain credible and objective.

11.3.3 Green Chemistry and Sustainable Practices

The most effective long-term policy strategy is pollution prevention at the source (Anastas & Zimmerman, 2018). Research and corporate investment must be intensified in the principles of Green Chemistry, specifically designing benign chemicals (Principle 4) and designing for safe degradation (Principle 10). This urges the industries to fulfill their fundamental duty in full by implementing sustainable practices to substantially reduce their ecological impact and boosting funds for scientific research for efficient EC governance (UNEP, 2022). In order to encourage decision-making that is informed and support the requisite stringent regulatory measures, it is essential for the public to be better educated on the matter of ECs.

12 Conclusion

Environmental toxicology has grown into a multidisciplinary field of research that combines biology, ecology and chemistry, and data-driven modeling to understand the detrimental effects of contaminants upon living systems and the integrity of ecosystems (Lei et al., 2015; Michelangeli & East, 2015). As previously said, the connections among chemical, physical, and biological contaminants demonstrate that contamination is an ongoing, interconnected process that is influenced by environmental dynamics, biological interactions, and human activities (Li et al., 2024).

It is essential to fully comprehend the ultimate aggregation, transit, and transformation of pollutants in order to assess their durability, bioaccumulation, and long-term ecological and health impacts (Mohan et al., 2020; Wang et al., 2024).

Recent developments in analytical chemistry, omics-based biology, and computer modeling have greatly improved our capacity to detect, characterize, and predict the behavior of novel contaminants (Lee et al., 2023; Li et al., 2024). The integration of bioinformatics, artificial intelligence (AI) tools, and New Approach Methodologies (NAMs) has enhanced risk assessment by providing mechanistic and predictive insights that outperform conventional toxicological testing (Wang et al., 2024). By emphasizing the interdependence of environmental, animal, and human health, the One Health concept, on the other hand, advocates for a coordinated approach to tackling pollution and their cumulative risks (Wang et al., 2024).

Environmental toxicology must continue to bridge the gap between regulatory practice, sustainable policy frameworks, and experimental research in the future (Wang et al., 2024). Promoting green chemistry, strengthening interdisciplinary collaboration, and putting proactive, prevention-based management strategies into practice will be crucial (Wang et al., 2024). By combining scientific innovation with effective governance and public awareness, the field may significantly contribute to preserving environmental quality and ensuring the welfare of present and future generations (Mohan et al., 2020).

References

Abonyi, H. N., et al. (2024). Nanotoxicology: Developments and new insights. *Nanomedicine, 20*(2), 225–241. https://doi.org/10.1080/17435889.2024.2443385

Akhtar, M., et al. (2025, July 31). Nanotechnology to nanotoxicology. *TIBAO Journal*. https://www.biomaterials.org.in/tibao/index.php/tibao/article/view/75

Ali, H., & Khan, E. (2018). Trophic transfer and biomagnification. *Human and Ecological Risk Assessment, 25*(6), 1353–1376. https://doi.org/10.1080/10807039.2018.1469398

Alum, E. U. (2025). The role of toxicology in climate change. *Sustainable Environment, 11*(1). https://doi.org/10.1080/27658511.2025.2467485

Anastas, P. T., & Zimmerman, J. B. (2018). The united nations sustainability goals: how can sustainable chemistry contribute? *Current Opinion in Green and Sustainable Chemistry, 13*, 150–153. https://doi.org/10.1016/j.cogsc.2018.04.017

Barton, C. C., & Ainerua, M. O. (2020). Environmental toxicology: Wildlife. *Elsevier eBooks, 337–344*. https://doi.org/10.1016/b978-0-12-813724-6.00030-x

Calabrese, E. J., & Baldwin, L. A. (2003). Hormesis: The dose-response revolution. *Annual Review of Pharmacology and Toxicology, 43*(1), 175–197. https://doi.org/10.1146/annurev.pharmtox.43.100901.140223

Chen, Q., & Liu, J. (2014). Development process and perspective on ecological risk assessment. *Acta Ecologica Sinica, 34*(5), 239–245. https://doi.org/10.1016/j.chnaes.2014.05.005

Clippinger, A. J., et al. (2018). Pathways-based predictive approaches. *Computational Toxicology, 7*, 77–93. https://doi.org/10.1016/j.comtox.2018.07.003

D'Amore, T., et al. (2025). Chemical food safety in Europe. *Foods, 14*(9), 1628. https://doi.org/10.3390/foods14091628

Deavall, D. G., et al. (2012). Drug-induced oxidative stress and toxicity. *Journal of Toxicology, 2012*, 645460. https://doi.org/10.1155/2012/645460

Deniz, G. Y. (2025). Chernobyl as a natural laboratory. *Mutation Research, 907*, 503898. https://doi.org/10.1016/j.mrgentox.2025.503898
Destoumieux-Garzón, D., et al. (2018). The one health concept. *Frontiers in Veterinary Science, 5*, 14. https://doi.org/10.3389/fvets.2018.00014
Ellos, D. M., & Bacosa, H. P. (2025). Oil spills and marine disasters in Asia. *Discover Oceans, 2*(1). https://doi.org/10.1007/s44289-025-00082-5
Geng, J., et al. (2023). Advances in environmental risk research. *Science of the Total Environment, 858*, 160137. https://doi.org/10.1016/j.scitotenv.2022.160137
Gill, D. A., & Mix, T. L. (2020). Love Canal: A classic case study. *Elsevier eBooks, 341–352.* https://doi.org/10.1016/b978-0-12-813602-7.00025-9
Han, P., et al. (2024). In vitro to in vivo extrapolation. *Environment & Health, 2*(7), 499–513. https://doi.org/10.1021/envhealth.4c00043
Hansen, B., & Blainey, M. (2008). Registration: The cornerstone of REACH. *RECIEL, 17*(1), 107–125. https://doi.org/10.1111/j.1467-9388.2008.00588.x
Harada, M. (1995). Minamata disease: Methylmercury poisoning in Japan caused by environmental pollution. *Critical Reviews in Toxicology, 25*(1), 1–24. https://doi.org/10.3109/10408449509089885
He, W., et al. (2025). Chemical ecological risk under climate change. *Reviews of Environmental Contamination and Toxicology, 263*(1). https://doi.org/10.1007/s44169-025-00085-5
Hei, T. K., & Filipic, M. (2004). Role of oxidative damage in the genotoxicity of arsenic. *Free Radical Biology and Medicine, 37*(5), 574–581. https://doi.org/10.1016/j.freeradbiomed.2004.02.003
Hollender, J., et al. (2017). Nontarget screening with HRMS. *Environmental Science & Technology, 51*(20), 11505–11512. https://doi.org/10.1021/acs.est.7b02184
Hu, X., Cheng, H., & Tao, S. (2021). Exposure to emerging contaminants. *Environmental Science & Technology, 55*(5), 2641–2663. https://doi.org/10.1021/acs.est.0c08278
Hutton, S. J., et al. (2024). Ecotoxicology challenges during climate change (pp. 147–165). Springer eBooks. https://doi.org/10.1007/978-3-031-53130-9_11
Kharrazian, D. (2021). Exposure to environmental toxins. https://pmc.ncbi.nlm.nih.gov/articles/PMC8325494/
Kirsch-Volders, M., et al. (2003). Indirect mechanisms of genotoxicity. *Toxicology Letters, 140–141*, 63–74. https://doi.org/10.1016/s0378-4274(02)00498-8
Kumar, N. (2025). The Bhopal gas tragedy. *International Journal of Law and Management*. https://doi.org/10.1108/ijlma-02-2025-0061
Lee, D., & Choi, K. (2019). Regulatory frameworks comparison. *Science of the Total Environment, 671*, 1026–1035. https://doi.org/10.1016/j.scitotenv.2019.03.372
Lee, J., et al. (2023). Microalgal bioassays review. *Environment International, 172*, 107802. https://doi.org/10.1016/j.envint.2023.107802
Lei, M., et al. (2015). Overview of emerging contaminants and associated human health effects. *BioMed Research International, 2015*, 404796. https://doi.org/10.1155/2015/404796
Li, X., et al. (2024). Review of emerging contaminants. *Ecotoxicology and Environmental Safety, 275*, 114489. https://doi.org/10.1016/j.ecoenv.2024.114489
Liu, N., et al. (2025). PPCPs in global surface waters. *Environmental Science & Technology, 59*(36), 19146–19159. https://doi.org/10.1021/acs.est.5c05659
Lynch, C., et al. (2024). High-throughput screening in in vitro toxicology. *Annual Review of Pharmacology and Toxicology, 64*(1), 191–209. https://doi.org/10.1146/annurev-pharmtox-112122-104310
Manke, A., Wang, L., & Rojanasakul, Y. (2013). Mechanisms of nanoparticle-induced oxidative stress and toxicity. *BioMed Research International, 2013*, 942916. https://doi.org/10.1155/2013/942916
Michelangeli, F., & East, J. M. (2015). Environmental toxicology. *FEBS Open Bio, 5*, 728–740. https://doi.org/10.1016/j.fob.2015.08.005

Mohan, M., Singh, K., & Verma, S. (2020). Emerging contaminants review. *International Journal of Advanced Science and Engineering, 6*(1), 1286–1298.

Noyes, P. D., et al. (2009). The toxicology of climate change. *Environment International, 35*(6), 971–986. https://doi.org/10.1016/j.envint.2009.02.006

OECD. (2018). *Users' handbook supplement to the guidance document for developing and assessing adverse outcome pathways*. OECD Series on Adverse Outcome Pathways, No. 1. OECD Publishing, Paris. https://doi.org/10.1787/5jlv1m9d1g32-en

Pamies, D., et al. (2022). Guidance document on good cell and tissue culture practice 2.0 (GCCP 2.0). *ALTEX, 39*(1), 30–70. https://doi.org/10.14573/altex.2111011

Patel, P., & Shah, J. (2017). Nanomedicines: Safety and toxicological considerations. *Current Clinical Pharmacology, 12*(2), 73–82. https://doi.org/10.2174/1574884712666170509161252

Peterson, E. K., et al. (2017). Integrative behavioral ecotoxicology. *Current Zoology, 63*(2), 185–194. https://doi.org/10.1093/cz/zox010

Raimondo, S., & Forbes, V. E. (2022). Moving beyond risk quotients. *Ecologies, 3*(2), 145–160. https://doi.org/10.3390/ecologies3020012

Rehman, M. U., et al. (2024). Effects of PPCPs. *Separation and Purification Technology, 342*, 126921. https://doi.org/10.1016/j.seppur.2024.126921

Renuka, N., & Patyal, D. (2025). Bioaccumulation and biomagnification. In *Global perspectives of toxic metals in bio environs* (Vol. 1, pp. 369–394). https://doi.org/10.1007/978-3-031-89725-2_14

Richardson, S. D., & Kimura, S. Y. (2020). Emerging contaminants. *Environmental Science & Technology, 54*(14), 8509–8524. https://doi.org/10.1021/acs.est.0c02194

Schmeisser, S., et al. (2023). New approach methodologies in human regulatory toxicology – Not if, but how and when! *Environment International, 178*, 108082. https://doi.org/10.1016/j.envint.2023.108082

Slob, W., & Setzer, R. W. (2013). Shape and steepness of toxicological dose–response relationships. *Critical Reviews in Toxicology, 44*(3), 270–297. https://doi.org/10.3109/10408444.2013.853726

Swenberg, J. A., et al. (2007). Biomarkers in toxicology and risk assessment. *Chemical Research in Toxicology, 21*(1), 253–265. https://doi.org/10.1021/tx700408t

Tralau, T., et al. (2015). Regulatory toxicology in the twenty-first century. *Archives of Toxicology, 89*(6), 823–850. https://doi.org/10.1007/s00204-015-1510-0

Turaga, R. M. R., & Sugathan, A. (2020). Environmental regulations in India. *Oxford Research Encyclopedia*. https://doi.org/10.1093/acrefore/9780199389414.013.417

UNEP. (2022). Global chemicals outlook II. https://www.unep.org/resources/report/global-chemicals-outlook-ii

Vinken, M. (2024). AOP networks for NAMs. *Current Opinion in Toxicology, 100504*. https://doi.org/10.1016/j.cotox.2024.100504

Wang, F., et al. (2021). Emerging contaminants: A One Health perspective. *The Innovation, 2*(4), 100157. https://doi.org/10.1016/j.xinn.2021.100157

Wang, R., et al. (2024). Emerging contaminants in water. *Water Science & Technology, 89*(10), 2763–2782. https://doi.org/10.2166/wst.2024.151

Wang, H., Xu, T., & Yin, D. (2025). Advancing environmental toxicology through NAMs. *Environment & Health*. https://doi.org/10.1021/envhealth.5c00323

Williams, E. S., Panko, J., & Paustenbach, D. J. (2009). The European Union's REACH regulation. *Critical Reviews in Toxicology, 39*(7), 553–575. https://doi.org/10.1080/10408440903036056

Yap, C. K., & Al-Mutairi, K. A. (2025). Chernobyl nuclear catastrophe. *Frontiers in Public Health, 13*. https://doi.org/10.3389/fpubh.2025.1552122

Chapter 12
Pharmacogenomics and Personalized Medicine: Applications in Environmental Toxicology

Vanktesh Kumar, Shivank Sharma, Stalin Arulsamy, Pankaj Wadhwa, and Rubal Kalra

Abstract Pharmacogenomics and Environmental Toxicology are two areas that are being increasingly integrated together in research on Precision Environmental Health through the use of pharmacogenomic approaches to identify how an individual's susceptibility to certain environmental toxicants may contribute to the unique effects that these environmental toxicants may have on an individual. The rationale for this integration is based upon studies that demonstrate the role of genetic variation for xenobiotic-metabolizing enzymes (e.g., cytochrome P450s, glutathione S-transferases, N-acetyltransferases), in addition to the transporters and receptors that mediate xenobiotic action and the ways in which genetic variation can alter the kinetics and dynamics of the environmental toxicants to which we all are exposed on a daily basis (e.g., heavy metals, air pollutants, pesticides, and chemical exposures). The interplay between genetics and environmental toxins is complex, as evidenced by the genomic composition of our population and by the differing responses of the members of the population to exposure to chemical toxicants. The combination of pharmacogenomic data with current research results yields a better understanding of the implications related to clinical and public health regarding the development of detoxifying metabolites, the conversion of chemicals to reactive intermediates and their interactions with receptors for both a clinical and public health perspective. The chapter will also discuss how emerging omics technologies can help to identify new biomarkers for individual risk of exposure to environmental hazards and assess their response to hazardous substances. A discussion of the difficulty and future pathways of this field exists as well and includes consideration of the integration of

V. Kumar · S. Sharma · S. Arulsamy · P. Wadhwa (✉)
Department of Pharmaceutical Chemistry, School of Pharmaceutical Sciences, Lovely Professional University, Phagwara, Jalandhar, Punjab, India
e-mail: pankajwadhwa88@gmail.com

V. Kumar · P. Wadhwa
Ummeed Institute of Pharmaceutical Sciences and Research Center, Badakhatlabana, Sri-Ganganagar, Rajasthan, India

R. Kalra
Department of Biomedical Engineering, Texas A&M University, College Station, TX, USA

K. Priya and S. K. Giri (eds.), *Environmental Toxicology: Xenobiotics and Effects on Human Genetic Material*, https://doi.org/10.1007/978-3-032-28619-2_12

pharmacogenomics into regulatory toxicology; the refinement of environmental risk assessment; and the associated ethical, legal, and social challenges and considerations stemming from these technologies.

Keywords Pharmacogenomics · Personalized medicine · Toxicology · Xenobiotic metabolism · Toxicogenomic · Gene-environment interaction

1 Introduction

The fields of Medicine and Public Health are undergoing a transitory, revolutionary change toward Personalized/Precision Intervention (Ratnani, 2021). Pharmacogenomics is a new and evolving field that studies the influence of a person's genetic profile on their response to medications (TumkurSattar et al., 2023). The field of genetics has identified many different genetic variations that can impact how medications are absorbed, distributed, metabolized, and excreted by the body. These genetic variations allow pharmacogenomics to be an evidence-based tool for enhancing the benefits of medications while simultaneously lowering the potential for adverse reactions (Bellanca et al., 2023). However, Pharmacogenomics does not limit itself to how Pharmacists can utilize pharmacogenomic principles and applications. All of our human biological systems utilize multiple processes to eliminate foreign objects and substances from our systems, including pharmaceutical drugs, which are often chemically identical to many other Xenobiotics (Chemical compounds that we come into contact with on a daily basis) (Banerjee et al., 2019). As such, an important foundation for Precision Environmental Health (PEH) exists at the intersection between Pharmacogenomics (or Precision Medicine) and Environmental Toxicology. What unites the two is a simple but powerful observation: People react differently to medications, but also have varying levels of susceptibility to the negative effects of environmental exposure (Varshavsky et al., 2023). Genetics largely determine the degree to which an individual is susceptible to specific diseases (Llaurens et al., 2017). Precision Environmental Health seeks to take the scientific knowledge of the human genome and apply it to gain an understanding of how the environment and genetic information interact with one another, thereby enabling Point (1) the prediction of future disease development; Point (2) identification of populations at increased risk; and Point (3) creation of tailored public health interventions directed specifically at those at the highest risk of developing disease (Virolainen et al., 2023). In Precision Environmental Health, the focus of an aetiology-based investigation does not focus on establishing the environmental toxicant of the disease. Instead, it concentrates on identifying the reasons that individuals do not develop the disease when other individuals with the same environmental exposures do. Thus, the focus is on how each individual's genetic make-up, when inter-related with their environmental exposures, contributes to the development of their respective diseases. This chapter will describe some of the scientific principles, bases of practice, societal issues related to Precision Environmental Health, and how pharmacogenomic technology/tools and principles may be applied to quantify and mitigate the risk to human health posed by our environmental exposures.

2 The Genetic Architecture of Xenobiotic Metabolism and Response

Xenobiotic metabolism is a complicated, eight-step process of managing the large influx of foreign chemicals entering the Human Body. It is primarily performed within the liver and divided, in accordance with traditional standards, into two phases that co-work together to make lipophilic (fat-soluble) xenobiotics into hydrophilic (water-soluble) metabolites that are easily excretable. Figure 1 is a graphical representation of the Phase I and Phase II processes and how they ultimately process the characteristics of a chemical in order to dispose of it. Phase I metabolism serves an important purpose in the determination of whether a chemical will become toxic or will not be Toxic. The Phase I process is concerned with introducing or revealing a polar Functional Group (PFG) to the xenobiotic via Hydroxyl (–OH) or Amine (–NH2) Functional Groups. The major enzymes involved with the Phase I metabolism process of xenobiotic metabolism are the Cytochrome P450 enzymes, which belong to the monooxygenase superfamily. These enzymes convert numerous substrates into substrates upon which Phase II enzymes can act effectively and therefore detoxify them from the body (Zhou et al., 2025). Phase I Metabolism is not exclusively a detoxification process. Many chemicals identified as carcinogens and/or toxic substances are considered pre-carcinogenic or pre-toxic in nature. For these chemicals, Phase I metabolism actually converts the parent (inactive) compound into an electrophilic (highly reactive) intermediate metabolite capable of interacting with the macromolecules (DNA) in the human body and damaging them (Khojasteh et al., 2018). After the Phase I metabolic transformation of a xenobiotic, Phase II will typically be involved through the addition of large and polar water-soluble endogenous compounds to create conjugates (conjugation). This aids in eliminating xenobiotics from the body by increasing their polarity (Chen, 2024). The principal family of enzymes in phase II are the glutathione S-transferases (GST) that catalyze the conjugation of xenobiotic compounds to glutathione, and N-acetyltransferases (NAT) which catalyze the transfer of an acetyl moiety to the xenobiotic compound. The combined effect of phases I and II illustrates how the metabolic elimination (metabolic gauntlet) determines the potential toxicity of a xenobiotic based on the exposure. However, the efficiency of the phase I and II metabolic systems varies across humans based on the polymorphic nature of gene encoding these two enzyme families, meaning that, in addition to being different within the population with respect to how the enzymes respond toward the xenobiotic, there are also multiple differences in the sequence and structure of the gene encoding the enzyme. This means that there are multiple different genotypes (genetic polymorphisms), from single nucleotide polymorphisms (SNPs) to complete gene deletions that affect enzyme activity, resulting in poor,

intermediate, extensive, and ultra-rapid metabolizer phenotypes (McCarthy et al., 2020). The genetic variability of how each person responds to the toxicity of xenobiotics is what constitutes the basis for the differential susceptibility of individuals to the toxic effects of xenobiotic substances. The context in which you find an individual will dictate the effects that a polymorphism will have on that individual. For example, if a chemical is detoxified by an enzyme like CYP and is being metabolized slowly, an individual carrying a "slow metabolizer" genotype may end up having a higher risk because of the longer he/she is exposed to the parent compound. On the other hand, if the chemical is benzene and requires bioactivation by a CYP to become toxic, an individual carrying a "rapid metabolizer" genotype would have a higher risk because he/she produces the toxic intermediate faster (as cited in Piekos, 2018). This concept of context-dependence illustrates the complexity of actual situations where individuals will be exposed to complex mixtures of chemicals rather than solely to one substance. An individual's total risk from any combination of exposures cannot be estimated based on evaluating a single gene-chemical interaction. A genotype that provides protection for one xenobiotic may not necessarily be protective for another (Ghosh et al., 2016). For example, a "rapid *CYP2E1*" genotype may pose increased risk from benzene in ambient air but also provide a lower risk from another co-pollutant that produces its toxic effects directly and requires *CYP2E1* for metabolism (Table 1).

3 Pharmacogenomics in Action: Case Studies in Environmental Toxicology

One way we can leverage genetic variability that has been documented to influence health following exposure to environmental toxins is through particular case studies representing examples of pharmacogenomics in environmental health. The case studies included in this report were chosen based upon their societal importance and the diversity of types of interactions between our genes and the environment. Three categories of gene-environment interactions have been considered: (1) Genetic modulation of detoxification processes; (2) Genetic regulation of metabolic activation; (3) Genetic basis for receptor-mediated toxicity. The findings presented in this report will help to develop broad concepts regarding how our genetics dictate how we respond to the environment through chemical exposures and provide examples of how to approach the toxicity of chemical mixtures containing unknown compounds.

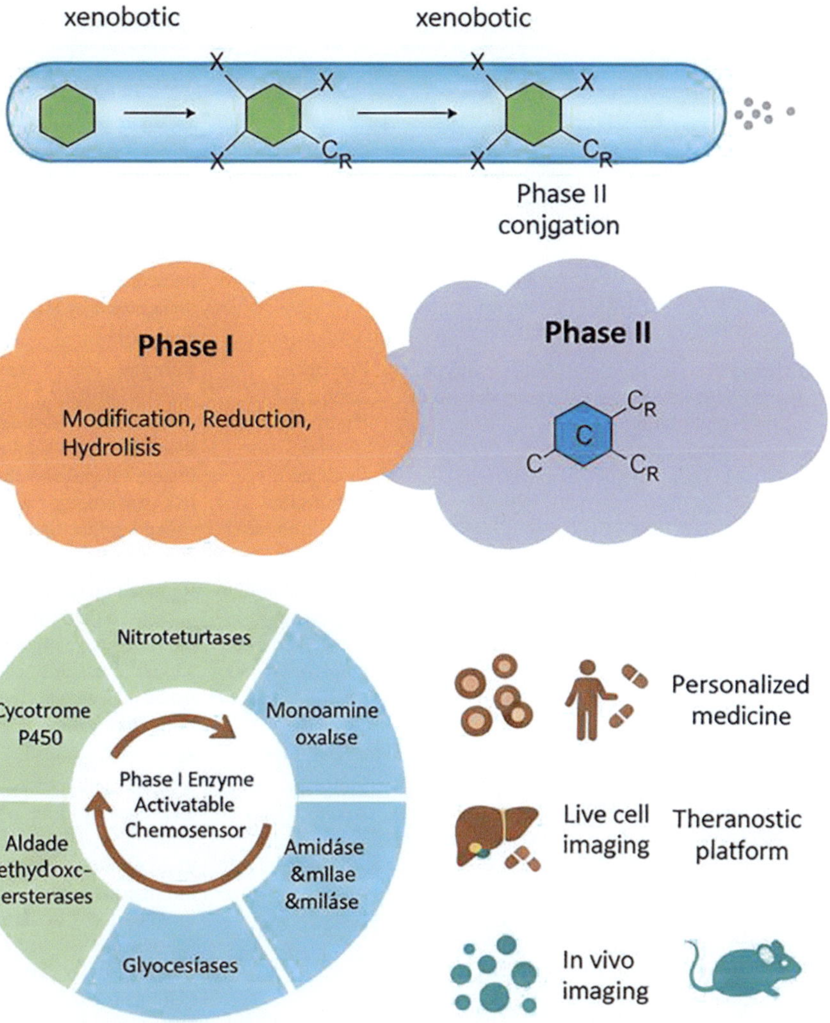

Fig. 1 The General Pathway of Xenobiotic Metabolism. This diagram illustrates the two-phase enzymatic process by which the human body metabolizes foreign chemical substances (xenobiotics). In Phase I, a lipophilic (fat-soluble) xenobiotic is modified by enzymes, primarily from the Cytochrome P450 (CYP) superfamily. This reaction introduces or exposes a polar functional group, which can either detoxify the compound or, in some cases, bioactivate it into a more reactive, toxic intermediate. In Phase II, conjugation enzymes such as Glutathione S-Transferases (GSTs) or N-acetyltransferases (NATs) attach a large, water-soluble molecule to the xenobiotic. This conjugation step significantly increases the compound's polarity, creating a hydrophilic (water-soluble) metabolite that can be easily eliminated from the body via urine or bile

Table 1 Key genetic polymorphisms in xenobiotic metabolizing enzymes and their toxicological significance

Gene symbol	Protein function	Common polymorphism(s)	Associated xenobiotics	Toxicological consequence
CYP2E1	Phase I oxidation of small molecules	RsaI/PstI polymorphisms leading to higher inducibility	Benzene, Paracetamol, N-nitrosamines	Increased bioactivation to toxic intermediates; higher risk of benzene-induced hematotoxicity and paracetamol-induced hepatotoxicity (Hartman et al., 2017)
GSTM1	Phase II detoxification (glutathione conjugation)	Null genotype (gene deletion)	Polycyclic Aromatic Hydrocarbons (PAHs), Benzene metabolites (e.g., quinones)	Complete loss of enzyme activity; impaired detoxification leading to increased DNA adduct formation and elevated risk of smoking-related cancers (Hausman-Cohen et al., 2020)
NAT2	Phase II detoxification (acetylation)	"Slow acetylator" alleles (multiple SNPs)	Aromatic amines (e.g., from tobacco smoke, industrial dyes)	Reduced enzyme activity; decreased detoxification rate leading to higher adduct levels in bladder tissue and increased risk of bladder cancer (Habil, 2022)
AS3MT	Arsenic methylation for detoxification	Met287Thr SNP and other variants affecting enzyme activity	Inorganic Arsenic (iAs)	Altered methylation efficiency, leading to accumulation of highly toxic monomethylated intermediates and increased risk of skin lesions, cancer, and cardiovascular disease (Thomas, 2021)

(continued)

Table 1 (continued)

Gene symbol	Protein function	Common polymorphism(s)	Associated xenobiotics	Toxicological consequence
AhR	Aryl hydrocarbon receptor (ligand-activated transcription factor)	Various SNPs affecting receptor affinity and function	Dioxins (e.g., TCDD), polychlorinated biphenyls (PCBs)	Modulated induction of downstream genes (e.g., *CYP1A1*, *CYP1A2*); variable susceptibility to carcinogenic, immunotoxic, and endocrine-disrupting effects (Liu et al., 2018)
NQO1	Phase II detoxification (quinone reduction)	C609T SNP leading to loss of enzyme activity	Benzene metabolites (benzoquinones)	Inability to detoxify reactive quinones; increased bone marrow toxicity and higher risk of benzene-induced leukemia, particularly when combined with rapid *CYP2E1* alleles (Lee et al., 2021)

3.1 Arsenic: A Case of Detoxification Efficiency

Inorganic arsenic (iAs) is found in drinking water and food, making it a worldwide environmental contaminant. It is known to be an effective human carcinogen and systemic toxicant, as it has been associated with cancer, skin lesions, diabetes, and cardiovascular disease (Tchounwou et al., 2019). In addition, the fate of iAs through metabolism determines its toxicity over time rather than being an instantaneous measurement. The first step in iAs detoxification involves methylation of iAs through the action of arsenite methyltransferase (encoded by AS3MT) (Stýblo et al., 2021). As such, methylation results in lower toxicity of iAs (dimethylated arsenic, DMA) and increased excretion from the body. Unfortunately, the methylation process generates an additional toxic metabolite (monomethylated arsenic, MMA) to the parent compound (Di et al., 2019). Thus, the ability of an individual to detoxify and metabolize arsenic is related to how efficiently the person can perform the two-step methylation reaction. Clinical pharmacogenomic research has established the significance of polymorphisms in AS3MT for regulating arsenic metabolic activity and toxicity (Raghavan, 2025). Specific SNPs, such as the Met287Thr variant, are strongly associated with an individual's methylation capacity, which can be measured by the ratio of different arsenic species in their urine (Xu et al., 2016). Figure 2 (Pánico et al., 2022) demonstrates that urinary excretion of MMA is greater and DMA is less in people with less functional AS3MT alleles due to the presence of a metabolic bottleneck resulting from impaired methylation processes on MMA. The impaired methylation process permits the accumulation and prolonged residence of an extremely toxic intermediary compound—MMA—thereby posing an elevated threat of developing

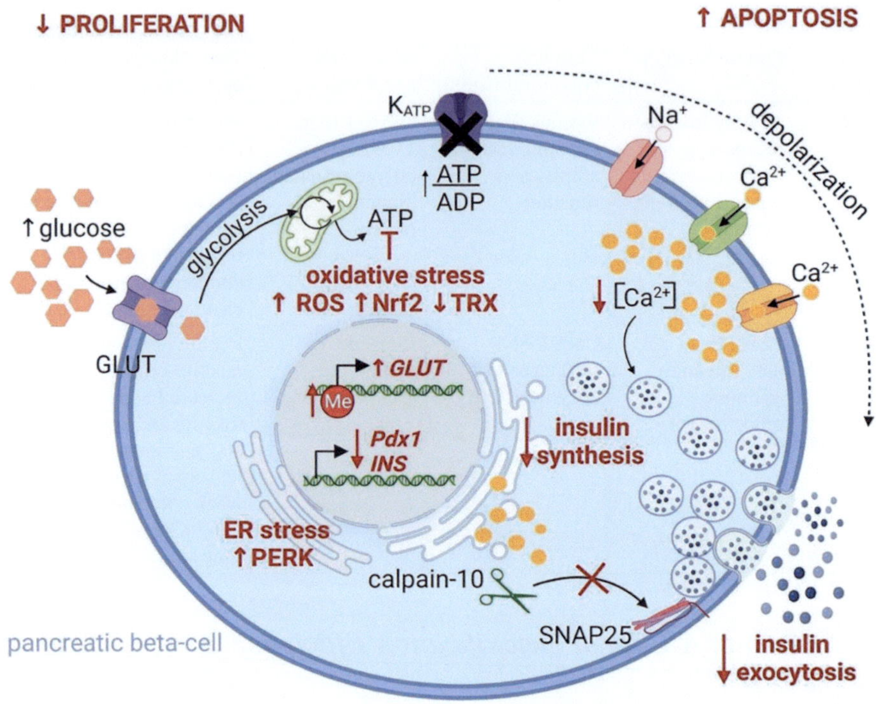

Fig. 2 The role of AS3MT polymorphisms in arsenic metabolism and toxicity. This figure depicts the metabolic pathway for inorganic arsenic (iAs) and highlights how genetic variation influences individual susceptibility

arsenic-related skin cancers and lesions for an individual with any degree of impairment from AS3MT (Chain et al., 2024). Arsenic, as described above, is an example of an adversarial human health risk stemming from knowledge of AS3MT genotypes and associated metabolite profiles that determine exposure-related susceptibility to certain health risks by exposing individuals to toxic substances.

3.2 *Benzene: A Case of Metabolic Activation*

Industrial processes, including the production of benzene as a solvent, will make up a significant portion of the production and use of benzene as an environmental contaminant, which is listed as a human carcinogen due to its capacity to cause multiple different types of blood-related diseases and other related diseases such as Acute Myeloid Leukemia (Cañas Portilla et al., 2024; Caldwell et al., 2020). While benzene is relatively inert, it becomes toxic through the metabolic activation of benzene in the human body to reactive metabolites (Fig. 3). The first and rate-limiting step in benzene metabolism is the oxidation of benzene to produce benzene oxide, which is catalyzed primarily by the enzyme *CYP2E1* (Thomas et al., 2025). After the production of benzene oxide, additional metabolism occurs, producing a wide range of toxic metabolites, including a variety of quinones that produce harmful effects on stem/progenitor cells in the bone marrow (Yusoff et al., 2023). A good example of the pharmacogenomic influences on benzene toxicity is found in both the activation and detoxification routes for benzene toxicity (Bauer et al., 2003). Polymorphisms in *CYP2E1* associated with the rapid metabolism "phenotype" confer increased risk for benzene toxicity. Individuals with these polymorphisms can generate more toxic metabolites than non-rapid metabolizers at the same benzene exposure level, which can lead to increased damage to the bone marrow (Yusoff et al., 2023). This risk is increased by genetic variations seen in Phase II detoxifying enzymes. Benzene's quinone metabolites are detoxified by NAD(P)H: quinone oxidoreductase 1 (*NQO1*) and glutathione S-transferase M1 (*GSTM1*). Those who possess the null genotype for *GSTM1* (gene deletion) or a low activity variant of NQO1 will not process these types of reactive metabolites as efficiently (Bhatia, 2015). The greatest risk profiles will be seen in those who carry a combination of "high risk" alleles, such as a *CYP2E1* rapidly metabolizing allele causing an increase in the production of these toxins, while at the same time they have either a null or slow-degrading *GSTM1/NQO1* allele that would cause the body's ability to clear the toxins from its system to be significantly decreased. This co-action demonstrates that an individual's total metabolic genotype, not just an individual gene, is what ultimately will determine their susceptibility to toxicants that require metabolizing before they can exert toxic effects.

3.3 Dioxins: A Case of Receptor-Mediated Toxicity

Dioxins, especially TCDD (2,3,7,8-Tetrachlorodibenzo-p-dioxin) (the most toxic dioxin congener), are highly persistent contaminants that have been shown to result from the processes of waste burning and a variety of industrial activities. Dioxins are extremely dangerous to human health and can cause a number of illnesses including but not limited to cancer through their carcinogenic properties, immunotoxin properties and their ability to inhibit normal developmental processes. The biological mechanism of TCDD toxicity differs from those caused by arsenic and benzene. In contrast to these agents which are metabolically activated before causing their adverse effects, TCDD exerts its toxic effects primarily through its binding to a specific intracellular receptor (the AhR) with high affinity (Grishanova & Perepechaeva, 2022). The AhR receptor functions as a transcription factor and is normally found in the cytoplasm bound to chaperone proteins in its native inactive form. When TCDD binds to the AhR receptor, it undergoes a conformational change allowing it to relocate from the cytoplasm to the nucleus, where it dimerizes with another cellular protein called ARNT, and interacts with specific sequences of DNA called DREs (dioxin responsive elements). Dioxin responsive elements initiate the transcription of several downstream genes (Fig. 4). For example, the genes *CYP1A1* and *CYP1A2*

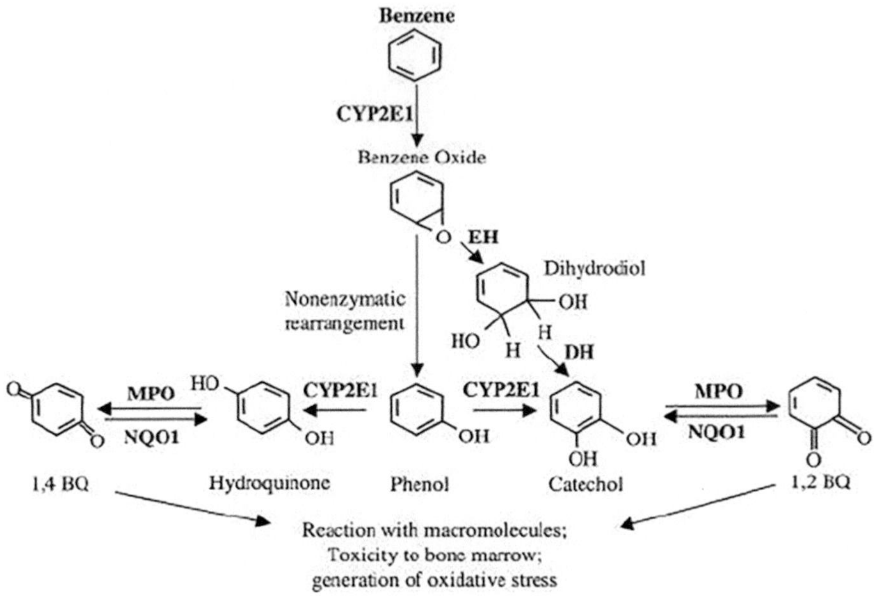

Fig. 3 Metabolic pathways determining benzene hematotoxicity. This diagram illustrates the critical balance between the bioactivation and detoxification of benzene, (a known human carcinogen. of toxins)

are particularly important genes involved in many cellular functions such as cell proliferation, differentiation and programmed cell death (Guarnieri, 2020). Dioxin's toxic properties are a product of dysfunctional and prolonged activation of the AhR pathway. Genetic differences in the AhR gene and related pathways create variability across individuals regarding the effects of dioxin exposure (Adegbola & Adetutu, 2024). Variations within AhR may lead to altered binding to TCDD or altered transcriptional activation of genes, causing a range of responses based on differences between individuals and other animal populations (Han et al, 2021). Accordingly, dioxin is an excellent model for studying receptor-based toxicology, with interindividual variability being caused by difference in the sensitivity of AhR signaling rather than differences in genes coding for metabolic enzymes.

4 High-Throughput "*Omics*" in Gene-Environment Research

Foundational studies into pharmacogenomics' environmental impacts began using a candidate-gene method where researchers examined certain genetic variants using prior knowledge to predict their biological activity. Although a very powerful strategy, candidate-gene studies are inherently constrained by what we already know

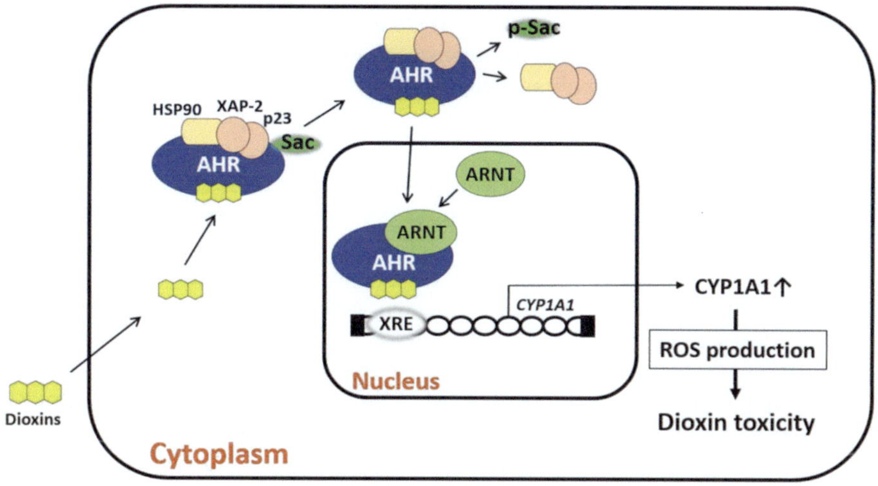

Fig. 4 The Aryl Hydrocarbon Receptor (AhR) signaling pathway activated by dioxin

about genetics. The emergence of high-throughput "omics" technologies transformed pharmacogenomic research to a hypothesis-free, data-driven study of entire genomes and their functional effects in response to environmental exposures (Morabito et al., 2025). Shifting from a local to a global perspective provides researchers with new levels of information regarding the intricate biological designs causing environmental disease (Manera, 2021). One major component of these innovative tools includes GWAS (Genome-Wide Association Study), where researchers scrutinized SNPs located throughout the genomes of many people to find SNPs that statistically correlate with specific diseases and traits (Hettiarachchi & Komar, 2022). Using a GWAS approach related to environmental health will help identify additional pathways and genes that contribute to susceptibility to disease but have not yet been identified. The GWEIS (Genome-Wide Environment Interaction Studies) approach utilizes advanced designs explicitly to engage each of the SNPs tested against a measured environmental exposure to systematically identify genetic variants where their impact on disease risk is only apparent when there is a specific environmental trigger (Werme et al., 2021). This has enabled the identification of several critical findings, such as novel loci for genetic susceptibility to pesticide exposure and neurodegenerative diseases such as Parkinson's disease (Tsalenchuk et al., 2023) and (Xue, 2025) and SNPs associated with the accumulation of toxic heavy metals and cancer risk (Motsinger-Reif et al., 2024).

As well as the genomic static blueprint, omics, or technologies such as transcriptomics (i.e., the complete array of all RNA transcripts produced by a cell at a given point in time) enable the analysis of the real-time response of cells to environmental challenges (Singh et al., 2025). By comparing the transcriptomes of exposed and unexposed individuals/cells in culture, researchers have been able to identify genes whose expression levels are either upregulated or downregulated by xenobiotics, and elucidate the underlying molecular pathways activated/repressed by these xenobiotic agents, thus providing insight into mechanisms of toxicity and potential biomarker(s) of exposure or effect in a timely manner (Jouannais & Pizzol, 2021).

The study of epigenomics demonstrates one of the most important regulators in the ongoing interaction between the environment and the expression of our genetic code (Metz & Faraji, 2026). Through two epigenetic chemical processes, DNA methylation and histone modification, there is an added layer of modification to the DNA that alters the way a gene is expressed, but does not alter the underlying structure of the DNA code. These epigenetic modifications can be altered in response to environmental influences like heavy metals, airborne pollutants, and endocrine disruptors. Some of the epigenetic modifications demonstrated to be caused by exposure to potentially harmful environmental exposures can remain stable and continue to exist indefinitely in the molecular memory of an individual's exposure history. Such processes provide insight into how an individual's exposure history can affect their susceptibility to chronic diseases, such as those that occur later in life. When combining the different types of data in genomics, epigenomics and transcriptomics to develop a comprehensive understanding of human health, it provides the ability to study biological responses at a different time scale than what we have traditionally used to study exposure influences on chronic health conditions, thereby providing a new framework to study chronic health outcomes caused by exposure to the environment. The RNA transcriptome and RNA expression reflect our immediate and ongoing response to any exposure defining the environment based on time from a few hours to a few days. Chronic diseases have been developed over decades. Therefore, we need to understand all of the time scales involved in the development of a chronic disease, including epigenetic and environmental triggers, to develop a better understanding of how chronic diseases develop. Knowing the history (or scarring) of the cellular processes through the epigenome can provide insight into how we can prevent chronic diseases from occurring, whereas knowing that a cell has not changed over time can help explain how an individual developed a disease (Statzer et al., 2023). The integration of these layers of data into a biological narrative allows for the connection of innate risk factors, cumulative environmental exposures, and the immediate response of a cell to acute stimuli. The integration of these data layers represents the foundation for how we can create environmentally-focused precision health for the next generation.

5 New Frontier for Risk Assessment and Regulatory Science

The integration of cutting-edge pharmacogenomics into public health policy and scientific standards is both an exciting opportunity and a serious challenge. There are regulatory agencies (including the U.S. EPA) that utilize the systematic framework of human health risk assessment to establish Safe Levels of Exposure for

Environmental Chemicals (Fitzpatrick et al., 2017). Historically, the risk assessment process has included four steps: hazard Identification, dose–response assessment, exposure assessment, and risk characterization. Therefore, one of the fundamental concepts of this framework, as required under statutes such as the TSCA, mandates the identification and protection of Potentially Exposed and Susceptible Populations (Gray). The definition of these "vulnerable groups" includes age, pre-existing health conditions, and (perhaps most importantly) genetic predisposition. Therefore, it is anticipated that the use of pharmacogenomics and toxicogenomic could ultimately transform this risk assessment process. Rather than applying generic (i.e., standard) uncertainty factors of 10 to account for all human-to-human variability, which is the current regulatory practice, the application of genomic data would enable regulators to establish and enforce data-based adjustments that accurately depict the actual distributions of susceptibility within the respective populations (Jang, 2023). Risk assessments can be designed specifically to protect the most sensitive individuals without artificially increasing the level of protection given to representatives of the general population, based on the knowledge gained from the identification of the presence of high-risk genotypes. There are several steps within the risk assessment process that may be informed by genomic data, including the determination of the chemical's mode of action for the purposes of hazard identification, as well as defining the dose–response relationship at low environmentally relevant exposure levels (SAARIMÄKI). Creation of this new scientific knowledge represents a major philosophical change in the current approach to public health policy. The current paradigm has focused on protecting the "average" or "medium" person. By considering the information provided by genomic data, the new approach will shift the focus to protecting the "most sensitive" individual. This represents a major economic and societal change, as it has previously been assumed that the average person would be the most likely to be exposed to hazards. If genomics data indicate that a subgroup of individuals, known as "poor metabolizers," make up 5% of the population and have 50 times greater sensitivity to the effects of chemical exposure than the average person, it becomes clear that the regulated allowable exposure limit, which is based on the established default uncertainty factor of 10, would be inadequate to protect the 5% of the population who are poor metabolizers. In order to provide meaningful protection to the 5%, the entire population's allowable exposure limit would have to be reduced by 5. As a result, the businesses and industries that are subject to existing regulations will incur significant economic costs as a result of the need to lower their allowable exposure limits. (Pérez Santín et al., 2021). This forces a difficult policy question: should a single, highly protective standard be set for everyone, or should a stratified approach be considered where susceptible individuals are given specific advice or protections? The latter path leads directly into the ethical minefield of genetic discrimination.

Although there is a great deal of potential associated with using genomic information for regulatory decision-making, there has been caution in applying this information to routine regulatory operations (Funk et al., 2019). The lack of incorporation of, and thus a large translation gap that exists between, the rapidly evolving academic literature related to genomics and the methodical and evidence-based decision-making processes utilized in regulatory science presents a challenge for both clinical and regulatory scientists. GWAS and "omics" studies allow for the identification of association and risk factors for diseases; however, regulatory decisions must be based on higher standards of evidence that demonstrate causation. Furthermore, transforming a probabilistic measure (e.g., an odds ratio for a particular SNP) into a deterministic regulatory value (e.g., Maximum Contaminant Levels for chemicals in drinking water) is a challenging scientific problem that currently lacks an established methodology. In addition, because of the vast amount and complexity of "omics" data, regulatory agencies require new analytical methodologies and resources. Although the EPA has initiated efforts such as "Next Generation of Risk Assessment" (NexGen) to advance the field of risk assessment, formal guidelines on how to use toxicogenomics data in risk assessment development remain in their infancy. Closing this translation gap by establishing validated, reproducible, and transparent methods for using genomic data is the central challenge that must be overcome to realize the full public health promise of precision environmental health.

6 The Societal Imperative: Ethical, Legal, and Social Implications (ELSI)

Pharmacogenomics will help develop knowledge and scientific understanding for Environmental Health; but there are many things to consider other than science with regard to ethical, legal and social implications (ELSI) of this technology that need to be addressed now to ensure that it is used correctly and responsibly as it moves forward. The knowledge derived from pharmacogenomics is likely to result in two areas of significant impact on individuals or groups with high genetic susceptibility to the adverse effects from exposure to environmental toxic agents. The first area of knowledge that will be developed through pharmacogenomics will be to give a better understanding of how at-risk or high-risk groups should be protected from exposure to toxic chemicals (Curran & Hamilton, 2017). The second area of knowledge to be developed will be to cause a serious dilemma for society—since while it supports the need for better protection of at-risk or high-risk populations because it provides greater detail about their vulnerability, it also presents the ability to isolate and discriminate against certain populations within our society. Therefore, for the successful establishment of the field of pharmacogenomics, it will be essential to establish an adequate, strong legal and ethical framework for its use, ensuring that the knowledge derived from pharmacogenomics will be used exclusively to protect those most vulnerable and not to create any situations of exclusion.

Genetic discrimination in the workplace arises as one of the largest concerns regarding employing predisposition testing. For example, if an employer discovers that a job candidate or employee has a "susceptibility genotype" for various workplace chemicals, the employer could choose to not hire the employee or terminate his/her employment rather than implement engineering controls to lower everyone's exposures (Ishikawa, 2017). The result is an unfair situation, in which the employer transfers the responsibility of ensuring a safe workplace to employees instead of shouldering that burden. Furthermore, this type of practice results in the emergence of a "genetic underclass," where workers are denied access to certain types of employment. The Genetic Information Nondiscrimination Act (GINA) creates some protections, but they only apply to specific instances regarding environmental and occupational exposure. In addition, the scope and enforcement of GINA are very ambiguous and complicated (Lefebvre, 2015). Privacy and confidentiality issues can arise because genomic information pertains to every person; therefore, a person's own genomic information is personal and sensitive in nature and could provide access to their family's health status based on their genetic information. As such, it is vital that genome information is securely stored and that there are policies detailing how that information will be used, shared, and protected so that victims do not suffer from unintended disclosure or exploitation of their sensitive information (Borgesius et al., 2015). This is further complicated by the challenge of obtaining truly informed consent for research, as genetic samples collected for one study may have unforeseen future uses as scientific knowledge expands.

The biggest challenge associated with health equity and health justice relates to the large amount of genomic research and pharmacogenomic discovery that has been carried out primarily using individuals of European ancestry (Ju et al., 2022; Srinivasan et al., 2016). Due to the lack of genomic diversity in the genomic data that researchers have used, the results of this research and the knowledge they have generated may prove to be significantly less accurate than they would be for individuals of other ancestry(s) (Hindorff et al., 2018). If precision environmental health tools were developed and validated for one specific group but then made available to individuals of all ethnicities, these tools would not only fail to protect non-European populations but also worsen the existing health disparities that exist for non-European individuals. Additionally, ethical considerations within the field of environmental genomics are more complex than those for environmental genomic research conducted for individuals receiving clinical care in terms of the interventions that environmental genomics can provide (e.g., regulation, remediation) versus those available for clinical care (e.g., prescribing medication) (Ching et al., 2018). Thus, instead of focusing on individual autonomy as the primary ethical consideration, the focus shifts to collective justice. Furthermore, identifying communities that are "genetically susceptible" can raise issues about resource allocation, stigmatization, and property values, which can be problematic even when targeted efforts are made to address environmental conditions that may be harmful to those individuals and groups (Allen et al., 2019; Sarwar, 2023).

Therefore, achieving equity in genomics by diversifying research cohorts, engaging with underrepresented communities, and ensuring that the benefits of this science are distributed fairly is not an ancillary goal but an absolute prerequisite for the ethical practice of precision environmental health.

7 Future Perspectives: A Novel Problem Statement for Advancing the Field

Pharmacogenomics for Environmental Toxicology is developed from recognizing individual gene alterations (polymorphisms) to understanding the genetic variations across the entire genome for environmental exposure susceptibility through performing genome-wide association studies (GWAS). Now the field needs an additional evolution and transformation to move beyond a static approach of measuring genetic susceptibility alone, toward developing predictive models describing how genetics continually interact with environmental factors throughout life to influence an individual's level of susceptibility. The biggest issue that needs to be addressed today is how to create predictive models that quantify the interaction between genetics and environmental factors based on an individual's history of diverse, time-dependent exposure to toxins (Pazzagli & Colleoni, 2018). Predictive models will continue to evolve, but the future of the field lies in developing continuous dialogue between genetics and the environment.

At present, a new and urgent challenge for researchers is to create and test a d-PESS. Today's polygenic risk scores (PRS) identify an individual's genetic potential only at the germinal level through examination of their DNA. An individual's risk for a health condition changes over time due to environmental exposures as well as age; it does not remain fixed over their lifetime. The d-PESS will be an innovative dynamic model that takes into account the dynamic nature of the genetic susceptibility as well as the effects of environmental insults over an individual's life, and can provide an individualized estimate of risk throughout the individual's lifetime. In order to accomplish this, the d-PESS will utilize three methods to gather biological information for its creation and use: Static Genetics: A baseline polygenic risk score (PRS) will be created using all known and potential genetic variants located within the genes of interest related to environmental response. This layer will represent an individual's inherited susceptibility to disease at a fixed point in time. Epigenomic History: An additional PRS will be developed based upon a longitudinal assessment of changes in the epigenetic markers (DNA methylation patterns) resulting from environmental exposure throughout the lifetime of an individual.

This layer would function as a molecular archive, representing the cumulative "memory" or biological "scarring" from an individual's past environmental exposures. The third and most dynamic layer would be the Transcriptomic Response, measured through real-time gene expression profiling from accessible tissues like peripheral blood mononuclear cells. This layer would capture the immediate cellular response to a recent or ongoing exposure, reflecting the current "state of alert" of the biological system. The development of the d-PESS would involve leveraging machine learning algorithms trained on data from large, longitudinal cohort studies that have collected rich genomic, epigenomic, transcriptomic, and detailed environmental exposure data over many years. Instead of producing an assigned "lifetime risk score," the resulting computer-generated solution would yield a changing profile representing risk which can change with every new set of variables added to the model as they become available for testing on public and clinical health levels. Currently the term "susceptibility" refers to a small portion of what determines whether or not someone becomes sick with disease due to genetic and environmental factors, yet it is possible to define a person's susceptibility as an ever-changing, evolving characteristic of the complicated system a person is making their way through life in (Fig. 5). If the d-PESS were to produce validated results, it would dramatically enhance the ability of many professionals today to promote population health and the prevention of chronic diseases. The validated d-PESS can provide information on when in a person's life they have been identified as being at a critical time, provide ongoing evidence to help use the clinical and scientific tools available to monitor their response to interventions designed to reduce their exposure to certain risk factors, and finally allow medical professionals to create highly individualized prevention approaches based on their assessment of the individual's continually developing susceptibility profile.

Fig. 5 A conceptual framework for the dynamic Polygenic Environmental Susceptibility Score (d-PESS)

References

Adegbola, P. I., & Adetutu, A. (2024). Genetic and epigenetic modulations in toxicity: The two-sided roles of heavy metals and polycyclic aromatic hydrocarbons from the environment. *Toxicology Reports, 12*, 502.

Allen, C. G., McBride, C. M., Engdawork, K., Ayode, D., & Tadele, G. (2019). Applying mental model methods to characterize understanding of gene-environment influences: The case of podoconiosis in ethiopia. *Critical Public Health, 29*, 84–99.

Banerjee, B. D., Kumar, R., Thamineni, K. L., Shah, H., Thakur, G. K., & Sharma, T. (2019). Effect of environmental exposure and pharmacogenomics on drug metabolism. *Current Drug Metabolism, 20*, 1103–1113.

Bauer, A., Faiola, B., Abernethy, D., Marchan, R., Pluta, L., Wong, V., Roberts, K., Jaiswal, A., Gonzalez, F., Butterworth, B., Borghoff, S., Parkinson, H., Everitt, J., & Recio, L. (2003).

Genetic susceptibility to benzene-induced toxicity: Role of NADPH: Quinone oxidoreductase-11. *Cancer Research, 63*, 929–935.

Bellanca, C. M., Augello, E., Cantone, A. F., di Mauro, R., Attaguile, G. A., di Giovanni, V., Condorelli, G. A., di Benedetto, G., Cantarella, G., & Bernardini, R. (2023). Insight into risk factors, pharmacogenetics/genomics, and management of adverse drug reactions in elderly: A narrative review. *Pharmaceuticals, 16*, 1542.

Bhatia, S. (2015). Genetic variation as a modifier of association between therapeutic exposure and subsequent malignant neoplasms in cancer survivors. *Cancer, 121*, 648–663.

Borgesius, F. Z., Gray, J., & van Eechoud, M. (2015). Open data, privacy, and fair information principles: Towards a balancing framework. *Berkeley Technology Law Journal, 30*, 2073–2131.

Caldwell, J., Sangster, S. A. & Sutton, J. D. (2020). The role of metabolic activation in target organ toxicity. In *Target organ toxicity* (vol I). CRC Press.

Cañas Portilla, A. I., Castaño, A. & Organization, W. H. (2024). *Human health effects of benzene, arsenic, cadmium, nickel, lead and mercury: Report of an expert consultation.* World Health Organization. Regional Office for Europe.

Chain, E. P. O. C. I. T. F., Schrenk, D., Bignami, M., Bodin, L., Chipman, J. K., Del Mazo, J., Grasl-Kraupp, B., Hogstrand, C., Hoogenboom, L. & Leblanc, J. C. (2024). Update of the risk assessment of inorganic arsenic in food. *EFSA Journal, 22*, e8488.

Chen, C.-H. (2024). Phase II detoxification enzymes. *Activation and detoxification enzymes: Functions and implications.* Springer.

Ching, T., Himmelstein, D. S., Beaulieu-Jones, B. K., Kalinin, A. A., Do, B. T., Way, G. P., Ferrero, E., Agapow, P.-M., Zietz, M., & Hoffman, M. M. (2018). Opportunities and obstacles for deep learning in biology and medicine. *Journal of the Royal Society Interface, 15*, 20170387.

Curran, W., & Hamilton, T. (2017). *Just green enough: Urban development and environmental gentrification.* Routledge.

Di, X., Beesley, L., Zhang, Z., Zhi, S., Jia, Y., & Ding, Y. (2019). Microbial arsenic methylation in soil and uptake and metabolism of methylated arsenic in plants: A review. *International Journal of Environmental Research and Public Health, 16*, 5012.

Fitzpatrick, J., Schoeny, R., Gallagher, K., Deener, K., Dockins, C., Firestone, M., Jordan, W., McDonough, M., Murphy, D., & Olsen, M. (2017). US environmental protection agency's framework for human health risk assessment to inform decision making. *International Journal of Risk Assessment and Management, 20*, 3–20.

Funk, W., Forester, B. R., Converse, S. J., Darst, C., & Morey, S. (2019). Improving conservation policy with genomics: A guide to integrating adaptive potential into US endangered species act decisions for conservation practitioners and geneticists. *Conservation Genetics, 20*, 115–134.

Ghosh, S., Ghosh, S., Bankura, B., Saha, M. L., Maji, S., Ghatak, S., Pattanayak, A. K., Sadhukhan, S., Guha, M., & Nachimuthu, S. K. (2016). Association of DNA repair and xenobiotic pathway gene polymorphisms with genetic susceptibility to gastric cancer patients in West Bengal, India. *Tumor Biology, 37*, 9139–9149.

Gray, D. *Environmental protection agency 40 CFR part 751.*

Grishanova, A. Y., & Perepechaeva, M. L. (2022). Aryl hydrocarbon receptor in oxidative stress as a double agent and its biological and therapeutic significance. *International Journal of Molecular Sciences, 23*, 6719.

Guarnieri, T. (2020). Aryl hydrocarbon receptor connects inflammation to breast cancer. *International Journal of Molecular Sciences, 21*, 5264.

Habil, M. (2022). Investigation of human n-acetyltransferases (NAT1 and NAT2) genetic polymorphisms in susceptibility to aromatic amine and alkylaniline genotoxicity.

Han, H., Safe, S., Jayaraman, A., & Chapkin, R. S. (2021). Diet–host–microbiota interactions shape aryl hydrocarbon receptor ligand production to modulate intestinal homeostasis. *Annual Review of Nutrition, 41*, 455–478.

Hartman, J. H., Miller, G. P., & Meyer, J. N. (2017). Toxicological implications of mitochondrial localization of CYP2E1. *Toxicology Research, 6*, 273–289.

Hausman-Cohen, S. R., Hausman-Cohen, L. J., Williams, G. E., & Bilich, C. E. (2020). Genomics of detoxification: How genomics can be used for targeting potential intervention and prevention strategies including nutrition for environmentally acquired illness. *Journal of the American College of Nutrition, 39*, 94–102.

Hettiarachchi, G. & Komar, A. A. (2022). GWAS to identify SNPs associated with common diseases and individual risk: genome wide association studies (GWAS) to identify SNPs associated with common diseases and individual risk. *Single nucleotide polymorphisms: Human variation and a coming revolution in biology and medicine.* Springer.

Hindorff, L. A., Bonham, V. L., Brody, L. C., Ginoza, M. E., Hutter, C. M., Manolio, T. A., & Green, E. D. (2018). Prioritizing diversity in human genomics research. *Nature Reviews Genetics, 19*, 175–185.

Ishikawa, S. (2017). 3 Promoting understanding of chemical sensitivity. *Chemical Sensitivity and Sick-Building Syndrome, 1.*

Jang, S. (2023). *Big data-enabled modeling approaches to address challenges in the chemical exposure to human risk continuum.*

Jouannais, P. & Pizzol, M. (2021). Ex-Ante LCA for microalgae-based veterinary molecules in finfish aquaculture: how to assess the environmental performance of unknown molecules? In *Setac Europe 31st Annual Meeting: Virtual Meeting.*

Ju, D., Hui, D., Hammond, D. A., Wonkam, A., & Tishkoff, S. A. (2022). Importance of including non-European populations in large human genetic studies to enhance precision medicine. *Annual Review of Biomedical Data Science, 5*, 321–339.

Khojasteh, S. C., Miller, G. P., Mitra, K., & Rietjens, I. M. (2018). Biotransformation and bioactivation reactions–2017 literature highlights. *Drug Metabolism Reviews, 50*, 221–255.

Lee, W.-S., Ham, W., & Kim, J. (2021). Roles of NAD (P) H: Quinone oxidoreductase 1 in diverse diseases. *Life, 11*, 1301.

Lefebvre, K. (2015). Genetic discrimination law in the United States: a socioethical & legal analysis of the Genetic information nondiscrimination act (GINA). *University of Massachusetts Amherst, School of Public Policy.*

Liu, D., Qin, S., Ray, B., Kalari, K. R., Wang, L., & Weinshilboum, R. M. (2018). Single nucleotide polymorphisms (SNPs) distant from xenobiotic response elements can modulate aryl hydrocarbon receptor function: SNP-dependent CYP1A1 induction. *Drug Metabolism and Disposition, 46*, 1372–1381.

Llaurens, V., Whibley, A., & Joron, M. (2017). Genetic architecture and balancing selection: The life and death of differentiated variants. *Molecular Ecology, 26*, 2430–2448.

Manera, M. (2021). Perspectives on complexity, chaos and thermodynamics in environmental pathology. *International Journal of Environmental Research and Public Health, 18*, 5766.

McCarthy, J. J., Graas, J., Leamon, M. H., Ward, C., Vasti, E. J., & Fassbender, C. (2020). The Use of the Methadone/Metabolite Ratio (MMR) to identify an individual metabolic phenotype and assess risks of poor response and adverse effects: Towards scientific methadone dosing. *Journal of Addiction Medicine, 14*, 431–436.

Metz, G. A. & Faraji, J. (2026). *Environmental epigenetics: New horizons in redefining biological and health outcomes.*

Morabito, A., de Simone, G., Pastorelli, R., Brunelli, L., & Ferrario, M. (2025). Algorithms and tools for data-driven omics integration to achieve multilayer biological insights: A narrative review. *Journal of Translational Medicine, 23*, 425.

Motsinger-Reif, A. A., Reif, D. M., Akhtari, F. S., House, J. S., Campbell, C. R., Messier, K. P., Fargo, D. C., Bowen, T. A., Nadadur, S. S. & Schmitt, C. P. (2024). Gene-environment interactions within a precision environmental health framework. *Cell Genomics, 4.*

Pánico, P., Velasco, M., Salazar, A. M., Picones, A., Ortiz-Huidobro, R. I., Guerrero-Palomo, G., Salgado-Bernabé, M. E., Ostrosky-Wegman, P. & Hiriart, M. (2022). Frontiers | is arsenic exposure a risk factor for metabolic syndrome? A review of the potential mechanisms. *Frontiers in Endocrinology, 13.*

Pazzagli, L., Linder, M., Zhang, M., Vago, E., Stang, P., Myers, D., Andersen, M., & Bahmanyar, S. (2018). Methods for time-varying exposure related problems in pharmacoepidemiology: An overview. *Pharmacoepidemiology and Drug Safety, 27*, 148–160.

Pérez Santín, E., Rodríguez Solana, R., González García, M., García Suárez, M. D. M., Blanco Díaz, G. D., Cima Cabal, M. D., Moreno Rojas, J. M. & López Sánchez, J. I. (2021). Toxicity prediction based on artificial intelligence: A multidisciplinary overview. *Wiley Interdisciplinary Reviews: Computational Molecular Science, 11*, e1516.

Piekos, S. C. (2018). *Consequences of antiepileptic drug exposure during postnatal liver development on cytochrome p450 expression and metabolism*, University of Connecticut.

Raghavan, R. (2025). *Environmental factors in carcinogenesis: Exposome-driven insights into cancer risk and prevention: Exposome-driven insights into cancer risk and prevention.* IGI Global.

Ratnani, I. (2021). Gap between practices of evidence-based medicine and personalized medicine: A barrier in learning.

Sarwar, E. (2019). Precision medicine and it's ethical and social implications: Public health and gobal persepctives.

Singh, S., Ahmed, T. & Chauhan, A. S. (2025). Beyond the genome: Exploring the multi-omics landscape. In *Multi-Omics in biomedical sciences and environmental sustainability: Applications and recent advances.* Springer.

Srinivasan, S., Clements, J. A., & Batra, J. (2016). Single nucleotide polymorphisms in clinics: Fantasy or reality for cancer? *Critical Reviews in Clinical Laboratory Sciences, 53*, 29–39.

Statzer, C., Park, J. Y. C., & Ewald, C. Y. (2023). Extracellular matrix dynamics as an emerging yet understudied hallmark of aging and longevity. *Aging and Disease, 14*, 670.

Stýblo, M., Venkatratnam, A., Fry, R. C., & Thomas, D. J. (2021). Origins, fate, and actions of methylated trivalent metabolites of inorganic arsenic: Progress and prospects. *Archives of Toxicology, 95*, 1547–1572.

Tchounwou, P. B., Yedjou, C. G., Udensi, U. K., Pacurari, M., Stevens, J. J., Patlolla, A. K., Noubissi, F., & Kumar, S. (2019). State of the science review of the health effects of inorganic arsenic: Perspectives for future research. *Environmental Toxicology, 34*, 188–202.

Thomas, D. J. (2021). Arsenic methylation–Lessons from three decades of research. *Toxicology, 457*, 152800.

Thomas, R., Kim, S., Lan, Q., Vermeulen, R., Zhang, L., Rothman, N., Smith, M. T., & Rappaport, S. M. (2025). Benzene metabolism is dominated by a high-affinity pathway at ambient exposures with implications for cancer risks. *International Journal of Molecular Sciences, 26*, 8550.

Tsalenchuk, M., Gentleman, S. M. & Marzi, S. J. (2023). Linking environmental risk factors with epigenetic mechanisms in Parkinson's disease. *npj Parkinson's Disease, 9*, 123.

Tumkursattar, A. H., Abushoumi, F., AL Dobayan, Z. & AL-Ghamdi, F. (2023). Investigate how genetic variations impact drug response and explore the integration of pharmacogenomics into clinical practice. *European Journal of Cardiovascular Medicine, 13*.

Varshavsky, J. R., Rayasam, S. D., Sass, J. B., Axelrad, D. A., Cranor, C. F., Hattis, D., Hauser, R., Koman, P. D., Marquez, E. C., & Morello-Frosch, R. (2023). Current practice and recommendations for advancing how human variability and susceptibility are considered in chemical risk assessment. *Environmental Health, 21*, 133.

Virolainen, S. J., Vonhandorf, A., Viel, K. C., Weirauch, M. T., & Kottyan, L. C. (2023). Gene–environment interactions and their impact on human health. *Genes & Immunity, 24*, 1–11.

Werme, J., van der Sluis, S., Posthuma, D., & de Leeuw, C. A. (2021). Genome-wide gene-environment interactions in neuroticism: An exploratory study across 25 environments. *Translational Psychiatry, 11*, 180.

Xu, X., Drobná, Z., Voruganti, V. S., Barron, K., González-Horta, C., Sánchez-Ramírez, B., Ballinas-Casarrubias, L., Cerón, R. H., Morales, D. V., & Terrazas, F. A. B. (2016). Association between variants in arsenic (+3 oxidation state) methyltranserase (AS3MT) and urinary metabolites of inorganic arsenic: Role of exposure level. *Toxicological Sciences, 153*, 112–123.

Xue, S. N. (2025). *Impact of air pollution on lung function in fibrotic interstitial lung disease and asthma: Insights from geospatial models to gene-environment interactions.* University of Pittsburgh.

Yusoff, N. A., Abd Hamid, Z., Budin, S. B. & Taib, I. S. (2023). Linking benzene, in utero carcinogenicity and fetal hematopoietic stem cell niches: A mechanistic review. *International Journal of Molecular Sciences, 24*, 6335.

Zhou, Z., Duan, Y., Li, Y., Zhang, P., Li, Q., Yu, L., Han, C., Huo, J., Chen, W. & Xiao, Y. (2025). CYP98A monooxygenases: a key enzyme family in plant phenolic compound biosynthesis. *Horticulture Research, 12*, uhaf074.

Chapter 13
The Toxicological Impact of Xenobiotics on Terrestrial and Aquatic Ecosystems

Pankaj Thakur, Vanshika Thakur, Radhika Sharma, Patricks Voua Otomo, and Reshma Sinha

Abstract Xenobiotics are a broader group of substances that include plant ingredients, industrial products, pharmaceutical substances, drugs, insecticides, personal care products and food additives. The chapter discusses their origins, forms, toxicokinetic routes and fate in both soil and water. Besides conventional xenobiotics, emergent forms such as microplastics and nanomaterials have also been discussed. It also records effects on the soil micro-communities, birds, fish and, lastly, the impact on human beings. Due to their broad use, they can enter environmental matrices directly and indirectly, releasing various metabolites and secondary products into the soil, water and air. They are more likely to bioaccumulate and can upset the food web. Upon interaction with biomolecules, they influence physiological mechanisms, targeting the endocrine, reproductive, immune and nervous systems, causing oxidative stress and inhibiting several enzymes, which may result in cancer. The intensity of the toxicity remains associated with the physicochemical characteristics of the medium, the chemical nature, solubility and charge of the xenobiotics, as well as the species under exposure, the concentration of toxicant and the duration of exposure. The chapter highlights the need to explore the xenobiotic toxicity so as to come up with efficient conservation and pollution management measures.

Keywords Bioaccumulation · Birds · Food web · Macroinvertebrates · Microplastics · Toxicokinetic

---

P. Thakur · V. Thakur · R. Sinha (✉)
Department of Animal Sciences, Central University of Himachal Pradesh, Kangra, India

R. Sharma
Department of Biology and Environmental Sciences, College of Basic Sciences, CSKHPKV, Palampur, India

P. V. Otomo
Department of Zoology and Entomology, Faculty of Natural and Agricultural Sciences, University of the Free State, Phuthaditjhaba, South Africa

Centre for Global Change, University of the Free State, Phuthaditjhaba, South Africa

K. Priya and S. K. Giri (eds.), *Environmental Toxicology: Xenobiotics and Effects on Human Genetic Material*, https://doi.org/10.1007/978-3-032-28619-2_13

1 Introduction: Xenobiotics, their Classification and Major Sources

Xenobiotics present in the environment as pollutants capable of causing biological effects are called environmental xenobiotics (Patel & Jyoti, 2013). These comprise a wide class of compounds including plant components, industrial commodities, medicinal compounds, narcotic drugs, insecticides, cosmetics, food additives and scents. They may enter environmental matrices either directly or indirectly and produce a variety of metabolites and secondary products (Singh et al., 2023).

Xenobiotics may be classified into industrial, agricultural, domestic, pharmaceutical and synthetic. Petrochemicals (oil, gas mixture), heavy metals (lead, mercury, cadmium, arsenic, chromium, zinc), aromatic/saturated hydrocarbons (phenanthrene, pyrene, naphthalene, crude oils, petroleum products), $N_2/S/O_2$, preservatives from food industries (like sodium benzoate, nitrites, parabens, tartrazine, erythrosine), organic and inorganic waste like industrial waste products and effluents from industrial activity such as phenols, hydrocarbons, dyes, surfactants, pharmaceuticals and plasticizers account for direct source of xenobiotics (Table 1) (Thakur et al., 2024). In the agricultural systems, extensive use of pesticides like thiophanate methyl, kasugamycin, mancozeb, etc., and fertilizers like urea, ammonium nitrate, superphosphate, potassium chloride, etc., contribute to their accumulation in the soil and water (Srivastav, 2020). Synthetic chemicals, such as phenanthrene, naphthalene, surfactant, triclosan, personal care products (PCPs) like UV filters, soaps, deodorants, cosmetics, perfumes and sunscreens are major environmental xenobiotics (Eriksson et al., 2003). PCPs contain naturally derived bioactive compounds, including cyanogenic glucosides, toxic alkaloids, terpenes, quinones and peroxides (Bucheli et al., 2018). Cosmetic formulations frequently contain high water content, necessitating the addition of preservatives and biocides, such as triclosan, to extend product shelf life.

Major types of pharmaceuticals found in the environment are analgesics, anti-inflammatory drugs (particularly non-steroid anti-inflammatory drugs or NSAIDs), antibiotics and psychiatric drugs. The abundance of a pharmaceutical is dependent on its spatial and temporal distribution (Zhang et al., 2021).

Microplastics and nanoparticles are novel xenobiotics because they are synthetic and persistent pollutants. Nanoparticles are engineered nanoscale substances with unique physicochemical properties not found in nature. Nanoparticles, including copper (CuNPs), copper oxide (CuONPs), zinc oxide (ZnONPs) and titanium dioxide (TiO_2NPs), are widely used, with applications spanning agriculture as fungicides and bactericides and cosmetics and medicines as active ingredients (Chauhan et al., 2021; Tortella et al., 2024). Thus, these xenobiotics have emerged as a significant and ubiquitous threat to ecosystems.

Table 1 Classification of xenobiotics with their major sources

Classification	Major source	Compound	Pathway to environment	References
Industrial xenobiotics	Chemical manufacturing, textile dyeing, petroleum refining, metal plating	Heavy metals, solvents, persistent organic pollutants, polycyclic aromatic hydrocarbons	Industrials effluents, inadequate treatment, rivers, lakes, soils	Thakur et al. (2024)
Agricultural xenobiotics	Pesticides and herbicides, veterinary pharmaceuticals and synthetic fertilizers	Organochlorine pesticides, glyphosate, neonicotinoids and animal antibiotics	Surface runoff and leaching, water bodies and groundwater	Gianfreda (2008)
Domestic xenobiotics	Household cleaning agents, detergents, cooking waste and discarded plastics	Surfactants, phosphates, microplastics and fragrances.	Wastewater discharge, treatment plant, effluent to aquatic systems	Eriksson et al. (2003)
Pharmaceutical	Hospitals, households and manufacturing waste	Antibiotics, analgesics, antidepressants, antipyretics	Disposal into sewage, partial removal in treatment, water contamination	Embrandiri et al. (2016), Zhang et al. (2021)
Personal care products	Cosmetics, sunscreens, fragrances, shampoos	UV filters, parabens, microbeads, soap and shampoo	Domestic wastewater, wastewater treatment plants, aquatic environment	Bester (2007)
Urban source	Road runoff, vehicle emissions, tire wear and construction activity	Heavy metals, hydrocarbons, microplastics	Stormwater runoff, urban drainage systems, rivers and coastal waters	Paul et al. (2023)
Water treatment plant discharges	Influenced by industrial, domestic and hospital	Mixed xenobiotic load (pharmaceuticals, heavy metals, synthetic polymers)	Treated effluent/ sludge release, aquatic and terrestrial systems	Schirmer et al. (2011)
Heavy metals	Industrial processes, mining and urban runoff	Lead, mercury, cadmium, arsenic	Industrial waste discharge, soil and sediment accumulation	Li et al. (2001)

(continued)

Table 1 (continued)

Classification	Major source	Compound	Pathway to environment	References
Synthetic polymers	Packaging, textiles, industrial materials	Polyethylene, polypropylene, polystyrene	Improper disposal, environmental fragmentation	Wang et al. (2023)
Microplastics and nanoplastics	Degradation of larger plastics, synthetic textiles and personal care products,	<5 mm polymer fragments, nanoplastics	Atmospheric deposition, water and soil transport, ingestion by biota	Bläsing and Amelung (2018), He et al. (2018)

2 Fate and Transport

2.1 Pathways into Terrestrial and Aquatic Environments

Rainwater runs off into storm drains straight into waterways, trapping sediment and pollutants (Hamlyn-Harris et al., 2019). Agricultural runoff often contains organic and xenobiotic compounds, including aromatics, plasticizers, fungicides and weedicides (Phale et al., 2019). The pesticides, herbicides and other agrochemicals reach the groundwater through leaching. Both laboratory column experiments and field studies have demonstrated that movement is significantly influenced by factors such as the type of soil, its absorption capacity, the chemical solubility of substances and the dynamics of local water (Sondhia, 2009; Toshiyuki Katagi, 2012). In addition to indirect entry into the environment, such as discharge through runoff, wastewater effluent and industrial releases, as well as atmospheric deposition, direct deposition of airborne substances, including herbicides and persistent organic pollutants, on soils and water bodies has been observed (Tran et al., 2018). Research studies on PFAS (per- and polyfluoroalkyl substances) indicates that direct discharges, like those from aqueous film forming foam (AFFF), are responsible for subsurface transport. Further, transport by sorption, hydrogeology, and degradation influences its interactions with plants (McGarr & Mbonimpa, 2023). Microplastics enter soil systems from sources such as residual agricultural mulch films, organic amendments, irrigation practices, surface runoff, incorporation of sewage sludge, mismanagement of plastic debris and deposition from the atmosphere (Bläsing & Amelung, 2018; He et al., 2018). Low plastic recycling rates contribute to the leakage of waste plastics into soils and landfills, where they gradually fragment into microplastics.

2.2 Properties Influencing the Environmental Fate and Transport of Xenobiotics

Persistence is among the various factors influencing the environmental fate and transport of xenobiotics. Persistence is often measured by degradation half-lives and compounds that have half-lives greater than 100 days, including most active pharmaceutical ingredients (APIs), are capable of persisting in the environment. During wastewater treatment, several APIs are transformed instead of being fully degraded, and they may continue to persist in the effluent discharge (Ortúzar et al., 2022). Another important aspect of the soil and groundwater pollution is through xenobiotics mobility. Sorption behaviour to a great extent controls mobility and is measured in terms of the 'partitioning coefficients', e.g. soil organic carbon–water partition coefficient. Those compounds with lower carbon–water partition coefficient values are weakly sorbed to soil particles and thus become very mobile and more prone to leaching into groundwater (Arp and Hale, 2022). Similarly, biofilm-coated microplastics display varying density and settling rates and can often spend more time in the benthic habitat (Mendrik et al., 2023). Thus, the physicochemical characteristics of xenobiotics, including water solubility, volatility, ionization, vapour pressure, and octanol-water partition coefficient, also affect the interaction of substances with environmental matrices. For example, highly volatile compounds may escape into the atmosphere, while low-volatility substances tend to accumulate in soils and sediments (Miglani et al., 2022; Xiang et al., 2022).

Temperature plays an important role, as it alters the aeration rate by increasing the uptake and absorption of pesticides or xenobiotic contaminants (Mishra et al., 2019; Ma et al., 2004; Meyer and Wania, 2008). Temperature positively affects degradation rates by altering the sorption dynamics (Knudsmark Sjøholm et al., 2021), control chemical speciation and mobility. Compounds become even more mobile at higher pH due to reduced sorption. pH shifts codetermine leaching, groundwater and plant uptake (Marín-Benito et al., 2019; Carter et al., 2020; Knudsmark Sjøholm et al., 2021).

3 Toxicity Mechanisms of Xenobiotics

Xenobiotics (pesticides and industrial chemicals, phthalates and brominated flame retardants, etc.) accumulate in the soil and are transferred to plants and animals, and concentrate at higher trophic levels due to difficulty in metabolism and excretion, leading to an increased concentration in animal tissues (bioaccumulation) over time (Gobas et al., 2015; Seghir et al., 2023; Rist et al., 2018). Further, upon release into the environment, they become part of food webs, bringing detrimental effects at every trophic level, causing biomagnification through food chains (Table 2) (Miglani et al., 2022; Borgå, 2013).

Table 2 Biomagnification of xenobiotics in food webs

Xenobiotic	Example ecosystem	Trophic transfer pathway	Impact on the food chain/ecosystem	Reference
Cadmium (heavy metal)	Aquatic	Algae, zooplankton, fish	Acute aquatic toxicity, kidney dysfunction in fish and humans, disrupts enzymatic systems	Ruangsomboon and Wongrat (2006), Lee et al. (2024)
Mercury (especially MeHg)	Aquatic	Small fish, large fish, top predators/ humans	Neurotoxin bioaccumulates in top predators, posing health hazards for humans due to seafood consumption	Lavoie et al. (2013)
Arsenic	Aquatic	Phytoplankton, zooplankton, fish	Chronic toxicity, population decline in invertebrates and fish, health hazard for consumers	Zhang et al. (2022), Chen and Folt (2000)
Lead (Pb)	Terrestrial	Plants, herbivores, carnivores	Neurotoxicity, anaemia, stunted growth in mammals and persistent soil contaminant	Kumar et al. (2020), Chen and Folt (2000)
Polychlorinated biphenyls (PCBs)	Aquatic/ terrestrial	Plankton/soil organisms, fish/ animals, mammals	Endocrine disruption, carcinogen, developmental issues, immune system effects in wildlife and humans	Wang et al. (2022)
Dioxins (PCDD/ PCDFs)	Aquatic/ terrestrial	Detritus, invertebrates, fish/birds/ mammals	Potent carcinogens, reproductive disorders, immunotoxicity, birth defects	Zhang et al. (2011), Picone et al. (2020)
Organochlorine pesticides (endrin, chlordane, HCB)	Aquatic/ terrestrial	Soil/water, plants/insects, birds/mammals	Neurological and reproductive damage, endocrine disorders, long-term soil and water contamination	Jayaraj et al. (2016), Kucklick and Baker (1998)
DDT (pesticide)	Aquatic/ terrestrial	Plankton, fish, birds/mammals	Causes eggshell thinning in birds, declining bird populations, potential carcinogen to mammals	Ray and Shaju (2023)

In a terrestrial system, root uptake from soil by plants is passive and influenced by the transpiration stream. After chemicals pass through the root membrane, they move within the plant through vascular tissue, where xylem carries dissolved material upward from roots to leaves at a rate of 10 cm/min (Helfter et al., 2007). Particles in the micro- and submicron range may penetrate plant roots via localized fissure-entry pathways, subsequently being translocated to the aerial parts of the plants (Yu et al.,

2023). Herbivores ingest xenobiotics by eating contaminated plants or soil particles (Wang et al., 2022).

Aboveground exposure includes absorption of xenobiotics via gas exchange, dry and wet deposition (for aerosol-adsorbed and water-soluble compounds) and cuticular absorption (Gobas et al., 2015). Further soil microbial communities are affected, with reduced nutrient cycling and altered soil chemistry. Microplastics can also bioaccumulate within food webs, posing risks to soil organisms across multiple trophic levels and potentially disrupting entire ecosystems functions (He et al., 2018; Wan et al., 2023).

Terrestrial air-breathing animals may absorb volatile xenobiotics from the air directly via their respiratory systems (Tison et al., 2024) and dermal absorption through the skin, cuticle or digestive organs (Chhabra, 1979; Tison et al., 2024).

Consequently, xenobiotics interference with metabolic and developmental processes in aquatic systems. Some xenobiotics (dyes and paints) interfere with the penetration of light in water, affecting photosynthesis and gas exchange. Further, the concentration of plastic in the floor (ocean and the sea) and bed (river and lake) results in the inhibition of the diffusion of gases between liquid and solid phases at the bottom of the water body (Wright et al., 2013). Particulate matter, besides acting as a carrier of xenobiotics and enhances the xenobiotic delivery in aquatic organisms, perpetuates the toxicity through systemic distribution and immune stimulation.

Among all xenobiotics, pharmaceuticals raise particular toxicological concerns because they are designed to act on biological processes. Even at lower environmental concentrations, these can disrupt physiological processes, promote antibiotic resistance and harm non-target species (Singh et al., 2023; Embrandiri et al., 2016). Active pharmacological ingredients (APIs), the main ingredients in drugs that cause the desired effect of the medicine, have been widely detected in terrestrial environments, particularly in areas with high human activity.

4 Toxicodynamic of Xenobiotics

The toxicity mechanisms of xenobiotics involve disruption in physiological, biochemical and ecological processes, primarily through bioaccumulation, altered metabolism, interference with cellular signalling and gene expression in the organism. Toxicodynamics refers to the biological effects occurring after xenobiotic interaction with molecular targets in the organism (Table 3) (Park et al., 2014).

4.1 Interaction with Biomolecules

The reactive xenobiotic metabolites can attach themselves to proteins, DNA and lipids, leading to dysfunction of cellular activity (Li et al., 2011). Xenobiotics that bind to amino acid side chains (i.e. cysteine, lysine, histidine) change the structure

Table 3 Toxicodynamics of xenobiotics in different physiological systems

Toxicity type	Key pathways and mechanisms	Key molecular players	Xenobiotic examples	Source of xenobiotics	Effects on organisms and ecosystem	References
Genotoxicity	ROS generation causes oxidative DNA damage, direct DNA adduct formation, inhibition of DNA repair enzymes and epigenetic modifications	Cytochrome P450 monooxygenases (CYPs), glutathione-S-transferase (GSTs), dioxygenases	Polycyclic aromatic hydrocarbons (PAHs, e.g. benzo[a]pyrene), heavy metals (cd, Pb and as), pesticides (DDT and chlordane), polychlorinated biphenyls (PCBs)	Industrial emission, fossil fuel combustion, pesticide application, agricultural runoff and mining activities	Mutation, chromosomal aberrations, impaired reproduction, developmental defects, altered gene expression	Miglani et al. (2022), Esteves et al. (2021)

(continued)

Table 3 (continued)

Toxicity type	Key pathways and mechanisms	Key molecular players	Xenobiotic examples	Source of xenobiotics	Effects on organisms and ecosystem	References
Immunotoxicity	Oxidative stress, induced disruption of immune signalling pathways (NF-kB, JNK, NRF2, STAT3), toll-like receptor (TLR) mediated innate immune modulation, altered cytokine production, apoptosis, chronic inflammation and immunosuppression	NF-kB, JNK, NRF2, STAT3, TLRs, ROS	Organophosphates pesticides, arsenic, BPA, heavy metals (Pb, hg), aromatic hydrocarbons	Pesticide spraying in agriculture, industrial wastewater, mining and atmospheric deposition	Reduced pathogen resistance, immune dysfunction, increased infection susceptibility, impaired organismal fitness in ecosystems	D'Souza et al. (2023)
Neurotoxicity	Acetylcholinesterase inhibition, ion channel blockade (Ca2+, K+), receptor binding interference, neuroinflammation (ROS, cytokines), endocrine-neural signalling disruption, lipophilic xenobiotics accumulate in nervous tissues	AChE, ion channels, neurotransmitter receptors, ROS, cytokines	Organophosphates (malathion), pyrethroids, heavy metals (Pb, cd, hg)	Agricultural pesticide use, industrial discharge, mining and combustion sources	Natural dysfunction, behavioural abnormalities, developmental neurotoxicity, neurodegeneration, reproductive impairments	Čadež et al. (2021), Manetti (2023), Patisaul (2021)

and functionality of proteins. Protein adduction leads to the disruption of cellular functions and triggers immune responses (Fang et al., 2009). When it binds to the DNA, it results in the formation of DNA adducts with the nucleophilic sites on DNA bases (especially guanine and adenine) (Yun et al., 2018), therefore causing mutations that can cause cancer (Liu et al., 2016). The binding of xenobiotics to lipids leads to the formation of reactive oxygen species (ROS), which promotes lipid peroxidation, resulting in the generation of toxic aldehydes that form further adducts with proteins and DNA, resulting in membrane and biomolecule damage, oxidative stress, apoptosis and cell injury (Gaschler and Stockwell, 2017; Mandelker, 2008).

4.2 *Disruption of Cell Signalling*

Xenobiotics can disrupt cellular signalling pathways by acting on hormone receptors and altering gene expression, thus developing physiological stress and developmental abnormalities (Miglani et al., 2022). Xenobiotics like organophosphates alter cellular communication by inhibiting several important enzymes and receptors (acetylcholinesterase) in soil invertebrates, resulting in neurotoxicity (Navenant et al., 2019; Ayoub et al., 2024). Xenobiotics can further dysregulate nuclear receptors (PXR, CAR) associated with xenobiotic metabolism genes and the rearrangement of phosphorylation cascades such as MAP kinase, which mediate stress and defence response.

4.3 *Oxidative Stress*

Most xenobiotics induce increased production of ROS including superoxide anions (O2–), hydrogen peroxide (H_2O_2) and hydroxyl radicals (.OH) that can cause oxidative stress and damage critical cellular constituents by lipid peroxidation of cellular membranes, protein oxidation, which results in enzyme inactivation, DNA strand breakage and base modification that disrupt genetic integrity (Liu et al., 2025). Xenobiotic stress can suppress photosynthesis and respiration in the organism. Organisms respond to stress using enzymatic and non-enzymatic antioxidants, such as superoxide dismutase (SOD), catalase (CAT) and glutathione peroxidases; and flavonoids, phenolic acids and vitamins C and E, respectively (Vicidomini et al., 2024; El-Demerdash et al., 2018). The resultant oxidative imbalance retards growth, compromises reproduction and may induce cell death, and the effect is propagated in the productivity and stability of ecosystems. Loss of mitochondrial activity has been evidenced upon interaction with cadmium in *Eisenia fetida* (Yi et al., 2022).

4.4 Endocrine Disruption

Some xenobiotics mimic, block or alter natural hormones—mainly oestrogens, androgens and thyroid hormones—by binding to receptors, disrupting synthesis or modifying gene expression, which can lead to impaired reproductive development and fertility (altered sex ratios, malformed reproductive organs), altered growth and immune responses, disrupted behaviour and population declines in wildlife (Patisaul et al., 2018). For example, Bisphenol A (BPA) in soil binds to oestrogen receptors in small mammals, reducing fertility (Siracusa et al., 2018). Industrial chemicals, pesticides, pharmaceuticals and persistent organic pollutants (such as PCBs and Dioxins) are prominent endocrine-disrupting chemicals reported in wildlife and terrestrial food webs (Lee et al., 2012).

4.5 Genotoxicity

Genotoxicity is the ability of xenobiotics to damage an organism's genetic material, either directly, by chemically interacting with DNA, or indirectly, via reactive oxygen species (ROS) and oxidative stress that oxidize and destabilize nucleic acids (Shukla et al., 2021; Miglani et al., 2022). Some xenobiotics, such as polycyclic aromatic hydrocarbons (PAHs), bind covalently to DNA adducts, alter the DNA structure and can block replication or transcription, leading to mutations if unrepaired (Miglani et al., 2022). Other mechanisms are often mediated by enzymes like cytochrome P450 monooxygenase, glutathione-S-transferase and other oxidoreductases that bio-transform xenobiotics, sometimes generating reactive intermediates that damage DNA (Esteves et al., 2021).

4.6 Immunotoxicity

Xenobiotics undermines the immune system by reducing immune response or inducing inappropriate inflammation (Azam, 2024). Overproduction of ROS induced by xenobiotics disturbs the immune signalling pathways, including JNK, NRF2, STAT3 and NF-kB, leading to the inappropriate activation of the immune system or immunosuppression. Although toll-like receptors (TLRs) and mitochondrial enzymes assist recognition of xenobiotic stress, prolonged exposure tends to increase oxidative damage and inflammation (D'Souza et al., 2023). It results in the production of distorted cytokines, impaired resistance to pathogens, apoptosis of immune cells and chronic inflammation, eventually lowering the fitness of organisms and disrupting terrestrial ecosystems (Zerdan et al., 2021; Pallardy et al., 1999).

4.7 Neurotoxicity

Xenobiotics can affect the central or peripheral nervous system, leading to neurotoxicity. Its lipophilic nature enables it to pass through biological membranes, concentrate in lipid-rich neural tissues and disrupt normal signalling (Iqubal et al., 2020). The toxicity mechanisms include enzyme inhibition, ion channel modulation, inhibiting receptor binding, neuroinflammation and endocrine-neural disruption. Organophosphates possess the capacity to irreversibly inactivate acetylcholinesterase (AChE), which disables neurotransmission in insects and soil invertebrates (Čadež et al., 2021). The heavy metals interfere with calcium or potassium channels, altering nerve excitability (Manetti, 2023). Concordantly, neurotoxicants occupy neurotransmitter receptors or membrane proteins, interfering with the transmission of signals (Silva-Adaya et al., 2021). Neural inflammation might be caused by the destruction of neuronal tissue by ROS and cytokine release (Mohan Kumar et al., 2023). Some pesticides can damage neuroendocrine disruptions, and thus cause abnormal behaviour, developmental anomalies and neuronal death (Apú et al., 2025; Patisaul, 2021).

5 Ecotoxicological Impact on Terrestrial and Aquatic Biota

Xenobiotics such as heavy metals, herbicides (e.g. glyphosate) and soil-persistent microplastics influence plants at physiological, morphological and genetic levels (Roy et al., 2022; Miglani et al., 2022). These undermine the health of plants, disrupting their essential physiological function by developing oxidative stress (Vicidomini et al., 2024; Neumann and Werner, 2000; Gomez-Sagasti and Marino, 2015). Nutrient and water uptake are also disrupted due to altered root-membrane integrity, transporter functionality and soil chemistry, consequently leading to Fe, K and Ca imbalance and poor water relations (Hafeez et al., 2023; Li et al., 2023). In addition, some chemicals cause hormonal perturbations, like disrupting the signalling of auxins, cytokines or abscisic acid (ABA), thus disrupting growth regulation, stress tolerance and development duration (Fuchs et al., 2022). All this adds up to reduce photosynthetic efficiency, respiration and nutrient assimilation, eventually holding back the plant's vigour, yield and ecological fitness, while morphologically causing reduced leaf area, chlorosis (yellowing of the leaf), necrosis (tissue death), stunted shoot, root, and root architecture (Hu et al., 2023; Singh and Shukla, 2025). Taken together, all these morphological damages contribute to the enhanced susceptibility of plants to secondary stresses (Hafeez et al., 2023). Xenobiotics also cause damage to plant DNA and alter gene expression, which can occur through oxidative stress or direct interaction with nucleic acids (Vicidomini et al., 2024). When exposed to heavy metals and hydrocarbon agents such as cadmium or fluoranthene, genes for root nodulation and growth are down-regulated. Furthermore, herbicides or heavy metals reduce soil microbial diversity, such as that of bacteria (*Pseudomonas, Bacillus*) and fungi (*Aspergillus*) involved in nutrient cycling, organic matter degradation and maintenance of soil structure and fertility (Mishra et al., 2021; Nwankwo et al., 2025; Yeboah et al., 2021). This leads, to ‘dysbiosis’, which is an imbalance of microorganisms, whereby some opportunistic or pathogenic groups increase in mass (Hrncir, 2022; Jyoti et al., 2024; Mishra et al., 2021; Chi et al., 2021). Interestingly, it has also been proposed that microbes in soil can become genetically and functionally adapted to xenobiotics, either through the development of novel degradation capacity or resistance (Ramakrishnan et al., 2018). *Burkholderia, Bradyrhizobium* and *Sphingobium* are microbes associated with xenobiotic degradation through enzymes such as phosphodiesterase, cytochrome P450s and laccases (Jha et al., 2015; Kumar and Singh, 2025). Changes in adaptation occur depending on the type of compound, concentration and soil conditions, where chronic exposure to toxins encourages the proliferation of microbial communities that are enriched with degradation genes (Aelion et al., 1987).

Xenobiotics also imparts toxic effects on soil invertebrates such as earthworms, nematodes and arthropods (Pelosi et al., 2013; Tison et al., 2024). For instance, annelids such as earthworms experience a lower rate of survival, growth and reproduction, as well as behavioural changes, including the loss of mobility and feeding capabilities upon consumption of xenobiotic-contaminated soil or absorbing it through the skin (Miglani & Bisht, 2019). Exposure to organophosphates (chlorpyrifos, diazinon), neonicotinoids (imidacloprid), carbamates (carbaryl) and other commonly used pesticides causes enzymatic inhibition, oxidative stress, DNA damage and other physiological dysfunctions in earthworms. The other effects include a reduction in cocoon production, delayed maturity, neurotoxicity (coiling, mucous secretion) and the disruption of gut microbiota that participate in nutrient recycling (Pelosi et al., 2013). Furthermore, they negatively affect nematodes and soil arthropods such as springtails and mites and consequently hamper soil biodiversity, soil aeration and soil quality (Sochová et al., 2005; Shehzad et al., 2024). Triclosan (antimicrobial), a global pollutant, poses a high risk to soil species and specifically earthworms, with the LC_{50} of 40 mg/kg soil, and the no observable effect concentration (NOEC) of 40 μg/kg (Ramires et al., 2021). The reporting of drugs from soil and biosolids samples has raised concern, as these drugs can be bioactive or active even at low concentrations. Some drugs cause microbial resistance (e.g. antibiotics), acute toxicity (e.g. diclofenac) and endocrine disruption (e.g. 172-estradiol) (De Aquino et al., 2021).

Birds are highly sensitive to xenobiotics due to their unique physiology and species-specific metabolic processes (Carrasco et al., 2025). Persistent organic pollutants like DDT exposure have exhibited chronic effects on birds (raptors and seabirds), leading to severe eggshell thinning, reproductive failure and population crashes. Some areas have recorded DDT residues even after decades of ban and linked them to low reproduction in birds (Kamata et al., 2020). Behavioural and physiological alterations included altered song, mating behaviour, predator avoidance and migration timing (Bean et al., 2014). Many pharmaceutical drugs, such as fluoxetine and painkillers, have also been reported to disrupt foraging behaviour, locomotion and reduce survival and reproduction (Whitlock et al., 2018). Secondary poisoning of insectivorous and piscivorous birds occurs with the consumption of contaminated prey, which causes organ damage, death and loss of population. Accordingly, an ecosystem imbalance occurs due to the loss of effective pest control and nutrient transfer mechanisms (Hernandez-Moreno et al., 2013). Xenobiotics can enter food webs through polluted water, soil and prey and become accumulated in fatty tissues of organisms (Montano et al., 2022). These can be toxic to apex predators because they are unable to metabolize or excrete them (Zenker et al., 2014; Danovaro et al., 2023). Further sediment–sediment interactions can trap microplastics in the pores or organic matrix and influence their bioavailability to benthic organisms (Bonifacio et al., 2022). The persistent compounds such as PCBs, organochlorine pesticides and heavy metals induce chronic ill effects (Afzal and Mahreen, 2024; Pain et al., 2019). Immune competence, cardiovascular and organ functions are physiologically altered and affected by drug-related toxicities (Mohammadkhani et al., 2023). Certain mutagenic or teratogenic xenobiotics can impair the microbiome of the gastrointestinal tract, affecting nutrient absorption, metabolic reactions and resistance to diseases (Rencuzogullar et al., 2018; Nogacka et al., 2019). Pesticides and certain neurotoxicants (methylmercury) cause neurological and behavioural defects, such as being unable to maintain coordination, disturbed foraging, poor predator avoidance and cognitive impairments, which ultimately decrease the survival and reproduction rate of mammals (Yamamoto et al., 2018). The majority of xenobiotics bypass the placenta, leading to developmental defects, a lower survival rate of the offspring and infertility (Kotta-Loizou et al., 2024). Conclusively, these effects at the population scale can result in a decline in reproductive output and increased mortality, which can cause local extinctions or declines, especially in long-lived predators such as marine mammals and large land-dwelling carnivores (Afzal and Mahreen, 2024).

6 Future Perspective and Knowledge Gaps and Priority Areas for Toxicological Research

In conclusion, even as toxicological research evolves, there exists a gap in knowledge that restricts our capacity to determine the risk of complex environmental exposure. These xenobiotics build up in the environment and pose various toxicological hazards. Conventional risk assessment using PEC/PNEC ratios and individual xenobiotic toxicity assays appears insufficient to reflect real-world scenarios involving mixture toxicity and stressor combinations. Future studies should focus on risk assessment of parent compounds and their transformation products, along with the interaction between several contaminants. With immediate interest, studies should specifically consider micro- and nanoplastics, as these are vectors of other toxicants and can substantially enhance the toxicological effects. It is necessary to bridge these gaps using interdisciplinary solutions and new methodologies to construct a more rigorous and safeguarding framework for the environment and human health.

References

Aelion, C. M., Swindoll, C. M., & Pfaender, F. K. (1987). Adaptation to and biodegradation of xenobiotic compounds by microbial communities from a pristine aquifer. *Applied and Environmental Microbiology, 53*(9), 2212–2217. https://doi.org/10.1128/aem.53.9.2212-2217.1987

Afzal, A., & Mahreen, N. (2024). Emerging insights into the impacts of heavy metals exposure on health, reproductive and productive performance of livestock. *Frontiers in Pharmacology, 15*. https://doi.org/10.3389/fphar.2024.1375137

Apú, N., et al. (2025). Endocrine-disrupting pesticide exposure relevant to reproductive health: a case study from Costa Rica. *Environmental Monitoring and Assessment, 197*(5). https://doi.org/10.1007/s10661-025-14011-8

Arp, H. P. H., & Hale, S. E. (2022). Assessing the persistence and mobility of organic substances to protect freshwater resources. *ACS Environmental Au, 2*(6), 482–509. https://doi.org/10.1021/acsenvironau.2c00024

Ayoub, L., et al. (2024). Exposure to organophosphate insecticides induces behavioral changes and acetylcholinesterase inhibition in Apis mellifera. *Ecotoxicology and Environmental Safety, 287*, 117279. https://doi.org/10.1016/j.ecoenv.2024.117279

Azam, H. (2024). Immunotoxicity: Understanding the impact of chemicals and drugs on the immune system. *World Journal of Pharmacology and Toxicology, 7*(6), 283. https://doi.org/10.4172/wjpt.1000283

Bean, T. G., et al. (2014). Behavioural and physiological responses of birds to environmentally relevant concentrations of an antidepressant. *Philosophical Transactions of the Royal Society B Biological Sciences, 369*(1656), 20130575. https://doi.org/10.1098/rstb.2013.0575

Bester, K. (2007). *Personal care compounds in the environment: Pathways, fate, and methods for determination*. Wiley-VCH Verlag GmbH & Co. KGaA. isbn:978-3-527-31567-3.

Bläsing, M., & Amelung, W. (2018). Plastics in soil: Analytical methods and possible sources. *Science of the Total Environment, 612*, 422–435. https://doi.org/10.1016/j.scitotenv.2017.08.086

Bonifacio, P. S. P., Metillo, E. B., & Romano, E. F. (2022). Microplastic in sediments and ingestion rates in three edible bivalve mollusc species in a southern Philippine estuary. *Water Air & Soil Pollution, 233*(11). https://doi.org/10.1007/s11270-022-05926-w

Borgå, K. (2013). *Ecotoxicology: bioaccumulation* (*Elsevier eBooks*). https://doi.org/10.1016/b978-0-12-409548-9.00765-x

Bucheli, T. D., Strobel, B. W., & Hansen, H. C. B. (2018). Personal care products are only one of many exposure routes of natural toxic substances to humans and the environment. *Cosmetics, 5*(1), 10. https://doi.org/10.3390/cosmetics5010010

Čadež, T., et al. (2021). Assessment of four organophosphorus pesticides as inhibitors of human acetylcholinesterase and butyrylcholinesterase. *Scientific Reports, 11*(1). https://doi.org/10.1038/s41598-021-00953-9

Carrasco, L., et al. (2025). Birds as Bioindicators: Revealing the widespread impact of microplastics. *Birds, 6*(1), 10. https://doi.org/10.3390/birds6010010

Carter, L. J., Wilkinson, J. L., & Boxall, A. B. A. (2020). Evaluation of Existing Models to Estimate Sorption Coefficients for Ionisable Pharmaceuticals in Soils and Sludge. *Toxics, 8*(1), 13. https://doi.org/10.3390/toxics8010013

Chauhan, R., et al. (2021). Advancing of Zinc Oxide Nanoparticles for Cosmetic Applications. In *Handbook of Consumer Nanoproducts* (pp. 1–16). Springer. https://doi.org/10.1007/978-981-15-6453-6_100-1

Chen, C. Y., & Folt, C. L. (2000). Bioaccumulation and diminution of arsenic and lead in a freshwater food web. *Environmental Science & Technology, 34*(18), 3878–3884. https://doi.org/10.1021/es991070c

Chhabra, R. S. (1979). Intestinal absorption and metabolism of xenobiotics. *Environmental Health Perspectives, 33*, 61–69. https://doi.org/10.1289/ehp.793361

Chi, L., et al. (2021). Studies of xenobiotic-induced gut microbiota dysbiosis: from correlation to mechanisms. *Gut Microbes, 13*(1). https://doi.org/10.1080/19490976.2021.1921912

D'Souza, L. C., et al. (2023). Environmental Chemical-Induced Reactive Oxygen Species Generation and Immunotoxicity: A Comprehensive review. *Antioxidants and Redox Signaling, 40*(10–12), 691–714. https://doi.org/10.1089/ars.2022.0117

Danovaro, R., et al. (2023). Bioaccumulation and biomagnification of heavy metals in marine micro-predators. *Communications Biology, 6*(1). https://doi.org/10.1038/s42003-023-05539-x

De Aquino, S. F., et al. (2021). Occurrence of pharmaceuticals and endocrine disrupting compounds in Brazilian water and the risks they may represent to human health. *International Journal of Environmental Research and Public Health, 18*(22), 11765. https://doi.org/10.3390/ijerph182211765

El-Demerdash, F. M., et al. (2018). Xenobiotics, oxidative stress, and antioxidants. *Oxidative Medicine and Cellular Longevity, 2018*(1). https://doi.org/10.1155/2018/9758951

Embrandiri, A., et al. (2016). Environmental Xenobiotics and Its Effects on Natural Ecosystem. In A. Singh, S. M. Prasad, & R. P. Singh (Eds.), *Plant Responses to Xenobiotics* (pp. 1–18). Springer. https://doi.org/10.1007/978-981-10-2860-1_1

Eriksson, E., et al. (2003). Household chemicals and personal care products as sources for xenobiotic organic compounds in grey wastewater. *Water SA, 29*(2). https://doi.org/10.4314/wsa.v29i2.4848

Esteves, F., Rueff, J., & Kranendonk, M. (2021). The Central Role of Cytochrome P450 in Xenobiotic Metabolism—A brief review on a fascinating enzyme family. *Journal of Xenobiotics, 11*(3), 94–114. https://doi.org/10.3390/jox11030007

Fang, J., Koen, Y. M., & Hanzlik, R. P. (2009). Bioinformatic analysis of xenobiotic reactive metabolite target proteins and their interacting partners. *BMC Chemical Biology, 9*(1). https://doi.org/10.1186/1472-6769-9-5

Fuchs, B., et al. (2022). A Glyphosate-Based herbicide in soil differentially affects hormonal homeostasis and performance of non-target crop plants. *Frontiers in Plant Science, 12*. https://doi.org/10.3389/fpls.2021.787958

Gaschler, M. M., & Stockwell, B. R. (2017). Lipid peroxidation in cell death. *Biochemical and Biophysical Research Communications, 482*(3), 419–425. https://doi.org/10.1016/j.bbrc.2016.10.086

Gianfreda, L. (2008). Interactions Between Xenobiotics and Microbial and Enzymatic Soil Activity. *Critical Reviews in Environmental Science and Technology, 38*(4), 269–310. https://doi.org/10.1080/10643380701413526

Gobas, F. A., et al. (2015). Review of existing terrestrial bioaccumulation models and terrestrial bioaccumulation modeling needs for organic chemicals. *Integrated Environmental Assessment and Management, 12*(1), 123–134. https://doi.org/10.1002/ieam.1690

Gomez-Sagasti, M. T., & Marino, D. (2015). PGPRs and nitrogen-fixing legumes: a perfect team for efficient Cd phytoremediation? *Frontiers in Plant Science, 6*. https://doi.org/10.3389/fpls.2015.00081

Hafeez, A., et al. (2023). Effect of heavy metals on growth, physiological and biochemical responses of plants. *Elsevier eBooks*, 139–159. https://doi.org/10.1016/b978-0-323-99978-6.00006-6

Hamlyn-Harris, D., McAlister, T., & Dillon, P. (2019). Chapter 9 - Water Harvesting Potential of WSUD Approaches. In A. K. Sharma, T. Gardner, & D. Begbie (Eds.), *Approaches to Water Sensitive Urban Design* (pp. 177–208). Woodhead Publishing. https://doi.org/10.1016/B978-0-12-812843-5.00009-5

He, D., et al. (2018). Microplastics in soils: Analytical methods, pollution characteristics and ecological risks. *TrAC Trends in Analytical Chemistry, 109*, 163–172. https://doi.org/10.1016/j.trac.2018.10.006

Helfter, C., et al. (2007). A noninvasive optical system for the measurement of xylem and phloem sap flow in woody plants of small stem size. *Tree Physiology, 27*(2), 169–179. https://doi.org/10.1093/treephys/27.2.169

Hernandez-Moreno, D., et al. (2013). Secondary poisoning of non-target animals in an Ornithological Zoo in Galicia (NW Spain) with anticoagulant rodenticides: a case report. *Veterinární Medicína, 58*(10), 553–559. https://doi.org/10.17221/7087-vetmed

Hrncir, T. (2022). Gut microbiota dysbiosis: Triggers, consequences, diagnostic and therapeutic options. *Microorganisms, 10*(3), 578. https://doi.org/10.3390/microorganisms10030578

Hu, Z., et al. (2023). Heavy metals can affect plant morphology and limit plant growth and photosynthesis processes. *Agronomy, 13*(10), 2601. https://doi.org/10.3390/agronomy13102601

Iqubal, A., et al. (2020). Environmental neurotoxic pollutants: review. *Environmental Science and Pollution Research, 27*(33), 41175–41198. https://doi.org/10.1007/s11356-020-10539-z

Jayaraj, R., Megha, P., & Sreedev, P. (2016). Review Article. Organochlorine pesticides, their toxic effects on living organisms and their fate in the environment. *Interdisciplinary Toxicology, 9*(3–4), 90–100. https://doi.org/10.1515/intox-2016-0012

Jha, S. K., Jain, P., & Sharma, H. P. (2015). Xenobiotic degradation by bacterial enzymes. *International Journal of Current Microbiology and Applied Sciences, 4*(6), 48–62. https://www.ijcmas.com (Retrieved from 14 Oct 2025

Jyoti, et al. (2024). *An assessment of degradation of xenobiotic compounds by soil borne bacteria and their adaptation to organic xenobiotic compounds: A review. J.Mycopathol.Res.* Indian Mycological Society, 231–240. https://imskolkata.org/article/003-06-24.pdf

Kamata, R., Shiraishi, F. and Nakamura, K. (2020) 'Avian eggshell thinning caused by transovarian exposure to *o,p′*-DDT: changes in histology and calcium-binding protein production in the oviduct uterus', *The Journal of Toxicological Sciences*, 45(3), pp. 131–136. https://doi.org/https://doi.org/10.2131/jts.45.131.

Knudsmark Sjøholm, K., et al. (2021). Determining the Temperature Dependency of Biodegradation Kinetics for 34 Hydrocarbons while Avoiding Chemical and Microbial Confounding Factors. *Environmental Science & Technology, 55*(16), 11091–11101. https://doi.org/10.1021/acs.est.1c02773

Kotta-Loizou, I., et al. (2024). Fetus exposure to drugs and chemicals: a holistic overview on the assessment of their transport and metabolism across the human placental barrier. *Diseases, 12*(6), 114. https://doi.org/10.3390/diseases12060114

Kucklick, J. R., & Baker, J. E. (1998). Organochlorines in Lake Superior's food web. *Environmental Science & Technology, 32*(9), 1192–1198. https://doi.org/10.1021/es970794q

Kumar, P., & Singh, J. (2025). Microbes in xenobiotics biodegradation. *Elsevier eBooks*, 581–612. https://doi.org/10.1016/b978-0-443-13615-3.00012-1

Kumar, A., Kumar, A., Cabral-Pinto, M. M. S., Chaturvedi, A. K., Shabnam, A. A., Subrahmanyam, G., Mondal, R., Gupta, D. K., Malyan, S. K, Kumar, S. S., Khan, S. A., Yadav. K. K. (2020). Lead toxicity: Health hazards, influence on food chain, and sustainable remediation approaches. *International Journal of Environmental Research and Public Health, 17*(7), 2179. https://doi.org/10.3390/ijerph17072179

Lavoie, R. A., et al. (2013). Biomagnification of Mercury in Aquatic Food Webs: A Worldwide Meta-Analysis. *Environmental Science & Technology, 47*(23), 13385–13394. https://doi.org/10.1021/es403103t

Lee, H., et al. (2012). Molecular mechanism(s) of endocrine-disrupting chemicals and their potent oestrogenicity in diverse cells and tissues that express oestrogen receptors. *Journal of Cellular and Molecular Medicine, 17*(1), 1–11. https://doi.org/10.1111/j.1582-4934.2012.01649.x

Lee, J.-W., et al. (2024). Review of cadmium bioaccumulation in fish exposed to cadmium. *Toxics, 13*(1), 7. https://doi.org/10.3390/toxics13010007

Li, F., Lu, J., & Ma, X. (2011). Profiling the reactive metabolites of xenobiotics using metabolomic technologies. *Chemical Research in Toxicology, 24*(5), 744–751. https://doi.org/10.1021/tx200033v

Li, G., et al. (2023). The effect of soil microplastics on Oryza sativa L. root growth traits under alien plant invasion. *Frontiers in Ecology and Evolution, 11*. https://doi.org/10.3389/fevo.2023.1172093

Li, X., Poon, C., & Liu, P. S. (2001). Heavy metal contamination of urban soils and street dusts in Hong Kong. *Applied Geochemistry, 16*(11), 1361–1368. https://doi.org/10.1016/S0883-2927(01)00045-2

Liu, B., et al. (2016). Mechanisms of mutagenesis: DNA replication in the presence of DNA damage. *Mutation Research/Reviews in Mutation Research, 768*, 53–67. https://doi.org/10.1016/j.mrrev.2016.03.006

Liu, S., et al. (2025). Oxidative stress: signaling pathways, biological functions, and disease. *MedComm, 6*(7). https://doi.org/10.1002/mco2.70268

Ma, J., Hung, H., & Blanchard, P. (2004). How Do Climate Fluctuations Affect Persistent Organic Pollutant Distribution in North America? Evidence from a Decade of Air Monitoring. *Environmental Science and Technology, 38*(9), 2538–2543. https://doi.org/10.1021/es0349610

Mandelker, L. (2008). Introduction to oxidative stress and mitochondrial dysfunction. *Veterinary Clinics of North America Small Animal Practice, 38*(1), 1–30. https://doi.org/10.1016/j.cvsm.2007.10.005

Manetti, P. (2023). Heavy Metal-Induced neurotoxicity and calcium channel involvement. *Toxicol Open Access, 9*(5), 236. https://doi.org/10.4172/2476-2067.1000236

Marín-Benito, J. M., et al. (2019). Previous degradation study of two herbicides to simulate their fate in a sandy loam soil: Effect of the temperature and the organic amendments. *Science of the Total Environment, 653*, 1301–1310. https://doi.org/10.1016/j.scitotenv.2018.11.015

McGarr, J.T. and Mbonimpa, E.G. (2023) *Fate and Transport of Per- and Polyfluoroalkyl Substances (PFAS) at Aqueous Film Forming Foam (AFFF) Discharge Sites: A Review*. https://www.mdpi.com/2571-8789/7/2/53 (Retrieved from 15 Aug 2025).

Mendrik, F., et al. (2023). Non-buoyant microplastic settling velocity varies with biofilm growth and ambient water salinity. *Communications Earth & Environment, 4*(1). https://doi.org/10.1038/s43247-023-00690-z

Meyer, T., & Wania, F. (2008). Organic contaminant amplification during snowmelt. *Water Research, 42*(8), 1847–1865. https://doi.org/10.1016/j.watres.2007.12.016

Miglani, R., & Bisht, S. S. (2019). World of earthworms with pesticides and insecticides. *Interdisciplinary Toxicology, 12*(2), 71–82. https://doi.org/10.2478/intox-2019-0008

Miglani, R., et al. (2022). Degradation of xenobiotic pollutants: an environmentally Sustainable approach. *Metabolites, 12*(9), 818. https://doi.org/10.3390/metabo12090818

Mishra, S., et al. (2021). Recent advanced technologies for the characterization of Xenobiotic-Degrading microorganisms and microbial communities. *Frontiers in Bioengineering and Biotechnology, 9*. https://doi.org/10.3389/fbioe.2021.632059

Mishra, V. K., Singh, G., & Shukla, R. (2019). Chapter 6 - Impact of Xenobiotics Under a Changing Climate Scenario. In K. K. Choudhary, A. Kumar, & A. K. Singh (Eds.), *Climate Change and Agricultural Ecosystems* (pp. 133–151). Woodhead Publishing. https://doi.org/10.1016/B978-0-12-816483-9.00006-2

Mohammadkhani, M. A., et al. (2023). Insights into Organochlorine Pesticides Exposure in the Development of Cardiovascular Diseases: A Systematic Review. *Archives of Iranian Medicine, 26*(10), 592–599. https://doi.org/10.34172/aim.2023.86

Mohan Kumar, S. M. J., et al. (2023). Role of cytokines and reactive oxygen species in brain aging. *Mechanisms of Ageing and Development, 214*, 111855. https://doi.org/10.1016/j.mad.2023.111855

Montano, L., et al. (2022). Polychlorinated biphenyls (PCBs) in the environment: occupational and exposure events, effects on human health and fertility. *Toxics, 10*(7), 365. https://doi.org/10.3390/toxics10070365

Navenant, A. L., et al. (2019). Metabolic mechanisms and acetylcholinesterase sensitivity involved in tolerance to chlorpyrifos-ethyl in the earwig Forficula auricularia. *Chemosphere, 227*, 416–424. https://doi.org/10.1016/j.chemosphere.2019.04.065

Neumann, H., & Werner, D. (2000). Gene Expression of Medicago sativa Inoculated with Sinorhizobium meliloti as Modulated by the Xenobiotics Cadmium and Fluoranthene. *Zeitschrift Für Naturforschung C, 55*(3–4), 222–232. https://doi.org/10.1515/znc-2000-3-414

Nogacka, A., et al. (2019). Xenobiotics formed during food processing: Their relation with the intestinal microbiota and colorectal cancer. *International Journal of Molecular Sciences, 20*(8), 2051. https://doi.org/10.3390/ijms20082051

Nwankwo, C. E. I., et al. (2025). Addressing emerging contaminants in agriculture affecting plant–soil interaction: a review on bio-based and nano-enhanced strategies for soil health and global food security (GFS). *Deleted Journal, 2*(1). https://doi.org/10.1007/s44339-025-00018-w

Ortúzar, M., et al. (2022). Pharmaceutical Pollution in Aquatic Environments: A Concise review of environmental impacts and bioremediation systems. *Frontiers in Microbiology, 13*. https://doi.org/10.3389/fmicb.2022.869332

Pain, D. J., Mateo, R., & Green, R. E. (2019). Effects of lead from ammunition on birds and other wildlife: A review and update. *Ambio, 48*(9), 935–953. https://doi.org/10.1007/s13280-019-01159-0

Pallardy, M., et al. (1999). Assessment of Apoptosis in Xenobiotic-Induced Immunotoxicity. *Methods, 19*(1), 36–47. https://doi.org/10.1006/meth.1999.0825

Park, Y.-C., Lee, S., & Cho, M.-H. (2014). The Simplest Flowchart Stating the Mechanisms for Organic Xenobiotics-induced Toxicity: Can it Possibly be Accepted as a 'Central Dogma' for Toxic Mechanisms? *Toxicological Research, 30*(3), 179–184. https://doi.org/10.5487/tr.2014.30.3.179

Patel, D. K., & Jyoti, S. D. (2013). Xenobiotics: An essential precursor for living system. *American Journal of Advanced Drug Delivery*. https://www.primescholars.com/articles/xenobiotics-an-essential-precursor-for-living-system.pdf

Patisaul, H. B. (2021). REPRODUCTIVE TOXICOLOGY: Endocrine disruption and reproductive disorders: impacts on sexually dimorphic neuroendocrine pathways. *Reproduction, 162*(5), F111–F130. https://doi.org/10.1530/rep-20-0596

Patisaul, H. B., Fenton, S. E., & Aylor, D. (2018). Animal models of endocrine disruption. *Best Practice & Research Clinical Endocrinology & Metabolism, 32*(3), 283–297. https://doi.org/10.1016/j.beem.2018.03.011

Paul, S., Ghosh, U., & Saha, M. (2023). Transport and Metabolism of Xenobiotics in the Urban Ecosystem. In R. Singh et al. (Eds.), *Xenobiotics in Urban Ecosystems: Sources, Distribution and Health Impacts* (pp. 69–85). Springer International Publishing. https://doi.org/10.1007/978-3-031-35775-6_4

Pelosi, C., et al. (2013). Pesticides and earthworms. A review. *Agronomy for Sustainable Development, 34*(1), 199–228. https://doi.org/10.1007/s13593-013-0151-z

Phale, P. S., Sharma, A., & Gautam, K. (2019). 11 - Microbial degradation of xenobiotics like aromatic pollutants from the terrestrial environments. In M. N. V. Prasad, M. Vithanage, & A. Kapley (Eds.), *Pharmaceuticals and Personal Care Products: Waste Management and Treatment Technology* (pp. 259–278). Butterworth-Heinemann. https://doi.org/10.1016/B978-0-12-816189-0.00011-1

Picone, M., et al. (2020). Bioaccumulation of Polychlorinated Dibenzo-p-Dioxins (PCDDs) and Dibenzofurans (PCDFs) in Hediste diversicolor (Polychaeta: Nereididae). *Frontiers in Ecology and Evolution, 8*. https://doi.org/10.3389/fevo.2020.00235

Ramakrishnan, B., et al. (2018). Local applications but global implications: Can pesticides drive microorganisms to develop antimicrobial resistance? *The Science of the Total Environment, 654*, 177–189. https://doi.org/10.1016/j.scitotenv.2018.11.041

Ramires, P. F., et al. (2021). Ecotoxicity of triclosan in soil: an approach using different species. *Environmental Science and Pollution Research, 28*(30), 41233–41241. https://doi.org/10.1007/s11356-021-13633-y

Ray, S., & Shaju, S. T. (2023). Bioaccumulation of pesticides in fish resulting toxicities in humans through food chain and forensic aspects. *Environmental Analysis Health and Toxicology, 38*(3), e2023017. https://doi.org/10.5620/eaht.2023017

Rencüzoğulları, E., & Aydın, M. (2018). Genotoxic and mutagenic studies of teratogens in developing rat and mouse. *Drug and Chemical Toxicology, 42*(4), 409–429. https://doi.org/10.1080/01480545.2018.1465950

Rist, S., et al. (2018). A critical perspective on early communications concerning human health aspects of microplastics. *Science of the Total Environment, 626*, 720–726. https://doi.org/10.1016/j.scitotenv.2018.01.092

Roy, T., Dey, T. K., & Jamal, M. (2022). Microplastic/nanoplastic toxicity in plants: an imminent concern. *Environmental Monitoring and Assessment, 195*(1). https://doi.org/10.1007/s10661-022-10654-z

Ruangsomboon, S., & Wongrat, L. (2006). Bioaccumulation of cadmium in an experimental aquatic food chain involving phytoplankton (Chlorella vulgaris), zooplankton (Moina macrocopa), and the predatory catfish Clarias macrocephalus × C. gariepinus. *Aquatic Toxicology, 78*(1), 15–20. https://doi.org/10.1016/j.aquatox.2006.01.015

Schirmer, M., et al. (2011). Mass fluxes of xenobiotics below cities: challenges in urban hydrogeology. *Environmental Earth Sciences, 64*(3), 607–617. https://doi.org/10.1007/s12665-010-0880-0

Seghir, B. B., et al. (2023). Food chain contamination and impact of xenobiotics on human health. *Springer eBooks*, 109–129. https://doi.org/10.1007/978-3-031-35775-6_6

Shehzad, K., et al. (2024). Arthropods in soil reclamation and bioremediation: Functional roles, mechanisms and future perspective. *Journal of Environmental Management, 370*, 122820. https://doi.org/10.1016/j.jenvman.2024.122820

Shukla, R. K., et al. (2021). Genotoxic potential of nanoparticles: structural and functional modifications in DNA. *Frontiers in Genetics, 12*. https://doi.org/10.3389/fgene.2021.728250

Silva-Adaya, D., Garza-Lombó, C., & Gonsebatt, M. E. (2021). Xenobiotic transport and metabolism in the human brain. *Neurotoxicology, 86*, 125–138. https://doi.org/10.1016/j.neuro.2021.08.004

Singh, A. K., & Shukla, S. K. (2025). Assessing air pollution tolerance, photosynthetic pigmentation, and dust capturing capacity of roadside trees in the mining region. *Journal of Sustainable Forestry, 44*(7), 572–598. https://doi.org/10.1080/10549811.2025.2529607

Singh, M., et al. (2023). *Xenobiotics: Sources, pathways, degradation, and risk associated with major emphasis on pharmaceutical compounds* (*Springer eBooks*) (pp. 87–106). https://doi.org/10.1007/978-3-031-35775-6_5

Siracusa, J. S., et al. (2018). Effects of bisphenol A and its analogs on reproductive health: A mini review. *Reproductive Toxicology, 79*, 96–123. https://doi.org/10.1016/j.reprotox.2018.06.005

Sochová, I., Hofman, J., & Holoubek, I. (2005). Using nematodes in soil ecotoxicology. *Environment International, 32*(3), 374–383. https://doi.org/10.1016/j.envint.2005.08.031

Sondhia, S. (2009). Leaching behaviour of metsulfuron in two texturally different soils. *Environmental Monitoring and Assessment, 154*(1), 111–115. https://doi.org/10.1007/s10661-008-0381-8

Srivastav, A. L. (2020). Chapter 6—Chemical fertilizers and pesticides: Role in groundwater contamination. In M. N. V. Prasad (Ed.), *Agrochemicals detection, treatment and remediation* (pp. 143–159). Butterworth-Heinemann. https://doi.org/10.1016/B978-0-08-103017-2.00006-4

Thakur, M., et al. (2024). How to deal with xenobiotic compounds through environment friendly approach? *Critical Reviews in Biotechnology*, 1–20. https://doi.org/10.1080/07388551.2024.2336527

Tison, L., et al. (2024). Transfer and bioaccumulation of pesticides in terrestrial arthropods and food webs: State of knowledge and perspectives for research. *Chemosphere, 357*, 142036. https://doi.org/10.1016/j.chemosphere.2024.142036

Tortella, G., et al. (2024). Copper nanoparticles as a potential emerging pollutant: Divergent effects in the agriculture, risk-benefit balance and integrated strategies for its use. *Emerging Contaminants, 10*(4), 100352. https://doi.org/10.1016/j.emcon.2024.100352

Toshiyuki Katagi (2012) Soil Column Leaching of Pesticides. SpringerLink. https://link.springer.com/chapter/10.1007/978-1-4614-4448-0_1. Retrieved from 15 Aug 2025.

Tran, N. H., Reinhard, M., & Gin, K. Y.-H. (2018). Occurrence and fate of emerging contaminants in municipal wastewater treatment plants from different geographical regions-a review. *Water Research, 133*, 182–207. https://doi.org/10.1016/j.watres.2017.12.029

Vicidomini, C., et al. (2024). Oxidative processes and xenobiotic metabolism in plants: mechanisms of defense and potential therapeutic implications. *Journal of Xenobiotics, 14*(4), 1541–1569. https://doi.org/10.3390/jox14040084

Wan, L., et al. (2023). Global meta-analysis reveals differential effects of microplastics on soil ecosystem. *The Science of the Total Environment, 867*, 161403. https://doi.org/10.1016/j.scitotenv.2023.161403

Wang, D., et al. (2023). Water-Soluble Synthetic Polymers: Their Environmental Emission Relevant Usage, Transport and Transformation, Persistence, and Toxicity. *Environmental Science & Technology, 57*(16), 6387–6402. https://doi.org/10.1021/acs.est.2c09178

Wang, X., et al. (2022). Pesticides xenobiotics in soil ecosystem and their remediation approaches. *Sustainability, 14*(6), 3353. https://doi.org/10.3390/su14063353

Whitlock, S. E., et al. (2018). Environmentally relevant exposure to an antidepressant alters courtship behaviours in a songbird. *Chemosphere, 211*, 17–24. https://doi.org/10.1016/j.chemosphere.2018.07.074

Wright, S. L., Thompson, R. C., & Galloway, T. S. (2013). The physical impacts of microplastics on marine organisms: A review. *Environmental Pollution, 178*, 483–492. https://doi.org/10.1016/j.envpol.2013.02.031

Xiang, L., et al. (2022). Integrating Biochar, Bacteria, and Plants for Sustainable Remediation of Soils Contaminated with Organic Pollutants. *Environmental Science & Technology*, 16546–16566. https://pubs.acs.org/doi/pdf/10.1021/acs.est.2c02976?ref=article_openPDF

Yamamoto, M., et al. (2018). Evaluation of neurobehavioral impairment in methylmercury-treated KK-Ay mice by dynamic weight-bearing test. *Journal of Applied Toxicology, 39*(2), 221–230. https://doi.org/10.1002/jat.3710

Yeboah, J. O., Shi, G., & Shi, W. (2021). Effect of heavy metal contamination on soil enzymes activities. *Journal of Geoscience and Environment Protection, 09*(06), 135–154. https://doi.org/10.4236/gep.2021.96008

Yi, L., et al. (2022). Cadmium-induced apoptosis of Leydig cells is mediated by excessive mitochondrial fission and inhibition of mitophagy. *Cell Death and Disease, 13*(11). https://doi.org/10.1038/s41419-022-05364-w

Yu, Z., et al. (2023). Uptake and transport of micro/nanoplastics in terrestrial plants: Detection, mechanisms, and influencing factors. *The Science of the Total Environment, 907*, 168155. https://doi.org/10.1016/j.scitotenv.2023.168155

Yun, B. H., et al. (2018). DNA adducts: Formation, biological effects, and new biospecimens for mass spectrometric measurements in humans. *Mass Spectrometry Reviews, 39*(1–2), 55–82. https://doi.org/10.1002/mas.21570

Zenker, A., et al. (2014). Bioaccumulation and biomagnification potential of pharmaceuticals with a focus to the aquatic environment. *Journal of Environmental Management, 133*, 378–387. https://doi.org/10.1016/j.jenvman.2013.12.017

Zerdan, M. B., et al. (2021). Mechanisms of immunotoxicity: Stressors and evaluators. *International Journal of Molecular Sciences, 22*(15), 8242. https://doi.org/10.3390/ijms22158242

Zhang, C., Barron, L., & Sturzenbaum, S. (2021). The transportation, transformation and (bio)accumulation of pharmaceuticals in the terrestrial ecosystem. *Science of the Total Environment, 781*, 146684. https://doi.org/10.1016/j.scitotenv.2021.146684

Zhang, Q., Yang, L., & Wang, W.-X. (2011). Bioaccumulation and trophic transfer of dioxins in marine copepods and fish. *Environmental Pollution, 159*(12), 3390–3397. https://doi.org/10.1016/j.envpol.2011.08.031

Zhang, W., et al. (2022). Arsenic bioaccumulation and biotransformation in aquatic organisms. *Environment International, 163*, 107221. https://doi.org/10.1016/j.envint.2022.107221

Chapter 14
Commercialization and Use and Its Regulatory Policies

Shivani Taya and Neeraj Rana

Abstract Xenobiotics are essential for stimulating economic growth and encouraging the development of new products and strategic positioning, yet their extensive spread across the globe is raising concerns about their effects on human genetic material, the health of future generations, and the environment. Studies suggest that long-term exposure to some xenobiotics may lead to oxidative stress, damage to the DNA, and epigenetic changes, thereby calling for informed regulatory oversight and commercialization practices. This chapter considers xenobiotics from a toxicological, environmental governance, business management, and regulatory policy perspective that is interdisciplinary. It discusses product development, market entry, and industrial application commercialization routes, as well as global and Indian regulatory systems (chemicals safety, environmental compliance, and protection of public health). The commentary also discusses the corporate social responsibility (CSR), environmental, social, and governance (ESG), and stakeholder accountability as it relates to the ethics burden of the industries. Examples from agrochemicals, pharmaceuticals, and consumer goods demonstrate the results of regulatory lacunae and governance collapse. The chapter overall underlines that industrial growth should be in tandem with ensuring human protection and the integrity of the genome with sustainable, transparent, and ethically founded commercialization of xenobiotics.

Keywords Xenobiotics · Environmental regulation · Risk management · Chemical safety · Public health policy

---

S. Taya · N. Rana (✉)
SCBM, Geeta University, Panipat, Haryana, India
e-mail: rananeeraj20@gmail.com

S. Taya
e-mail: taya.shivani910@gmail.com

K. Priya and S. K. Giri (eds.), *Environmental Toxicology: Xenobiotics and Effects on Human Genetic Material*, https://doi.org/10.1007/978-3-032-28619-2_14

1 Introduction

Xenobiotics, generally considered to be chemical entities that are alien to biological systems, are now indispensable in a range of modern industrial, pharmaceutical, agricultural, and consumer applications. These compounds are synthetic pesticides, active pharmaceutical ingredients, plasticizers, additives in cosmetics, dyes, solvents, preservatives, and a multitude of industrial chemicals that are not part of either human metabolic processes or environmental cycles (Štefanac et al., 2021). Their rising acceptance in commerce closely parallels industrial development and technological change, as well as growing global consumption tendencies making xenobiotics, a hallmark of contemporary economies. But while xenobiotics are central to global commerce, they have also raised serious concerns about potential long-term impacts on human health and genotoxicity, resilience of the natural environment, and the risk of being captured in biological systems, leading to heightened regulation in recent years.

In biological term, xenobiotics can affect human genome and stability of metabolism through several route including oxidative stress, DNA adduct formation, chromosomal abnormalities, lipid peroxidation, and modulation in cell-signaling pathways (Lang et al., 2025). While the technical machinery of these processes is in the domain of toxicology and molecular biology, the societal consequences are potentially far removed from the laboratory. Governments view xenobiotics as a health hazard that requires stringent regulation. For businesses, these substances represent both an expense and a revenue stream, with potential for reputational harm, legal issues, and financial setbacks if safety protocols are not adhered to. Consumers, on the other hand, are concerned about product safety, the risks associated with prolonged exposure, and the sustainability of manufacturing practices.

The economic importance of xenobiotics stems from their wide range of functional benefits. In the pharmaceutical sector, xenobiotic compounds serve as active ingredients essential for treating diseases, managing infections, and improving health outcomes worldwide. Similarly, agrochemical xenobiotics, like herbicides and insecticides, play a crucial role in boosting agricultural output, controlling pest infestations, and ensuring stable food supply chains. The production of plastics and polymers depends on chemicals such as bisphenol-A (BPA) and phthalates to enhance material strength, flexibility, and performance (Huang et al., 2021). Additionally, the cosmetics and FMCG industries use xenobiotic preservatives and additives to maintain product quality and appeal. Collectively, these industries contribute billions to the global GDP, highlighting that xenobiotics are not only commercially viable but also a significant economic driver. However, the increasing presence of xenobiotics in global markets has raised public health concerns (Huang et al., 2024).

Scientific evidence indicates that long-duration or repeated exposure to certain xenobiotics via polluted air, water, food, or consumer products may contribute to an increased risk of genotoxicity, endocrine disruption, carcinogenicity, and metabolic diseases (World Health Organization [WHO], 2020a, 2020b). These worries have increased consumers' consciousness and have subjected industries to more scrutiny to implement good manufacturing practices, minimize hazardous chemicals, and meet stricter regulations. They have also contributed to urges for governments to update chemical safety systems to reflect emerging scientific understandings of toxicogenomic, bioaccumulation, and long-term, low-dose risks of exposure. The promotion of xenobiotics to the status of a market good, thus necessitates an analysis that considers knowledge from science, public policy, business strategy, and regulatory governance. Publishing requires scientific discovery, R&D spending, toxicological assessment, regulatory approval, intellectual property management, market analysis, risk communication, and post-market surveillance and control. In each of these stages, there are ethical and managerial issues. "Decisions regarding product design, labeling, sourcing from the supply chain, and consumer marketing need to comply with national and international legislation including the REACH legislation in the European Union; the Toxic Substances Control Act (TSCA) in the United States; and the Environmental Protection Act and associated regulations in India (European Chemicals Agency [ECHA], 2023)."

Regulatory authorities, in this case, have a vital responsibility to monitor the safe introduction of xenobiotics into the market. Few of the most powerful ones are: Environmental Protection Agency (EPA), the European Chemicals Agency (ECHA), the US Food and Drug Administration (FDA), the World Health Organization (WHO), the Organization for Economic Co-operation and Development (OECD), and India's Central Drugs Standard Control Organization (CDSCO). These organizations impose rules related to allowable exposure limits, chemicals registration, safety assessments, product labeling, releases to environment, and monitoring after the market. While these entities are executed in disparate legal and political environments, their fundamental purpose is the same: To ensure that Xenobiotics do not present intolerable risks to human health or the environment.

As global supply chains become more intricate, companies are confronted with a maze of regulatory compliance requirements while facing pressures from markets and innovation. Scientific/regulatory issues represent only one facet of the xenobiotic commercialization landscape; ethical debates, consumer expectations, corporate social responsibility (CSR), and developing environmental, social, and governance (ESG) paradigms are playing an increasing role and driving innovation. Companies that fail to integrate sustainability and transparency into their xenobiotic-related processes risk of losing stakeholder trust and their market position (Roszkowska-Menkes, 2021). Conversely, firms demonstrating proactive regulatory compliance, sustained investment in safer alternatives, and environmentally responsible business practices tend to achieve improved market credibility and greater long-term viability.

Overall, commercialization and application of xenobiotics involves a complex dynamic with factors of economic development, scientific risk, regulatory governance, and ethical responsibility. This chapter addresses these facets through analysis of industry trends, commercialization routes, global and Indian regulatory models, challenges to compliance, CSR and ESG requirements, and practical cases. The high-level goal is to show how the commercialization of xenobiotics can be conducted in a responsible manner that protects human health—especially genetic integrity—and preserves ecosystem sustainability.

2 Industrial Landscape and Economic Importance

2.1 Xenobiotics in the Pharma Industry

The pharmaceutical industry is one of the largest users of xenobiotics at global level. Synthetic active pharmaceutical ingredients (APIs), solvents, stabilizers, and excipients are at the heart of drug manufacture and are what makes therapeutic innovation possible and what underpins global healthcare systems. The increased need for a variety of novel xenobiotic molecules for the treatment of chronic diseases, antimicrobial resistances, and disorders of the lifestyle accelerated the development of drugs (Behl & Sharma, 2025). As a leading global producer of generic medicines, India has a crucial part to play in xenobiotic-based pharmaceuticals production and catering to the international markets with her enormous capacity and export prowess. Thus, the commercialization of pharmaceutical xenobiotics has a large economic impact but presents regulatory and security concerns regarding toxicity, genetic effects, and residual contamination in drug formulations.

2.2 Agrochemical Xenobiotics and Agricultural Performance

To escalate yield, manage pests, and assure food safety, agriculture highly depends on xenobiotics such pesticides, herbicides, fungicides, and insecticides. To enable large production, decrease crop losses, and maintaining global food supply lines, these chemicals have entirely transformed modern agriculture. Their ecologic and health effects, inclusive soil erosion, water pollution, and genotoxicity in exposed populations, have raised concerns that call for more stringent regulatory approaches. At the same time, the agrochemical sector is witnessing robust growth, with global demand forecast to reach approximately USD 253 billion by 2030, intensifying concerns related to regulatory oversight and environmental impact (BioSpace, 2024). The economic reliance on these xenobiotics underscores their continued importance in global food systems, even as policymakers increasingly promote safer alternatives and integrated pest management (IPM) strategies (Table 1).

2.3 Xenobiotics in the Production of Plastic and Polymer

One of the largest consumers of xenobiotics is the plastics and polymer industry, by utilizing substances like, bisphenol-A (BPA), phthalates, and flame retardants, among others, to enhance flexibility, durability, and heat resistance of the material. These compounds are integrated in client goods, medical devices, electronics, vehicles parts, and packaging materials. Although many polymer-associated xenobiotics essential for product performance have been linked to endocrine disruption, developmental

Table 1 Major industrial sectors using xenobiotics and their economic significance

Industry sector	Key xenobiotics used	Economic contribution	Major concerns	Regulatory bodies
Pharmaceutical	APIs, solvents, stabilizers	Global drug supply, exports	Genotoxicity, residues	USFDA, CDSCO
Agriculture	Pesticides, herbicides	Crop yield, food security	Soil/water toxicity	WHO, FAO
Plastics and polymers	BPA, phthalates, flame retardants	Manufacturing, packaging	Endocrine disruption	EPA, EU REACH
Cosmetics	Parabens, fragrances	Global beauty market	Bioaccumulation	EU SCCS, FDA
Food industry	Preservatives, colorants	Processed food sector	Migration risks	FSSAI, EFSA
Textiles	Azo dyes, fixatives	Global apparel trade	Water contamination	OEKO-TEX®BIS

toxicity, and even long-term genetic effects resulting in heightened scientific attention and lay concern (Jung et al., 2023). The need to comply with regulatory limitations in some countries has driven the industry to seek greener polymers, BPA-free products, and more sustainable technologies for additives and so forth, providing a signal of renewed interest in sustainable commercialization.

2.4 *Xenobiotics in the Cosmetics and Personal Care Products Industry*

Xenobiotics such as parabens, synthetic fragrances, dyes, and UV filters are used in cosmetic formulations to provide antimicrobial activity, enhance product stability, increase shelf life, and improve sensory properties. The high-performance cosmetics consumer has made these chemicals more widely used. Nonetheless, growing concerns for potential endocrine-disrupting and bio accumulative effects have triggered dramatic changes in legislation and consumer attitude (Wirtu, 2024). Consequently, the cosmetic world is seeing a move toward "clean beauty," transparency when it comes to disclosing ingredients and reformulations to minimize or remove toxic xenobiotics. This transition has had an impact on competition and global market strategies.

2.5 *Xenobiotics in Food Processing and Packaging*

In the food industry, xenobiotics are employed as preservatives, synthetic colorants, flavoring agents, and packaging substances to increase the shelf life and to ensure the quality of the product. Packaging materials based on polymers are indispensable to protect food but are also associated with the risk of chemical migration upon application of heat or during extended storage (Bhunia et al., 2013). This is spurring calls from regulatory agencies and consumers for more scrutiny of food-contact materials and chemical residues in processed foods. New regulations and food safety standards, such as those by the Food Safety and Standards Authority of India (FSSAI), are also gradually placing more stringent requirements on the adoption of safe packaging technologies and providing clear labels.

2.6 Xenobiotics in the Manufacture of Textiles and Dyes

The textile and dye industries use xenobiotic substances such as azo dyes, bleaches, mordants, and finishing materials to achieve desired levels of colorfastness', texture, and design effects. These chemicals are part and parcel of the world's textile economy, but have massive environmental ramifications. Pollution from treated fabric waste, typically laced with xenobiotics, leads to the contamination of water supplies and endangers the lives of people living in the vicinity (Negi, 2025). Consequently, environmental labels like OEKO-TEX® and more stringent discharge regulations are now the key to industry compliance and global competitiveness.

2.7 Economic Importance of Xenobiotic-Dependent Industries

Xenobiotics are vital in generating revenue for all industries, whether agriculture, medicine, or other sectors of human life. Xenobiotics still play a large part in the output of industrial productivity, technological progress, global supply chain maintenance, and positive gross domestic product across all sectors. Furthermore, the specialty chemicals market which contains a considerable proportion of xenobiotic-based products is witnessing growth due to increasing demand from developing nations and rising levels of industrialization (OECD, 2022). With robust chemical production in India combined with a strong export potential and competitive labor cost, India has become one of the world's chief producers of xenobiotics. But along with economic returns come ever-higher expectations that these be responsibly managed, produced safely, and comply with environmental and public health regulations.

3 Commercialization Pathways of Xenobiotics

The production and sale of xenobiotics have associated commercial opportunities and implications that are described below. The development of xenobiotics is a tightly regulated, costly, and time-consuming process due to the high risk of developing xenobiotic-based products, which may cause human harm. Unlike regular consumer goods, innovations based on xenobiotics have innate risks associated with their chemical composition and related to human and environmental implications (Chin, 2019). Therefore, the route of commercialization is more stringent and involves laboratory research, testing and trials, regulation, approval, protection of intellectual property, manufacturing, distribution, and surveillance on the product, and after this, it is introduced in the market. Such pathways help to facilitate ensuring that xenobiotics introduced into the market are safe for use while also being economically viable and competitive.

3.1 Research, Development, and Pre-commercial Testing

When a xenobiotic compound is discovered and synthesized in the course of laboratory research, the process of commercialization is initiated. Science teams recognize molecules with favorable characteristics (e.g., having pesticidal activity, pharmaceutical utility, or useful material characteristics) and screen for their compatibility with applications. This phase includes initial toxicological examination, genotoxicity study, and physiochemical testing as well as investigation of the mechanism-of-action (MacDonald & Robertson, 2009). Preclinical studies for drugs examine absorption, distribution, metabolism, excretion, and toxicity (ADMET) properties and are used to decide whether a xenobiotic drug test can continue to clinical study. This phase encompasses laboratory and greenhouse evaluations of efficacy, environmental fate, and ecotoxicity (U.S. Environmental Protection Agency [EPA], 2025). Preliminary screening is important to evaluate whether the molecule is likely to be commercially attractive and meet future expectations regarding safety.

3.2 Regulatory Review and Approval Stages

Control of surveillance has long been a crucial gatekeeper for xenobiotic markets. To produce and market a product with xenobiotic components, a complete dossier on toxicity data, environmental impact data, patterns of exposure, and overall risk assessments needs to be provided to the national and international regulators (U.S. Environmental Protection Agency [EPA], 2025). In the USA, animal drugs and feed additives are regulated under specific schemes by the Food and Drug Administration (FDA), and drugs and chemicals used in animal and crop production are regulated under the Toxic Substances Control Act (TSCA), the Federal Insecticide, Fungicide and Rodenticide Act (FIFRA), and others by the Environmental Protection Agency (EPA). Similarly, the Registration, Evaluation, Authorization, and Restriction of Chemicals (REACH) regulation in the European Union, which is administered by the European Chemicals Agency (ECHA), requires submission of safety data as well as the authorization to place chemical substances on the market (Fig. 1). That's not so easy to say, because these treatments' barriers must make sure that no xenobiotics ended up in the market.

3.3 Technology Transfer and Proof of Industry—Academia Engagement

When a xenobiotic molecule has advanced to a stage where it is likely to be used in a commercial product, the tools of technology transfer are what bring the research institution-derived innovation to the marketplace. Universities and research institutions are often partners with industry in joint research programs, incubation arrangements, and licensing agreements. These alliances are particularly important in medicine, biotechnology, and agrochemistry research, accelerating the translation of science-based discoveries into commercial products. Technology transfer offices (TTOs) are responsible for managing the intellectual property, negotiating licensing agreements, and ensuring compliance with the rules. Better innovation output and commercialization feasibility of xenobiotics with safety, efficacy, and market demand can be realized through collaboration.

3.4 Intellectual Property Rights and Patenting Strategies

Intellectual property (IP) protection is essential for maintaining a competitive edge and for R&D expenditures in innovations resulting from xenobiotics. Patenting enables the protection of novel chemical entities, manufacturing processes and uses. By allocating a period of market exclusivity, robust IP protection can stimulate private sector investment (World Intellectual Property Organization [WIPO], 2025). In the pharma sector, the patents on the xenobiotic molecules frequently determine the commercial success of therapeutic development and their pricing strategies. At the same pace, patents protection also raises important ethical and public policy questions related to access, affordability, and monopolistic behavior. In agriculture and chemicals sectors, the patent system also plays a critical role in stimulates innovations in safer pesticides, biocides, and polymer additives.

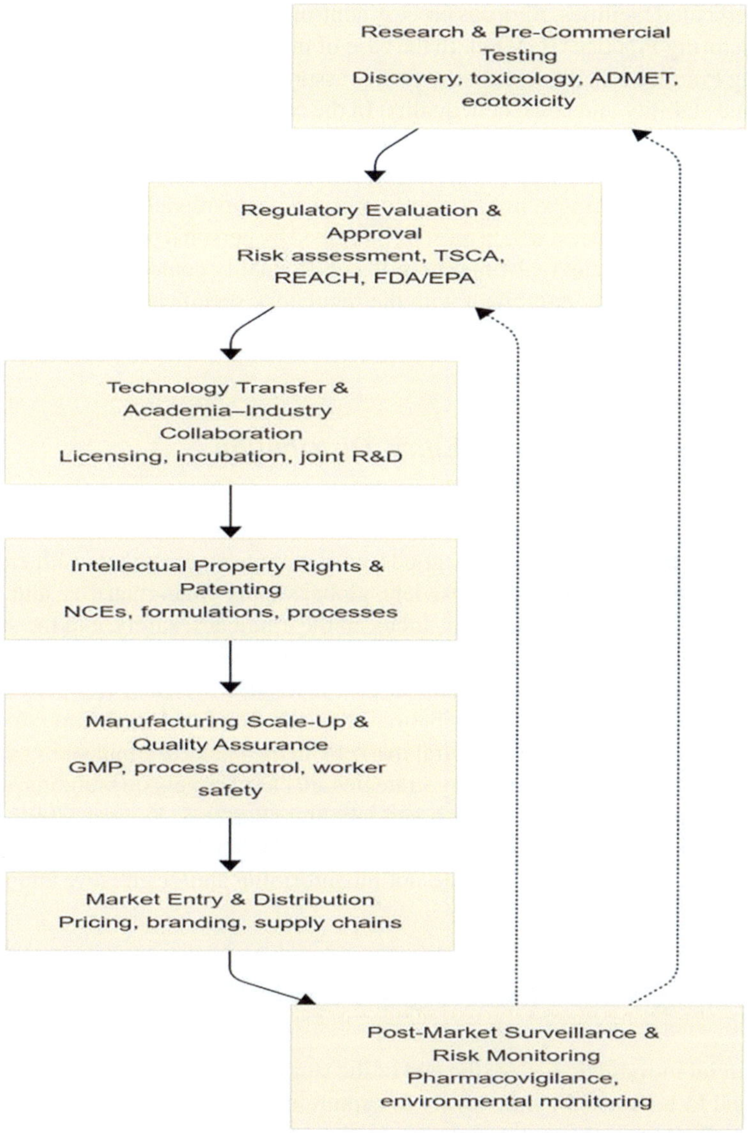

Fig. 1 Commercialization stages of xenobiotics

3.5 Optimization of the Manufacturing Process and Quality Control

Once the relevant regulatory authorities and the patent offices have been notified, the next critical stage is production. Manufacturing of xenobiotics at commercial scale needs dedicated facilities, rigorous process controls, and strict compliance with Good Manufacturing Practices (GMPs). In the case of pharmaceutical sector, Good Manufacturing Practice (GMP) serves as a quality assurance framework to ensure product purity and stability and consistent quality. In the case of chemicals materials, clearly defined quality control (QC) specifications are important to prevent unintended toxicity and contamination and patient's harm (U.S. Food and Drug Administration [FDA], 2021). Producing xenobiotics also presents occupational health issues related to exposure of employees, which must be mitigated by personal protective equipment, engineering control, and real-time surveillance. A quality control system guarantees the final product being compliant with the regulatory specifications in term of each batch production.

3.6 Strategies for Market Entry, Distribution, and Commercialization

Once produced, xenobiotic-based goods must compete in markets influenced by customer preferences, regulatory barriers, global supply chain matrices, and pricing mechanisms. Promotional messages focus on the efficiency, safety, and the status of the product's registration. Market access for pharmaceuticals includes the healthcare providers, hospital purchasing organizations, insurance systems, and public health organizations. Agrochemical distribution channels are developed through wholesalers, farming co-operatives and rural markets. In the business of plastic, cosmetics, and food packaging, marketing entry strategies are based largely on branding, sustainability claims, and compliance with eco-labeling (Porter & Kramer, 2006). To be valid, such market efforts must be guided by a commercial sense but also a sense of responsibility that the eco claims are not misinforming and/or greenwashing.

3.7 Oversight and Monitoring of Risks in the Post-market

Commercial introduction is not the end of the story for market introduction; xenobiotics need to be tracked continuously to capture new information on their potential harmful effects, on their accumulation in the environment or on genetic toxicity which may be hidden in the controlled testing. Environmental sampling, analyses of consumer feedback, safety reviews at certain intervals, inspections of compliance by regulatory body are post-market surveillance. In pharmaceuticals, pharmacovigilance systems are the source of information on ADRs, whereas in agrochemicals, outcomes of ecological and human exposures are evaluated through observations conducted at the field level (World Health Organization [WHO], 2020a, 2020b). Continued surveillance of products based on xenobiotics is essential to ensure their continued safety and to provide the scientific basis for revision of permissible exposure limits, restrictions on use, or recommendations for safer substitutes.

4 Global Regulatory Framework for Xenobiotics

The global xenobiotics regulatory framework is influenced by varying historical legislations, scientific risk assessment methodologies, and intents for public health protection. Countries establish product regulations to control for potential risk to human health and the environment and to promote responsible commerce in chemical products. While individual regimes vary in the structures of their administration, they rest on a shared underpinning of toxicological/scientific evaluation, environmental risk assessment, labeling, and post-market monitoring. International regulatory regimes act as gatekeepers, deciding which xenobiotics will be allowed to enter commercial markets, and on what terms they may be manufactured, distributed, and used (Bax & Green, 2015).

4.1 Regulatory Approaches in the USA

The US's policy on xenobiotic regulation is highly concentrated on two large federal organizations: the Food and Drug Administration (FDA) and the Environmental Protection Agency (EPA). The EPA regulates the Toxic Substances Control Act (TSCA), which mandates that companies report by submitting pre-manufacture notices, perform risk evaluations, and hold chemical inventories for new and existing materials. The Frank R. Lautenberg Chemical Safety for the 21st Century Act amendments in 2016 bolstered the TSCA by granting the EPA additional authority to obtain safety information, limit the use of dangerous chemicals, and prioritize chemicals on the basis of potential risks. Also, the Federal Insecticide, Fungicide, and Rodenticide Act (FIFRA) oversees the regulation of pesticides and agrochemicals in USA, which pertains to the registration, labeling, and allowable exposure limits for such products to maintain the safety within agricultural and environmental concerns (Frank, 2016). The FDA oversees pharmaceutical and cosmetic xenobiotics under the Federal Food, Drug, and Cosmetic Act (FD&C Act), ensuring product safety, efficacy, and quality before approval. Collectively, these regulatory instruments create a structured, science-driven system that governs the lifecycle of xenobiotics in the USA (U.S. Food and Drug Administration [FDA], 2025).

4.2 European Union Regulation Through REACH and CLP

The EU has something like the strongest chemical regulation in the world. The European Chemicals Agency-based (ECHA) Registration, Evaluation, Authorization, and Restriction of Chemicals (REACH) regulation obliges companies to register chemical substances, provide extensive safety dossiers, and prove that their products are not associated with substantial risks to human health or the environment (Deubner et al., 2024). REACH places the burden of proof on industry, requiring manufacturers and importers to produce and submit safety data prior to market access. Alongside REACH, the Classification, Labeling, and Packaging (CLP) Regulation ensures a harmonized chemical hazard communication in all EU member states through the adoption of the Globally Harmonized System (GHS) of classification and labeling. Together these regulations provide for strict regulation of xenobiotics, ensuring transparency, accountability, and harmonized safety standards throughout Europe.

4.3 International Standards and Guidelines: WHO and OECD

Various international organizations have a bearing on world chemical management through recommendations, best practices, and joint initiatives. The World Health Organization (WHO) provides health risk evaluations, exposure standards, and toxicological assessments for pesticides, pharmaceuticals, and other environmental pollutants. The Organization for Economic Co-operation and Development (OECD) is the key body, through its Chemicals Safety Program, in elaborating harmonized methodologies for testing, risk assessment, and data (mutual acceptance of data, MAD) which apply to its member countries (Organization for Economic Co-operation and Development [OECD], 2021–2024). The OECD test guidelines have gained high credibility within the international regulatory community and therefore are used for the evaluation of new and existing xenobiotics. These bodies make an important contribution to scientific harmonization, and thus to evidence-based decisions on chemical safety on a global scale.

4.4 Regulatory Systems in East Asia: Japan, China, and South Korea

East Asia has become a major center for chemical production and use and the development of strong regulatory systems responded (Fig. 2). The Japanese Chemical Substances Control Law (CSCL) is the legislation that governs the production, importation, and use of chemical substances with regard to its harmfulness and environmental persistence. China's Environmental Management of Novel Chemical Substances Establishing the New Chemical Substances Schedule the Act on the Registration, Evaluation, etc. of Chemicals' Requirements for Pre-market Notification Hazard Classification and Risk Assessment (Fukushima, 2021), (K-REACH) The Chemical Act (also known as K-REACH), which mandates the registration, assessment, and post-registration management of industrial chemicals for consumer safety, is another piece of legislation introduced in South Korea. This shows a dedication to both harmonization with international standards and systematic xenobiotic control as a recurrent subject for the area.

4.5 Chemical Safety Regulations in Latin America and Africa

Latin American countries, notably Brazil and Mexico, have developed more stringent regulations for chemicals in recent years. National agencies in Brazil's government: the National Health Surveillance Agency (ANVISA) and the Brazilian Institute of Environment and Renewable Natural Resources (IBAMA) are responsible for regulating agrochemicals, industrial chemicals, and pharmaceutical xenobiotics according to comprehensive authorization procedures. Mexico's system of chemical safety, based on NOMs, includes toxicological assessment and labeling processes. In Africa, regulatory systems are developing with South Africa's National Environmental Management Act (NEMA) being used as the basis for chemical risk management and a number of countries working together in the African Chemicals Convention and other regional initiatives under the auspices of the United Nations Environment Program. These systems reveal inequalities in regulatory capacity on a global scale, but they do so through a shared lens of concern for the safety of Xenobiotics.

4.6 Comparative Insights and Challenges of Global Harmonization

While the regulatory regimes are not the same for each region, there are various shared outlooks on xenobiotics management, and these aspects include risk assessment, hazard classification, labeling requirements, exposure control, and market or post-market surveillance. Nevertheless, global harmonization is difficult to achieve considering the contrasts in national priorities, scientific capacity, legal traditions, and enforcement mechanisms. Developed countries, particularly high-income areas such as the European Union and the USA, have well established regulatory systems, but developing countries may be unable to provide the fruits of extensive toxicological testing, monitoring, and enforcement. Global harmonization initiatives, such as

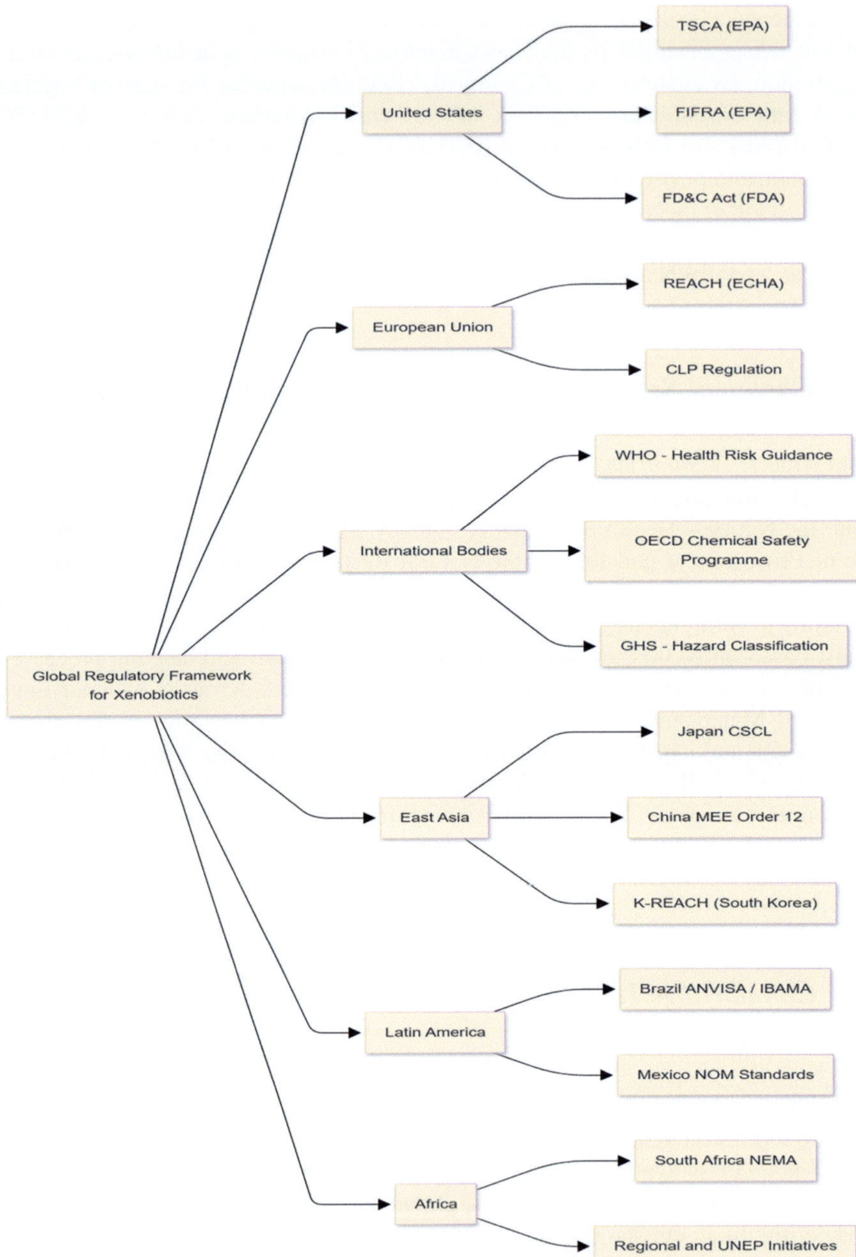

Fig. 2 Global regulatory framework for xenobiotics

the Globally Harmonized System (GHS), for classification and labeling of chemicals and the OECD harmonized test guidelines for chemical testing are important in filling regulatory gaps and ensuring coherence in chemical safety standards. With the globalization of xenobiotic supply chains, effective inter-country collaboration will become increasingly important to ensure safety, transparency, and sustainability.

5 Indian Regulatory Framework for Xenobiotics

Over the past decades, India's regulations concerning xenobiotics have made dramatic changes with the pace of heavy industrialization, growth in pharmaceutical as well as agrochemical industries, increasing environmental concerns, and the need for safeguarding public health. In contrast to more consolidated arrangements like the European Union's REACH, the Indian system of regulation is spread over multiple ministries, agencies, and laws and is still taking shape. Such a multiagency apparatus is indicative of the varied uses of xenobiotics—pharmaceuticals and pesticides, yes, but also food additives, plastics, dyes, and industrial chemicals. While regulatory division may create problems, recent reforms suggest an increasing national interest in enhancing oversight of chemical safety, increasing transparency, and harmonizing with international best practices in Indian policy.

5.1 Ministry of Environment, Forest, and Climate Change (MoEFCC) and the Environmental Protection Act

The MoEFCC is the highest body responsible for regulation of environment in India. The 1986 Environmental Protection Act (EPA) is the key instrument in the Indian environmental law that allows it to deal with chemical substances and industrial emissions, hazardous waste management, and environmental monitoring. MoEFCC prescribes standards for air and water quality, for allowable levels of discharges, and for the management of hazardous xenobiotic materials under the EPA. A few of the rules under the EPA, such as the Hazardous and Other Wastes Rules, 2016, the Chemical Accidents Rules, 1996, and MSIHC Rules, 1989 [Manufacture, Storage and Import of Hazardous Chemicals/(MSIHC) Rules], regulate the management of xenobiotics explicitly (Ahuja, 2018). These laws and regulations call for risk assessments, safety audits, emergency planning, and reporting requirements for industries that use hazardous chemicals.

5.2 Central Pollution Control Board (CPCB) and Environmental Monitoring

The Central Pollution Control Board (CPCB), operating under Ministry of Environment, Forest, and Climate Change (MoEFCC), plays a pivotal role in implementing environmental standards for xenobiotic emissions, effluents, and industrial waste in India. The CPCB is responsible for establishing permissible pollutant limits, monitoring environmental quality, and enforcing compliance through inspections, regulations, and directives. CPCB regulations for wastewater treatment, stack emissions, chemical storage, and waste disposal must be followed by industries that release xenobiotics, whether through effluent wastewater discharges, air emissions, or solid waste (Mohapatra et al., 2025). Sectors such as pharmaceuticals, dyes, textiles, and petrochemicals are subject to rigorous monitoring to ensure that xenobiotic contaminants do not exceed prescribed environmental limits. CPCB also collaborates with State Pollution Control Boards (SPCBs) to oversee local enforcement, strengthen data collection, and manage industrial clusters with high chemical loads (Table 2).

5.3 Central Drugs Standard Control Organization (CDSCO) and Pharmaceutical Xenobiotics

In India, pharmaceutical xenobiotics are regulated by the Central Drugs Standard Control Organization (CDSCO) under the Ministry of Health and Family Welfare. CDSCO implements the provisions of Drugs and Cosmetics Act 1940 and its subsequent rules that regulate the approval process, manufacturing, import, distribution, and post-market surveillance of pharmaceuticals. Xenobiotic-derived drugs are subject to rigorous preclinical and clinical testing, including toxicology studies, pharmacokinetic evaluations, and safety trials, prior to clinical availability. The approval process involves expert committees, inspection of manufacturing facilities for Good Manufacturing Practices (GMPs), and a series of periodic safety updates. Through the Pharmacovigilance Program of India, CDSCO also tracks adverse drug reactions and continues to evaluate xenobiotic drugs for genetic toxicity, long-term risk, and safety issues post market. CDSCO's role in regulating is particularly crucial for global pharmaceutical safety given India's vast generic drug market.

5.4 Food Safety and Standards Authority of India (FSSAI) and Food-Related Xenobiotics

Food xenobiotics in India are regulated by the Food Safety and Standards Authority of India (FSSAI) set up under the Food Safety and Standard Act (FSS Act) 2006. The FSSAI is also responsible to regulate the standards for food-additives, preservatives,

Table 2 Key regulatory bodies and legislative instruments governing xenobiotics in India

Regulatory body/ ministry	Relevant legislation/ rules	Scope of regulation	Key functions related to xenobiotics
Ministry of Environment, Forest and Climate Change (MoEFCC)	Environmental Protection Act, 1986; MSIHC Rules, 1989; Hazardous and Other Wastes Rules, 2016; Chemical Accidents Rules, 1996	Environmental protection, hazardous chemicals, industrial pollution	Sets emission/ discharge standards; regulates handling of hazardous chemicals; mandates risk assessment, safety audits, emergency response
Central Pollution Control Board (CPCB)	EPA Rules, Water Act (1974), Air Act (1981)	Environmental monitoring and pollution control	Monitors xenobiotic emissions/effluents; enforces environmental norms; coordinates with SPCBs; regulates industrial discharges
Central Drugs Standard Control Organization (CDSCO)	Drugs and Cosmetics Act, 1940 and Rules	Pharmaceutical xenobiotics: drugs, APIs, excipients, cosmetics	Approves drugs; evaluates toxicology and clinical safety; GMP inspections; pharmacovigilance; post-market safety monitoring
Food Safety and Standards Authority of India (FSSAI)	Food Safety and Standards Act, 2006	Food additives, packaging materials, contaminants, MRLs	Sets standards for additives and preservatives; regulates contamination; controls packaging migration; conducts food surveillance
Directorate of Plant Protection, Quarantine and Storage (DPPQS) under MoA&FW	Insecticides Act, 1968; Insecticides Rules, 1971	Pesticides and agrochemicals	Registers pesticides; evaluates toxicity, residues, environmental impact; labeling, licensing, quality checks
Bureau of Indian Standards (BIS)	BIS Act, 2016; BIS Specifications and Standards	Industrial chemicals, plastics, dyes, consumer products	Specifies quality and chemical composition limits; regulates impurities, labeling, safety standards; supports export compliance

(continued)

Table 2 (continued)

Regulatory body/ ministry	Relevant legislation/ rules	Scope of regulation	Key functions related to xenobiotics
State Pollution Control Boards (SPCBs)	EPA Rules, Water and Air Acts (state enforcement)	Regional enforcement of environmental laws	Issues consents; inspects facilities; monitors compliance; handles local environmental grievances
Emerging Policy Bodies (MoEFCC, NITI Aayog)	Proposed National Chemicals Policy; draft Chemical Safety frameworks	Chemical management modernization	Strengthening oversight; addressing endocrine disruptors, microplastics, nanomaterials; enhancing toxicovigilance

colorants, contaminants, and packaging materials. Xenobiotics in food processing such as chemical preservatives or pesticides residues need to adhere to maximal residue limits (MRLs), microbiological criteria, and labeling rules as set by FSSAI (Food Safety and Standards Authority of India [FSSAI], 2025). It also sets limits on migration from packaging to prevent leaching of chemicals into food products. Enforcement is backed by continuous surveillance, sampling, and national monitoring programs, and xenobiotic content in food is maintained at a level deemed to be considered as safe. These regulations are developed in alignment with the Codex Alimentarius to enhance international harmonization and trade of food products.

5.5 *Pesticides and Agrochemicals Under the Insecticides Act*

Application, production, and sale of agrochemical xenobiotics in India are regulated under the Provisions of the Insecticides Act, 1968, and the Insecticides Rules, 1971. These regulations impose requirements for product registration, labeling, quality assurance, and licensing on pesticide manufacturers and distributors. Pesticides are evaluated for effectiveness, toxicity, residue dynamics, environmental impact, and risk mitigation procedures prior to registration. Despite the existence of a regulatory framework, the persistence of challenges associated with illegal pesticide imports, counterfeit products, and inappropriate usage indicates a continued need for stronger enforcement measures and legislative reforms. The long-awaited Pesticide Management Bill is expected to overhaul regulatory control to include more stringent licensing, digital monitoring, and greater transparency for pesticide registration.

5.6 Bureau of Indian Standards (BIS) and Industrial Chemical Quality

The Bureau of Indian Standards (BIS) prescribes the quality and safety standards that need to be complied with for certain industrial chemicals, polymers, dyes and also for certain types of consumer products. Be it mandatory or voluntary, BIS certification confirms the adherence following standardized specifications which disallows a higher content of hazardous xenobiotics in the product. Chemical composition, allowed impurities, labeling, packaging, and test methods are described in the BIS standards. In industries like plastic, textile, and consumer product, BIS is crucial in ensuring the quality of the products and preventing any hazard to public life due to release of materials containing toxic xenobiotic. The rising international trade has additionally prompted India to harmonize BIS standards with international standards to enhance their export competitiveness.

5.7 Emerging Policies and Regulatory Gaps

Although India has established an extensive regulatory framework, significant challenges remain to persist. Inconsistencies in the governance of xenobiotics are caused by fragmentation between overlapping institutional agencies and scientific deficiencies. Enforcement outcome also differs across states, reflecting differences in administrative capabilities and resources availability. In addition, many existing regulations are not designed to cover issues arising today, such as microplastics, endocrine disruptions, nanomaterials, or exposure to chemicals at low doses for long periods of time. Recognizing this limitation, India has initiated to pursue reform to strengthen national chemical management systems, including the development of a National Chemicals Policy and toxicovigilance system infrastructure. Together, these initiatives exemplify a broader shift toward regulatory approaches grounded on scientific evidence, global best practice, and principles of sustainability.

6 Compliance Management and Corporate Governance Framework in Xenobiotic Regulation

Because the marketing and use of xenobiotics, which can endanger human health, genetic integrity, and environmental sustainability, these activities are subject to stringent regulatory oversight and require effective corporate governance mechanisms. Industries that rely heavily on xenobiotics, such as pharmaceuticals, agrochemicals, plastics, textiles, and food processing, operate within complex and multi-layer regulatory frameworks that address product safety, environment protection, quality assurance, and ethical conduct. Failure in compliance can have serious risks to public health and safety and may result in product recalls, market withdrawals, litigation, financial penalties, and long-term damage to a company's reputation. As global awareness chemical safety continues to expand, compliance and governance have become defining elements of an organization's strength, credibility, resilience, and sustained market participation.

6.1 Regulatory Compliance, Risk Management, and Strategic Resilience

Regulatory compliance in xenophobic-producing industries has evolved beyond a narrow legal requirement to become a strategic importance with direct implications for organizational performance and legitimacy. National and international regulation requires companies to adhere to requirements related to chemicals registration, permissible exposure limits, environmental discharge, product labeling, and post-marketing surveillance. Fulfilling these requirements involves a maintenance of systematic documentation, transparent reporting practices, and continuous internal surveillance to ensure sustained compliance the time (Chatterjee et al., 2024). Those that are able to successfully reintegrate compliance into their business models can mitigate operational risk and enhance their competitive positioning. On the other hand, breaches of compliance can lead to drastic consequences—a PFMEA can be accepted, a license to operate can be revoked, and in worst cases, there could be an initiation of legal prosecution, especially in tightly regulated industries such as pharmaceuticals and agrochemicals.

6.2 Operational Risk Controls and Safety Protocols

Robust operational controls that reduce the potential for occupational exposure and the release of hazardous materials to the environment are needed to manage xenobiotics in production facilities. Industry should provide engineering controls, containment, air filtration, safe storage, and treatment of wastewater to protect workers and the surrounding communities from exposure. Integrated risk management systems include hazard identification, exposure assessment, emergency preparedness, and reporting systems (Elmadhoun et al., 2025). In drug and chemical manufacturing facilities, Good Manufacturing Practices (GMPs) and Good Laboratory Practices (GLPs) guarantee the uniformity, cleanliness, and security of the product throughout the process. Good risk controls not only ensure compliance, but also improve operational performance and decrease exposure to potential and costly business interruptions.

6.3 Supply Chain Governance for Xenobiotic-Based Industries

The safety of xenobiotics is not confined to the internal processes of production, and it also covers the whole supply chain, including the sourcing of raw materials and the distribution and end-of-life management. There is growing scrutiny on the origin and quality of materials related to xenobiotics, as companies contend with global supply chains wherein complex transactions may hide safety risks. Robust supply chain management includes supplier auditing, disclosure of chemicals, quality verification processes, environmental, and social compliance in the supply chain (Alfaouri et al., 2025). Companies operating in sectors such as pesticide, pharmaceutical, and food packaging companies are increasingly required to demonstrate that their suppliers comply with acceptable manufacturing standards, impurity thresholds, and documentation requirement. As regulatory expectations increase, those companies implementing transparent, traceable supply chains will be best positioned to stay in compliance and maintain stakeholder confidence.

6.4 Corporate Governance and Ethical Responsibility

Corporate governance structures can at times employ mechanisms that they execute intentionally or not and can create irresponsible decision-making process that encourage irresponsible corporate behavior. In sectors involving xenobiotics, such actions are regulated through scientific and regulatory frameworks to prevent actions or practices that could compromise environmental or human health or environmental integrity. Responsibility chemical safety governance rests primarily with boards of directors and senior management, who must ensure full regulatory compliance with all regulatory requirements and that environmental and social issues are incorporated into strategic planning. Ethical corporate governance therefore requires to take into account not just the business interests, but also the long-term knock-on effects on public health and ecological systems. Stakeholder communication is open employees, consumers, regulators, and communities are all part of the conversation and that's what builds trust, particularly around high-risk xenobiotics (Trump et al., 2023). Failures of governance of the chemical industries are often catastrophic, with environmental disasters, public backlash, and investor flight.

6.5 Environmental, Social, and Governance (ESG) Reporting

The Environmental, Social, and Governance (ESG) narrative has become a compelling lens through which to analyze corporate responsibility, particularly in sectors associated with toxic chemicals. Environmental impacts, emissions, waste management, use of chemicals, and strategies to mitigate risks should be disclosed by firms in their ESG reports. ESG ratings are increasingly sought out by investors and consumers to judge organizational responsibility and sustainability beyond the short-term. In industries that make use of xenobiotics, robust performance in ESG indicators reflects compliance with chemical safety standards, the limitation of hazardous substances, and adherence to ethical production standards (Ghag, 2024). Several multinational companies have started substituting hazardous xenobiotics with safer ones or by integrating principles of green chemistry to enhance their ESG ratings and to be in tune with the global sustainability trends.

6.6 Challenges in Compliance and Governance

Although there are increasing demands for chemical safety, the xenobiotic industries are faced with a multitude of challenges when it comes to ensuring compliance and the application of good governance. The complexity of regulation, lack of sufficient toxicological data, and differing international standards can make compliance difficult. In high monitoring cost industries, such as manufacturing, small and medium enterprises (SMEs) find it difficult to comply, and compliance levels may be diverse across sectors and subsectors. Moreover, new scientific knowledge constantly uncovers new hazards associated with xenobiotics (e.g., endocrine-disrupting properties, microplastic pollution, toxicity of nanomaterials) that are not adequately considered by existing regulations. Meeting these challenges will depends on firms' ability to develop adaptive governance strategies, participate in continuous learning, and collaborate across sectors.

6.7 Responsible Commercialization and Risk Governance

Robust compliance and governance mechanisms are essential to ensuring xenobiotics are responsibly commercialized. Companies must integrate chemical safety into their core business models, allocate appropriate resources toward safer alternatives, implement transparent reporting systems, and engage constructively with regulators and affected communities. As global attention focuses on the toxic risks and environmental sustainability, those companies with well-developed compliance and governance will be best positioned to remain competitive markets and contribute positively to public health protection and the preservation of our planet.

7 The Ethical, Environmental, and Social Impacts of Xenobiotic Trade

The trade and marketing of xenobiotics raise ethical, environmental, and social implications beyond the simple prospecting for economic gain. Although these industries depend on xenobiotics to facilitate industrial innovation, operational efficiency, and market competitiveness, the pervasive use of such chemicals often generative adverse impacts that are externalized, most severely affecting public health, marginalized communities, and the natural ecosystem. Consequently, the ethical trade-off requires industries, regulators, and policymakers to weigh economic considerations with moral responsibility, scientific rigor, and environmental stewardship. Achieving this balance is particularly critical in traditional areas where xenobiotics are associated with genotoxic effects, be phenotypically persistent in the environment, or cause irreversible damage to ecosystems.

7.1 Ethics: The Good of the Genus Xenobiotic?

Ethical principles are the ground rules of chemicals management. Companies must bear the responsibility to make sure that the production, use, and disposal of xenobiotics do not cause damage to human health or the environment. Ethical responsibility entails full transparency about the constituency of chemicals, transparency regarding risk and products of the chemicals, full reliance on scientific evidence, and no use of misleading communications on chemicals. Of course, ethical transgressions in chemical companies—covering up toxicity data or pushing dangerous products—have historically caused harm on a mass scale, resulting in litigation and public mistrust. Ethical and safety concerns also make it imperative for companies to prioritize safety rather than temporary financial gain, invest in safer alternatives to hazardous chemicals, and respect the rights of consumers and communities potentially vulnerable to chemical exposure or risk elements.

7.2 Environmental Impacts and Ecological Considerations

Xenobiotics have an impact on the ecosystem due to their mobility, persistence, and bioaccumulation. Many xenobiotics, such as pesticides, industrial solvents, heavy metals, and plasticizers, are released into soil, water, and air where they persist and come into contact with biota. Their environmental effects involve the loss of biodiversity, changes to the ecosystem, pollution of lakes and rivers, and magnification within the food (Embrandiri et al., 2016). Persistent organic pollutants (POPs) and endocrine-disrupting chemicals (EDCs) attract special attention as they have long half-lives and can interfere with hormone systems of wild life and human. In

India and elsewhere in the developing world, poor waste disposal practices, lack of adequate wastewater treatment facilities, and industry operating in the informal sector compound the environmental hazard posed by the release of xenobiotics.

7.3 Social Equity and Vulnerable Populations

Exposure to xenobiotics tends to disproportionately impact susceptible populations, thereby posing concerns for public equity and environmental justice. Due to occupational exposure, contaminated resources, or living near industrial areas, agricultural workers, industrial workers, children, pregnant women, and poor communities have a greater risk of being exposed to hazardous levels of xenobiotics. These groups are also more susceptible to genotoxicity, respiratory disease, cancer, and developmental defects. There is more socioeconomic disparity because these xenobiotics are more readily consumed and disposed of in underserved areas where regulatory systems are less strictly enforced. To protect the rural and vulnerable discount ages in their air and water, ethical commercialization means that governance and industry must protect those most at risk through better enforcement, health monitoring, and community engagement.

7.4 Corporate Social Responsibility (CSR) in Xenobiotic-Using Industries

CSR refers to a company's discretionary or compliance-related activities that contribute to social welfare, environment conservation, and sustainability. For instance, India's Companies Act of 2013's mandatory CSR regulations have assisted in directing corporate funds into public health, pollution prevention, environmental preservation, and rural development initiatives. Chemical-intensive enterprises also frequently deploy CSR programs to mitigate some of the negative effects of their operations, such as labor safety programs, environmental rehabilitation projects, and donations to health camps (Thakur, 2024). However, corporate social responsibility (CSR) must be genuine and significant rather than merely symbolic or promotional. This means that businesses must incorporate responsible practices into their daily operations rather than viewing CSR as an afterthought.

7.5 *Environmental, Social, and Governance (ESG) Considerations*

ESG frameworks are now embedded in assessments of corporate performance and ethical conduct, including corporate conduct in relation to toxic substances. Xenobiotic-related ESG aspects take into account the environmental aspects of xenobiotics, including emission rates, waste management, energy use, and chemical safety policies. ESG frameworks assess social issues such as worker health, community well-being, and supply chain transparency, alongside governance factors including board oversight, ethical conduct, and the effectiveness compliance mechanisms. Those with strong ESG credentials are better placed to attract investment, limit liability, and retain consumer faith. As global markets place growing emphasis sustainable development, adherence to ESG standards has shifted from a voluntary signal of responsibility to a practical requirement for long-term survival in xenobiotic-utilizing industries.

7.6 *Public Perception, Consumer Awareness, and Risk Communication*

Public perception plays a decisive role in shaping the commercial trajectory of xenobiotics. Greater access to scientific information, combined with sustained environmental campaigning and the rise of social media activism has heightened consumers' awareness over possible hazards related to chemicals in food, cosmetics, plastic, and everyday household items. Poorly managed risk communication or conflicting withhold signals can erode public trust, resulting in consumer reaction, product boycott or regulatory intervention. To communicate risk effectively, companies and regulators must deliver information about xenobiotic ingredients, potential hazards, safety measures, and regulatory adherence that is accurate, understandable, sufficiently detailed, and based on the best available science. Enhancing public participation and science literacy is key to transitioning consumer behavior to sustainable trajectories, and holding industry accountable.

7.7 *Ethical Marketing and the Problem of Greenwashing*

With "sustainability" as the new marketing angle, a growing number of companies are presenting their chemical products as "green," "non-toxic," or "natural." But greenwashing, or the exaggeration or the outright lie about environmental credentials, is just as much an ethical issue in xenobiotic-related fields. By concealing harmful behavior under a marketing-friendly façade, Greenwash deceives consumers and diverts attention from true sustainability (Vangeli et al., 2023). Greenwashing, in response, regulatory agencies in the EU, and other regions have issued guidance to combat greenwashing and have stressed the need for substantiation of claims and for ingredient disclosure. The ethics of marketing would dictate that the firms gave up on selling their lies and instead funded true safer, science-based alternatives to the products they sell.

7.8 Toward Ethical and Sustainable Practices

The political, ecological, and societal consequences of xenobiotic trade illustrate the importance of diligent oversight underpinned by scientific validity, social responsibility, and environmental care. Ethical issues need to be considered at all stages in the life cycle of a xenobiotic, including the research and development and disposal stages, by companies. Enforcement of regulations, public health surveillance, and protective interventions for at-risk populations must be enhanced by policymakers. With increasing global attention on the hazards of chemicals, those industries which demonstrate ethical, sustainable behavior will be more likely to continue to have legitimacy, to be responsive to regulatory demands, and ultimately to support the long-term fortunes of human society.

8 Future Prospects in Commercialization, Regulation, Market Dynamics, and Ethics

Potential pathways for xenobiotic sustainability lie directly within the larger context of commercialization, regulations, market dynamics, and ethics. With world industry expanding use of xenobiotics in pharmaceuticals, agriculture, manufacturing, and consumer goods, the pressure to redefine commercialization pathways is growing. The pending trends in green chemistry, predictive toxicology, circular economy models, regulatory harmonization, improved surveillance, transparency, and investing in safer alternatives, together reshape the production, marketing, and governance of xenobiotics.

8.1 Green Chemistry and Regulatory Policy Modernization

Advances in green chemistry, together with ongoing revisions of regulatory policy, are driving a major shift toward sustainable manufacturing and safer consumer products. Development in green chemistry has expanded the availability of safer alternatives to harmful substances, and it affects decision-making in the regulatory arena. As authorities impose more stringent toxicological standards and safety thresholds, industry faces the challenge to introduce product that are more biodegradable, less persistent, and less likely to pose genetic risks. These shifts align closely with ethical commitments to the protection of both ecological and human health. Firms that successfully produce green xenobiotics often benefit better market access and a reduced compliance risk, illustrating how sustainability are now shapes competitive dynamics.

8.2 *Predictive Toxicology and Ethical Commercialization*

AI-enabled predictive toxicology is increasingly reshaping regulation, enabling prompt detection of hazards, reducing animal testing, and accelerating safety assessment processes. The activity of such tools toward the ethical commercialization is to keep detrimental xenobiotics away from the consumers and the ecosystems. In the market, predictive analytics speeds up research and development (R&D) efficiency, decreases product failures, and builds consumer trust—important factors contributing to market acceptance and long-term brand credibility.

8.3 *Circular Economy Approaches and Waste Reduction*

Circular economy-based system addresses a different set of concerns than the typical linear "take-make-dispose" industrial system, by focusing on waste prevention, resource efficiency, and material reutilization. In relation to xenobiotics, circular economy practices will include creating products for recyclability, utilizing closed-loop manufacturing systems, and reclaiming chemical byproducts for use. To give an example, recycling schemes on plastic packaging prevent the buildup of xenobiotics-containing wastes in landfills and oceans. Sustainable textile production technologies have been developed to reduce the release of xenobiotic dyes such as dry processing, dyeing without water, natural pigment systems, and water-treatment advanced systems. Circular economy advancements enable industries to diminish their xenobiotic impact and meet the expectations of a growing economy and a changing regulatory environment.

8.4 *Global Harmonization of Regulatory Frameworks*

As xenobiotics flow through world trade, there is a growing need to harmonize the regulatory systems of different countries. International systems such as the GHS for classification and the OECD's Mutual Acceptance of Data (MAD) mechanism are harmonizing testing schemes, risk assessment approaches, and chemicals' safety communication. Nevertheless, considerable differences remain between national regulatory frameworks, especially when comparing high-income and low-income countries. Strengthening global collaboration through shared databases, cross-border scientific partnerships, and multilateral agreements can enhance regulatory consistency and collective capacity to address xenobiotic risk. Regulatory harmonization also eliminates duplication in testing, facilitates international trade, and support the sound management of chemicals.

8.5 Strengthening Environmental Surveillance and Public Health Monitoring

Future xenobiotic regulation will increasingly rely on extensive environmental monitoring, biomonitoring, and the application of public health surveillance systems. Environmental monitoring can enable xenobiotic contamination in air, water, soil, and food, while biomonitoring provides measures to the extent of exposure to chemicals within human population. Strategic investments in monitoring infrastructure, laboratory capacity, and real-time data analytics will yield disproportionate benefits, strengthening early-warning systems and enabling more effective regulatory enforcement. In parallel, systems for public health surveillance that are able to monitor disease patterns associated with xenobiotic exposure such as cancers, endocrine, and reproductive disorders will be essential in assessing long-term risk and in guiding changes in policy.

8.6 Enhancing Transparency and Consumer Empowerment

Consumers are increasingly demanding more transparency around chemical ingredients, safety information, and environmental claims. As a result, future commercialization pathways will depend on companies being more transparent about the potential risks associated with their products through labeling and responsible marketing. Emerging digital technologies, such as smartphone-based information platforms and block-chain-enabled traceability system, offer new options for informing consumers about product ingredients in real time and product safety. Empowering consumers with more power encourage them to make informed choices and to push companies to maintain high safety standards. As public awareness rises, market-driven demand for safer xenobiotics is expected to accelerate industry to switch to safer Xenobiotics and sustainable production practices.

8.7 Investment in Safer Substitutes and Innovation Incentives

Governments, and in some cases industry, are increasingly investing in the R&D aimed at identifying safer xenobiotic substitutes. Innovation incentives, such as tax credits, public grants, and public–private partnerships, support companies to develop cleaner production technology and more environmentally friendly chemical alternatives. In parallel, regulatory authorities worldwide are also increasingly implementing substitution policies that require hazardous substances to be replaced when safer options are exist. These policies aim to incentivize industry to phase out its use of chemicals associated with genotoxicity, risk of contamination, and persistence in the environment. A competition for innovation-driven transformation continuously enables faster market shifts to safer and more sustainable xenobiotic products.

8.8 A Vision for Sustainable Xenobiotic Governance

Looking ahead, the future of xenobiotics commercialization will depend on successfully integrating advancements, ethical standards, and sustainable practices within robust governance frameworks to protect both public health and the environment. Moving forward, the regulatory landscape must shift from a reliance on narrowly reactive, compliance-driven approaches toward proactive mindsets that prioritize safety and long-term environmental and public health protection. Accordingly, regulatory systems have to advance to better identify emerging risks, strengthen enforcement mechanisms, and promote global harmonization. Ultimately, sustainable oversight of xenobiotics that is sustainable can only be achieved through a collaborative engagement of scientists, policymakers, industry stakeholders, and the public which achieves an appropriate balance between economic development and ecological and genetic security. As societies shifting to greener, more resilient economic models, the effective management of xenobiotics will play a pivotal role in determining public health and sustainable development outcomes.

9 Conclusion

The commercialization and application of xenobiotics occupy a critical intersection of science, economics, public policy, and ethics. As this chapter has demonstrated, xenobiotics are the foundation of many industries—pharmaceuticals, agrochemical, plastics, cosmetics, and processed foods—that drive technological development, productivity growth, and economic growth. Their indispensable role guarantees that xenobiotics will remain integral component to the modern industrial landscape. At the same, their proliferating presence gives rise to multifaceted issues of genetic toxicity, ecological disruption, public health risks, and social and environmental justice considerations. Addressing these issues calls for a more holistic, integrative approach extends beyond traditional regulatory control mechanisms.

The discussion presented in this chapter states it clear that the commercializing of xenobiotics is most effectively placed under strong governance, robust regulations, and apparently ethical corporate conduct. International regulatory regimes such as REACH, TSCA, and WHO guidelines offer valuable models of integrated chemical governance, while India's multi agency regulatory structure illustrates the need for coordination and systemic reform in the context of rapidly growing chemical markets. To safeguard human health and the environment integrity, and to maintain public trust, industries' actors need to commit responsible compliance by ensuring quality assurance and transparency at every point in the supply chain. Ethical considerations, including CSR, ESG obligation, and equitable protection of at-risk groups, are fundamental to determining the production, sale, and regulation of xenobiotics. Evidence from case studies of glyphosate, BPA, talc-based cosmetics, microplastics, and pharmaceutical contamination demonstrates that failures in governance, transparency, or scientific rigor can result in devastating outcomes for both the corporate and society. Collectively, these are good examples that illustrate the need to manage risk proactively, identify hazards early, and have honest conversations with consumers and regulatory authorities.

Looking ahead, the opportunities to commercialize xenobiotics will be shaped by advances in green chemistry, predictive toxicology, circular economy approaches, global regulatory harmonization, as well as improved environmental monitoring systems. In an environment characterized by increasing regulation and environmental awareness will be those industries that respond to the market by making safer chemical substitutions, by implementing sustainable production practices, and engage in transparent communication with their stakeholders. At the same time, governments, however, should also continue to enhance policy frameworks, build scientific capacity, and encourage international co-operation to promote effective governance of the risks associated to xenobiotics.

Ultimately, achieving an appropriate balance in the middle of xenobiotic commercialization means balancing scientific proof, economic feasibility, ethical considerations, and regulatory control. In this way, the positive impacts of xenobiotics can be realized with minimizing harm done to human genetic integrity and the ecological systems. Responsible commercialization, therefore, is not simply a business ideal, but it is a social one influence the health of the public, the sustainability of the environment, and the well-being of future generations.

References

Ahuja, V. (2018). Regulation of emerging gene technologies in India. *BMC Proceedings*, *12*(Suppl. 8), 14. https://doi.org/10.1186/s12919-018-0106-0

Alfaouri, M., Jaaron, A. A. M., & Igudia, E. (2025). Pharmaceutical supply chain management challenges in developing countries: A systematic literature review. *Journal of African Business*, *26*(4), 798–841. https://doi.org/10.1080/15228916.2025.2532943

Bax, R., & Green, S. (2015). Antibiotics: The changing regulatory and pharmaceutical industry paradigm. *Journal of Antimicrobial Chemotherapy*, *70*(5), 1281–1284. https://doi.org/10.1093/jac/dku572

Behl, A., & Sharma, K. K. (2025). Xenobiotics mediated modulation of gut microbiota and its role in lifestyle diseases: A critical appraisal on exposomics. *Letters in Applied Microbiology*, *78*(5), Article ovaf067. https://doi.org/10.1093/lambio/ovaf067

Bhunia, K., Sablani, S. S., Tang, J., & Rasco, B. (2013). Migration of chemical compounds from packaging polymers during microwave, conventional heat treatment, and storage. *Comprehensive Reviews in Food Science and Food Safety*, *12*(5), 523–545. https://doi.org/10.1111/1541-4337.12028

BioSpace. (2024, April 16). *Agricultural biotechnology market is projected to achieve sustainable growth 8.6%*. BioSpace. https://www.biospace.com/agricultural-biotechnology-market-is-projected-to-achieve-sustainable-growth-8-6-percent

Chatterjee, R. P., Mane, V. K., & Pajni, N. S. (2024). Effect of xenobiotics on human health. In *Futuristic trends in agriculture engineering & food sciences* (Vol. 3, Book 1, Chapter 41, p. 532). IIP Series. e-ISBN: 978-93-5747-515-0.

Chin, K. (2019, March 14). *Frameworks and strategies for commercialization success in the biopharmaceutical ecosystem*. BioProcess International. https://www.bioprocessintl.com/intellectual-property/frameworks-and-strategies-for-commercialization-success-in-the-biopharmaceutical-ecosystem

Deubner, H.-L., Walendzik, G., Lüdeke, A., & Schlüter, U. (2024). Assessing the impact of authorisation process as a regulatory tool in the European REACH regulation: A study on improving occupational safety for applying companies. *Annals of Work Exposures and Health*, *68*, 617–625. https://api.semanticscholar.org/CorpusID:269633585

ECHA. (2023). *REACH regulation: Guidance for industry*. European Chemicals Agency. https://echa.europa.eu

Elmadhoun, B., Alsaidalani, R., & Burczynski, F. (2025). Quality risk management in the final operational stage of sterile pharmaceutical manufacturing: A case study highlighting the management of sustainable related risks in product sterilization, inspection, labeling, packaging, and storage processes. *Sustainability*, *17*(4), 1670. https://doi.org/10.3390/su17041670

Embrandiri, A., Kiyasudeen, K., Rupani, P. F., & Ibrahim, M. (2016). Environmental xenobiotics and its effects on natural ecosystem. In *Plant responses to xenobiotics* (pp. 1–18). https://doi.org/10.1007/978-981-10-2860-1_1

Food Safety and Standards Authority of India. (2025, April 02). *Food safety and standards (contaminants, toxins and residues) regulations, 2011: Compendium version VIII*. https://fssai.gov.in/upload/uploadfiles/files/Comp_Contaminants_Regulations_2_4_2025_VIII.pdf

Frank R. (2016). *Lautenberg chemical safety for the 21st century act* (Pub. L. No. 114-182, 130 Stat. 448).

Fukushima, T. (2021). CSCL and industrial chemical management framework in Japan. *ACS Chemical Health & Safety*. https://doi.org/10.1021/acs.chas.1c00037

Ghag, M. K. (2024). Determinants of environmental, social and corporate governance (ESG) disclosure: A study of pharmaceutical companies in India. *African Journal of Biomedical Research*, *27*(5S), 709–718. https://doi.org/10.53555/AJBR.v27i5S.6769

Huang, H., Jin, Y., Chen, C., Feng, M., Wang, Q., Li, D., Chen, W., Xing, X., Yu, D., & Xiao, Y. (2021). A toxicity pathway-based approach for modeling the mode of action framework of lead-induced neurotoxicity. *Environmental Research, 199*, 111328. https://doi.org/10.1016/j.envres.2021.111328

Huang, Z., Gustave, W., Bai, S., Li, Y., Li, B., Elçin, E., Jiang, B., Jia, Z., Zhang, X., Shaheen, S. M., & He, F. (2024). Challenges and opportunities in commercializing whole-cell bioreporters in environmental application. *Environmental Research, 262*(Part 1), 119801. https://doi.org/10.1016/j.envres.2024.119801

Jung, H., Shin, G., Kwak, H., Hao, L. T., Jegal, J., Kim, H. J., Jeon, H., Park, J., & Oh, D. X. (2023). Review of polymer technologies for improving the recycling and upcycling efficiency of plastic waste. *Chemosphere, 320*, 138089. https://doi.org/10.1016/j.chemosphere.2023.138089

Lang, T., Lipp, A. M., & Wechselberger, C. (2025). Xenobiotic toxicants and particulate matter: Effects, mechanisms, impacts on human health, and mitigation strategies. *Journal of Xenobiotics, 15*(4), 131. https://doi.org/10.3390/jox15040131

MacDonald, J. S., & Robertson, R. T. (2009). Toxicity testing in the 21st century: A view from the pharmaceutical industry. *Toxicological Sciences, 110*(1), 40–46. https://doi.org/10.1093/toxsci/kfp088

Mohapatra, B., Dutta, A., Phale, P. S., & Kumar, U. (2025). Xenobiotics and emerging contaminants in ecosystems: Innovative geo-microbial strategies for prevention, efficient clean-up and biosafety. *Frontiers in Bioengineering and Biotechnology, 13*, 1619769. https://doi.org/10.3389/fbioe.2025.1619769

Negi, A. (2025). Environmental impact of textile materials: Challenges in fiber–dye chemistry and implication of microbial biodegradation. *Polymers*, *17*(7), Article 871. https://doi.org/10.3390/polym17070871

Organisation for Economic Co-operation and Development. (2021–2024). *OECD work on chemical safety and biosafety 2021–24*. https://www.oecd.org/content/dam/oecd/en/topics/policy-issues/chemical-safety-and-biosafety/work-on-chemical-safety-and-biosafety-2021-24.pdf

Porter, M. E., & Kramer, M. R. (2006). Strategy and society: The link between competitive advantage and corporate social responsibility. *Harvard Business Review, 84*(12), 78–92.

Roszkowska-Menkes, M. (2021). Porter and Kramer's (2006) Shared value. In *[Title of the reference work]* (Entry 393-1). Springer. https://doi.org/10.1007/978-3-030-02006-4_393-1

Štefanac, T., Grgas, D., & Landeka Dragičević, T. (2021). Xenobiotics—Division and methods of detection: A review. *Journal of Xenobiotics*, *11*(4), 130–141. https://doi.org/10.3390/jox11040009

Thakur, P. D. (2024). Contribution of pharmaceutical industries in corporate social responsibility (CSR). *International Journal of Advanced Research, 12*, 671–675. https://doi.org/10.21474/IJAR01/18929

Trump, B., Cummings, C., Klasa, K., Galaitsi, S., & Linkov, I. (2023). Governing biotechnology to provide safety and security and address ethical, legal, and social implications. *Frontiers in Genetics, 13*, 1052371. https://doi.org/10.3389/fgene.2022.1052371

U.S. Environmental Protection Agency. (2025, September 30). *Test guidelines for pesticides and toxic substances*. https://www.epa.gov/test-guidelines-pesticides-and-toxic-substances

U.S. Food and Drug Administration. (2021, February 11). *Office of generic drugs 2020 annual report*. https://www.fda.gov/drugs/generic-drugs/office-generic-drugs-2020-annual-report

U.S. Food and Drug Administration. (2025, April 17). *FDA's regulation of plant and animal biotechnology products*. https://www.fda.gov/safety/fdas-regulation-plant-and-animal-biotechnology-products

Vangeli, A., Małecka, A., Mitręga, M., & Pfajfar, G. (2023). From greenwashing to green B2B marketing: A systematic literature review. *Industrial Marketing Management, 115*, 281–299. https://doi.org/10.1016/j.indmarman.2023.10.002

Wirtu, Y. D. (2024). A review of environmental and health effects of synthetic cosmetics. *Frontiers in Environmental Science*, *12*, Article 1402893. https://doi.org/10.3389/fenvs.2024.1402893

World Health Organization. (2020a). *Guidance on post-market surveillance of medical devices including in vitro diagnostics*. https://www.who.int/publications/i/item/9789240015319

World Health Organization (WHO). (2020b). *Human health and chemical exposure: Global risk assessment*. https://www.who.int

World Intellectual Property Organization. (2025). *A primer on technology transfer in the field of biotechnology*. WIPO. https://www.wipo.int/web-publications/a-primer-on-technology-transfer-in-the-field-of-biotechnology/en/5-role-of-intellectual-property-in-biotechnology-commercialization.html

MIX
Papier aus verantwortungsvollen Quellen
Paper from responsible sources
FSC® C105338

If you have any concerns about our products,
you can contact us on
ProductSafety@springernature.com

In case Publisher is established outside the EU,
the EU authorized representative is:
Springer Nature Customer Service Center GmbH
Europaplatz 3, 69115 Heidelberg, Germany

Printed by Libri Plureos GmbH
in Hamburg, Germany